3 0011 00772 0246

AUTOTOOLS

COMPUTERS

005.3
CAL

1/3/2020

Calcote, John

Autotools

DISCARD

D1224304

COMPUTERS

JAN 0 3 2020

246

AUTOTOOLS

2ND EDITION

A Practitioner's Guide to GNU Autoconf, Automake, and Libtool

by John Calcote

no starch press

San Francisco

FAIRPORT PUBLIC LIBRARY
1 VILLAGE LANDING
FAIRPORT, NY 14450

AUTOTOOLS, 2ND EDITION. Copyright © 2020 by John Calcote.

All rights reserved. No part of this work may be reproduced or transmitted in any form or by any means, electronic or mechanical, including photocopying, recording, or by any information storage or retrieval system, without the prior written permission of the copyright owner and the publisher.

Printed in USA

First printing

23 22 21 20 19   1 2 3 4 5 6 7 8 9

ISBN-10: 1-59327-972-8
ISBN-13: 978-1-59327-972-1

Publisher: William Pollock
Production Editor: Janelle Ludowise
Cover and Interior Design: Octopod Studios
Developmental Editor: Ellie Bru
Technical Reviewer: Eric Blake
Copyeditor: Barton D. Reed
Compositors: Janelle Ludowise and Danielle Foster
Proofreader: Paula L. Fleming
Indexer: Beth Nauman-Montana

For information on distribution, translations, or bulk sales, please contact No Starch Press, Inc. directly:
No Starch Press, Inc.
245 8th Street, San Francisco, CA 94103
phone: 1.415.863.9900; info@nostarch.com
www.nostarch.com

The Library of Congress has catalogued the first edition as follows:

Calcote, John, 1964-
   Autotools : a practitioner's guide to GNU Autoconf, Automake, and Libtool / by John Calcote.
       p. cm.
   ISBN-13: 978-1-59327-206-7 (pbk.)
   ISBN-10: 1-59327-206-5 (pbk.)
  1. Autotools (Electronic resource) 2. Cross-platform software development. 3. Open source software.
 4. UNIX (Computer file) I. Title.
   QA76.76.D47C335 2010
   005.3--dc22

                                                    2009040784

No Starch Press and the No Starch Press logo are registered trademarks of No Starch Press, Inc. Other product and company names mentioned herein may be the trademarks of their respective owners. Rather than use a trademark symbol with every occurrence of a trademarked name, we are using the names only in an editorial fashion and to the benefit of the trademark owner, with no intention of infringement of the trademark.

The information in this book is distributed on an "As Is" basis, without warranty. While every precaution has been taken in the preparation of this work, neither the author nor No Starch Press, Inc. shall have any liability to any person or entity with respect to any loss or damage caused or alleged to be caused directly or indirectly by the information contained in it.

For Michelle

But to see her was to love her;
Love but her, and love forever.
—Robert Burns

About the Author

John Calcote is currently a senior software engineer at *Hammerspace.com*, a software company specializing in cloud data management. He's been writing portable networking and storage software for over 25 years and is active in developing, debugging, and analyzing diverse open source software packages. John is an advocate for the open source movement, participating in a number of open source communities.

About the Technical Reviewer

Eric Blake has contributed to a variety of open source projects since 2000. He became maintainer for GNU M4 in 2006, and GNU Autoconf in 2007. Eric currently works at Red Hat Inc. since 2010, where his primary focus has been on the qemu and libvirt virtualization software. He is also an active participant in the Austin Group which maintains the POSIX specification.

BRIEF CONTENTS

CONTENTS IN DETAIL

5
MORE FUN WITH AUTOCONF: CONFIGURING USER OPTIONS   113

6
AUTOMATIC MAKEFILES WITH AUTOMAKE   145

9
UNIT AND INTEGRATION TESTING WITH AUTOTEST 235

10
FINDING BUILD DEPENDENCIES WITH PKG-CONFIG 271

11
INTERNATIONALIZATION 293

12
LOCALIZATION 331

18
A CATALOG OF TIPS AND REUSABLE SOLUTIONS FOR CREATING GREAT PROJECTS 499

FOREWORD FOR THE FIRST EDITION

When I was asked to do a technical review on a book about the Autotools, I was rather skeptical. Several online tutorials and a few books already introduce readers to the use of GNU Autoconf, Automake, and Libtool. However, many of these texts are less than ideal in at least some ways: they were either written several years ago and are starting to show their age, contain at least some inaccuracies, or tend to be incomplete for typical beginner's tasks. On the other hand, the GNU manuals for these programs are fairly large and rather technical, and as such, they may present a significant entry barrier to learning your ways around the Autotools.

John Calcote began this book with an online tutorial that shared at least some of the problems facing other tutorials. Around that time, he became a regular contributor to discussions on the Autotools mailing lists, too. John kept asking more and more questions, and discussions with him uncovered some bugs in the Autotools sources and documentation, as well as some issues in his tutorial.

Since that time, John has reworked the text a lot. The review uncovered several more issues in both software and book text, a nice mutual benefit. As a result, this book has become a great introductory text that still aims to be accurate, up to date with current Autotools, and quite comprehensive in a way that is easily understood.

Always going by example, John explores the various software layers, portability issues and standards involved, and features needed for package build development. If you're new to the topic, the entry path may just have become a bit less steep for you.

Ralf Wildenhues
Bonn, Germany
June 2010

FOREWORD FOR THE
SECOND EDITION

The GNU Autotools have been around for a long time. My own introduction to Autoconf was back in 1999, around the same time I began my first foray into the world of Open Source Software. I had obtained employment in a research lab using the Java language to control FPGA hardware, and one of my co-workers showed me a program he had found, jikes, which compiled our project nearly 10 times faster than javac from Sun. But it crashed on one corner case of string concatenation, and I was assigned to figure out why. The jikes compiler was an Open Source project from IBM written in C++, and I soon found and fixed the problem, and got the pleasure of seeing my patch accepted upstream.

What's more, since jikes used GNU Autoconf to allow building it on multiple platforms, after I had fixed the initial bug using the Solaris 7 machines at my work, I was able to experiment with further fixes to jikes at home on my Microsoft Windows 95 personal machine, using the Cygwin project that I found on the internet to provide gcc. I did not immediately appreciate the magnitude of the portability problems that had already been solved to be able to build software across vastly different systems, because the Autotools already hid so much of that complexity. But over time, my interests shifted away from Java and more into cross-platform compatibility, where I started contributing patches for GNU M4, and later GNU Autoconf, and eventually reached the point where I took over as the maintainer for both projects. Although my employment and my active project contributions have changed, in the meantime, I still find myself using the Autotools on a regular basis.

The point of all this? While the Java programming language is still around, it looks much different than 20 years ago. Between such changes as the introduction of generic typing, a change in ownership when Oracle bought Sun, and my own career taking me down a different path focused on the C language, I find it hard to compare modern Java projects to the work I did back then. And the jikes project that I worked on? It is now obsolete, unmaintained because it never kept up with the changes in the language, and because the javac compiler itself improved in quality and speed, where jikes was no longer a strong competition. But in all that time, the GNU Autotools have still been in active use, as a backbone behind the many GNU/Linux distributions. Even if most Linux users these days have never personally interacted with running a configure script, and instead rely on the distribution to pre-package binaries, they are beneficiaries of the power of the Autotools in making it easy for the distributions to bundle so many Open Source products.

John Calcote has taken on the huge task of introducing the power of the Autotools to a new user. And while I may not necessarily be the target audience (after all, I am no longer a new user), I definitely learned some things as I read through this book. The documentation shipped with the various Autotools tends to feel more like a brain dump, trying to cover every last feature with no regards to which features will be more useful to the beginning user, and where it seems like you have to already know what feature exists before you can search to find out the specifics of that feature. In contrast, John has done an an excellent job of breaking down the tasks at hand into a series of logical steps, guiding the reader through a typical evolutionary project that uses progressively more of the Autotools along the way, focusing on the most commonly-used aspects, and providing good explanations as to why each newly-introduced feature is worth using. I hope that you learn enough of the Autotools from reading this book to feel comfortable using them with your project. Whether or not you feel like an Autotools expert, this book is a success if it allows your end user to run ./configure && make without any further thought as to the portability problems solved by the Autotools on their behalf.

Eric Blake
GNU Autoconf Maintainer

PREFACE

I've often wondered during the last ten years how it could be that the *only* third-party book on the GNU Autotools that I've been able to discover is *GNU AUTOCONF, AUTOMAKE, and LIBTOOL* by Gary Vaughan, Ben Elliston, Tom Tromey, and Ian Lance Taylor, affectionately known by the community as *The Goat Book* (so dubbed for the front cover—an old-fashioned photo of goats doing acrobatic stunts).[1]

I've been told by publishers that there is simply no market for such a book. In fact, one editor told me that he himself had tried unsuccessfully to entice authors to write this book a few years ago. His authors wouldn't finish the project, and the publisher's market analysis indicated that there was very little interest in the book. Publishers believe that open source software developers tend to disdain written documentation. Perhaps they're

---

1. Vaughan, Elliston, Tromey, and Taylor, *GNU Autoconf, Automake, and Libtool* (Indianapolis: Sams Publishing, 2000).

right. Interestingly, books on IT utilities like Perl sell like Perl's going out of style—which is actually somewhat true these days—and yet people are still buying enough Perl books to keep their publishers happy. All of this explains why there are ten books on the shelf with animal pictures on the cover for Perl, but literally nothing for open source software developers.

I've worked in software development for 25 years, and I've used open source software for quite some time now. I've learned a lot about open source software maintenance and development, and most of what I've learned, unfortunately, has been by trial and error. Existing GNU documentation is more often reference material than solution-oriented instruction. Had there *been* other books on the topic, I would have snatched them all up immediately.

What we need is a cookbook-style approach with the recipes covering real problems found in real projects. First the basics are covered, sauces and reductions, followed by various cooking techniques. Finally, master recipes are presented for culinary wonders. As each recipe is mastered, the reader makes small intuitive leaps—I call them *minor epiphanies*. Put enough of these under your belt and overall mastery of the Autotools is ultimately inevitable.

Let me give you an analogy. I'd been away from math classes for about three years when I took my first college calculus course. I struggled the entire semester with little progress. I understood the theory, but I had trouble with the homework. I just didn't have the background I needed. So the next semester, I took college algebra and trigonometry back to back as half-semester classes. At the end of that semester, I tried calculus again. This time I did very well—finishing the class with a solid *A* grade. What was missing the first time? Just basic math skills. You'd think it wouldn't have made *that* much difference, but it really does.

The same concept applies to learning to properly use the Autotools. You need a solid understanding of the tools upon which the Autotools are built in order to become proficient with the Autotools themselves.

Why Use the Autotools?

In the early 1990s, I was working on the final stages of my bachelor's degree in computer science at Brigham Young University. I took an advanced computer graphics class where I was introduced to C++ and the object-oriented programming paradigm. For the next couple of years, I had a love-hate relationship with C++. I was a pretty good C coder by that time, and I thought I could easily pick up C++, as close in syntax as it was to C. How wrong I was! I fought with the C++ compiler more often than I'd care to recall.

The problem was that the most fundamental differences between C and C++ are not obvious to the casual observer, because they're buried deep within the C++ language specification rather than on the surface in the language syntax. The C++ compiler generates an amazing amount of code beneath the covers, providing functionality in a few lines of C++ code that require dozens of lines of C code.

Just as programmers then complained of their troubles with C++, so likewise programmers today complain about similar difficulties with the GNU Autotools. The differences between make and Automake are very similar to the differences between C and C++. The most basic single-line *Makefile.am* generates a *Makefile.in* (an Autoconf template) containing 300–400 lines of parameterized make script, and it tends to increase with each revision of the tool as more features are added.

Thus, when you use the Autotools, you have to understand the underlying infrastructure managed by these tools. You need to take the time to understand the open source software distribution, build, test, and installation philosophies embodied by—in many cases even enforced by—these tools, or you'll find yourself fighting against the system. Finally, you need to learn to agree with these basic philosophies because you'll only become frustrated if you try to make the Autotools operate outside of the boundaries set by their designers.

Source-level distribution relegates to the end user a particular portion of the responsibility of software development that has traditionally been assumed by the software developer—namely, building products from source code. But end users are often not developers, so most of them won't know how to properly build the package. The solution to this problem, from the earliest days of the open source movement, has been to make the package build and installation processes as simple as possible for the end user so that he could perform a few well-understood steps to have the package built and installed cleanly on his system.

Most packages are built using the make utility. It's very easy to type make, but that's not the problem. The problem crops up when the package doesn't build successfully because of some unanticipated difference between the user's system and the developer's system. Thus was born the ubiquitous configure script—initially a simple shell script that configured the end user's environment so that make could successfully find the required external resources on the user's system. Hand-coded configuration scripts helped, but they weren't the final answer. They fixed about 65 percent of the problems resulting from system configuration differences—and they were a pain in the neck to write properly and to maintain. Dozens of changes were made incrementally over a period of years, until the script worked properly on most of the systems anyone cared about. But the entire process was clearly in need of an upgrade.

Do you have any idea of the number of build-breaking differences there are between existing systems today? Neither do I, but there are a handful of developers in the world who know a large percentage of these differences. Between them and the open source software community, the GNU Autotools were born. The Autotools were designed to create configuration scripts and makefiles that work correctly and provide significant chunks of valuable end-user functionality under most circumstances, and on most systems—even on systems not initially considered (or even conceived of) by the package maintainer.

With this in mind, the primary purpose of the Autotools is not to make life simpler for the package maintainer (although it really does in the long run). *The primary purpose of the Autotools is to make life simpler for the end user.*

Acknowledgments for the First Edition

I could not have written a technical book like this without the help of a lot of wonderful people. I would like to thank Bill Pollock and the editors and staff at No Starch Press for their patience with a first-time author. They made the process interesting and fun.

Additionally, I'd like to thank the authors and maintainers of the GNU Autotools. Specifically, Ralf Wildenhues, who believed in this project enough to spend hundreds of hours of his personal time in technical review. His comments and insight were invaluable in taking this book from mere wishful thinking to an accurate and useful text.

I would also like to thank my friend Cary Petterborg for encouraging me to "just go ahead and do it," when I thought it would probably never happen.

Finally, I'd like to thank my wife Michelle and my children; Ethan, Mason, Robby, Haley, Joey, Nick, and Alex for allowing me to spend all of that time away from them while I worked on the book. A novel would have been easier (and more lucrative), but the world has plenty of novels and not enough books about the Autotools.

Acknowledgments for the Second Edition

I would like to thank my technical reviewer, Eric Blake, whose insights and comments lead to a much better work than I'd originally envisioned. He's been an institution on the Autotools mailing lists; I was extremely lucky to have his editorial contributions.

I would also like to thank Tony Mobily, editor and chief of *Free Software Magazine*. Tony gave me my first chance at publishing this material. Without his guidance and direction, I'd never have taken this project to paper format. I should have made this acknowledgment in the first edition of this book but, much to my own embarrassment, it was omitted.

I Wish You the Very Best

I spent a long time and a lot of effort learning what I now know about the Autotools. Most of this learning process was more painful than it really had to be. I've written this book so that you won't have to struggle to learn what should be a core set of tools for the open source programmer. Please feel free to contact me, and let me know your experiences with learning the Autotools. I can be reached at my personal email address at *john.calcote@gmail.com*. Good luck in your quest for a better software development experience!

John Calcote
Elk Ridge, Utah
September 2019

INTRODUCTION

Few open source software developers would deny that GNU Autoconf, Automake, and Libtool (the *Autotools*) have revolutionized the open source software world. However, although there are many thousands of Autotools advocates, there are also many software developers who *hate* the Autotools—with a passion. I believe the reason for this dread of the Autotools is that when you use them without understanding the underlying infrastructure they manage, you find yourself fighting against the system.

This book solves this problem by first providing a framework for understanding the underlying infrastructure of the Autotools and then building on that framework with a tutorial-based approach to teaching Autotools concepts in a logically ordered fashion.

Who Should Read This Book

This book is primarily for the open source software package maintainer who wants to become an Autotools expert. That said, this book also provides instructions to end users who wish to understand what's happening during the process of downloading, unpacking, and building software packages whose build processes are managed by the Autotools. Existing material on the subject is limited to the GNU Autotools manuals and a few internet-based tutorials. For years, most real-world questions have been answered on the Autotools mailing lists, but mailing lists are an inefficient form of teaching because the same answers to the same questions are given time and again. This book provides a cookbook-style approach, covering real problems found in real projects.

How This Book Is Organized

The book starts with an end-user perspective on the Autotools and then moves from high-level development build concepts to mid-level use cases and examples, finally finishing with more advanced details and examples. As though you were learning arithmetic, we'll begin with some basic math—algebra and trigonometry—and then move on to analytic geometry and calculus.

Chapter 1 provides an end-user perspective on the Autotools. It covers topics that a Linux power user, who is not necessarily a software developer, needs to understand in order to take full advantage of the features of Autotools-managed source packages (so-called "tarballs") downloaded from project websites containing perhaps the latest beta version of some software the user would like to try out. Often, Linux users find that the solution to a software problem involves updating to a version that contains the fix for that problem, only to discover that the version they need is so new there is no RPM or Debian package for that version in any of the package repositories for their Linux distribution of choice. Chapter 1 provides relief to the newbie who needs to know what to do with that *tar.gz* file containing that configure script and all those *.c* source files.

Chapter 2 begins the discussion of concepts of interest to software developers. It presents a general overview of the packages considered part of the GNU Autotools. This chapter describes the interaction between these packages and the files consumed by and generated by each one. In each case, figures depict the flow of data from hand-coded input to final output files.

Chapter 3 covers open source software project structure and organization. This chapter also goes into some detail about the *GNU Coding Standards (GCS)* and the *Filesystem Hierarchy Standard (FHS)*, both of which have played vital roles in the design of the GNU Autotools. This chapter presents some fundamental tenets upon which the design of each of the Autotools is based. With these concepts, you'll better understand the theory behind the architectural decisions made by the Autotools designers.

In this chapter, we'll also design a simple project, Jupiter, from start to finish using hand-coded makefiles. We'll add to Jupiter in a stepwise fashion as we discover functionality that we can use to simplify tasks and to provide features that open source software users have come to expect.

Chapters 4 and 5 present the framework designed by the GNU Autoconf engineers to ease the burden of creating and maintaining portable, functional project configuration scripts. The GNU Autoconf package provides the basis for creating complex configuration scripts with just a few lines of information provided by the project maintainer.

In these chapters, we'll quickly convert our hand-coded makefiles into Autoconf *Makefile.in* templates and then begin adding to them in order to gain some of the most significant Autoconf benefits. Chapter 4 discusses the basics of generating configuration scripts, while Chapter 5 moves on to more advanced Autoconf topics, features, and uses.

Chapter 6 begins by converting the Jupiter project *Makefile.in* templates into Automake *Makefile.am* files. Here, you'll discover that Automake is to makefiles what Autoconf is to configuration scripts. This chapter presents the major features of Automake in a manner that will not become outdated as new versions of Automake are released.

Chapters 7 and 8 explain the basic concepts behind shared libraries and show how to build shared libraries with Libtool—a standalone abstraction for shared-library functionality that can be used with the other Autotools. Chapter 7 begins with a shared-library primer and then covers some basic Libtool extensions that allow Libtool to be a drop-in replacement for the more basic library generation functionality provided by Automake. Chapter 8 covers library versioning and the runtime dynamic module management abstraction provided by Libtool.

Chapter 9 presents a relatively new addition to the Autotools—autotest. The autotest functionality in Autoconf allows you to easily create and manage integration test execution frameworks for your projects. In previous chapters, we will have covered unit testing in individual makefiles. Autotest provides a mechanism for adding more global testing that depends on multiple components in your project. Honestly, autotest can be used to do about any sort of testing you want. We'll focus on adding autotest suites that ensure your project works the way you believe it should—automatically.

Chapter 10 discusses the concepts of finding compile- and link-time dependencies and adding the appropriate references to build tool command lines. Specifically, this chapter introduces `pkg-config`, which has become a de facto standard in Linux software development, providing the framework for easily finding and consuming components that your package depends on. This chapter shows you how to both consume `pkg-config` *.pc* files to find your dependencies and how to play nicely in the sandbox by providing *.pc* files for your projects.

Chapters 11 and 12 discuss internationalization (abbreviated *i18n*) and localization (*l10n*), respectively—the ability to easily manage text strings and other locale-specific attributes (such as references to numbers, money, and dates) within your project that should be different for localized releases of your project.

Chapter 13 talks about obtaining maximum portability in your projects by using Gnulib.

Chapters 14 and 15 illustrate the transformation of an existing, fairly complex open source project (FLAIM) from using a hand-built build system to using an Autotools build system. This example will help you to understand how you might *autoconfiscate* one of your own existing projects.

Chapter 16 provides an overview of the features of the M4 macro processor that are relevant to obtaining a solid understanding of Autoconf. This chapter also considers the process of writing your own Autoconf macros.

Chapter 17 discusses using the Autotools to build software designed to run on Microsoft Windows platforms. I'll show you how to cross compile on Linux for Windows, and how to install and use the three most popular Windows-based POSIX platforms—Cygwin, Msys2, and MinGW—to build Windows software using GNU tools, including the Autotools.

Microsoft has a great set of free tools for building Windows software, but if your package is already working on Linux and being built with POSIX build tools, using the Autotools to build for Windows can be a great way to get you up and running there fast. From there, you can decide whether what you have is good enough for your project or if you need to provide a native build environment for Windows.

Chapter 18 is a compilation of tips, tricks, and reusable solutions to Autotools problems. The solutions in this chapter are presented as a set of individual topics or items. Each can be understood without context from the surrounding items.

Chapters 3 through 9 are built around the Jupiter project. Chapters 14 and 15 cover the FLAIM project. Chapters 11 and 12 cover the gettext project and Chapter 13 covers the b64 project. These projects are found on the GitHub NSP-Autotools site at *https://github.com/NSP-Autotools*.

Except for the FLAIM project, each of these repositories are tagged at commits representing topic transitions. The tags are called out within the margins of the chapters pertaining to these projects. You can easily follow along by checking out the tagged commit in the repository when you see one in the book.

Conventions Used in This Book

This book contains hundreds of program listings in roughly two categories: console examples and file listings. Console examples have no captions, and user input is **bolded**.

Often, I'll use the Linux ls command with various options to show the contents of a directory before or after changes are made. Different Linux distributions often ship with an alias for ls enabled by default. I'm using Linux Mint 18 with the Cinnamon desktop to write this book; my pre-defined alias for ls is:

```
$ alias
--snip--
alias ls='ls --color=auto'
$
```

You may find you have an ls alias on your system that provides a different default set of functionality that will defeat your attempts to exactly duplicate my console examples. Just be aware of the reasons for these possible differences.

File listings contain full or partial listings of the files discussed in the text. All named listings are provided in the associated git repositories. I've tried to provide enough context around modified portions of partial listings so that you can easily see where lines are added or changed. However, there are a few listings where lines are deleted. In these cases, I've called out the deleted lines in the text near the listing.

Listings without filenames are entirely contained in the printed listing itself, are meant to be considered independently without context, and are not part of the provided source repositories. In general, text that remains the same as a previously listed version of the file will be grayed out, whereas modified areas will be in black text.

For listings that do relate to the Jupiter and FLAIM projects, the caption first specifies the path of the file relative to the project root directory and then provides a description of the changes made to that file in the listing.

Throughout this book, I refer to the GNU/Linux operating system simply as *Linux*. It should be understood that by the use of the term *Linux*, I'm referring to GNU/Linux, its actual official name. I use *Linux* simply as shorthand for the official name.

Autotools Versions Used in This Book

The Autotools are always being updated—on average, a significant update of each of the three most important tools, Autoconf, Automake, and Libtool, is released every year and a half, and minor updates are released every three to six months. The Autotools developers attempt to maintain a reasonable level of backward compatibility with each new release, but occasionally something significant is broken, and older documentation simply becomes out-of-date. More recently, the Autotools have been considered mature and complete; release cycles have slowed and major changes seldom happen anymore. This is good for the community and for you, the reader, as it means that the material you find in this book will remain relevant for a long time to come.

Although I describe new, major features of recent releases of the Autotools, in my efforts to make this book more evergreen, I've tried to stick to descriptions of Autotools features (Autoconf macros, for instance) that have been in widespread use for several years. Minor details change occasionally, but the general use has stayed the same through many releases.

At appropriate places in the text, I mention the versions of the Autotools I used for this book, but I'll summarize here. I used version 2.69 of Autoconf, version 1.15 of Automake (the latest version as of this writing is actually 1.16.1), and version 2.4.6 of Libtool. Through the publication process, I was able to make minor corrections and update to new releases as they became available.

Additionally, I used version 0.19.7 of GNU gettext (the latest is 0.20.1) and version 0.29.1 of pkg-config (the latest is 0.29.2). The GNU portability library, Gnulib, is not distributed as a package but rather as a set of code snippets that are downloaded directly from the GNU website (*https://www .gnu.org/software/gnulib/*).

1

AN END USER'S PERSPECTIVE ON THE GNU AUTOTOOLS

I am not afraid of storms, for I am
learning how to sail my ship.
—*Louisa May Alcott,* Little Women

If you're not a software developer, either by trade or by hobby, you may still have a need or desire at some point to build open source software to be installed on your computer. Perhaps you're a graphic artist who wishes to use the latest version of GIMP, or maybe you're a video enthusiast and you need to build a late version of FFmpeg. This chapter, therefore, may be the only one you read in this book. I hope that is not the case, because even a power user can gain so much more by striving to understand what goes on under the covers. Nevertheless, this chapter is designed for you. Here, I'll discuss what to do with that so-called *tarball* you downloaded from that project website. I'll use the Autoconf package to illustrate, and I'll try to provide enough context so that you can follow the same process for any package you download[1] from a project website.

If you are a software developer, there's a good chance the material in this chapter is too basic for you; therefore, I'd recommend skipping right to the next chapter, where we'll jump into a more developer-centric discussion of the Autotools.

Software Source Archives

Open source software is distributed as single-file source archives containing the source and build files necessary to build the software on your system. Linux distributions remove much of the pain for end users by prebuilding these source archives and packaging the built binaries into installation packages ending in extensions like *.rpm* (for Red Hat–based systems) and *.deb* (for Debian/Ubuntu-based systems). Installing software using your system package manager is relatively easy, but sometimes you need the latest feature set of some software and it hasn't yet been packaged for your particular flavor of Linux. When this happens, you need to download the source archive from the project website's download page and then build and install it yourself. Let's begin by downloading version 2.69 of the Autoconf package:

```
$ wget https://ftp.gnu.org/gnu/autoconf/autoconf-2.69.tar.gz
```

Source archive names generally follow a de facto standard format supported by the Autotools. Unless the project maintainer has gone out of their way to modify this format, the Autotools will automatically generate a source archive file named according to the following template: *pkgname -version.format*. Here, *pkgname* is the short name of the software, *version* is the version of the software, and *format* represents the archive format, or file extensions. The *format* portion may contain more than one period, depending on the way the archive was built. For instance, *.tar.gz* represents two encodings in the format—a tar archive that has been compressed with the gzip utility, as is the case with the Autoconf source archive:

```
$ ls -1
autoconf-2.69.tar.gz
automake-1.16.1.tar.gz
gettext-0.19.8.1.tar.gz
libtool-2.4.6.tar.gz
pkg-config-0.29.2.tar.gz
$
```

---

1. It's important to distinguish the difference between a distribution archive—often found on the project website's download page—and a GitHub tarball downloaded from the green "Clone or download" button on a GitHub project page. In fact, any online source repository presenting a download button on the source browsing page is likely to send you an archive containing the raw contents of the repository, rather than a distribution archive designed to be unpacked and built by users.

Unpacking a Source Archive

By convention, source archives contain a single root directory as the top-level entry. You should feel safe unpacking a source archive to find only a single new directory in the current directory, named the same as the archive file minus the *format* portion. Source archives packaged using Autotools-based build systems never unpack the contents of the original top-level directory into the current directory.

Nevertheless, occasionally, you'll download an archive and unpack it to find dozens of new files in the current directory. It's therefore prudent to unpack a source archive of unknown origin into a new, empty subdirectory. You can always move it up a level if you need to. Additionally, you can see what will happen by using the tar utility's t option (instead of x), which lists the contents of the archive without unpacking it. The unzip utility supports the -l option to the same effect.

Source archives can take many shapes, each ending in a unique file extension: *.zip*, *.tar*, *.tar.gz* (or *.tgz*), *.tar.bz2*, *.tar.xz*, *tar.Z*, and so on. The files contained in these source archives are the source code and build files used to build the software. The most common of these formats are *.zip*, *.tar.gz* (or *.tgz*), and *.tar.bz2*. Newer formats that are gaining in popularity include *.xz* (for which the latest Autotools even have native support) and *.zstd*.

ZIP files use compression techniques developed decades ago by Phil Katz on Microsoft DOS systems. ZIP was a proprietary multifile compressed archive format that was eventually released into the public domain. Since then, versions have been written for Microsoft Windows and Linux as well as other Unix-like operating systems. In later versions of Windows, a user can unpack a *.zip* file merely by right-clicking it in Windows Explorer and selecting an **Extract** menu option. The same is true of the Nautilus (Nemo on Mint's Cinnamon desktop) file browser on Linux Gnome desktops.

ZIP files can be unpacked at the Linux command line using the more or less ubiquitous unzip program,[2] like so:

```
$ unzip some-package.zip
```

ZIP files are most often intended by project maintainers to be used on Microsoft Windows systems. A much more common format used on Linux platforms is the compressed *.tar* file. The name *tar* comes from *tape archive*. The tar utility was originally designed to stream the contents of online storage media, such as hard disk drives, to more archival storage formats, such as magnetic tape. Because it's not a random-access format, magnetic tape doesn't have a hierarchical filesystem. Rather, data is written to tape in one long string of bits, with these archive files appended end to end. To find

---

2. Reading the man page, you might get the impression that the gunzip utility can handle *.zip* files as well as *.gz* files, but this feature is intended only to convert *.tar.zip* files into *.tar.gz* files. Essentially, the gunzip utility cannot understand compressed archives that contain more than one file. If you do have a *.tar.zip* file, you can uncompress it to a *.tar* file using a command like gunzip < file.tar.zip. It doesn't recognize the *.zip* extension, so piping it from stdin is the only way to get it to work.

a particular file on tape, you have to read from the beginning of the tape through to the file you're interested in. Hence, it's better to store fewer files on tape to reduce search time.

The tar utility was designed to convert a set of files in a hierarchical filesystem into just such a long string of bits—an archive. The tar utility was specifically *not* designed to compress this data in a manner that would reduce the amount of space it takes up, as there are other utilities to do that sort of thing—remember, a founding principle of Unix is that of a single responsibility per tool. In fact, a *.tar* file is usually slightly larger than the sum of the sizes of the files it contains because of the overhead of storing the hierarchy, names, and attributes of the archived files.

Occasionally, you'll find a source archive that ends only in a *.tar* extension. This implies that the file is an uncompressed *.tar* archive. More often, however, you'll see extensions such as *.tar.gz*, *.tgz*, and *.tar.bz2*. These are compressed *.tar* archives. An archive is created from the contents of a directory tree using the tar utility, and then the archive is compressed using the gzip or bzip2 utility. A file with an extension of *.tar.gz* or *.tgz* is a *.tar* archive that has been compressed with the gzip utility. Technically, you can extract the contents of a *.tar.gz* file by using a pipeline of commands to first uncompress the *.gz* file with gunzip and then unpack the remaining *.tar* file with tar, in the following manner:

```
$ gunzip -c autoconf-2.69.tar.gz | tar xf -
```

However, the tar utility has evolved since it was used for creating tape data streams. Nowadays, it's used as a general-purpose archive file management tool. It understands, based on file extensions and sometimes the initial bytes of an archive, how to execute the correct tools to uncompress a compressed *.tar* archive before unpacking the files. For example, the following command recognizes *autoconf-2.69.tar.gz* as a *.tar* archive that was subsequently compressed with the *gzip* utility:

```
$ tar xf autoconf-2.69.tar.gz
```

This command first executes the gunzip program (or the gzip program with the -d option) to uncompress the archive, and then it uses internal algorithms to convert the archive into its original multifile directory structure, complete with original timestamps and file attributes.

Building the Software

Once you've unpacked the source archive, the next step usually involves examining the contents of the unpacked directory tree in an effort to determine how the software should be built and installed. A few patterns have become pervasive in the open source world, and GNU and the Autotools try to promote the use of these patterns as the default behavior of an Autotools-based project.

First, look for a file named *INSTALL* in the root directory of the unpacked archive. This file usually contains step-by-step instructions for how to build and install the software, or it tells you how to find those instructions—perhaps via a URL reference to a project web page.

The *INSTALL* file for GNU packages such as Autoconf is pretty verbose. The GNU project tends to try to set an example for the rest of the open source world. Nevertheless, it does carefully outline the steps required to build the Autoconf package. I'd recommend reading a GNU project *INSTALL* file completely at least once, because it contains details about how most GNU projects are built and installed. In fact, the one bundled with the Autoconf package is actually a generic one that GNU bundles with many of its packages—which in itself is a testament to the consistency of Autotools-generated build systems. Let's dive in and see what it tells us about building Autoconf:

```
$ tar xf autoconf-2.69.tar.gz
$ cd autoconf-2.69
$ more INSTALL
--snip--
```

The instructions indicate that you should use the cd command to change to the directory containing the project's source code and then type ./configure to configure the package for your system. However, it should be clear that if you're reading the *INSTALL* file, you're probably already in the directory containing configure.

Running configure can take a while if the package is large and complex. For the Autoconf package, it takes only a couple of seconds and spews a single page of text to the screen in the process. Let's take a closer look at what gets displayed during a successful Autoconf configuration process:

```
$ ./configure
checking for a BSD-compatible install... /usr/bin/install -c
checking whether build environment is sane... yes
--snip--
configure: creating ./config.status
config.status: creating tests/Makefile
--snip--
config.status: creating bin/Makefile
config.status: executing tests/atconfig commands
$
```

There are basically two parts to configure's output. The first part contains lines that start with checking (though there are a few in the middle that start with configure:). These lines indicate the status of the features that configure was programmed to look for. If a feature is not found, the trailing text will be no. On the other hand, if the feature is found, the trailing text will sometimes be yes but will often be the filesystem location of the tool or feature that was discovered.

It's not uncommon for configure to fail due to missing tools or utilities, especially if this is a newly installed system or if you haven't downloaded and built a lot of software on this system. A new user will often start posting questions to online forums at this point—or just give up.

It's important to understand the contents of this section because it can help you figure out how to solve problems. Addressing a failure is often as simple as installing a compiler using your system's package manager. For the Autoconf package, not much is required that isn't installed by default on most Linux systems. There are a few exceptions, however. For example, here's the output of configure on a system that doesn't have M4 installed:

```
$ ./configure
checking for a BSD-compatible install... /usr/bin/install -c
checking whether build environment is sane... yes
--snip--
checking for GNU M4 that supports accurate traces... configure: error: no
    acceptable m4 could be found in $PATH.
GNU M4 1.4.6 or later is required; 1.4.16 or newer is recommended.
GNU M4 1.4.15 uses a buggy replacement strstr on some systems.
Glibc 2.9 - 2.12 and GNU M4 1.4.11 - 1.4.15 have another strstr bug.
$
```

Here, you'll notice the last few lines show an error. The Autoconf package is a GNU software tool, and, true to form, it provides a lot of information to help you figure out what's wrong. You need to install an M4 macro processor, and we'll do that with our package manager. My system is a Linux Mint system, based on Ubuntu, so I'll use the apt utility. If you're using a Red Hat–based system, you may use yum to accomplish the same thing or just use the graphical user interface (GUI) for your system package manager from the GUI desktop. The key here is that we're installing the m4 package:

```
$ sudo apt install m4
```

Now configure can complete successfully:

```
$ ./configure
checking for a BSD-compatible install... /usr/bin/install -c
checking whether build environment is sane... yes
--snip--
❶ configure: creating ./config.status
config.status: creating tests/Makefile
config.status: creating tests/atlocal
--snip--
config.status: creating bin/Makefile
config.status: executing tests/atconfig commands
$
```

The second section is a set of lines beginning with `config.status:`. This section starts with the line `configure: creating ./config.status`, at ❶. The last thing `configure` does is create another script called *config.status* and then execute this script. The lines that start with `config.status:` are actually displayed by *config.status*. The primary task of *config.status* is to generate the build system based on the findings of `configure`. The lines output by this script merely tell you the names of the files being generated.

You can also run `configure` from a different directory, if you wish, by using a relative path to the `configure` command. This is useful if, for example, the project source code comes to you on a CD or via a read-only NFS mount. You could, at this point, create a build directory in your home directory and, using a relative or absolute path, execute `configure` from the read-only source directory. The `configure` script will create the entire build tree for the project in the current directory, including makefiles and any other files needed to build the project with `make`.

Once `configure` has completed, it's possible to run `make`. Before this point, there are no files in the directory tree named *Makefile*. Running `make` after `configure` yields the following:

```
$ make
make  all-recursive
make[1]: Entering directory '/.../autotools/autoconf-2.69'
Making all in bin
make[2]: Entering directory '/.../autotools/autoconf-2.69/bin'
--snip--
make[2]: Leaving directory '/.../autotools/autoconf-2.69/man'
make[1]: Leaving directory '/.../autotools/autoconf-2.69'
$
```

The primary task of `configure` is to ensure that `make` will succeed, so it's not likely that `make` will fail. If it does, the problem will probably be very specific to your system, so I can't provide any guidelines here except to suggest a careful reading of the `make` output in order to determine what caused the failure. If you can't discover the problem by reading the output, you can check the Autoconf mailing list archives, ask on the mailing list directly, and finally post a bug report to the Autoconf project website.

Testing the Build

Once we've built the software using `make`, it would be nice to exercise any tests the project maintainers might have added to the build system to provide some level of assurance that the software will run correctly on our system.

When we built the software, we ran `make` without any command line arguments. This caused `make` to assume we wanted to build the *default target*, which by convention is the `all` target. Therefore, running `make all` is the same as running `make` without any arguments. However, Autotools build systems have many targets that can be directly specified on the `make` command line. The one we're interested in at this point is the `check` target.

Running make check within the source directory will build and execute any test programs that were included by the project maintainers (this takes several minutes to complete for Autoconf):

```
$ make check
if test -d ./.git; then \
  cd . && \
  git submodule --quiet foreach test '$(git rev-parse $sha1)' \
    = '$(git merge-base origin $sha1)' \
    || { echo 'maint.mk: found non-public submodule commit' >&2; \
  exit 1; }; \
else \
  : ; \
fi
make  check-recursive
make[1]: Entering directory '/home/jcalcote/Downloads/autotools/autoconf-2.69'
Making check in bin
--snip--
/bin/bash ./testsuite
## -------------------------- ##
## GNU Autoconf 2.69 test suite. ##
## -------------------------- ##
Executables (autoheader, autoupdate...).
  1: Syntax of the shell scripts                    skipped (tools.at:48)
  2: Syntax of the Perl scripts                     ok
--snip--
501: Libtool                                        FAILED (foreign.at:61)
502: shtool                                         ok
Autoscan.
503: autoscan                                       FAILED (autoscan.at:44)
## ------------- ##
## Test results. ##
## ------------- ##
ERROR: 460 tests were run,
6 failed (4 expected failures).
43 tests were skipped.
## ------------------------- ##
## testsuite.log was created. ##
## ------------------------- ##
Please send `tests/testsuite.log' and all information you think might help:
  To: <bug-autoconf@gnu.org>
  Subject: [GNU Autoconf 2.69] testsuite: 501 503 failed
You may investigate any problem if you feel able to do so, in which
case the test suite provides a good starting point.  Its output may
be found below `tests/testsuite.dir'.
--snip--
make: *** [check] Error 2
$
```

NOTE   *Your output may differ slightly in minor ways from mine. Different Linux distributions and tool versions display differently, so don't be too concerned about minor differences. The number of tests skipped or failed may also differ from system to system due to differences in the tools installed.*

As you can see, the Autoconf package provides 503 tests; 460 of those were run and 43 were purposely skipped. Of the 460 tests that were executed, six failed, but four of those were expected failures, so we have only two problems: test 501 and test 503.

With only two failures out of 460, I'd personally call this a whopping success, but if you would like to dig a little deeper to see what's causing these problems, there are two approaches you can take. The first is to go to the Autoconf mailing list archives and either search for a similar question with answers or ask the list directly; notice the request in the preceding output to send the *tests/testsuite.log* file to *bug-autoconf@gnu.org*.

The other option requires a bit more programming skill. These tests are run by Autoconf's *autotest* framework, which automatically creates a directory for each failed test under *tests/testsuite.dir*. Each directory found under *testsuite.dir* is named after the number of the failed test. If you look there, you'll see six directories, including directories for the four expected failures. Each of these numbered directories contains a run script that will re-execute the failed test, displaying output to stdout rather than to a log file. This allows you to experiment with your system (perhaps by installing a different version of Libtool for test 501, for example) and then try running the test again.

There is also the possibility, however slight, that the project maintainers are aware of these test failures. In this case, they would likely respond to your email with a comment to this effect (or a quick search of the archives may also turn up the same answer), at which point you can simply ignore the failed tests.

Installing the Built Software

Running make usually leaves built software products—executables, libraries, and data files—scattered throughout the build directory tree. Take heart, you're almost there. The final step is installing the built software onto your system so you can use it. Thankfully, most build systems, including those managed by the Autotools, provide a mechanism for installing built software.

A complex build system is only useful to non-experts if it assumes a lot of basic defaults; otherwise, the poor user would be required to specify dozens of command line options for even the simplest build. The location of software installation is one such assumption; by default, the build system assumes you want to install built software into the */usr/local* directory tree.

The */usr/local* directory tree mirrors the */usr* directory tree; it's the standard location for software that is built locally. The */usr* directory tree, on the other hand, is where Linux distribution packages get installed. For instance, if you installed the Autoconf package using the command sudo apt-get install autoconf (or sudo yum install autoconf), the package binaries would be installed into the */usr/bin* directory. When you install your hand-built Autoconf binaries, they'll go into */usr/local/bin*, by default.

It's most often the case that */usr/local/bin* is positioned in your PATH environment variable before */usr/bin*. This allows your locally built and installed programs to override the ones installed by your distribution's package manager.

If you wish to override this default behavior and install your software into a different location, you can use the --prefix option on configure's command line,[3] as shown here:

```
$ ./configure --prefix=$HOME
```

This will cause configure to generate the build scripts such that executable binaries will be installed into your $HOME/*bin* directory.[4] If you don't have root access on your system, this is a good compromise that will allow you to install built software without asking your system administrator for extra rights.

Another reason for choosing a different --prefix location is to allow yourself to install the software into an isolated location. You can then examine the location after installation to see exactly what got installed and where it went, relative to --prefix.

Let's first install into a private installation location so we can see what the Autoconf project installs onto our system:

```
$ ./configure --prefix=$PWD/private-install
--snip--
$ make
--snip--
$ make install
--snip--
$ tree --charset=ascii private-install
private-install
├── bin
│   ├── autoconf
│   ├── autoheader
│   ├── autom4te
│   ├── autoreconf
│   ├── autoscan
│   ├── autoupdate
│   `── ifnames
`── share
    ├── autoconf
    │   ├── autoconf
    │   │   ├── autoconf.m4
--snip--
    `── man
        `── man1
            ├── autoconf.1
--snip--
            `── ifnames.1

11 directories, 61 files
$
```

---

3. Another way of saying $HOME is ~, so the command ./configure --prefix ~ has the same effect.

4. The existence and relative position of the $HOME/*bin* entry in the PATH is distribution dependent. Some distros put it before the */usr/bin* directory, some put it after, and some don't even add it to PATH by default. You may need to update your PATH to make it work the way you want.

NOTE *As with the earlier build process, the number of files and directories on your system may differ slightly from mine, based on the difference in tool availability between our systems. If you have additional documentation tools installed, for example, you may see more directories than I do, as Autoconf will build more documentation if the tools are available.*

Note that I specified the installation location on `configure`'s command line using a full path—the `PWD` environment variable contains the absolute path of the current directory in the shell. It's important to always use a full path in `--prefix`. In many cases, using a relative path will cause installation failures because the `--prefix` argument is referenced from different directories during the installation process.[5]

I used the tree command on the *private-install* directory in order to get a visual picture of what Autoconf installs.[6] There were 61 files installed into 11 directories within *private-install*.

Now, let's install Autoconf into the default location in */usr/local/bin*:

```
$ ./configure
--snip--
$ make
--snip--
$ sudo make install
--snip--
$
```

It's important to note the use of `sudo` on this command line to run `make install` with root privileges. When you install software outside of your home directory, you'll need higher privileges. If you set the `--prefix` directory to somewhere within your home directory, then you can omit the use of `sudo` in the command.

Summary

At this point, you should understand what a source archive is and how to download, unpack, build, test, and install it. I hope I've also given you the impetus to dig further and discover more about open source build systems. Those generated by the Autotools follow common patterns so pedantically that they're reasonably predictable. For hints on the sorts of things you can do, try running `./configure --help`.

There are other build systems out there. Most of them follow a reasonable set of patterns, but once in a while you'll run into one that's significantly different from all the rest. All open source build systems tend to follow some very fundamental, high-level concepts—the idea of a configuration

---

5. Modern Autoconf generates `configure` scripts that require a full path for `--prefix` anyway. An error is generated if you try to use a relative path.

6. The tree utility is not often installed by default; you can install it using your system's package manager—the package name is usually simply tree.

process, followed by a build step, is one such principle. However, the nature of the configuration process as well as the command used to build the software might not align very closely with what we've discussed here. One of the benefits of the Autotools is the consistent nature of the build systems they generate.

If you want to understand how all this magic works, keep reading.

2

A BRIEF INTRODUCTION TO THE GNU AUTOTOOLS

We shall not cease from exploration
And the end of all our exploring
Will be to arrive where we started
And know the place for the first time.
—*T.S. Eliot, "Quartet No. 4: Little Gidding"*

As stated in the preface to this book, the purpose of the GNU Autotools is to make life simpler for the end user, not the maintainer. Nevertheless, using the Autotools will make your job as a project maintainer easier in the long run, although maybe not for the reasons you suspect. The Autotools framework is as simple as it can be, given the functionality it provides. The real purpose of the Autotools is twofold: it serves the needs of your users, and it makes your project incredibly portable—even to systems on which you've never tested, installed, or built your code.

Throughout this book, I will often use the term *Autotools*, although you won't find a package in the GNU archives with this label. I use this term to signify the following three GNU projects, which are considered by the community to be part of the GNU build system:

- Autoconf, which is used to generate a configuration script for a project
- Automake, which is used to simplify the process of creating consistent and functional makefiles
- Libtool, which provides an abstraction for the portable creation of shared libraries

Other build tools, such as the open source projects CMake and SCons, attempt to provide the same functionality as the Autotools but in a more user-friendly manner. However, because these tools attempt to hide much of their complexity behind GUI interfaces and script builders, they actually end up being less functional, and more difficult to manage, because the build system is not as transparent. In the final analysis, this transparency is what makes the Autotools both simpler to use and simpler to understand. Initial frustration with the Autotools, therefore, comes not from their complexity—for they are truly very simple—but from their extensive use of less well understood tools and subsystems, such as the Linux command shell (Bash), the make utility, and the M4 macro processor and accompanying macro libraries. Indeed, the meta-language provided by Automake is so simple it can be entirely digested and comprehended within a few hours of perusing the manual (though the ramifications of this meta-language may take a bit longer to thoroughly internalize).

Who Should Use the Autotools?

If you're writing open source software that targets Unix or Linux systems, you should absolutely be using the GNU Autotools, and even if you're writing proprietary software for Unix or Linux systems, you'll still benefit significantly from using them. The Autotools provide you with a build environment that allows your project to build successfully on future versions or distributions with virtually no changes to the build scripts. This is useful even if you only intend to target a single Linux distribution, because—let's be honest—you really *can't* know in advance whether or not your company will want your software to run on other platforms in the future.

When Should You Not Use the Autotools?

About the only time it makes sense not to use the Autotools is when you're writing software that will *only* run on non-Unix platforms, such as Microsoft Windows.

Autotools support for Windows requires an Msys[1] environment in order to work correctly, because Autoconf-generated configuration scripts are Bourne-shell scripts, and Windows doesn't provide a native Bourne shell.[2] Unix and Microsoft tools are just different enough in command line options and runtime characteristics that it's often simpler to use Windows ports of GNU tools, such as Cygwin, Msys2, or MinGW, to build Windows programs with an Autotools build system.

For these reasons, I'll focus mostly on using the Autotools on POSIX-compliant platforms. Nevertheless, if you're interested in trying out the Autotools on Windows, check out Chapter 17 for an in-depth overview.

NOTE *I'm not a typical Unix bigot. While I love Unix (and especially Linux), I also appreciate Windows for the areas in which it excels.[3] For Windows development, I highly recommend using Microsoft tools. The original reasons for using GNU tools to develop Windows programs are more or less academic nowadays because Microsoft has made the better part of its tools available for download at no cost. For download information, see Visual Studio Community at* https://visualstudio.microsoft.com/vs/express/.

Apple Platforms and Mac OS X

The Macintosh operating system has been POSIX compliant since 2007 when the "Leopard" release of macOS version 10 (OS X) was published. OS X is derived from NeXTSTEP/OpenStep, which is based on the Mach kernel, with parts taken from FreeBSD and NetBSD. As a POSIX-compliant operating system, OS X provides all the infrastructure required by the Autotools. The problems you'll encounter with OS X will most likely involve Apple's graphical user interface and package management systems, both of which are specific to the Mac.

The user interface presents the same issues you encounter when dealing with the X Window system on other Unix platforms, and then some. The primary difference is that the X Window system is used exclusively on most Unix systems, but macOS has its own graphical user interface called *Cocoa*. While the X Window system can be used on the Mac (Apple provides

---

1. See MinGW, Minimalist GNU for Windows at *http://www.mingw.org/* for more information on the Msys concept.

2. Windows 10 actually supports a Linux environment called the Windows Subsystem for Linux (WSL). The integration between the Windows host and the Linux subsystem is much tighter than that of, say, a virtual machine running Linux on a Windows host. It's well worth exploring if you're interested in running Linux but don't want to entirely give up your Windows applications. Be aware, however, that open source software programs built using the Autotools will not run as native Windows applications but will instead interface with the WSL kernel components. Perhaps these days the distinction simply isn't that important.

3. Hard-core gamers will agree with me, I'm sure. I wrote the original edition of this book on a laptop running Windows 7, but I used OpenOffice as my content editor, and I wrote the book's sample code on a 3GHz 64-bit dual-processor openSUSE 11.2 Linux workstation. Lately I've been running the Ubuntu-based Linux Mint distribution and using LibreOffice 5.3.

a window manager that makes X applications look a lot like native Cocoa apps), Mac programmers will sometimes wish to take full advantage of the native user interface features provided by the operating system.

The Autotools skirt the issue of package management differences between Unix platforms by simply ignoring them. Instead, they create packages that are little more than compressed source archives using the tar and gzip utilities, and they install and uninstall products from the make command line. The macOS package management system is an integral part of installing an application on an Apple system, and projects like Fink (*http://www.finkproject.org/*) and MacPorts (*http://www.macports.org/*) help make existing open source packages available on the Mac by providing simplified mechanisms for converting Autotools packages into installable Mac packages.

The bottom line is that the Autotools can be used quite effectively on Apple Macintosh systems running OS X or later, as long as you keep these caveats in mind.

The Choice of Language

Your choice of programming language is another important factor to consider when deciding whether to use the Autotools. Remember that the Autotools were designed by GNU people to manage GNU projects. In the GNU community, two factors determine the importance of a computer programming language:

- Are there any GNU packages written in the language?
- Does the GNU compiler tool set support the language?

Autoconf provides native support for the following languages based on these two criteria (by *native support*, I mean that Autoconf will compile, link, and run source-level feature checks in these languages):

- C
- C++
- Objective C
- Objective C++
- Fortran
- Fortran 77
- Erlang
- Go

Therefore, if you want to build a Java package, you can configure Automake to do so (as you'll see in Chapters 14 and 15), but you can't ask

Autoconf to compile, link, or run Java-based checks,[4] because Autoconf simply doesn't natively support Java. However, you can find Autoconf macros (which I will cover in more detail in later chapters) that enhance Autoconf's ability to manage the configuration process for projects written in Java.

The general feeling is that Java has plenty of its own build environments and tools that work very well (maven, for instance); therefore, adding full support for Java seems like a wasted effort. This is especially true since Java and its build tools are themselves highly portable—even to non-Unix/Linux platforms such as Windows.

Rudimentary support does exist in Automake for Java compilers and JVMs. I've used these features myself on projects, and they work well, as long as you don't try to push them too far.

If you're into Smalltalk, ADA, Modula, Lisp, Forth, or some other non-mainstream language, you're probably not too interested in porting your code to dozens of platforms and CPUs. However, if you *are* using a non-mainstream language and you're concerned about the portability of your build systems, consider adding support for your language to the Autotools yourself. This is not as daunting a task as you may think, and I guarantee that you'll be an Autotools expert when you're finished.[5]

Generating Your Package Build System

The GNU Autotools framework includes three main packages: Autoconf, Automake, and Libtool. The tools in these packages can depend on utilities and functionality from the gettext, M4, sed, make, and Perl packages, among others; however, the build systems generated by these packages rely only on a Bourne shell and the make utility.

With respect to the Autotools, it's important to distinguish between a *maintainer's* system and an *end user's* system. The design goals of the Autotools specify that an Autotools-generated build system should rely only on tools that are readily available and preinstalled on the end user's machine (assuming the end user's system has rudimentary support for building programs from source code). For example, the machine a maintainer uses to create distributions requires a Perl interpreter, but a machine on which an end user builds products from release distribution source archives should not require Perl (unless, of course, the project sources are written in Perl).

A corollary is that an end user's machine doesn't need to have the Autotools installed—an end user's system only requires a reasonably

---

4. This statement is not strictly true: I've seen third-party macros that use the Java virtual machine (JVM) to execute Java code within checks, but these are usually very special cases. None of the built-in Autoconf checks rely on a JVM in any way. Chapters 14 and 15 outline how you might use a JVM in an Autoconf check. Additionally, the portable nature of Java and the Java virtual machine specification make it fairly unlikely that you'll need to perform a Java-based Autoconf check in the first place.

5. For example, native Erlang support made it into the Autotools because members of the Erlang community thought it was important enough to add it themselves.

POSIX-compliant version of make and some variant of the Bourne shell that can execute the generated configuration script. And, of course, any package will also require compilers, linkers, and other tools needed to convert source files into executable binary programs, help files, and other runtime resources.

Configuration

Most developers understand the purpose of the make utility, but what's the point of configure? While Unix systems have followed the de facto standard Unix kernel interface for decades, most software has to stretch beyond these boundaries.

Originally, configuration scripts were hand-coded shell scripts designed to set environment variables based on platform-specific characteristics. They also allowed users to configure package options before running make. This approach worked well for decades, but as the number of Linux distributions and Unix-like systems grew, the variety of features and installation and configuration options exploded, so it became very difficult to write a decent portable configuration script. In fact, it was much more difficult to write a portable configuration script than it was to write makefiles for a new project. Therefore, most people just created configuration scripts for their projects by copying and modifying the script for a similar project.

In the early 1990s, it was apparent to many open source software developers that project configuration would become painful if something wasn't done to ease the burden of writing massive complex shell scripts to manage configuration options. The number of GNU project packages had grown to hundreds, and maintaining consistency across their separate build systems had become more time-consuming than simply maintaining the code for these projects. These problems had to be solved.

Autoconf

Autoconf[6] changed this paradigm almost overnight. David MacKenzie started the Autoconf project in 1991, but a look at the *AUTHORS* file in the Savannah Autoconf project[7] repository will give you an idea of the number of people who had a hand in making the tool. Although configuration scripts were long and complex, users needed to specify only a few variables when executing them. Most of these variables were simply choices about components, features, and options, such as *Where can the build system find libraries and header files? Where do I want to install my finished products? Which optional components do I want to build into my products?*

---

6. For more on Autoconf origins, see the GNU web page on the topic at *http://www.gnu.org /software/autoconf/*.

7. See *http://savannah.gnu.org/projects/autoconf/*.

Instead of modifying and debugging hundreds of lines of supposedly portable shell script, developers can now write a short metascript file using a concise, macro-based language, and Autoconf will generate a perfect configuration script that is more portable, more accurate, and more maintainable than a hand-coded one. In addition, Autoconf often catches semantic or logic errors that could otherwise take days to debug. Another benefit of Autoconf is that the shell code it generates is portable between most variations of the Bourne shell. Mistakes made in portability between shells are very common and, unfortunately, are the most difficult kinds of mistakes to find, because no one developer has access to all Bourne-like shells.

NOTE *While portable scripting languages like Perl and Python are now more pervasive than the Bourne shell, this was not the case when the idea for Autoconf was first conceived.*

Autoconf-generated configuration scripts provide a common set of options that are important to all portable software projects running on POSIX systems. These include options to modify standard locations (a concept I'll cover in more detail in Chapter 3), as well as project-specific options defined in the *configure.ac* file (which I'll discuss in Chapter 5).

The autoconf package provides several programs, including the following:

- autoconf
- autoreconf
- autoheader
- autoscan
- autoupdate
- ifnames
- autom4te

The autoconf program is a simple Bourne shell script. Its main task is to ensure that the current shell contains the functionality necessary to execute the m4 macro processor. (I'll discuss Autoconf's use of M4 in detail in Chapter 4.) The remainder of the script parses command line parameters and executes autom4te.

autoreconf

The autoreconf utility executes the configuration tools in the autoconf, automake, and libtool packages as required by the project. This utility minimizes the amount of regeneration required to address changes in timestamps, features, and project state. It was written as an attempt to consolidate existing maintainer-written, script-based utilities that ran all the required Autotools in the right order. You can think of autoreconf as a sort of smart Autotools bootstrap utility. If all you have is a *configure.ac* file, you can run autoreconf to execute all the tools you need, in the correct order, so that configure will be properly generated. Figure 2-1 shows how autoreconf interacts with other utilities in the Autotools suite.

FAIRPORT PUBLIC LIBRARY
1 VILLAGE LANDING
FAIRPORT, NY 14450   246

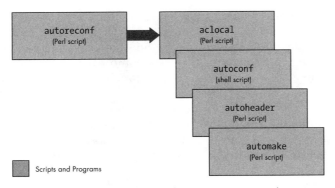

Figure 2-1: A dataflow diagram for the autoreconf utility

Nevertheless, there are times when a project requires more than simply bootstrapping the Autotools to get a developer up and running on a newly checked-out repository work area. In these cases, a small shell script that runs autoreconf, along with any non-Autotools-related processes, is appropriate. Many projects name such a script autogen.sh, but this is often confusing to developers because there is a GNU Autogen project. A better name would be something like bootstrap.sh.

Additionally, when used with the -i option, autoreconf will bootstrap a project into a distributable state by adding missing files that are recommended or required by GNU for proper open source projects. These include a proper *ChangeLog* and template *INSTALL, README,* and *AUTHORS* files and so on.

autoheader

The autoheader utility generates a C/C++ compatible header file template from various constructs in *configure.ac*. This file is usually called *config.h.in*. When the end user executes configure, the configuration script generates *config.h* from *config.h.in*. As maintainer, you'll use autoheader to generate the template file you will include in your distribution package. (We'll examine autoheader in greater detail in Chapter 4.)

autoscan

The autoscan program generates a default *configure.ac* file for a new project; it can also examine an existing Autotools project for flaws and opportunities for enhancement. (We'll discuss autoscan in more detail in Chapters 4 and 14.) autoscan is very useful as a starting point for a project that uses a non-Autotools-based build system, but it may also be useful for suggesting features that might enhance an existing Autotools-based project.

autoupdate

The autoupdate utility is used to update *configure.ac* or the template (*.in*) files to match the syntax supported by current versions of the Autotools.

ifnames

The `ifnames` program is a small and generally underused utility that accepts a list of source file names on the command line and displays a list of C-preprocessor definitions. This utility was designed to help maintainers determine what to put into the *configure.ac* and *Makefile.am* files to make them portable. If your project was written with some level of portability in mind, `ifnames` can help you determine where those attempts at portability are located in your source tree and give you the names of potential portability definitions.

autom4te

The `autom4te` utility is a Perl-based intelligent caching wrapper for `m4` that is used by most of the other Autotools. The `autom4te` cache decreases the time successive tools spend accessing *configure.ac* constructs by as much as 30 percent.

I won't spend a lot of time on `autom4te` (pronounced *automate*) because it's primarily used internally by the Autotools. The only sign that it's working is the *autom4te.cache* directory that appears in your top-level project directory after you run `autoconf` or `autoreconf`.

Working Together

Of the previously listed tools, `autoconf` and `autoheader` are the only ones project maintainers use when generating a `configure` script, and `autoreconf` is the only one that the developer needs to directly execute. Figure 2-2 shows the interaction between input files and `autoconf` and `autoheader` that generates the corresponding product files.

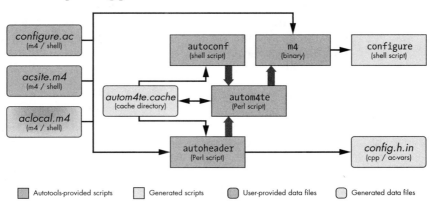

Figure 2-2: A data flow diagram for autoconf and autoheader

NOTE   *I use the data flow diagram format shown in Figure 2-2 throughout this book. Dark boxes represent objects provided either by the user or by an Autotools package. Light boxes represent generated objects. Boxes with square corners are scripts and programs, and boxes with rounded corners are data files. The meaning of most of* *the labels here should be obvious, but at least one deserves an explanation: the term* ac-vars *refers to Autoconf-specific replacement text. I'll explain the gradient shading of the* aclocal.m4 *box shortly.*

The primary task of this suite of tools is to generate a configuration script that can be used to configure a project build directory for a target platform (not necessarily the local host). This script does not rely on the Autotools themselves; in fact, autoconf is designed to generate configuration scripts that will run on all Unix-like platforms and in most variations of the Bourne shell. This means that you can generate a configuration script using autoconf and then successfully execute that script on a machine that does not have the Autotools installed.

The autoconf and autoheader programs are executed either directly by you or indirectly by autoreconf. They take their input from your project's *configure .ac* file and various Autoconf-flavored M4 macro definition files (which, by convention, have a *.m4* extension), using autom4te to maintain cache information. The autoconf program generates a configuration script called configure, a very portable Bourne shell script that enables your project to offer many useful configuration capabilities. The program autoheader generates the *config.h.in* template based on certain macro definitions in *configure.ac*.

Automake

Once you've done it a few times, writing a basic makefile for a new project is fairly simple. But problems may occur when you try to do more than just the basics. And let's face it—what project maintainer has ever been satisfied with just a basic makefile?

Attention to detail is what makes an open source project successful. Users lose interest in a project fairly easily—especially when functionality they expect is missing or improperly written. For example, power users have come to expect makefiles to support certain standard targets or goals, specified on the make command line, like this:

```
$ make install
```

Common make targets include all, clean, and install. In this example, install is the target. But you should realize that none of these are *real* targets: a *real target* is a filesystem object that is produced by the build system—usually a file (but sometimes a directory or a link). When building an executable called doofabble, for instance, you'd expect to be able to enter:

```
$ make doofabble
```

For this project, doofabble is a real target, and this command works for the doofabble project. However, requiring the user to enter real targets on the make command line is asking a lot of them, because each project must be built differently—make doofabble, make foodabble, make abfooble, and so on. Standardized targets for make allow all projects to be built in the same way using commonly known commands like make all and make clean. But *commonly known* doesn't mean *automatic,* and writing and maintaining makefiles that support these targets is tedious and error prone.

Automake's job is to convert a simplified specification of your project's build process into boilerplate makefile syntax that always works correctly the first time *and provides all the standard functionality expected.* Automake creates projects that support the guidelines defined in the *GNU Coding Standards* (discussed in Chapter 3).

Just like autoconf produces a configure script that is portable to many flavors of the Bourne shell, automake produces make script that is portable to many flavors of make.

The automake package provides the following tools in the form of Perl scripts:

- automake
- aclocal

automake

The automake program generates standard makefile templates (named *Makefile.in*) from high-level build specification files (named *Makefile.am*). These *Makefile.am* input files are essentially just regular makefiles. If you were to put only the few required Automake definitions in a *Makefile.am* file, you'd get a *Makefile.in* file containing several hundred lines of parameterized make script.

If you add additional make syntax to a *Makefile.am* file, Automake will move this code to the most functionally correct location in the resulting *Makefile.in* file. In fact, you can write your *Makefile.am* files so all they contain is ordinary make script, and the resulting makefiles will work just fine. This pass-through feature gives you the ability to extend Automake's functionality to suit your project's specific requirements.[8]

aclocal

In the *GNU Automake Manual,* the aclocal utility is documented as a temporary workaround for a certain lack of flexibility in Autoconf. Automake enhances Autoconf by adding an extensive set of macros, but Autoconf was not really designed with this level of enhancement in mind.

The original documented method for adding user-defined macros to an Autoconf project was to create a file called *aclocal.m4,* place the user-defined macros in this file, and place the file in the same directory as *configure.ac.* Autoconf then automatically includes this set of macros while processing *configure.ac.* The designers of Automake found this extension mechanism too useful to pass up; however, users would have been required to add an m4_include statement to a possibly unnecessary *aclocal.m4* file in order to

---

8. Other metabuild tools like CMake also generate makefiles but do not allow you to directly specify what ends up in these files. Rather, you have to find the correct approach in CMake's macro language in order to coerce it into writing make script that does what you want it to.

include the Automake macros. Since both user-defined macros and the use of M4 itself are considered advanced concepts, this was deemed too harsh a requirement.

The aclocal script was designed to solve this problem. This utility generates an *aclocal.m4* file for a project that contains both user-defined macros and all required Automake macros.[9] Instead of adding user-defined macros directly to *aclocal.m4*, project maintainers should now add them to a new file called *acinclude.m4*.

To make it clear to readers that Autoconf doesn't depend on Automake (and perhaps due to a bit of stubbornness), the *GNU Autoconf Manual* doesn't make much mention of the aclocal utility. The *GNU Automake Manual* originally suggested that you rename *aclocal.m4* to *acinclude.m4* when adding Automake to an existing Autoconf project, and this approach is still commonly used. The flow of data for aclocal is depicted in Figure 2-3.

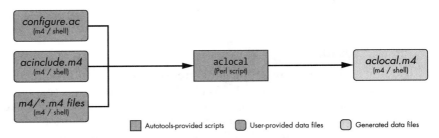

Figure 2-3: A data flow diagram for `aclocal`

However, the latest documentation for both Autoconf and Automake suggests that the entire paradigm is now obsolete. Developers should now specify a directory that contains a set of M4 macro files. The current recommendation is to create a directory in the project root directory called *m4* and add macros as individual *.m4* files to it. All files in this directory will be gathered into *aclocal.m4* before Autoconf processes *configure.ac*.[10]

It may now be more apparent why the *aclocal.m4* box in Figure 2-2 couldn't decide which color it should be. When you're using it without Automake and Libtool, you write *aclocal.m4* by hand. However, when you're using it with Automake, the file is generated by the aclocal utility, and you provide project-specific macros either in *acinclude.m4* or in an *m4* directory.

Libtool

How do you build shared libraries on different Unix platforms without adding a lot of very platform-specific conditional code to your build system and source code? This is the question that the Libtool project tries to address.

---

9. Automake macros are copied into this file, but the user-written *acinclude.m4* file is merely referenced with an m4_include statement at the end of the file.

10. As with *acinclude.m4*, this gathering is virtual; *aclocal.m4* merely contains m4_include statements that reference these other files in place.

There's a significant amount of common functionality among Unix-like platforms. However, one very significant difference has to do with how shared libraries are built, named, and managed. Some platforms name their libraries *lib*name.*so*, others use *lib*name.*a* or even *lib*name.*sl*. The Cygwin system for Windows names Cygwin-generated shared libraries *cyg*name.*dll*. Still others don't even provide native shared libraries. Some platforms provide *libdl.so* to allow software to dynamically load and access library functionality at runtime, while others provide different mechanisms, and some platforms don't provide this functionality at all.

The developers of Libtool have carefully considered all of these differences. Libtool supports dozens of platforms, not only providing a set of Autoconf macros that hide library-naming differences in makefiles but also offering an optional library of dynamic loader functionality that can be added to programs. This functionality allows maintainers to make their runtime, dynamic shared-object management code more portable and easier to maintain.

The libtool package provides the following programs, libraries, and header file:

- libtool (program)
- libtoolize (program)
- *ltdl* (static and shared libraries)
- *ltdl.h* (header file)

libtool

The libtool shell script that ships with the libtool package is a generic version of the custom script that libtoolize generates for a project.

libtoolize

The libtoolize shell script prepares your project to use Libtool. It generates a custom version of the generic libtool script and adds it to your project directory. This custom script is shipped with the project along with the Automake-generated makefiles, which execute the script on the user's system at the appropriate time.

ltdl, the Libtool C API

The libtool package also provides the *ltdl* library and associated header files, which provide a consistent runtime shared-object manager across platforms. The *ltdl* library may be linked statically or dynamically into your programs, giving them a consistent runtime shared-library access interface between platforms.

Figure 2-4 illustrates the interaction between the automake and libtool scripts, and the input files used to create products that configure and build your projects.

Automake and Libtool are both standard pluggable options that can be added to *configure.ac* with just a few simple macro calls.

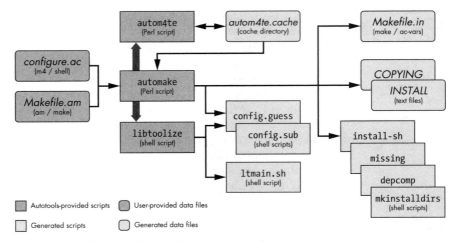

- Autotools-provided scripts
- Generated scripts
- User-provided data files
- Generated data files

Figure 2-4: A data flow diagram for automake and libtool

Building Your Package

As maintainer, you probably build your software packages fairly often, and you're also probably intimately familiar with your project's components, architecture, and build system. However, you should make sure that your users' build experiences are much simpler than your own. One way to do this is to give users a simple, easy-to-understand pattern to follow when building your software packages. In the following sections, I'll show you the build pattern supported by the Autotools.

Running configure

After running the Autotools, you're left with a shell script called configure and one or more *Makefile.in* files. These files are intended to be shipped with your project release distribution packages.[11] Your users will download these packages, unpack them, and enter ./configure && make from the top-level project directory. The configure script will generate makefiles (called *Makefile*) from the *Makefile.in* templates created by automake and a *config.h* header file from the *config.h.in* template generated by autoheader.

Automake generates *Makefile.in* templates rather than makefiles because without makefiles, your users can't run make; you don't want them to run make until after they've run configure, and this functionality guards against them doing so. *Makefile.in* templates are nearly identical to makefiles you might

---

11. GPL licensing also requires *configure.ac* and *Makefile.am* to be shipped with your package, and the Autotools ensure that these files are in the distribution tarball. The reasoning is that the GPL requires the full source of a project to be distributed in preferred-editing form. A user obtaining the distribution tarball would not be able to edit anything without the base source files for the build system. However, end users need not touch or interact with these files unless they wish to customize the program in a manner not supported by project configuration options.

write by hand, except that you didn't have to. They also do a lot more than most people are willing to hand-code. Another reason for not shipping ready-to-run makefiles is that it gives `configure` the chance to insert platform characteristics and user-specified optional features directly into the makefiles. This makes them a better fit for their target platforms and the end user's build preferences. Finally, the makefiles can also be generated outside the source tree, which means you can create custom build systems in different directories for the same source directory tree. I'll discuss this topic in greater detail in "Building Outside the Source Directory" on page 28.

Figure 2-5 illustrates the interaction between `configure` and the scripts it executes during the configuration process in order to create the makefiles and the *config.h* header file.

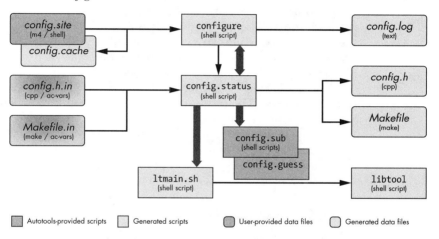

Figure 2-5: A data flow diagram for `configure`

The `configure` script has a bidirectional relationship with another script called `config.status`. You may have thought that your `configure` script generated your makefiles. But actually, the only file (besides a log file) that `configure` generates is `config.status`.

The `configure` script is designed to determine platform characteristics and features available on the user's system, as specified in the maintainer-written *configure.ac*. Once it has this information, it generates `config.status`, which contains all of the check results, and then it executes this script. The `config.status` script, in turn, uses the check information embedded within it to generate platform-specific *config.h* and makefiles, as well as any other template-based output files specified in *configure.ac*.

NOTE *As the double-ended fat arrow in Figure 2-5 shows, `config.status` can also call `configure`. When used with the `--recheck` option, `config.status` will call `configure` using the same command line options used to originally generate `config.status`.*

The `configure` script also generates a log file called *config.log*, which will contain very useful information in the event that an execution of `configure` fails on the user's system. As the maintainer, you can use this information for debugging. The *config.log* file also logs how configure was executed. (You can run config.status --version to discover the command line options used to generate config.status.) This feature can be particularly handy when, for example, a user returns from vacation and can't remember which options they used to originally generate the project build directory.

NOTE *To regenerate makefiles and the* config.h *header files, just enter* ./config.status *from within the project build directory. The output files will be generated using the same options originally used to generate* config.status.

The *config.site* file can be used to customize the way configure works based on the --prefix option passed to it. The *config.site* file is a script, but it's not meant to be executed directly. Rather, configure looks for *$(prefix)/share/config .site* and "sources" it (incorporates it as part of its own script) before executing any of its own code. This can be a handy way of specifying the same set of options for many packages, all destined to be built and installed the same way. Since configure is just a shell script, *config.site* should just contain shell code.

The *config.cache* file is generated by configure when the -C or --config-cache options are used. The results of configuration tests are cached in this file and are reusable by subdirectory configure scripts or by future runs of configure. By default, *config.cache* is disabled because it can be a potential source of configuration errors. If you're confident with your configuration process, *config.cache* can really speed up the configuration process between executions of configure.

Building Outside the Source Directory

A little-known feature of Autotools build environments is that they don't need to be generated within a project source tree. That is, if a user executes configure from a directory other than the project source directory, they can generate a full build environment within an isolated build directory.

In the following example, the user downloads *doofabble-3.0.tar.gz*, unpacks it, and creates two sibling directories called *doofabble-3.0.debug* and *doofabble-3.0.release*. They change into the *doofabble-3.0.debug* directory; execute doofabble's configure script, using a relative path, with a doofabble-specific debug option; and then run make from within this same directory. Then they switch over to the *doofabble-3.0.release* directory and do the same thing, this time running configure without the debug option:

```
$ gzip -dc doofabble-3.0.tar.gz | tar xf -
$ mkdir doofabble-3.0.debug
$ mkdir doofabble-3.0.release
$ cd doofabble-3.0.debug
$ ../doofabble-3.0/configure --enable-debug
--snip--
$ make
--snip--
$ cd ../doofabble-3.0.release
$ ../doofabble-3.0/configure
```

```
--snip--
$ make
--snip--
```

Users generally don't care about remote build functionality, because all they usually want to do is configure, build, and install your code on their platforms. Maintainers, on the other hand, find remote build functionality very useful, as it allows them to not only maintain a reasonably pristine source tree but also to maintain multiple build environments for their project, each with complex configuration options. Rather than reconfigure a single build environment, a maintainer can simply switch to another build directory that has been configured with different options.

There is one case, however, where a user might wish to use remote-build. Consider the case where one obtains the full unpacked source code of a project on CD or has access to it via a read-only NFS mount. The ability to build outside the source tree can grant the ability to build the project without having to copy it to writable media.

Running make

Finally, you run plain old make. The designers of the Autotools went to a *lot* of trouble to ensure that you didn't need any special version or brand of make. Figure 2-6 depicts the interaction between make and the makefiles that are generated during the build process.

NOTE   *There has been some discussion on the Autotools mailing lists during the last few years about supporting only GNU make, as modern GNU make is so much more functional than other make utilities. Almost all Unix-y platforms (and even Microsoft Windows) have a version of GNU make today, so the rationale for continuing to support other brands of make is no longer as important as it once was.*

As you can see, make runs several generated scripts, but these are all really ancillary to the make process. The generated makefiles contain commands that execute these scripts under the appropriate conditions. These scripts are part of the Autotools, and they are either shipped with your package or generated by your configuration script.

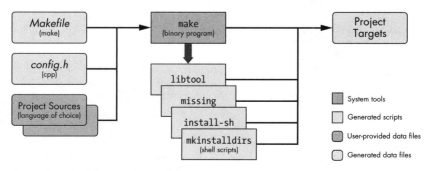

Figure 2-6: A data flow diagram for make

Installing the Most Up-to-Date Autotools

If you're running a variant of Linux and you've chosen to install the compilers and tools used for developing C-language software, you probably already have some version of the Autotools installed on your system. To determine which versions of Autoconf, Automake, and Libtool you're using, simply open a terminal window and type the following commands (if you don't have the which utility on your system, try **type -p** instead):

```
$ which autoconf
/usr/local/bin/autoconf
$
$ autoconf --version
autoconf (GNU Autoconf) 2.69
Copyright (C) 2012 Free Software Foundation, Inc.
License GPLv3+/Autoconf: GNU GPL version 3 or later
<http://gnu.org/licenses/gpl.html>, <http://gnu.org/licenses/exceptions.html>
This is free software: you are free to change and redistribute it.
There is NO WARRANTY, to the extent permitted by law.

Written by David J. MacKenzie and Akim Demaille.
$
$ which automake
/usr/local/bin/automake
$
$ automake --version
automake (GNU automake) 1.15
Copyright (C) 2014 Free Software Foundation, Inc.
License GPLv2+: GNU GPL version 2 or later <http://gnu.org/licenses/gpl-
2.0.html>
This is free software: you are free to change and redistribute it.
There is NO WARRANTY, to the extent permitted by law.

Written by Tom Tromey <tromey@redhat.com>
        and Alexandre Duret-Lutz <adl@gnu.org>.
$
$ which libtool
/usr/local/bin/libtool
$
$ libtool --version

libtool (GNU libtool) 2.4.6
Written by Gordon Matzigkeit, 1996

Copyright (C) 2014 Free Software Foundation, Inc.
This is free software; see the source for copying conditions.  There is NO
warranty; not even for MERCHANTABILITY or FITNESS FOR A PARTICULAR PURPOSE.
$
```

NOTE   *If you have the Linux-distribution varieties of these Autotools packages installed on your system, the executables will probably be found in* /usr/bin *rather than* /usr/local/bin, *as you can see from the output of the* which *command here.*

If you choose to download, build, and install the latest released version of any one of these packages from the GNU website, you must do the same for all of them, because the Automake and Libtool packages install macros into the Autoconf macro directory. If you don't already have the Autotools installed, you can install them using your system package manager (for example, yum or apt), or from source, using their GNU distribution source archives. The latter can be done with the following commands (be sure to change the version numbers as necessary):

```
$ mkdir autotools && cd autotools
$ wget -q https://ftp.gnu.org/gnu/autoconf/autoconf-2.69.tar.gz
$ wget -q https://ftp.gnu.org/gnu/autoconf/autoconf-2.69.tar.gz.sig
$ gpg autoconf-2.69.tar.gz.sig
gpg: assuming signed data in `autoconf-2.69.tar.gz'
gpg: Signature made Tue 24 Apr 2012 09:17:04 PM MDT using RSA key ID 2527436A
gpg: Can't check signature: public key not found
$
$ gpg --keyserver keys.gnupg.net --recv-key 2527436A
gpg: requesting key 2527436A from hkp server keys.gnupg.net
gpg: key 2527436A: public key "Eric Blake <eblake@redhat.com>" imported
gpg: key 2527436A: public key "Eric Blake <eblake@redhat.com>" imported
gpg: no ultimately trusted keys found
gpg: Total number processed: 2
gpg:               imported: 2  (RSA: 2)$ gpg autoconf-2.69.tar.gz.sig
gpg: assuming signed data in `autoconf-2.69.tar.gz'
gpg: Signature made Tue 24 Apr 2012 09:17:04 PM MDT using RSA key ID 2527436A
gpg: Good signature from "Eric Blake <eblake@redhat.com>"
gpg:                 aka "Eric Blake (Free Software Programmer) <ebb9@byu.net>"
gpg:                 aka "[jpeg image of size 6874]"
gpg: WARNING: This key is not certified with a trusted signature!
gpg:          There is no indication that the signature belongs to the owner.
Primary key fingerprint: 71C2 CC22 B1C4 6029 27D2  F3AA A7A1 6B4A 2527 436A
$
$ gzip -cd autoconf* | tar xf -
$ cd autoconf*/
$ ./configure && make all check
        # note - a few tests (501 and 503, for example) may fail
        # - this is fine for this release)
--snip--
$ sudo make install
--snip--
$ cd ..
$ wget -q https://ftp.gnu.org/gnu/automake/automake-1.16.1.tar.gz
$ wget -q https://ftp.gnu.org/gnu/automake/automake-1.16.1.tar.gz.sig
$ gpg automake-1.16.1.tar.gz.sig
gpg: assuming signed data in `automake-1.16.1.tar.gz'
gpg: Signature made Sun 11 Mar 2018 04:12:47 PM MDT using RSA key ID 94604D37
gpg: Can't check signature: public key not found
$
$ gpg --keyserver keys.gnupg.net --recv-key 94604D37
gpg: requesting key 94604D37 from hkp server keys.gnupg.net
gpg: key 94604D37: public key "Mathieu Lirzin <mthl@gnu.org>" imported
gpg: no ultimately trusted keys found
gpg: Total number processed: 1
```

```
gpg:                   imported: 1  (RSA: 1)
$
$ gpg automake-1.16.1.tar.gz.sig
gpg: assuming signed data in `automake-1.16.1.tar.gz'
gpg: Signature made Sun 11 Mar 2018 04:12:47 PM MDT using RSA key ID 94604D37
gpg: Good signature from "Mathieu Lirzin <mthl@gnu.org>"
gpg:                 aka "Mathieu Lirzin <mthl@openmailbox.org>"
gpg:                 aka "Mathieu Lirzin <mathieu.lirzin@openmailbox.org>"
gpg: WARNING: This key is not certified with a trusted signature!
gpg:          There is no indication that the signature belongs to the owner.
Primary key fingerprint: F2A3 8D7E EB2B 6640 5761  070D 0ADE E100 9460 4D37
$
$ gzip -cd automake* | tar xf -
$ cd automake*/
$ ./configure && make all check
--snip--
$ sudo make install
--snip--
$ cd ..
$ wget -q https://ftp.gnu.org/gnu/libtool/libtool-2.4.6.tar.gz
$ wget -q https://ftp.gnu.org/gnu/libtool/libtool-2.4.6.tar.gz.sig
$ gpg libtool-2.4.6.tar.gz.sig
gpg: assuming signed data in `libtool-2.4.6.tar.gz'
gpg: Signature made Sun 15 Feb 2015 01:31:09 PM MST using DSA key ID 2983D606
gpg: Can't check signature: public key not found
$
$ gpg --keyserver keys.gnupg.net --recv-key 2983D606
gpg: requesting key 2983D606 from hkp server keys.gnupg.net
gpg: key 2983D606: public key "Gary Vaughan (Free Software Developer) <gary@vaughan.pe>"
imported
gpg: key 2983D606: public key "Gary Vaughan (Free Software Developer) <gary@vaughan.pe>"
imported
gpg: no ultimately trusted keys found
gpg: Total number processed: 2
gpg:                   imported: 2  (RSA: 1)
$
$ gpg libtool-2.4.6.tar.gz.sig
gpg: assuming signed data in `libtool-2.4.6.tar.gz'
gpg: Signature made Sun 15 Feb 2015 01:31:09 PM MST using DSA key ID 2983D606
gpg: Good signature from "Gary Vaughan (Free Software Developer) <gary@vaughan.pe>"
gpg:                   aka "Gary V. Vaughan <gary@gnu.org>"
gpg:                   aka "[jpeg image of size 9845]"
gpg: WARNING: This key is not certified with a trusted signature!
gpg:          There is no indication that the signature belongs to the owner.
Primary key fingerprint: CFE2 BE70 7B53 8E8B 2675  7D84 1513 0809 2983 D606
$
$ gzip -cd libtool* | tar xf -
$ cd libtool*/
$ ./configure && make all check
--snip--
$ sudo make install
--snip--
$ cd ..
$
```

The preceding example shows how to use the associated *.sig* files to validate the signature on GNU packages. The example assumes you have not configured a gpg key server on your system and that you have not installed the public key for any of these packages. If you have already configured a preferred key server, you can skip the gpg command line --keyserver options. Once you've imported the public keys for these packages, you need not do it again.

You may also wish to install in a manner that does not require root access via sudo. To do this, execute configure with a --prefix option such as --prefix=$HOME/autotools and then add *~/autotools/bin* to your PATH environment variable.

You should now be able to successfully execute the version-check commands from the previous example. If you still see older versions, ensure your PATH environment variable properly contains */usr/local/bin* (or wherever you installed to) before */usr/bin*.

Summary

In this chapter, I presented a high-level overview of the Autotools to give you a feel for how everything ties together. I also showed you the pattern to follow when building software from distribution tarballs created by Autotools build systems. Finally, I showed you how to install the Autotools and how to tell which versions you have installed.

In Chapter 3, we'll step away from the Autotools briefly and begin creating a hand-coded build system for a toy project called *Jupiter*. You'll learn the requirements of a reasonable build system, and you'll become familiar with the rationale behind the original design of the Autotools. With this background knowledge, you'll begin to understand why the Autotools do things the way they do. I can't really emphasize this enough: *Chapter 3 is one of the most important chapters in this book, because it will get you past any emotional stigma you may have associated with the Autotools due to misconceptions.*

3

UNDERSTANDING THE GNU CODING STANDARDS

I don't know what's the matter with people: they don't
learn by understanding, they learn by some other way—
by rote or something. Their knowledge is so fragile!
—Richard Feynman, "Surely You're Joking,
Mr. Feynman!"

In Chapter 2, I gave an overview of the
GNU Autotools and some resources that
can help reduce the learning curve required
to master them. In this chapter, we're going to
step back a little and examine project organization
techniques that you can apply to any project, not just
one that uses the Autotools.

When you're done reading this chapter, you should be familiar with the
common make targets and why they exist. You should also have a solid understanding of why projects are organized the way they are. You will, in fact, be
well on your way to becoming an Autotools expert.

The information provided in this chapter comes primarily from two sources:

- The *GNU Coding Standards (GCS)*[1]
- The *Filesystem Hierarchy Standard (FHS)*[2]

If you'd like to brush up on your make syntax, you may also find the *GNU Make Manual*[3] very useful. If you're particularly interested in portable make syntax, then check out the POSIX man page for make.[4] Note, however, there are current discussions on the Autotools mailing lists around making GNU make the target standard because it's so widely available today. Therefore, portable make script isn't as important as it used to be.

Creating a New Project Directory Structure

You need to ask yourself two questions when you're setting up the build system for an open source software project:

- Which platforms will I target?
- What do my users expect?

The first is an easy question—you get to decide which platforms to target, but you shouldn't be too restrictive. Open source software projects are only as good as their communities, and arbitrarily limiting the number of platforms reduces the potential size of your community. However, you might consider supporting only current versions of your target platforms. You can check with user groups and communities to determine which versions of each are relevant.

The second question is more difficult to answer. First, let's narrow the scope to something manageable. What you really need to ask is, *What do my users expect of my build system?* Experienced open source software developers become familiar with these expectations by downloading, unpacking, building, and installing hundreds of packages. Eventually, they come to know intuitively what users expect of a build system. But, even so, the processes of package configuration, build, and installation vary widely, so it's difficult to define any solid norm.

Rather than taking a survey of every build system out there yourself, you can consult the Free Software Foundation (FSF), sponsor of the GNU project, which has done a lot of the legwork for you. The FSF provides some of the best definitive sources of information on free, open source software, including the *GCS*, which covers a wide variety of topics related to writing,

---

1. See the Free Software Foundation's *GNU Coding Standards* at *http://www.gnu.org/prep/standards/*.

2. See Daniel Quinlan's overview at *http://www.pathname.com/fhs/*.

3. See the Free Software Foundation's *GNU Make Manual* at *http://www.gnu.org/software/make/manual/*.

4. See the "Open Group Base Specifications," Issue 6, at *http://www.opengroup.org/online pubs/009695399/utilities/make.html*.

publishing, and distributing free, open source software. Even many non-GNU open source software projects align themselves with the *GCS*. Why? Well, the FSF invented the concept of free software, and the ideas make sense, for the most part.[5] There are dozens of issues to consider when designing a system that manages packaging, building, and installing software, and the *GCS* takes most of them into account.

WHAT'S IN A NAME?

You probably know that open source software projects generally have quirky names—they might be named after some device, an invention, a Latin term, a past hero, an ancient god, or they might be named after some small, furry animal that has (vaguely) similar characteristics to the software. Some names are just made-up words or acronyms that are catchy and easy to pronounce. Another significant characteristic of a good project name is uniqueness—it's important that your project be easy to distinguish from others. You also want your project name to be easy to distinguish from any other uses of the name in a search engine. Additionally, you should ensure that your project's name does not have negative connotations in any language or culture.

Project Structure

We'll start with a basic sample project and build on it as we continue our exploration of source-level software distribution. We'll call our project *Jupiter* and create a project directory structure using the following commands:

```
$ cd projects
$ mkdir -p jupiter/src
$ cd jupiter
$ touch Makefile src/Makefile jupiter/src/main.c
$
```

We now have one source code directory called *src*, one C source file called *main.c*, and a makefile for each of the two directories in our project. Minimal, yes, but this is a new endeavor and the key to a successful open source software project is evolution. Start small and grow as needed—and as you have the time and inclination.

Let's start by adding support for building and cleaning our project. We'll need to add other important capabilities to our build system later on, but these two will get us going. The top-level makefile does very little at this point; it merely passes requests down to *src/Makefile*, recursively. This

---

5. In truth, it's likely that the standards that came about from the BSD project were written much earlier than the standards of the FSF, but the FSF had a big hand in spreading the information to many different platforms and non-system-specific software projects. Thus, it had a large part in making these standards publicly visible and widely used.

constitutes a fairly common type of build system, known as a *recursive build system*, so named because makefiles recursively invoke make on subdirectory makefiles.[6] We'll spend a little time at the end of this chapter considering how to convert our recursive system into a nonrecursive system.

Listings 3-1 through 3-3 show the contents of each of these three files, thus far.

Git tag 3.0

```
all clean jupiter:
        cd src && $(MAKE) $@

.PHONY: all clean
```

Listing 3-1: Makefile: An initial draft of a top-level makefile for Jupiter

```
all: jupiter

jupiter: main.c
        gcc -g -O0 -o $@ main.c
clean:
        -rm jupiter

.PHONY: all clean
```

Listing 3-2: src/Makefile: The first draft of Jupiter's src directory makefile

```
#include <stdio.h>
#include <stdlib.h>

int main(int argc, char * argv[])
{
    printf("Hello from %s!\n", argv[0]);
    return 0;
}
```

Listing 3-3: src/main.c: The first version of the only C source file in the Jupiter project

NOTE *As you read this code, you will probably notice places where a makefile or a source code file contains a construct that is not written in the simplest manner or is perhaps not written the way you would have chosen to write it. There is a method to my madness: I've tried to use constructs that are portable to many flavors of the* make *utility.*

Now let's discuss the basics of make. If you're already pretty well versed in it, you can skip the next section. Otherwise, give it a quick read, and we'll return our attention to the Jupiter project later in the chapter.

---

6. Peter Miller's seminal paper "Recursive Make Considered Harmful" (*http://aegis.source forge.net/auug97.pdf*), published over 20 years ago, discusses some of the problems recursive build systems can cause. I encourage you to read this paper and understand the issues Miller presents. While the issues are valid, the sheer simplicity of implementing and maintaining a recursive build system makes it, by far, the most widely used form of build system.

Makefile Basics

If you don't use make on a regular basis, it's often difficult to remember exactly what goes where in a makefile, so here are a few things to keep in mind. Besides comments, which begin with a hash mark (#), there are only two basic types of entities in a makefile:

- Rule definitions
- Variable assignments

While there are several other types of constructs in a makefile (including conditional statements, directives, extension rules, pattern rules, function variables, and include statements, among others), for our purposes, we'll just touch lightly on them as needed instead of covering them all in detail. This doesn't mean they're unimportant. On the contrary, they're very useful if you're going to write your own complex build system by hand. However, our purpose is to gain the background necessary for understanding the GNU Autotools, so I'll only cover the aspects of make you need to know to accomplish that goal.

If you want a broader education on make syntax, refer to the *GNU Make Manual*. For strictly portable syntax, the POSIX man page for make is an excellent reference. If you want to become a make expert, be prepared to spend a good deal of time studying these resources—there's much more to the make utility than is initially apparent.

Rules

Rules follow the general format shown in Listing 3-4.

```
targets: [dependencies][; command-0]
[<tab>command-1
<tab>command-2
--snip--
<tab>command-N]
```

Listing 3-4: The syntax of a rule within a makefile

In this syntax definition, square brackets ([and]) denote optional portions of a rule and <tab> represents a TAB (CTRL-I) character.

Except for the TAB characters and the line feeds, all other whitespace is optional and ignored. When a line in a makefile begins with a TAB character, make generally considers it a command (with the exception of continuation lines, discussed later). Indeed, one of the most frustrating aspects of makefile syntax to neophytes and experts alike is that commands must be prefixed with an essentially invisible character. The error messages generated by the legacy UNIX make utility when a required TAB is missing (or has been converted to spaces by your editor), or when an unintentional TAB is inserted at the start of a line that follows something that could be interpreted as a rule, are obscure at best. GNU make does a better job with such error messages. Nonetheless, be careful to use leading TAB characters properly in your makefiles—always and only before commands.[7]

Note that almost everything in a rule is optional; the only required aspect of a rule is the *targets* portion and its colon (:) character. Use of the first command, *command-0* and its preceding semicolon (;), is an optional form that's generally discouraged by the Autotools, but is perfectly legitimate make syntax if you have a single command to execute. You may even combine *command-0* with additional commands, but this almost never done.

In general, *targets* are objects that need to be built, and *dependencies* are objects that provide source material for targets. Thus, targets are said to depend upon the dependencies. Dependencies are essentially *prerequisites* of the targets, and therefore they should be updated first.[8]

Listing 3-5 shows the general layout of a makefile.

```
var1 = val1
var2 = val2
--snip--
target1 : t1_dep1 t1_dep2 ... t1_depN
<tab>shell-command1a
<tab>shell-command1b
--snip--
target2 : t2_dep1 t2_dep2 ... t2_depN
<tab>shell-command2a
<tab>shell-command2b
--snip--
```

Listing 3-5: The general layout of a makefile

The contents of a makefile comprise a *declarative language* wherein you define a set of desired goals and make decides the best way to accomplish those goals. The make utility is a rule-based command engine, and the rules at work indicate which commands should be executed and when. When you define commands within rules, you're telling make that you want it to execute each of the following statements from a shell whenever the preceding target should be built. Presumably, the commands actually do create or update the target. The existence and timestamps of the files mentioned in the targets and dependencies of rules indicate whether the commands should be executed and in what order.

As make processes the text in a makefile, it builds a web of dependency chains (technically called a *directed acyclic graph*, or *DAG*). When building a particular target, make must walk backward through the entire graph to the beginning of each "chain." While traversing a chain, make executes the commands for each rule, beginning with the rule farthest from the target and

---

7. In the spirit of full disclosure, it is possible to use TAB characters at the start of some other lines, but you have to be careful not to accidentally trick make into thinking you're building a rule, so I find it easiest to simply not use TAB except at the start of commands.

8. You'll often hear dependencies referred to as *prerequisites* for this reason. In fact, the *GNU Make Manual* calls dependencies "prerequisites" and commands "recipes."

working forward to the rule for the desired target. As make discovers targets that are older than their dependencies, it must execute the associated set of commands to update those targets before it can process the next rule in the chain. As long as the rules are written correctly, this algorithm ensures that make will build a completely up-to-date product using the least number of operations possible. Indeed, as we'll see shortly, when the rules in a makefile are written properly, it's rather a joy to watch it run after various changes to files in the project.

Variables

Lines in a makefile containing an equal sign (=) are variable definitions. Variables in makefiles are somewhat similar to shell or environment variables, but there are some key differences.

In Bourne-shell syntax, you'd reference a variable in this manner: ${my_var}. Equally viable, without the curly brackets, is $my_var. The syntax for referencing variables in a makefile is nearly identical, except that you have the choice of using curly brackets or parentheses: $(my_var). To minimize confusion, it has become somewhat of a convention to use parentheses rather than curly brackets when dereferencing make variables. For single-character make variables, using these delimiters is optional, but you should use them in order to avoid ambiguity. For example, $X is functionally equivalent to $(X) or ${X}, but $(my_var) would require parentheses so make does not interpret the reference as $(m)y_var.

NOTE   *To dereference a shell variable inside a make command, escape the dollar sign by doubling it—for example, $${shell_var}. Escaping the dollar sign tells make not to interpret the variable reference but rather to treat it as literal text in the command. The variable reference is thus left to be interpolated by the shell when the command is executed.*

By default, make reads the process environment into its variable table before processing the makefile; this allows you to access most environment variables without explicitly defining them in the makefile. Note, however, that variables set inside the makefile will override those obtained from the environment.[9] It's generally not a good idea to depend on the existence of environment variables in your build process, although it's okay to use them conditionally. In addition, make defines several useful variables of its own, such as the MAKE variable, the value of which is the command used to invoke make for the current process.

You can assign variables at any point in the makefile. However, you should be aware that make processes a makefile in two passes. In the first pass, it gathers variables and rules into tables and internal structures. In the second pass, it resolves dependencies defined by the rules, invoking those rules as necessary to rebuild the dependencies based on the filesystem

---

9. You can use the -e option on the make command line to reverse this default behavior so that variables defined within the environment override those defined within the makefile. However, relying on this option can lead to problems caused by subtle environmental differences between systems.

timestamps gathered during the first pass. If a dependency in a rule is newer than the target or if the target is missing, then make executes the commands of the rule to update the target. Some variable references are resolved immediately during the first pass while processing rules, and others are resolved later during the second pass while executing commands.

A Separate Shell for Each Command

As it processes rules, make executes each command independently of those around it. That is, each individual command under a rule is executed in its own shell. This means that you cannot export a shell variable in one command and then try to access its value in the next.

To do something like this, you would have to string commands together on the same command line with command separator characters (for example, semicolons in Bourne-shell syntax). When you write commands like this, make passes the set of concatenated commands as one command line to the same shell. To avoid long command lines and increase readability, you can wrap them using a backslash at the end of each line—by convention, after the semicolon.[10] The wrapped portion of such commands may also be preceded by a TAB character. POSIX specifies that make remove all leading TAB characters (even those following escaped newlines) before processing commands, but be aware that some implementations of make do output—usually harmlessly—the TAB characters embedded within wrapped commands.[11]

Listing 3-6 shows a few simple examples of multiple commands that will be executed by the same shell.

```
❶ foo: bar.c
        sources=bar.c; \
        gcc -o foo $${sources}
❷ fud: baz.c
        sources=baz.c; gcc -o fud $${sources}
❸ doo: doo.c
        TMPDIR=/var/tmp gcc -o doo doo.c
```

Listing 3-6: A makefile with some examples of multiple commands executed by the same shell

In the first example at ❶, both lines are executed by the same shell because the backslash escapes the newline character between the lines. The make utility will remove any escaped newline characters before passing a single, multi-command statement to the shell. The second example at ❷ is identical to the first, from make's perspective.

---

10. Some shell commands may naturally be wrapped without semicolons, such as within parts of if or case statements. In these cases, you still need the backslash before the newline, but you don't need the semicolon.

11. Experiments have shown that many make implementations generate cleaner output if you don't use TAB characters after escaped newlines. Nevertheless, the community seems to have settled on the consistent use of TAB characters in all command lines, whether wrapped or not.

The third example at ❸ is a bit different. In this case, I've defined the TMPDIR variable only for the child process that will run gcc.[12] Note the missing semicolon; as far as the shell is concerned, this is a single command.[13]

NOTE *If you choose to wrap commands with a trailing backslash, be sure that there are no spaces or other invisible characters after it. The backslash escapes the newline character, so it must immediately precede that character.*

Variable Binding

Variables referenced in commands may be defined after the command in the makefile because such references are not bound to their values until just before make passes the command to the shell for execution—long after the entire makefile has been read. In general, make binds variables to values as late as it possibly can.

Since commands are processed at a later stage than rules, variable references in commands are bound later than those in rules. Variable references found in rules are expanded when make builds the directed graph from the rules in the makefile. Thus, a variable referenced in a rule must be fully defined in a makefile before the referencing rule. Listing 3-7 shows a portion of a makefile that illustrates both of these concepts.

```
--snip--
mytarget = foo
❶ $(mytarget): $(mytarget).c
        ❷ gcc -o $(mytarget) $(mytarget).c
mytarget = bar
--snip--
```

Listing 3-7: Variable expansion in a makefile

In the rule at ❶, both references to $(mytarget) are expanded to foo because they're processed during the first pass, when make is building the variable list and directed graph. However, the outcome is probably not what you'd expect, because both references to $(mytarget) in the command at ❷ are not expanded until much later, long after make has already assigned bar to mytarget, overwriting the original assignment of foo.

Listing 3-8 shows the same rule and command the way make sees them after the variables are fully expanded.

```
--snip--
foo: foo.c
        gcc -o bar bar.c
--snip--
```

Listing 3-8: The results after variable expansion of the code in Listing 3-7

---

12. The gcc compiler uses the value of the TMPDIR variable to determine where to write temporary intermediate files between tools such as the C-preprocessor and the compiler.

13. You cannot dereference TMPDIR on the command line when it's defined in this manner. Only the child process has access to this variable; the current shell does not.

The moral of this story is that you should understand where variables will be expanded in makefile constructs so you're not surprised when make refuses to act in a sane manner when it processes your makefile. It is good practice (and a good way to avoid headaches) to always assign variables before you intend to use them. For more information on immediate and deferred expansion of variables in makefiles, refer to "How make Reads a Makefile" in the *GNU Make Manual.*

Rules in Detail

The rules used in my examples, known as *common* make rules, contain a single colon character (:). The colon separates targets on the left from dependencies on the right.

Remember that targets are products—that is, filesystem entities that can be produced by running one or more commands, such as a C or C++ compiler, a linker, or a documentation generator like Doxygen or LaTeX. Dependencies, on the other hand, are source objects, or objects from which targets are created. These may be computer language source files, intermediate products built by a previous rule, or anything else that can be used by a command as a resource.

You can specify any target defined within a makefile rule directly on the make command line, and make will execute all the commands necessary to generate that target.

NOTE *If you don't specify any targets on the make command line, make will use the default target—the first one it finds in the makefile.*

For example, a C compiler takes dependency *main.c* as input and generates target *main.o.* A linker then takes dependency *main.o* as input and generates a named executable target—program, in this case.

Figure 3-1 shows the flow of data as it might be specified by the rules defined in a makefile.

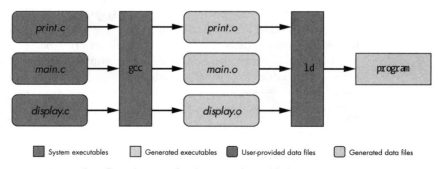

Figure 3-1: A data flow diagram for the compile and link processes

The make utility implements some fairly complex logic to determine when a rule should be run, based on whether a target exists and whether it is older than its dependencies. Listing 3-9 shows a makefile containing rules, some of which execute the actions in Figure 3-1.

```
program: main.o print.o display.o
    ❶ ld main.o print.o display.o ... -o program

main.o: main.c
        gcc -c -g -O2 -o main.o main.c

print.o: print.c
        gcc -c -g -O2 -o print.o print.c

display.o: display.c
        gcc -c -g -O2 -o display.o display.c
```

Listing 3-9: Using multiple make rules to compile and link a program

The first rule in this makefile says that program depends on *main.o*, *print.o*, and *display.o*. The remaining rules say that each *.o* file depends on the corresponding *.c* file. Ultimately, program depends on the three source files, but the object files are necessary as intermediate dependencies because there are two steps to the process—compile and link—with a result in between. For each rule, make uses an associated list of commands to build the rule's target from its list of dependencies.

Unix compilers are designed as higher-level tools than linkers. They have built-in, low-level knowledge about system-specific linker requirements. In the makefile in Listing 3-9, the ellipsis in the line at ❶ is a placeholder for a list of system-specific, low-level objects and libraries required to build all programs on this system. The compiler can be used to call the linker, silently passing these system-specific objects and libraries. (It's so effective and widely used that it's often difficult to discover how to manually execute the linker on a given system.) Listing 3-10 shows how you might rewrite the makefile from Listing 3-9 to use the compiler to compile the sources and call the linker in a single rule.

```
sources = main.c print.c display.c

program: $(sources)
        gcc -g -O2 -o program $(sources)
```

Listing 3-10: Using a single make rule to compile sources into an executable

NOTE
Using a single rule and command to process both steps is possible in this case because the example is very basic. For larger projects, skipping from source to executable in a single step is usually not the wisest way to manage the build process. However, in either case, using the compiler to call the linker can ease the burden of determining the many system objects that need to be linked into an application, and, in fact, this very technique is used quite often. More complex examples, wherein each file is compiled separately, use the compiler to compile each source file into an object file and then use the compiler to call the linker to link them all together into an executable.

In this example, I've added a make variable (sources) that allows us to consolidate all product dependencies into one location. We now have a list of source files captured in a variable definition that is referenced in two places: in the dependency list and on the command line.

Automatic Variables

There may be other kinds of objects in a dependency list that are not in the sources variable, including precompiled objects and libraries. These other objects would have to be listed separately, both in the rule and on the command line. Wouldn't it be nice if we had a shorthand notation for referencing the rule's entire dependency list in the commands?

As it happens, various *automatic* variables can be used to reference portions of the controlling rule during the execution of a command. Unfortunately, most of these are all but useless if you care about portability between implementations of make. The $@ variable (which references the current target) happens to be portable and useful, but most of the other automatic variables are too limited to be very useful.[14] The following is a complete list of portable automatic variables defined by POSIX for make:

- $@ refers to the full target name of the current target or the archive filename part of a library archive target. This variable is valid in both explicit and implicit rules.
- $% refers to a member of an archive and is valid only when the current target is an archive member—that is, an object file that is a member of a static library. This variable is valid in both explicit and implicit rules.
- $? refers to the list of dependencies that are newer than the current target. This variable is valid in both explicit and implicit rules.
- $< refers to the member of the dependency list whose existence allowed the rule to be chosen for the target. This variable is only valid in implicit rules.
- $* refers to the current target name with its suffix deleted. This variable is guaranteed by POSIX to be valid only in implicit rules.

GNU make dramatically extends the POSIX-defined list, but since GNU extensions are not portable, it's unwise to use any of these except $@.

Dependency Rules

Let us now assume that *print.c* and *display.c* each have a header file of the same name, ending in *.h*. Each of these source files includes its own header file, but *main.c* includes both *print.h* and *display.h*. Given the makefiles of Listings 3-9 and 3-10, what do you suppose would happen if you executed

---

14. This is because POSIX is not so much a specification for the way things *should be* done as it is a specification for the way things *are* done. Essentially, the purpose of the POSIX standard is to keep Unix implementations from deviating any further from the norm than necessary. Unfortunately, most make implementations had wide acceptance within their own communities long before the idea for a POSIX standard was conceived.

make to build program, then modified one of the header files—say *print.h*—and then re-executed make? Nothing would happen because make is unaware even of the existence of these header files. As far as make is concerned, you didn't touch anything related to program.

In Listing 3-11, I've replaced the sources variable with an objects variable and replaced the list of source files with a list of object files. This version of the makefile in Listing 3-10 also eliminates redundancy by making use of both standard and automatic variables.

```
objects = main.o print.o display.o

main.o: main.c print.h display.h
print.o: print.c print.h
display.o: display.c display.h

program: $(objects)
        gcc -g -O2 -o $@ $(objects)
```

Listing 3-11: Using automatic variables in a command

I've also added three *dependency rules*, which are rules without commands that clarify the relationships between compiler output files and dependent source and header files. Because *print.h* and *display.h* are (presumably) included by *main.c, main.c* must be recompiled if either of those files changes; however, make has no way of knowing that these two header files are included by *main.c*. Dependency rules allow the developer to tell make about such backend relationships.

Implicit Rules

If you attempt to mentally follow the dependency graph that make would build from the rules within the makefile in Listing 3-11, you'll find what appears to be a hole in the web. According to the last rule in the file, the program executable depends on *main.o, print.o,* and *display.o.* This rule also provides the command to link these objects into an executable (using the compiler merely to call the linker this time). The object files are tied to their corresponding C source and header files by the three dependency rules. But where are the commands that compile the *.c* files into *.o* files?

We could add these commands to the dependency rules, but there's really no need because make has a *built-in* rule that knows how to build *.o* files from *.c* files. There's nothing magic about make—it only knows about the relationships you describe to it through the rules you write. But make does have certain built-in rules that describe the relationships between, for example, *.c* files and *.o* files. This particular built-in rule provides commands for building anything with a *.o* extension from a file of the same base name with a *.c* extension. These built-in rules are called *suffix rules* or, more generally, *implicit rules*, because the name of the dependency (source file) is implied by the name of the target (object file).

To make the built-in implicit rules more widely usable, their commands often consume well-known make variables. If you set those variables, overriding the default values, you can wield some control over the execution of a built-in rule. For instance, the command in the standard POSIX definition of the built-in implicit rule for converting *.o* files to *.c* files is:[15]

```
$(CC) $(CPPFLAGS) $(CFLAGS) -c
```

Here, you can override just about every aspect of this built-in rule by setting your own values for CC, the compiler; CPPFLAGS, options passed to the C preprocessor; and CFLAGS, options passed to the C compiler.

You can write implicit rules yourself, if you wish. You can even override the default implicit rules with your own versions. Implicit rules are a powerful tool, and they shouldn't be overlooked, but for the purposes of this book, we won't go into any more detail. You can learn more about writing and using implicit rules within makefiles in "Using Implicit Rules" in the *GNU Make Manual.*

To illustrate this implicit functionality, I created trivial C source and header files to accompany the sample makefile from Listing 3-11. Here's what happened when I executed make on this makefile:

```
❶ $ make
cc     -c -o main.o main.c
$
❷ $ make program
cc     -c -o print.o print.c
cc     -c -o display.o display.c
gcc -g -O2 -o program main.o print.o display.o
$
```

As you can see, cc was magically executed with -c and -o options to generate *main.o* from *main.c.* This is common command line syntax used to make a C-language compiler build objects from sources—it's so common, in fact, that the functionality is built into make. If you look for cc on a modern GNU/Linux system, you'll find that it's a soft link in */usr/bin* that refers to the system's GNU C compiler. On other systems, it refers to the system's native C compiler. Calling the system C compiler *cc* has been a de facto standard for decades.[16]

The extra spaces between cc and -c in that output under ❶ represent the spaces between the uses of the CPPFLAGS and CFLAGS variables, which are defined as empty by default.

But why did the make utility build only *main.o* when we typed make at ❶? Simply because the dependency rule for *main.o* provided the first (and thus, the default) target for the makefile. In this case, to build program, we needed to execute make program, as we did in ❷. Remember that when you enter make

---

15. See "Catalogue of Built-In Rules" in the *GNU Make Manual.*

16. POSIX has standardized the program (or link) names *c89* and *c99* to refer to 1989 and 1999 C-language standard compatible compilers, respectively. Since these commands can refer to the same compiler with different command line options, they're often implemented as binary programs or shell scripts rather than merely as soft links.

on the command line, the make utility attempts to build the first explicitly defined target within the file called *Makefile* in the current directory. If we wanted to make program the default target, we could rearrange the rules so the program rule would be the first one listed in the makefile.

To see the dependency rules in action, touch one of the header files and then rebuild the program target:

```
$ touch display.h
$ make program
cc -c -o main.o main.c
cc -c -o display.o display.c
gcc -g -O0 -o program main.o print.o display.o
$
```

After *display.h* was updated, only *display.o*, *main.o*, and program were rebuilt. The *print.o* object didn't need to be rebuilt because *print.c* doesn't depend on *display.h*, according to the rules specified in the makefile.

Phony Targets

Targets are not always files. They can also be so-called *phony targets*, as in the case of all and clean. These targets don't refer to true products in the filesystem but rather to particular outcomes or actions—when you make these targets, the project is *cleaned*, *all* products are built, and so on.

Multiple Targets

In the same way that you can list multiple dependencies within a rule on the right side of a colon, you can combine rules for multiple targets with the same dependencies and commands by listing the targets on the left side of a colon, as shown in Listing 3-12.

```
all clean:
        cd src && $(MAKE) $@
```

Listing 3-12: Using multiple targets in a rule

While it may not be immediately apparent, this example contains two separate rules: one for each of the two targets, all and clean. Because these two rules have the same set of dependencies (none, in this case) and the same set of commands, we're able to take advantage of a short-hand notation supported by make that allows us to combine their rules into one specification.

To help you understand this concept, consider the $@ variable in Listing 3-12. Which target does it refer to? Well, that depends on which rule is currently executing—the one for all or the one for clean. Since a rule can only be executed on a single target at any given time, $@ can only ever refer to one target, even when the controlling rule specification contains several.

Resources for Makefile Authors

GNU make is significantly more powerful than the original AT&T UNIX make utility, although GNU make is completely backward compatible, as long as you avoid GNU extensions. The *GNU Make Manual*[17] is available online, and O'Reilly has published an excellent book on the original AT&T UNIX make utility[18] and all of its many nuances. While you can still find this title, the publisher has since merged its content into a new edition that also covers GNU make extensions.[19]

This concludes the general discussion of makefile syntax and the make utility, although we will look at additional makefile constructs as we encounter them throughout the rest of this chapter. With this general information behind us, let's return to the Jupiter project and begin adding some more interesting functionality.

Creating a Source Distribution Archive

In order to actually get source code for Jupiter to our users, we're going to have to create and distribute a source archive—a tarball. We could write a separate script to create the tarball, but since we can use phony targets to create arbitrary sets of functionality in makefiles, let's design a make target to perform this task instead. Building a source archive for distribution is usually relegated to the dist target.

When designing a new make target, we need to consider whether its functionality should be distributed among the makefiles of the project or handled in a single location. Normally, the rule of thumb is to take advantage of a recursive build system's nature by allowing each directory to manage its own portions of a process. We did just this in Listing 3-1 when we passed control of building the jupiter program down to the *src* directory, where the source code is located. However, building a compressed archive from a directory structure isn't really a recursive process.[20] This being the case, we'll have to perform the entire task in one of the two makefiles.

Global processes are often handled by the makefile at the highest relevant level in the project directory structure. We'll add the dist target to our top-level makefile, as shown in Listing 3-13.

---

17. See the Free Software Foundation's *GNU Make Manual* at *http://www.gnu.org/software /make/manual/*.

18. Andy Oram and Steve Talbott, *Managing Projects with make,* Second Edition (Sebastopol, CA: O'Reilly Media, 1991), *http://oreilly.com/catalog/9780937175903/*.

19. Robert Mecklenburg, *Managing Projects with GNU Make,* Third Edition: *The Power of GNU make for Building Anything* (Sebastopol, CA: O'Reilly Media, 2004), *http://www.oreilly.com/ catalog/9780596006105/*.

20. Well, okay, it is a recursive process, but the recursive portions of the process are tucked away inside the tar utility.

```
❶ package = jupiter
   version = 1.0
   tarname = $(package)
   distdir = $(tarname)-$(version)

   all clean jupiter:
           cd src && $(MAKE) $@

❷ dist: $(distdir).tar.gz

❸ $(distdir).tar.gz: $(distdir)
           tar chof - $(distdir) | gzip -9 -c > $@
           rm -rf $(distdir)

❹ $(distdir):
           mkdir -p $(distdir)/src
           cp Makefile $(distdir)
           cp src/Makefile src/main.c $(distdir)/src

❺ .PHONY: all clean dist
```

Listing 3-13: Makefile: *Adding the* dist *target to the top-level makefile*

Besides the addition of the dist target at ❷, I've also made several other modifications. Let's look at them one at a time. I've added the dist target to the .PHONY rule at ❺. The .PHONY rule is a special kind of built-in rule called a *dot-rule* or *directive*. The make utility understands several dot-rules. The purpose of .PHONY is simply to tell make that certain targets don't generate filesystem objects. Normally, make determines which commands to run by comparing the timestamps of the targets to those of their dependencies in the filesystem—but phony targets don't have associated filesystem objects. Using .PHONY ensures that make won't go looking for nonexistent product files named after these targets. It also ensures that if a file or directory named *dist* somehow inadvertently gets added to the directory, make will still treat the dist target as non-real.

Adding a target to the .PHONY rule has another effect. Since make won't be able to use timestamps to determine whether the target is up-to-date (that is, newer than its dependencies), make has no recourse but to *always* execute the commands associated with phony targets whenever these targets either are requested on the command line or appear in a dependency chain.

I've separated the functionality of the dist target into three separate rules (❷, ❸, and ❹) for the sake of readability, modularity, and maintenance. This is a great rule of thumb to follow in any software engineering process: *build large processes from smaller ones and reuse the smaller processes where it makes sense.*

The dist target at ❷ depends on the existence of the ultimate goal—in this case, a source-level compressed archive package, *jupiter-1.0.tar.gz.* I've used one variable to hold the version number (which makes it easier to update the project version later) and another variable for the package name at ❶, which will make it easier to change the name if I ever decide to reuse this makefile for another project. I've also logically split the functions of package name and tarball name; the default tarball name is the package name, but we do have the option of making them different.

The rule that builds the tarball at ❸ indicates how this should be done with a command that uses the gzip and tar utilities to create the file. But, notice that the rule has a dependency—the directory to be archived. The directory name is derived from the tarball name and the package version number; it's stored in yet another variable called distdir.

We don't want object files and executables from our last build attempt to end up in the archive, so we need to build an image directory containing exactly what we want to ship—including any files required in the build and install processes and any additional documentation or license files. Unfortunately, this pretty much mandates the use of individual copy (cp) commands.

Since there's a rule in the makefile (at ❹) that tells how this directory should be created, and since that rule's target is a dependency of the tarball, make runs the commands for that rule *before* running the commands for the tarball rule. Recall that make processes rules to build dependencies recursively, from the bottom up, until it can run the commands for the requested target.[21]

Forcing a Rule to Run

There's a subtle flaw in the $(distdir) target that may not be obvious right now, but it will rear its ugly head at the worst of times. If the archive image directory (*jupiter-1.0*) already exists when you execute make dist, then make won't try to create it. Try this:

```
$ mkdir jupiter-1.0
$ make dist
tar chof - jupiter-1.0 | gzip -9 -c > jupiter-1.0.tar.gz
rm -rf jupiter-1.0
$
```

Notice that the dist target didn't copy any files—it just built an archive out of the existing *jupiter-1.0* directory, which was empty. Our users would get a real surprise when they unpack this tarball! Worse still, if the image directory from the previous attempt to archive happened to still be there, the new tarball would contain the now-outdated sources from our last attempt to create a distribution tarball.

The problem is that the $(distdir) target is a real target with no dependencies, which means that make will consider it up-to-date as long as it exists in the filesystem. We could add the $(distdir) target to the .PHONY rule to force make to rebuild it every time we make the dist target, but it's not a phony target—it's a real filesystem object. The proper way to ensure that $(distdir) is always rebuilt is to ensure that it doesn't exist before make attempts to build it. One way to accomplish this is to create a true phony target that will always execute and then add that target to the dependency list for the $(distdir) target. A common name for this kind of target is FORCE, and I've implemented this concept in Listing 3-14.

---

21. This process is formally called *post-order recursion.*

```
--snip--
$(distdir).tar.gz: $(distdir)
        tar chof - $(distdir) | gzip -9 -c > $@
        rm -rf $(distdir)

❶ $(distdir): FORCE
        mkdir -p $(distdir)/src
        cp Makefile $(distdir)
        cp src/Makefile $(distdir)/src
        cp src/main.c $(distdir)/src

❷ FORCE:
        -rm $(distdir).tar.gz >/dev/null 2>&1
        rm -rf $(distdir)

.PHONY: FORCE all clean dist
```

Listing 3-14: Makefile: *Using the FORCE target*

The FORCE rule's commands (at ❷) are executed every time because FORCE is a phony target. Since we made FORCE a dependency of the $(distdir) target (at ❶), we have the opportunity to delete any previously created files and directories *before* make begins to evaluate whether it should execute the commands for $(distdir).

Leading Control Characters

A leading dash character (-) on a command tells make not to care about the status code of the command it precedes. Normally, when make encounters a command that returns a nonzero status code to the shell, it will stop execution and display an error message, but if you use a leading dash, it will just ignore the error and continue. I use a leading dash on the first rm command in the FORCE rule because I want to delete previously created product files that may *or may not* exist, and rm will return an error if I attempt to delete a nonexistent file.

In general, a better option is to use the -f flag on the rm command line, which causes rm to ignore missing file errors. Another benefit of using -f is that we no longer need to redirect error messages to */dev/null*, as we really care about other errors—permission errors, for example. From this point on, we'll remove the leading dash in front of any rm commands and ensure we use -f.

Another leading character that you may encounter is the at sign (@). A command prefixed with an at sign tells make not to perform its normal behavior of printing the command to the stdout device as it executes it. It is common to use a leading at sign on echo statements. You don't want make to print echo statements, because then your message will be printed twice: once by make and then again by the echo statement itself.

NOTE *You may also combine these leading characters (@, -, and +) in any order. The plus (+) character is used to force a command to execute that would otherwise not be executed due, for example, to a -n command line option, which tells* make *to perform a so-called dry run. Some commands make sense even in a dry run.*

It's best to use the at sign judiciously. I usually reserve it for commands I *never* want to see, such as echo statements. If you like quiet build systems, consider using the global .SILENT directive in your makefiles. Or better still, simply do nothing, thereby allowing the user the option of adding the -s option to their make command lines. This enables the user to choose how much noise they want to see.

Automatically Testing a Distribution

The rule for building the archive directory is probably the most frustrating rule in this makefile because it contains commands to copy individual files into the distribution directory. Every time we change the file structure in our project, we have to update this rule in our top-level makefile, or we'll break the dist target. But there's nothing more we can do—we've made the rule as simple as possible. Now we just have to remember to manage this process properly.

Unfortunately, though, even worse things than breaking the dist target could happen if you forget to update the distdir rule's commands. It may *appear* that the dist target is working, but it may not actually be copying all of the required files into the tarball. In fact, it is far more likely that this, rather than an error, will occur, because adding files to a project is a more common activity than moving them around or deleting them. New files will not be copied, but the dist rule won't notice the difference.

There is a way to perform a sort of self-check on the dist target. We can create another phony target, called distcheck, that does exactly what our users will do: unpack the tarball and build the project. We can have this rule's commands perform this task in a temporary directory. If the build process fails, then the distcheck target will break, telling us that we forgot something crucial in our distribution.

Listing 3-15 shows the modifications to our top-level makefile that are required to implement the distcheck target.

Git tag 3.3
```
--snip--
$(distdir): FORCE
        mkdir -p $(distdir)/src
        cp Makefile $(distdir)
        cp src/Makefile src/main.c $(distdir)/src

distcheck: $(distdir).tar.gz
        gzip -cd $(distdir).tar.gz | tar xvf -
        cd $(distdir) && $(MAKE) all
        cd $(distdir) && $(MAKE) clean
        rm -rf $(distdir)
        @echo "*** Package $(distdir).tar.gz is ready for distribution."
--snip--
.PHONY: FORCE all clean dist distcheck
```

Listing 3-15: Makefile: Adding a distcheck target to the top-level makefile

The `distcheck` target depends on the tarball itself, so the rule that builds the tarball is executed first. The make utility then executes the `distcheck` commands, which unpack the tarball just built and then recursively run `make` on the `all` and `clean` targets within the resulting directory. If that process succeeds, the `distcheck` target prints out a message indicating that your users will likely not have a problem with this tarball.

Now all you have to do is remember to execute `make` `distcheck` *before* you post your tarballs for public distribution!

Unit Testing, Anyone?

Some people insist that unit testing is evil, but the only honest rationale they can come up with for not doing it is laziness. Proper unit testing is hard work, but it pays off in the end. Those who do it have learned a lesson (usually in childhood) about the value of delayed gratification.

A good build system should incorporate proper unit testing. The most commonly used target for testing a build is the `check` target, so we'll go ahead and add it in the usual manner. The actual unit test should probably go in *src/Makefile* because that's where the `jupiter` executable is built, so we'll pass the `check` target down from the top-level makefile.

But what commands do we put in the `check` rule? Well, `jupiter` is a pretty simple program—it prints the message *Hello from* some/path/*jupiter!* where *some/path* depends on the location from which `jupiter` was executed. I'll use the `grep` utility to test that `jupiter` actually outputs such a string.

Listings 3-16 and 3-17 illustrate the modifications to our top-level and *src* directory makefiles, respectively.

Git tag 3.4

```
--snip--
all clean check jupiter:
        cd src && $(MAKE) $@
--snip--
.PHONY: FORCE all clean check dist distcheck
```

Listing 3-16: Makefile: Passing the check target to src/Makefile

```
--snip--
src/jupiter: src/main.c
        $(CC) $(CFLAGS) $(CPPFLAGS) -o $@ src/main.c

check: all
        ./jupiter | grep "Hello from .*jupiter!"
        @echo "*** ALL TESTS PASSED ***"
--snip--

.PHONY: all clean check
```

Listing 3-17: src/Makefile: Implementing the unit test in the check target

Note that `check` depends on `all`. We can't really test our products unless they are up-to-date, reflecting any recent source code or build system changes that may have been made. It makes sense that if the user wants to test the products, they also want the products to exist and be up-to-date. We can ensure they exist and are current by adding all to check's dependency list.

There's one more enhancement we can make to our build system: we can add check to the list of targets executed by make in our distcheck rule, between the commands to make all and clean. Listing 3-18 shows where this is done in the top-level makefile.

Git tag 3.5

```
--snip--
distcheck: $(distdir).tar.gz
        gzip -cd $(distdir).tar.gz | tar xvf -
        cd $(distdir) && $(MAKE) all
        cd $(distdir) && $(MAKE) check
        cd $(distdir) && $(MAKE) clean
        rm -rf $(distdir)
        @echo "*** Package $(distdir).tar.gz is ready for distribution."
--snip--
```

Listing 3-18: Makefile: Adding the check target to the $(MAKE) command

Now when we run make distcheck, it will test the entire build system shipped with the package.

Installing Products

We've reached the point where our users' experiences with Jupiter should be fairly painless—even pleasant—as far as building the project is concerned. Users will simply unpack the distribution tarball, change into the distribution directory, and type make. It really can't get any simpler than that.

But we still lack one important feature—installation. In the case of the Jupiter project, this is fairly trivial. There's only one program, and most users would guess correctly that to install it, they should copy jupiter into either their */usr/bin* or */usr/local/bin* directory. More complex projects, however, could cause users real consternation over where to put user and system binaries, libraries, header files, and documentation, including man pages, info pages, PDF files, and the more or less obligatory *README, AUTHORS, NEWS, INSTALL,* and *COPYING* files generally associated with GNU projects.

We don't really want our users to have to figure all that out, so we'll create an install target to manage putting things where they go once they're built properly. In fact, why not just make installation part of the all target? Well, let's not get carried away. There are actually a few good reasons for not doing this.

First, build and installation are separate logical concepts. The second reason is a matter of filesystem rights. Users have rights to build projects in their own home directories, but installation often requires *root*-level rights to copy files into system directories. Finally, there are several reasons why a user may wish to build but not install a project, so it would be unwise to tie these actions together.

While creating a distribution package may not be an inherently recursive process, installation certainly is, so we'll allow each subdirectory in our project to manage installation of its own components. To do this, we need to modify both the top-level and the *src*-level makefiles. Changing the top-level makefile is easy: since there are no products to be installed in the top-level directory, we'll just pass the responsibility on to *src/Makefile* in the usual way.

The modifications for adding an install target are shown in Listings 3-19 and 3-20.

Git tag 3.6

```
--snip--
all clean check install jupiter:
        cd src && $(MAKE) $@
--snip--

.PHONY: FORCE all clean check dist distcheck install
```

Listing 3-19: Makefile: *Passing the* install *target to* src/Makefile

```
--snip--
check: all
        ./src/jupiter | grep "Hello from .*jupiter!"
        @echo "*** All TESTS PASSED"

install:
        cp jupiter /usr/bin
        chown root:root /usr/bin/jupiter
        chmod +x /usr/bin/jupiter
--snip--

.PHONY: all clean check install
```

Listing 3-20: src/Makefile: *Implementing the* install *target*

In the top-level makefile shown in Listing 3-19, I've added install to the list of targets passed down to *src/Makefile*. The installation of files is handled by the *src*-level makefile shown in Listing 3-20.

Installation is a bit more complex than simply copying files. If a file is placed in the */usr/bin* directory, then *root* should own it so that only *root* can delete or modify it. Additionally, the jupiter binary should be flagged executable, so I've used the chmod command to set the mode of the file as such. This is probably redundant, as the linker ensures that jupiter is created as an executable file, but some types of executable products are not generated by a linker—shell scripts, for example.

Now our users can just type the following sequence of commands and the Jupiter project will be built, tested, and installed with the correct system attributes and ownership on their platforms:

```
$ gzip -cd jupiter-1.0.tar.gz | tar xf -
$ cd jupiter-1.0
$ make all check
--snip--
```

```
$ sudo make install
Password: ******
--snip--
$
```

Installation Choices

All of this is well and good, but it could be a bit more flexible with regard to where things are installed. Some users may be okay with having jupiter installed into the */usr/bin* directory. Others are going to ask why it isn't installed into the */usr/local/bin* directory—after all, this is a common convention. We could change the target directory to */usr/local/bin*, but then users may ask why they don't have the option of installing into their home directories. This is the perfect situation for a little command line–supported flexibility.

Another problem with our current build system is that we have to do a lot of stuff just to install files. Most Unix systems provide a system-level program—sometimes simply a shell script—called install that allows a user to specify various attributes of the files being installed. The proper use of this utility could simplify things a bit for Jupiter's installation, so while we're adding location flexibility, we might as well use the install utility, too. These modifications are shown in Listings 3-21 and 3-22.

Git tag 3.7
```
package = jupiter
version = 1.0
tarname = $(package)
distdir = $(tarname)-$(version)

prefix=/usr/local
❶ export prefix

all clean check install jupiter:
        cd src && $(MAKE) $@
--snip--
```

Listing 3-21: Makefile: Adding a prefix variable

```
--snip--
install:
    ❷ install -d $(prefix)/bin
        install -m 0755 jupiter $(prefix)/bin
--snip--
```

Listing 3-22: src/Makefile: Using the prefix variable in the install target

Notice that I only declared and assigned the prefix variable in the top-level makefile, but I referenced it in *src/Makefile*. I can do this because I used the export modifier at ❶ in the top-level makefile—this modifier exports the variable to the shell that make spawns when it executes itself in the *src* directory. This feature of make allows us to define all of our user variables in one obvious location—at the beginning of the top-level makefile.

NOTE *GNU make allows you to use the export keyword on the assignment line, but this syntax is not portable between GNU make and other versions of make. Technically, POSIX doesn't support the use of export at all, but most make implementations support it.*

I've now declared the prefix variable to be */usr/local*, which is very nice for those who want to install jupiter in */usr/local/bin* but not so nice for those who want it in */usr/bin*. Fortunately, make allows you to define make variables on the command line, in this manner:

```
$ sudo make prefix=/usr install
--snip--
```

Remember that variables defined on the command line override those defined within the makefile.[22] Thus, users who want to install jupiter into the */usr/bin* directory now have the option of specifying this on the make command line.

With this system in place, our users may install jupiter into a *bin* directory beneath any directory they choose, including a location in their home directory (for which they do not need additional rights). This is, in fact, the reason we added the install -d $(prefix)/bin command at ❷ in Listing 3-22—this command creates the installation directory if it doesn't already exist. Since we allow the user to define prefix on the make command line, we don't actually know where the user is going to install jupiter; therefore, we have to be prepared for the possibility that the location may not yet exist. Give this a try:[23]

```
$ make all
$ make prefix=$PWD/inst install
$
$ ls -1p
inst/
Makefile
src/
$
$ ls -1p inst
bin/
$
$ ls -1p inst/bin
jupiter
$
```

---

22. Unfortunately, some make implementations do not propagate such command line variables to recursive $(MAKE) processes. To alleviate this potential problem, variables that might be set on the command line can be passed as var="$(var)" on sub-make command lines. My simple examples ignore this issue because it's a corner case, but you should at least be aware of this problem.

23. In the examples throughout this book, for the sake of simplicity, and to keep from becoming sidetracked on irrelevant issues, I simply assume paths don't contain any whitespace—files and directories with spaces in their names. You can write makefiles that handle these conditions properly, but it involves the judicious use of quoting around various variable references both on the command line and within the makefiles themselves. All of these problems disappear with the use of the Autotools, as they handle all the cases where file and directory names may contain whitespace.

Uninstalling a Package

What if a user doesn't like our package after they've installed it, and they just want to get it off their system? This is a fairly likely scenario for the Jupiter project, as it's rather useless and takes up valuable space in the *bin* directory. In the case of *your* projects, however, it's more likely that a user would want to do a clean install of a newer version of the project or replace the test build they downloaded from the project website with a professionally packaged version that comes with their Linux distribution. Support for an uninstall target would be very helpful in situations like these.

Listings 3-23 and 3-24 show the addition of an uninstall target to our two makefiles.

Git tag 3.8

```
--snip--
all clean check install uninstall jupiter:
        cd src && $(MAKE) $@
--snip--

.PHONY: FORCE all clean check dist distcheck install uninstall
```

Listing 3-23: Makefile: *Adding the* uninstall *target to the top-level makefile*

```
--snip--
install:
        install -d $(prefix)/bin
        install -m 0755 jupiter $(prefix)/bin

uninstall:
        rm -f $(prefix)/bin/jupiter
        -rmdir $(prefix)/bin >/dev/null 2>&1
--snip--

.PHONY: all clean check install uninstall
```

Listing 3-24: src/Makefile: *Adding the* uninstall *target to the src-level makefile*

As with the install target, this target requires root-level rights if the user is using a system prefix, such as */usr* or */usr/local*. You should be very careful about how you write your uninstall targets; unless a directory belongs specifically to your package, you shouldn't assume you created it. If you do, you may end up deleting a system directory like */usr/bin*!

On the other hand, we did create the directory in the install target if it was originally missing, so we should remove it if possible. Here, we can use the rmdir command, whose job it is to remove empty directories. Even if the directory is a system directory such as */usr/bin*, removing it is harmless if it's empty, but rmdir will fail if it's not empty. Recalling that command failure stops the make process, we'll also prefix it with a dash character. And we don't really want to see such a failure, so we'll redirect it's output to */dev/null*.

The list of things to maintain in our build system is getting out of hand. There are now two places we need to update when we change our installation processes: the install and uninstall targets. Unfortunately, this is really about the best we can hope for when writing our own makefiles, unless we resort to fairly complex shell script commands. But hang in there—in Chapter 6, I'll show you how to rewrite this makefile in a much simpler way using GNU Automake.

Testing Install and Uninstall

Now let's add some code to our `distcheck` target to test the functionality of the `install` and `uninstall` targets. After all, it's fairly important that both of these targets work correctly from our distribution tarballs, so we should test them in `distcheck` before declaring the tarball release worthy. Listing 3-25 illustrates the necessary changes to the top-level makefile.

Git tag 3.9

```
--snip--
distcheck: $(distdir).tar.gz
        gzip -cd $(distdir).tar.gz | tar xvf -
        cd $(distdir) && $(MAKE) all
        cd $(distdir) && $(MAKE) check
        cd $(distdir) && $(MAKE) prefix=$${PWD}/_inst install
        cd $(distdir) && $(MAKE) prefix=$${PWD}/_inst uninstall
        cd $(distdir) && $(MAKE) clean
        rm -rf $(distdir)
        @echo "*** Package $(distdir).tar.gz is ready for distribution."
--snip--
```

Listing 3-25: Makefile: Adding `distcheck` tests for the `install` and `uninstall` targets

Note that I used a double dollar sign on the $${PWD} variable references, ensuring that make passes the variable reference to the shell with the rest of the command line, rather than expanding it inline before executing the command. I wanted this variable to be dereferenced by the shell rather than by the make utility.[24]

What we're doing here is testing to ensure the `install` and `uninstall` targets don't generate errors—but this isn't very likely because all they do is install files into a temporary directory within the build directory. We could add some code immediately after the make install command that looks for the products that are supposed to be installed, but that's more than I'm willing to do. One reaches a point of diminishing returns, where the code that does the checking is just as complex as the installation code—in which case, the check becomes pointless.

But there is something else we can do: we can write a more or less generic test that checks to see if everything we installed was properly removed. Since the stage directory was empty before our installation, it had better be in a similar state after we uninstall. Listing 3-26 shows the addition of this test.

---

24. Technically, I didn't have to do this because the PWD make variable was initialized from the environment, but it serves as a good example of this process. Additionally, there are corner cases where the PWD make variable is not quite as accurate as the PWD shell variable. It may be left pointing to the parent directory on a subdirectory make invocation.

```
--snip--
distcheck: $(distdir).tar.gz
        gzip -cd $(distdir).tar.gz | tar xvf -
        cd $(distdir) && $(MAKE) all
        cd $(distdir) && $(MAKE) check
        cd $(distdir) && $(MAKE) prefix=$${PWD}/_inst install
        cd $(distdir) && $(MAKE) prefix=$${PWD}/_inst uninstall
      ❶ @remaining="`find $(distdir)/_inst -type f | wc -l`"; \
        if test "$${remaining}" -ne 0; then \
      ❷ echo "*** $${remaining} file(s) remaining in stage directory!"; \
          exit 1; \
        fi
        cd $(distdir) && $(MAKE) clean
        rm -rf $(distdir)
        @echo "*** Package $(distdir).tar.gz is ready for distribution."
--snip--
```

Listing 3-26: Makefile: Adding a test for leftover files after uninstall finishes

The test first generates a numeric value at ❶ in a shell variable called remaining, which represents the number of regular files found in the stage directory we used. If this number is not zero, the test prints a message to the console at ❷ indicating how many files were left behind by the uninstall commands and then it exits with an error. Exiting early leaves the stage directory intact so we can examine it to find out which files we forgot to uninstall.

NOTE   *This test code represents a good use of multiple shell commands passed to a single shell. I had to do this here so that the value of remaining would be available for use by the if statement. Conditionals don't work very well when the closing if is not executed by the same shell as the opening if!*

I don't want to alarm people by printing the embedded echo statement unless it really should be executed, so I prefixed the entire test with an at sign (@) so that make wouldn't print the code to stdout. Since make considers these five lines of code a single command, the only way to suppress printing the echo statement is to suppress printing the entire command.

Now, this test isn't perfect—not by a long shot. This code only checks for regular files. If your installation procedure creates any soft links, this test won't notice if they're left behind. The directory structure that's built during installation is purposely left in place because the check code doesn't know whether a subdirectory within the stage directory belongs to the system or to the project. The uninstall rule's commands can be aware of which directories are project specific and properly remove them, but I don't want to add project-specific knowledge into the distcheck tests—it's that problem of diminishing returns again.

The Filesystem Hierarchy Standard

You may be wondering by now where I'm getting these directory names. What if some Unix system out there doesn't use */usr* or */usr/local*? For one thing, this is another reason for providing the prefix variable—to allow the user some choice in these matters. However, most Unix-like systems nowadays follow the *Filesystem Hierarchy Standard (FHS)* as closely as possible. The *FHS* defines a number of standard places, including the following root-level directories:

/bin	*/etc*	*/home*
/opt	*/sbin*	*/srv*
/tmp	*/usr*	*/var*

This list is by no means exhaustive. I've only mentioned the directories that are most relevant to our study of open source project build systems. In addition, the *FHS* defines several standard locations beneath these root-level directories. For instance, the */usr* directory should contain the following subdirectories:

/usr/bin	*/usr/include*	*/usr/lib*
/usr/local	*/usr/sbin*	*/usr/share*
/usr/src		

The */usr/local* directory should contain a structure very similar to that of the */usr* directory. The */usr/local* directory provides a location for software installation that overrides versions of the same packages installed in the */usr* directory structure, because system software updates often overwrite software in */usr* without prejudice. The */usr/local* directory structure allows a system administrator to decide which version of a package to use on their system because */usr/local/bin* may be (and usually is) added to the PATH before */usr/bin*. A fair amount of thought has gone into designing the *FHS*, and the GNU Autotools take full advantage of this consensus of understanding.

Not only does the *FHS* define these standard locations, but it also explains in detail what they're for and what types of files should be kept there. All in all, the *FHS* leaves you, as project maintainer, just enough flexibility and choice to keep your life interesting but not enough to make you wonder whether you're installing your files in the right places.[25]

---

25. Before I discovered the *FHS*, I relied on my personal experience to decide where files should be installed in my projects. Mostly I was right, because I'm a careful guy, but after I read the *FHS* documentation, I went back to some of my past projects with a bit of chagrin and changed things around. I heartily recommend you become thoroughly familiar with the *FHS* if you seriously intend to develop Unix software.

Supporting Standard Targets and Variables

In addition to those I've already mentioned, the *GNU Coding Standards* lists some important targets and variables that you should support in your projects—mainly because your users will expect support for them.

Some of the chapters in the *GCS* document should be taken with a grain of salt (unless you're actually working on a GNU-sponsored project). For example, you probably won't care much about the C source code formatting suggestions in Chapter 5 of the *GCS*. Your users certainly won't care, so you can use whatever source code formatting style you wish.

That's not to say that all of Chapter 5 is worthless to non-GNU open source projects. The "Portability between System Types" and "Portability between CPUs" subsections, for instance, provide excellent information on C source code portability. Also, the "Internationalization" subsection gives some useful tips on using GNU software to internationalize your projects. We'll consider internationalization in greater detail in Chapter 11 of this book.

While Chapter 6 of the *GCS* discusses documentation the GNU way, some sections of Chapter 6 describe various top-level text files commonly found in projects, such as the *AUTHORS*, *NEWS*, *INSTALL*, *README*, and *ChangeLog* files. These are all bits of information that the well-indoctrinated open source software user expects to see in any reputable project.

The *really* useful information in the *GCS* document begins in Chapter 7, "The Release Process." This chapter is critical to you as a project maintainer because it defines what your users will expect of your projects' build systems. Chapter 7 contains the de facto standards for the user options that packages provide in source-level distributions.

Standard Targets

The "How Configuration Should Work" subsection of Chapter 7 of the *GCS* defines the configuration process, which I cover briefly in "Configuring Your Package" on page 77. The "Makefile Conventions" subsection of the *GCS* covers all of the standard targets and many of the standard variables that users have come to expect in open source software packages. Standard targets defined by the *GCS* include the following:

all	install	install-html
install-dvi	install-pdf	install-ps
install-strip	uninstall	clean
distclean	mostlyclean	maintainer-clean
TAGS	info	dvi
html	pdf	ps
dist	check	installcheck
installdirs		

You don't need to support all of these targets, but you should consider supporting the ones that make sense for your project. For example, if you build and install HTML pages, you should probably consider supporting the html and install-html targets. Autotools projects support these and more. Some targets are useful to end users, while others are useful only to project maintainers.

Standard Variables

Variables you should support as you see fit include those listed in the following table. In order to provide flexibility for the end user, most of these variables are defined in terms of a few of them and, ultimately, only one of them: prefix. For lack of a more standard name, I call these *prefix variables*. Most of these could be classified as *installation directory variables* that refer to standard locations, but there are a few exceptions, such as srcdir.

These variables are meant to be fully resolved by make, so they're defined in terms of make variables, using parentheses rather than curly brackets. Table 3-1 lists these prefix variables and their default values.

Table 3-1: Prefix Variables and Their Default Values

Variable	Default Value
prefix	/usr/local
exec_prefix	$(prefix)
bindir	$(exec_prefix)/bin
sbindir	$(exec_prefix)/sbin
libexecdir	$(exec_prefix)/libexec
datarootdir	$(prefix)/share
datadir	$(datarootdir)
sysconfdir	$(prefix)/etc
sharedstatedir	$(prefix)/com
localstatedir	$(prefix)/var
includedir	$(prefix)/include
oldincludedir	/usr/include
docdir	$(datarootdir)/doc/$(package)
infodir	$(datarootdir)/info
htmldir	$(docdir)
dvidir	$(docdir)
pdfdir	$(docdir)
psdir	$(docdir)
libdir	$(exec_prefix)/lib
lispdir	$(datarootdir)/emacs/site-lisp
localedir	$(datarootdir)/locale

(continued)

Table 3-1 (continued)

Variable	Default Value
mandir	$(datarootdir)/man
man*N*dir	$(mandir)/man*N* (*N* = 1..9)
manext	.1
man*N*ext	.*N* (*N* = 1..9)
srcdir	The source-tree directory corresponding to the current directory in the build tree

Autotools-based projects support these and other useful variables automatically, as needed; Automake provides full support for them, while Autoconf's support is more limited. If you write your own makefiles and build systems, you should support as many of these as you use in your build and installation processes.

Adding Location Variables to Jupiter

To support the variables that we've used so far in the Jupiter project, we need to add the bindir variable, as well as any variables that it relies on—in this case, the exec_prefix variable. Listings 3-27 and 3-28 show how to do this in the top-level and *src* directory makefiles.

Git tag 3.11

```
--snip--
prefix = /usr/local
exec_prefix = $(prefix)
bindir = $(exec_prefix)/bin
export prefix
export exec_prefix
export bindir
--snip--
```

Listing 3-27: Makefile: *Adding the* bindir *variable*

```
--snip--
install:
        install -d $(bindir)
        install -m 0755 jupiter $(bindir)

uninstall:
        rm -f $(bindir)/jupiter
        -rmdir $(bindir) >/dev/null 2>&1
--snip--
```

Listing 3-28: src/Makefile: *Adding the* bindir *variable*

Even though we only use bindir in *src/Makefile*, we have to export prefix, exec_prefix, and bindir because bindir is defined in terms of exec_prefix, which is itself defined in terms of prefix. When make runs the install commands, it will first resolve bindir to $(exec_prefix)/*bin*, then to $(prefix)/*bin*, and finally to */usr/local/bin*. Thus, *src/Makefile* needs to have access to all three variables during this process.

How do such recursive variable definitions make life better for the end user? After all, the user can change the root install location from */usr/local* to */usr* by simply typing the following:

```
$ make prefix=/usr install
--snip--
```

The ability to change prefix variables at multiple levels is particularly useful to a Linux distribution packager (an employee or volunteer at a Linux company whose job it is to professionally package your project as a *.deb* or *.rpm* package) who needs to install packages into very specific system locations. For example, a distro packager could use the following command to change the installation prefix to */usr* and the system configuration directory to */etc*:

```
$ make prefix=/usr sysconfdir=/etc install
--snip--
```

Without the ability to change prefix variables at multiple levels, configuration files would end up in */usr/etc* because the default value of $(sysconfdir) is $(prefix)*/etc*.

Getting Your Project into a Linux Distro

When a Linux distro picks up your package for distribution, your project magically moves from the realm of tens of users to that of tens of thousands of users—almost overnight. Some people will be using your software without even knowing it. Since one great value of open source software for the developer is free help in making your software better, this can be seen as a good thing—a dramatic increase in community size.

By following the *GCS* within your build system, you remove many of the barriers to including your project in a Linux distro. If your tarball follows all the usual conventions, distro packagers will immediately know what to do with it. These packagers generally get to decide, based on needed functionality and their feelings about your package, whether it should be included in their flavor of Linux. Since they have a fair amount of power in this process, it behooves you to please them.

Section 7 of the *GCS* contains a small subsection that talks about supporting *staged installations*. It is easy to support this concept in your build system, but if you neglect to support it, it will almost always cause problems for packagers.

Packaging systems such as the Red Hat Package Manager (RPM) accept one or more tarballs, a set of patch files, and a specification file. The so-called *spec file* describes the process of building and packaging your project for a particular system. In addition, it defines all of the products installed into the target installation directory structure. The package manager software uses this information to install your package into a temporary directory, from which it then pulls the specified products, storing them in a special binary archive that the package installation program (for example, rpm) understands.

To support staged installation, all you need is a variable named DESTDIR that acts as a sort of super-prefix to all of your installed products. To show you how this is done, I'll add staged installation support to the Jupiter project. This is so trivial that it requires only four changes to *src/Makefile*. The required changes are highlighted in Listing 3-29.

Git tag 3.12

```
--snip--
install:
        install -d $(DESTDIR)$(bindir)
        install -m 0755 jupiter $(DESTDIR)$(bindir)

uninstall:
        rm -f $(DESTDIR)$(bindir)/jupiter
        -rmdir $(DESTDIR)$(bindir) >/dev/null 2>&1
--snip--
```

Listing 3-29: src/Makefile: Adding staged build functionality

As you can see, I've added the $(DESTDIR) prefix to the $(bindir) references in the install and uninstall targets that refer to installation paths. You don't need to define a default value for DESTDIR, because when it is left undefined, it expands to an empty string, which has no effect on the paths to which it's prepended.

> **NOTE** *Do not add a slash after $(DESTDIR), which is usually empty. The prefix variables ultimately resolve to something starting with a slash; adding a slash after $(DESTDIR) is therefore redundant and, in some situations, can cause unintended side effects.*

I didn't need to add $(DESTDIR) to the uninstall rule's rm command for the sake of the package manager, because package managers don't care how your package is uninstalled. They only install your package so they can copy the products from a stage directory. To uninstall the stage directory, package managers simply delete it. Package manager programs such as rpm use their own rules for removing products from a system, and these rules are based on a package manager database rather than your uninstall target.

However, for the sake of symmetry, and to be complete, it doesn't hurt to add $(DESTDIR) to uninstall. Besides, we need it to be complete for the sake of the distcheck target, which we'll now modify to take advantage of our staged installation functionality. This modification is shown in Listing 3-30.

Git tag 3.13

```
--snip--
distcheck: $(distdir).tar.gz
        gzip -cd $(distdir).tar.gz | tar xvf -
        cd $(distdir) && $(MAKE) all
        cd $(distdir) && $(MAKE) check
        cd $(distdir) && $(MAKE) DESTDIR=$${PWD}/inst install
        cd $(distdir) && $(MAKE) DESTDIR=$${PWD}/inst uninstall
```

```
            @remaining="`find $(distdir)/inst -type f | wc -l`"; \
            if test "$${remaining}" -ne 0; then \
              echo "*** $${remaining} file(s) remaining in stage directory!"; \
              exit 1; \
            fi
            cd $(distdir) && $(MAKE) clean
            rm -rf $(distdir)
            @echo "*** Package $(distdir).tar.gz is ready for distribution."
--snip--
```

Listing 3-30: Makefile: Using DESTDIR in the distcheck target

Changing prefix to DESTDIR in the install and uninstall commands allows us to properly test a complete installation directory hierarchy, as we'll see shortly.

At this point, an RPM spec file could provide the following text as the installation commands for the Jupiter package:

```
%install
make prefix=/usr DESTDIR=%BUILDROOT install
```

Don't worry about package manager file formats. Instead, just focus on providing staged installation functionality through the DESTDIR variable.

You may be wondering why the prefix variable couldn't provide this functionality. For one thing, not every path in a system-level installation is defined relative to the prefix variable. The system configuration directory (sysconfdir), for instance, is often defined as */etc* by packagers. You can see in Table 3-1 that the default definition of sysconfdir is $(prefix)/*etc*, so the only way sysconfdir would resolve to */etc* would be if you explicitly set it to do so on the configure or make command line. If you configured it that way, only a variable like DESTDIR would affect the base location of sysconfdir during staged installation. Other reasons for this will become clearer as we talk about project configuration later on in this chapter, and then again in the next two chapters.

Build vs. Installation Prefix Overrides

At this point, I'd like to digress slightly to explain an elusive (or at least nonobvious) concept regarding prefix and other path variables defined in the *GCS*. In the preceding examples, I used prefix overrides on the make install command line, like this:

```
$ make prefix=/usr install
--snip--
```

The question I wish to address is: what is the difference between using a prefix override for make all and for make install? In our small sample makefiles, we've managed to avoid using prefixes in any targets not related to installation, so it may not be clear to you at this point that a prefix is *ever* useful during the build stage. However, prefix variables can be very useful during the build stage to substitute paths into source code at compile time, as shown in Listing 3-31.

```
program: main.c
        gcc -DCFGDIR="\"$(sysconfdir)\"" -o $@ main.c
```

Listing 3-31: Substituting paths into source code at compile time

In this example, I'm defining a C-preprocessor variable called CFGDIR on the compiler command line for use by *main.c*. Presumably, there's some code in *main.c* like that shown in Listing 3-32.

```
#ifndef CFGDIR
# define CFGDIR "/etc"
#endif
const char *cfgdir = CFGDIR;
```

Listing 3-32: Substituting CFGDIR at compile time

Later in the code, you might use the C global variable cfgdir to access the application's configuration file.

Linux distro packagers often use different prefix overrides for build and install command lines in RPM spec files. During the build stage, the actual runtime directories are hardcoded into the executable using commands like the ./configure command shown in Listing 3-33.

```
%build
%setup
./configure prefix=/usr sysconfdir=/etc
make
```

Listing 3-33: The portion of an RPM spec file that builds the source tree

Note that we have to explicitly specify sysconfdir along with prefix, because, as I mentioned earlier, the system configuration directory is usually outside of the prefix directory structure. The package manager installs these executables into a stage directory so it can then copy them out of their installed locations when it builds the binary installation package. The corresponding installation commands might look like those shown in Listing 3-34.

```
%install
make DESTDIR=%BUILDROOT% install
```

Listing 3-34: The installation portion of an RPM spec file

Using DESTDIR during installation will temporarily override *all* installation prefix variables, so you don't have to remember which variables you've overridden during configuration. Given the configuration command shown in Listing 3-33, using DESTDIR in the manner shown in Listing 3-34 has the same effect as the code shown in Listing 3-35.

```
%install
make prefix=%BUILDROOT%/usr sysconfdir=%BUILDROOT%/etc install
```

Listing 3-35: Overriding the default sysconfdir during installation

WARNING *The key point here is one that I touched on earlier. Never write your install target to build all or even part of your products in your makefiles. Installation functionality should be limited to copying files, if possible. Otherwise, your users won't be able to access your staged installation features if they are using prefix overrides.*

Another reason for limiting installation functionality in this way is that it allows the user to install sets of packages as a group into an isolated location and then create links to the actual files in the proper locations. Some people like to do this when they are testing out a package and want to keep track of all its components.[26]

One final point: if you're installing into a system directory hierarchy, you'll need *root* permissions. People often run make install like this:

```
$ sudo make install
```

If your install target depends on your build targets, and you've neglected to build them beforehand, make will happily build your program before installing it—but the local copies will all be owned by *root*. This inconvenience is easily avoided by having make install fail for lack of things to install, rather than jumping right into a build while running as *root*.

User Variables

The *GCS* defines a set of variables that are sacred to the user. These variables should be *referenced* by a GNU build system but never *modified* by a GNU build system. These so-called *user variables* include those listed in Table 3-2 for C and C++ programs.

Table 3-2: Some User Variables and Their Purposes

Variables	Purpose
CC	A reference to the system C compiler
CFLAGS	Desired C compiler flags
CXX	A reference to the system C++ compiler

(continued)

---

26. Some Linux distributions provide a way of installing multiple versions of common packages. Java is a great example; to support packages using multiple versions or brands of Java (perhaps Oracle Java versus IBM Java), some Linux distributions provide a script set called the *alternatives* scripts. These allow a user (running as *root*) to swap all of the links in the various system directories from one grouped installation to another. Thus, both sets of files can be installed in different auxiliary locations, but links in the expected installation locations can be changed to refer to each group at different times with a single *root*-level command.

Table 3-2 (continued)

Variables	Purpose
CXXFLAGS	Desired C++ compiler flags
LDFLAGS	Desired linker flags
CPPFLAGS	Desired C/C++ preprocessor flags
--snip--	

This list is by no means comprehensive, and interestingly, there isn't a comprehensive list to be found in the *GCS*. In fact, most of these variables come from the documentation for the make utility itself. These variables are used in the built-in rules of the make utility—they're somewhat hardcoded into make, so they are effectively defined by make. You can find a fairly complete list of program name and flag variables in the "Variables Used by Implicit Rules" section of the *GNU Make Manual*.

Note that make assigns default values for many of these variables based on common Unix utility names. For example, the default value of CC is cc, which (at least on Linux systems) is a soft link to the GCC C compiler (gcc). On other systems, cc is a soft link to the system's own compiler. Thus, we don't need to set CC to gcc, which is good, because GCC may not be installed on non-Linux platforms. There may be times when you do wish to set CC on the make command line, such as when using an alternative compiler like clang or when using the ccache utility to cache gcc results for faster recompilation.

For our purposes, the variables shown in Table 3-2 are sufficient, but for a more complex makefile, you should become familiar with the larger list outlined in the *GNU Make Manual*.

To use these variables in our makefiles, we'll just replace gcc with $(CC). We'll do the same for CFLAGS and CPPFLAGS, although CPPFLAGS will be empty by default. The CFLAGS variable has no default value either, but this is a good time to add one. I like to use -g to build objects with symbols and -O0 to disable optimizations for debug builds. The updates to *src/Makefile* are shown in Listing 3-36.

Git tag 3.14

```
CFLAGS = -g -O0
--snip--
jupiter: main.c
        $(CC) $(CPPFLAGS) $(CFLAGS) -o $@ main.c
--snip--
```

Listing 3-36: src/Makefile: Adding appropriate user variables

This works because the make utility allows such variables to be overridden by options on the command line. For example, to switch compilers and set some compiler command line options, a user need only type the following:

```
$ make CC=ccache CFLAGS='-g -O2' CPPFLAGS=-Dtest
```

In this case, our user has decided to use the ccache utility instead of gcc, generate debug symbols, and optimize their code using level-two optimizations. They've also decided to enable the test option through the use of a C-preprocessor definition. Note that these variables are set on the make command line; this apparently equivalent Bourne-shell syntax will not work as expected:

```
$ CC=ccache CFLAGS='-g -O2' CPPFLAGS=-Dtest make
```

The reason is that we're merely setting environment variables in the local environment passed to the make utility by the shell. Remember that environment variables do not automatically override those set in the makefile. To get the functionality we want, we could use a little GNU make–specific syntax in our makefile, as shown in Listing 3-37.

```
--snip--
CFLAGS ?= -g -O0
--snip--
```

Listing 3-37: Using the GNU make–specific query-assign operator (?=) in a makefile

The ?= operator is a GNU make–specific operator, which will only set the variable in the makefile if it hasn't already been set elsewhere. This means we can now override these particular variable settings by setting them in the environment. But don't forget that this will only work in GNU make. In general, it's better to set make variables on the make command line.

Nonrecursive Build Systems

Now that we've spent all this time creating the perfect build system for our project, let's take a look at a *more perfect* solution—a nonrecursive system. I mentioned at the start of this chapter that there was a problem with recursive builds that we'd discuss at a later point.

The fundamental problem with recursive build systems is that they artificially introduce flaws into make's directed graph—the set of rules make uses to determine what depends on what and when something needs to be rebuilt. For Jupiter, very little can go wrong because there's one top-level makefile invoking make on a single subdirectory makefile, but let's consider a more complex project where multiple submodules, nested arbitrarily deeply, are interdependent upon each other in more complex ways.

With a single makefile, the one make process can "see the big picture." That is, it can see and understand all of the interdependencies in the system, and it can create a DAG that properly represents all of the interdependencies among all of the filesystem objects within the project. With multiple makefiles, each child make process executed by parent make can see only a portion of the dependency graph. Ultimately, this can cause make to build products out of order so that a product that depends on prerequisites not within its own purview is built before those prerequisites are updated.

The preceding problem is compounded when you use *parallel make* by adding -j to the make command line. The -j option tells make to examine its DAG and find places where portions of the DAG do not depend on each other, then execute those portions at the same time. On a multiprocessor system, this can dramatically speed up the build process for large projects. However, this causes problems from two different angles. First, since make can't see the whole picture, it can make incorrect assumptions about what things can be done in parallel. Second, as far as the top-level make is concerned, child make processes are all independent and can be run in parallel, which we can easily see is simply not true. For an example that does not even rely on the differences between recursive and nonrecursive build systems, consider the following command line:

```
$ make -j clean all
```

As far as make is concerned, clean and all are 100 percent independent of each other, so make will happily run them both at the same time. Even a novice can see the problems with this assumption. The point is, make doesn't understand the high-level relationship between clean and all. That relationship is understood only by the author of the makefile. Similar barriers to make's understanding of the big picture are artificially introduced at the boundaries between parent and child make invocations in a recursive build system.

So, how hard is it to turn Jupiter's recursive build system into a nonrecursive system? We want to maintain modularity, so we still want a *Makefile* in each directory that essentially manages the tasks of that directory. This is easily accomplished by using another feature of common make—the include directive. The include directive allows us to break up our single, parent-level makefile into chunks of directory-specific rules and then include just those snippets in the top-level makefile. Listing 3-38 shows what the complete updated top-level makefile looks like.

Git tag 3.15

```
package = jupiter
version = 1.0
tarname = $(package)
distdir = $(tarname)-$(version)

prefix = /usr/local
exec_prefix = $(prefix)
bindir = $(exec_prefix)/bin

❶ #export prefix
#export exec_prefix
#export bindir

❷ all jupiter: src/jupiter

dist: $(distdir).tar.gz

$(distdir).tar.gz: $(distdir)
        tar chof - $(distdir) | gzip -9 -c > $@
        rm -rf $(distdir)
```

```
$(distdir): FORCE
        mkdir -p $(distdir)/src
        cp Makefile $(distdir)
        cp src/Makefile $(distdir)/src
        cp src/main.c $(distdir)/src

distcheck: $(distdir).tar.gz
        gzip -cd $(distdir).tar.gz | tar xvf -
        cd $(distdir) && $(MAKE) all
        cd $(distdir) && $(MAKE) check
        cd $(distdir) && $(MAKE) DESTDIR=$${PWD}/_inst install
        cd $(distdir) && $(MAKE) DESTDIR=$${PWD}/_inst uninstall
        @remaining="`find $${PWD}/$(distdir)/_inst -type f | wc -l`"; \
        if test "$${remaining}" -ne 0; then \
          echo "*** $${remaining} file(s) remaining in stage directory!"; \
          exit 1; \
        fi
        cd $(distdir) && $(MAKE) clean
        rm -rf $(distdir)
        @echo "*** Package $(distdir).tar.gz is ready for distribution."

FORCE:
        -rm -f $(distdir).tar.gz >/dev/null 2>&1
        -rm -rf $(distdir) >/dev/null 2>&1
```

❸ `include src/Makefile`

```
.PHONY: FORCE all clean check dist distcheck install uninstall
```

Listing 3-38: Makefile: A nonrecursive version of the top-level makefile

Three changes were made here, but please note that the only really significant change made to this makefile was the replacement of the rule at ❷ where recursion was done with a single rule for `all`, `clean`, `check`, `install`, `uninstall`, and an explicit `jupiter` target. Even this replacement could have been a simple deletion if we hadn't cared that the new default target would have become `dist`, had we not added the `all` target at this location. I've also added an explicit `jupiter` target that maps to `src/jupiter` to maintain feature parity with the previous system.

The second change made was to include the *src*-level makefile at ❸. Finally, I also commented out the `export` statements at ❶ because we no longer need to export variables to child `make` processes; they're left as comments simply for illustration.

Now, let's examine what changed in the *src*-level makefile. The complete, updated version is shown in Listing 3-39.

```
CFLAGS = -g -O0

src/jupiter: src/main.c
        $(CC) $(CFLAGS) $(CPPFLAGS) -o $@ src/main.c

check: all
```

```
        ./src/jupiter | grep "Hello from .*jupiter!"
        @echo "*** All TESTS PASSED"

install:
        install -d $(DESTDIR)$(bindir)
        install -m 0755 src/jupiter $(DESTDIR)$(bindir)

uninstall:
        rm -f $(DESTDIR)$(bindir)/jupiter
        -rmdir -f $(DESTDIR)$(bindir) >/dev/null 2>&1

clean:
        rm -f src/jupiter
```

Listing 3-39: src/Makefile: *A nonrecursive version of the src-level makefile*

First, the all target was removed. We don't need one here now because this makefile is not intended to be executed directly but, rather, included by the parent makefile. Hence, we do not need a default target. Second, all references to objects in the *src* directory are now referenced by paths that are relative to the parent directory. Again, this is because make is executed only once from the parent directory, so references to objects in the *src* directory must be considered relative to where make is running—the parent directory.

We also removed the .PHONY directive at the bottom because this directive contained a proper subset of the .PHONY directive in the parent makefile, making the directive redundant. In short, we merely converted this makefile into a snippet that could be included in the parent makefile, removed redundancies, and ensured that all filesystem references are now made relative to the parent directory. I hope you can see that these changes actually constitute a simplification of what we had before. Intuitively, it seems more complicated but it is actually simpler.

This makefile is a more accurate and faster version of our recursive system. I say "this makefile" because there is really only one makefile here—the included file can be pasted directly into the parent makefile at the point of inclusion (at ❷ in Listing 3-38), just as with inclusions of header files in C-language source files. Ultimately, after all the inclusions are resolved, there are only one makefile and one make process that executes commands based on the rules in that makefile.

One apparent drawback of nonrecursive build systems is that you cannot simply enter make while sitting in the *src* directory and build the portion of the project related to that directory. Instead, you have to change into the parent directory and run make, which builds everything. But this, too, is a fallacious concern because you've always had the ability to execute any portion of the build system you wished by specifying exactly the target you desired on the make command line. The difference is that now what gets built is actually what should get built because make understands the entire set of dependencies for any given target you command it to build.

As we'll see in the coming chapters, Automake has full support for nonrecursive build systems. I encourage you to start writing your next project build system in a nonrecursive fashion because it can seem like an overwhelming task to retrofit an existing system, even though, as we've seen here, it's not really all that difficult.

Configuring Your Package

The *GCS* describes the configuration process in the "How Configuration Should Work" subsection of Section 7. Up to this point, we've been able to do about everything we've wanted to with Jupiter using only makefiles, so you might be wondering what configuration is actually for. The opening paragraphs of this subsection in the *GCS* answer our question:

> Each GNU distribution should come with a shell script named configure. This script is given arguments which describe the kind of machine and system you want to compile the program for. The configure script must record the configuration options so that they affect compilation.
>
> The description here is the specification of the interface for the configure script in GNU packages. Many packages implement it using GNU Autoconf (see "Introduction" in Autoconf) and/ or GNU Automake (see "Introduction" in Automake), but you do not have to use these tools. You can implement it any way you like; for instance, by making configure be a wrapper around a completely different configuration system.
>
> Another way for the configure script to operate is to make a link from a standard name such as *config.h* to the proper configuration file for the chosen system. If you use this technique, the distribution should *not* contain a file named *config.h*. This is so that people won't be able to build the program without configuring it first.
>
> Another thing that configure can do is to edit the *Makefile*. If you do this, the distribution should *not* contain a file named *Makefile*. Instead, it should include a file *Makefile.in* which contains the input used for editing. Once again, this is so that people won't be able to build the program without configuring it first.[27]

So then, the primary tasks of a typical configuration script are as follows:

- Generate files from templates containing replacement variables.
- Generate a C-language header file (*config.h*) for inclusion by project source code.

27. See Section 7.1, "How Configuration Should Work," in the *GNU Coding Standards* document at *http://www.gnu.org/prep/standards/html_node/Configuration.html#Configuration*. GNU documentation changes quite often. This text came from the January 14, 2019 version of the *GCS* document.

- Set user options for a particular make environment (debug flags and so on).
- Set various package options as environment variables.
- Test for the existence of tools, libraries, and header files.

For complex projects, configuration scripts often generate the project makefiles from one or more templates maintained by project developers. These templates contain configuration variables in a format that is easy to recognize (and substitute). The configuration script replaces these variables with values determined during the configuration process—either from command line options specified by the user or from a thorough analysis of the platform environment. This analysis entails such things as checking for the existence of certain system or package header files and libraries, searching various filesystem paths for required utilities and tools, and even running small programs designed to indicate the feature set of the shell, C compiler, or desired libraries.

The tool of choice for variable replacement has, in the past, been the sed stream editor. A simple sed command can replace all the configuration variables in a makefile template in a single pass through the file. However, Autoconf versions 2.62 and newer prefer awk to sed for this process. The awk utility is almost as pervasive as sed these days, and it provides more functionality to allow for efficient replacement of many variables. For our purposes on the Jupiter project, either of these tools would suffice.

Summary

We have now created a complete project build system by hand, with one important exception: we haven't designed a configure script according to the design criteria specified in the *GNU Coding Standards*. We could do this, but it would take a dozen more pages of text to build one that even comes close to conforming to these specifications. Still, there are a few key build features related specifically to the makefiles that the *GCS* indicates are desirable. Among these is the concept of vpath building. This is an important feature that can be properly illustrated only by actually writing a configuration script that works as specified by the *GCS*.

Rather than spend the time and effort to do this now, I'd like to simply move on to a discussion of Autoconf in Chapter 4, which will allow us to build one of these configuration scripts in as little as two or three lines of code. With that behind us, it will be trivial to add vpath building and other common Autotools features to the Jupiter project.

4

CONFIGURING YOUR PROJECT WITH AUTOCONF

Come my friends,
'Tis not too late to seek a newer world.
—Alfred, Lord Tennyson, "Ulysses"

The Autoconf project has had a long history, starting in 1992 when David McKenzie, while volunteering for the Free Software Foundation, was looking for a way to simplify the process of creating the complex configuration scripts necessary to support the target platforms that were being added daily at that time to the GNU project. At the same time, he was working on his bachelor's degree in computer science at the University of Maryland, College Park.

After McKenzie's initial work on Autoconf, he continued to be a strong contributor to the project through 1996, at which point Ben Elliston took over project maintenance. Since then, maintainers and primary contributors have included Akim Demaille, Jim Meyering, Alexandre Oliva, Tom Tromey, Lars J. Aas (inventor of the name *autom4te*, among others), Mo DeJong, Steven G. Johnson, Matthew D. Langston, Paval Roskin, and Paul Eggert (the list of contributors is much longer—see the Autoconf *AUTHORS* file for more history).

Today's maintainer, Eric Blake, began making strong contributions to Autoconf in 2012. He's been maintainer of the project ever since while working for Red Hat. Because Automake and Libtool are essentially add-on components to the original Autoconf framework, it's useful to spend some time focusing on using Autoconf without Automake and Libtool. This will provide a fair amount of insight into how Autoconf operates by exposing aspects of the tool that are often hidden by Automake.

Before Automake came along, Autoconf was used alone. In fact, many legacy open source projects never made the transition from Autoconf to the full GNU Autotools suite. As a result, it's not unusual to find a file called *configure.in* (the original Autoconf naming convention), as well as handwritten *Makefile.in* templates, in older open source projects.

In this chapter, I'll show you how to add an Autoconf build system to an existing project. I'll spend most of this chapter talking about the more basic features of Autoconf, and in Chapter 5 I'll go into much more detail about how some of the more complex Autoconf macros work and how to properly use them. Throughout this process, we'll continue using the Jupiter project as our example.

Autoconf Configuration Scripts

The input to the autoconf program is Bourne shell script sprinkled with macro calls. The input data stream must also include the definitions of all referenced macros—both those that Autoconf provides and those that you write yourself.

The macro language used in Autoconf is called *M4*. (The name means *M, plus 4 more letters*, or the word *Macro*.[1]) The m4 utility is a general-purpose macro language processor originally written by Brian Kernighan and Dennis Ritchie in 1977.

While you may not be familiar with it, you can find some form of M4 on every Unix and Linux variant (as well as other systems) in use today. The ubiquitous nature of this tool is the main reason it's used by Autoconf, as the original design goals of Autoconf stated that it should be able to run on all systems without the addition of complex tool chains and utility sets.[2]

Autoconf depends on the existence of relatively few tools: a Bourne shell, M4, and a Perl interpreter. The configuration scripts and makefiles it generates rely on the existence of a different set of tools, including a Bourne shell, grep, ls, and sed or awk.[3]

---

1. As a point of interest, this naming convention is a fairly common practice in some software-engineering domains. For example, the term *internationalization* is often abbreviated *i18n*, for the sake of brevity (or perhaps just because programmers love acronyms).

2. In fact, whatever notoriety M4 may have today is likely due to the widespread use of Autoconf.

3. Autoconf versions 2.62 and later generate configuration scripts that require awk in addition to sed on the end user's system.

NOTE *Do not confuse the requirements of the Autotools with the requirements of the scripts and makefiles they generate. The Autotools are maintainer tools, whereas the resulting scripts and makefiles are end user tools. We can reasonably expect a higher level of installed functionality on development systems than we can on end user systems.*

The configuration script ensures that the end user's build environment is configured to properly build your project. This script checks for installed tools, utilities, libraries, and header files, as well as for specific functionality within these resources. What distinguishes Autoconf from other project configuration frameworks is that Autoconf tests also ensure that these resources can be properly consumed by your project. You see, it's important not only that your users have *libxyz.so* and its public header files properly installed on their systems but also that they have compatible versions of these files. Autoconf is pathological about such tests. It ensures that the end user's environment is in compliance with the project requirements by compiling and linking a small test program for each feature—a quintessential example, if you will, that does what your project source code does on a larger scale.

Can't I just ensure that libxyz.2.1.0.so *is installed by searching library paths for the filename?* The answer to this question is debatable. There are legitimate situations where libraries and tools get updated quietly. Sometimes, the specific functionality upon which your project relies is added in the form of a security bug fix or enhancement to a library, in which case vendors aren't even required to bump up the version number. But it's often difficult to tell whether you've got version 2.1.0.r1 or version 2.1.0.r2 unless you look at the file size or call a library function to make sure it works as expected.

Additionally, vendors often backport bug fixes and features from newer products onto older platforms without bumping the version number. Hence, you can't tell even by looking at the version number whether the library supports a feature that was added *after* that version of the library was published.

However, the most significant reason for not relying on library version numbers is that they do not represent specific marketing releases of a library. As we will discuss in Chapter 8, library version numbers indicate binary interface characteristics on a particular platform. This means that library version numbers for the same feature set can be different from platform to platform. As a result, you may not be able to tell—short of compiling and linking against the library—whether or not a particular library has the functionality your project needs.

Finally, there are several important cases where the same functionality is provided by entirely different libraries on different systems. For example, you may find cursor manipulation functionality in *libtermcap* on one system, *libncurses* on another, and *libcurses* on yet another system. But it's not critical that you know about all of these side cases, because your users will tell you when your project won't build on their system because of such a discrepancy.

What can you do when such a bug is reported? You can use the Autoconf AC_SEARCH_LIBS macro to test multiple libraries for the same functionality. Simply add a library to the search list, and you're done. Since this fix is so easy, it's likely the user who noticed the problem will simply send a patch to your *configure.ac* file.

Because Autoconf tests are written in shell script, you have a lot of flexibility as to how the tests operate. You can write a test that merely checks for the existence of a library or utility in the usual locations on your user's system, but this bypasses some of the most significant features of Autoconf. Fortunately, Autoconf provides dozens of macros that conform to Autoconf's feature-testing philosophy. You should carefully study and use the list of available macros, rather than write your own, because they're specifically designed to ensure that the desired functionality is available on the widest variety of systems and platforms.

The Shortest configure.ac File

The input file for autoconf is called *configure.ac*. The simplest possible *configure.ac* file has just two lines, as shown in Listing 4-1.

```
AC_INIT([Jupiter], [1.0])
AC_OUTPUT
```

Listing 4-1: The simplest configure.ac *file*

To those new to Autoconf, these two lines appear to be a couple of function calls, perhaps in the syntax of some obscure programming language. Don't let their appearance throw you—these are M4 macro invocations. The macros are defined in files distributed with the autoconf package. You can find the definition of AC_INIT, for example, in *general.m4* in Autoconf's installation directory (usually */usr/(local/)share/autoconf/autoconf*). AC_OUTPUT is defined in *status.m4* in the same directory.

Comparing M4 to the C Preprocessor

M4 macros are similar in many ways to the C-preprocessor (CPP) macros defined in C-language source files. The C preprocessor is also a text replacement tool, which isn't surprising: both M4 and the C preprocessor were designed and written by Kernighan and Ritchie around the same time.

Autoconf uses square brackets around macro parameters as a quoting mechanism. Quotes are necessary only for cases in which the context of the macro call could cause an ambiguity that the macro processor may resolve incorrectly (usually without telling you). We'll discuss M4 quoting in much more detail in Chapter 16. For now, just use square brackets around every argument to ensure that the expected macro expansions are generated.

As with CPP macros, you can define M4 macros to accept a comma-delimited list of arguments enclosed in parentheses. With CPP, macros are defined using a *preprocessor directive*: #define name(args) expansion, while in M4, macros are defined with a built-in macro: define(name, expansion). Another significant difference is that in CPP, the arguments specified in the macro definition are required,[4] while in M4, the arguments to parameterized macros are optional and the caller may simply omit them. If no arguments are passed, you can also omit the parentheses. Extra arguments passed to M4 macros are simply ignored. Finally, M4 does not allow intervening whitespace between a macro name and the opening parenthesis in a macro invocation.

The Nature of M4 Macros

If you've been programming in C for many years, you've no doubt run across a few C-preprocessor macros from the dark regions of the lower realm. I'm talking about those truly evil macros that expand into one or two pages of C code. They should have been written as C functions, but their authors were either overly worried about performance or just got carried away, and now it's your turn to debug and maintain them. But, as any veteran C programmer will tell you, the slight performance gains you get by using a macro where you should have used a function do not justify the trouble you cause maintainers trying to debug your fancy macros. Debugging such macros can be a nightmare because the source code generated by macros is usually inaccessible from within a symbolic debugger.[5]

Writing such complex macros is viewed by M4 programmers as a sort of macro nirvana—the more complex and functional they are, the "cooler" they are. The two Autoconf macros in Listing 4-1 expand into a file containing almost 2,400 lines of Bourne-shell script that total more than 70KB! But you wouldn't guess this by looking at their definitions. They're both fairly short—only a few dozen lines each. The reason for this apparent disparity is simple: they're written in a modular fashion, with each macro expanding several others, which in turn expand several others, and so on.

For the same reasons that programmers are taught not to abuse the C preprocessor, the extensive use of M4 causes a fair amount of frustration for those trying to understand Autoconf. That's not to say Autoconf shouldn't use M4 this way; quite the contrary—this is the domain of M4. But there is a school of thought that says M4 was a poor choice for Autoconf because of the problems with macros mentioned earlier. Fortunately, being able to use Autoconf effectively usually doesn't require a deep understanding of the inner workings of the macros that ship with it.[6]

---

4. I'm ignoring the newer CPP variadic macros in modern preprocessors. M4 has always had optional arguments; all M4 macro arguments are optional without special syntax to make optional arguments work.

5. A technique I've used in the past for debugging large macros involves manually generating source code using the C preprocessor and then compiling this generated source. Symbolic debuggers can only work with the source code you provide. By providing source with the macros fully expanded, you enable the debugger to allow you to step through the generated source.

6. There are a few exceptions to this rule. Poor documentation can sometimes lead to a misunderstanding about the intended use of some of the published Autoconf macros. This book highlights a few of these situations, but a degree of expertise with M4 is the only way to work your way through most of these problems.

Executing autoconf

Running Autoconf is simple: just execute autoconf in the same directory as your *configure.ac* file. While I could do this for each example in this chapter, I'm going to use the autoreconf program instead of the autoconf program, because running autoreconf has exactly the same effect as running autoconf, except that autoreconf will also do the right thing when you start adding Automake and Libtool functionality to your build system. That is, it will execute all of the Autotools in the right order based on the contents of your *configure.ac* file.

The autoreconf program is smart enough to execute only the tools you need, in the order you need them, with the options you want (with one caveat that I'll mention shortly). Therefore, running autoreconf is the recommended method for executing the Autotools tool chain.

Let's start by adding the simple *configure.ac* file from Listing 4-1 to our project directory. The top-level directory currently contains only a *Makefile* and a *src* directory that contains its own *Makefile* and a *main.c* file. Once you've added *configure.ac* to the top-level directory, run autoreconf:

Git tag 4.0

```
$ autoreconf
$
$ ls -1p
autom4te.cache/
configure
configure.ac
Makefile
src/
$
```

First, notice that autoreconf operates silently by default. If you want to see something happening, use the -v or --verbose option. If you want autoreconf to execute the Autotools in verbose mode as well, then add -vv to the command line.[7]

Next, notice that autoconf creates a directory called *autom4te.cache*. This is the autom4te cache directory. This cache speeds up access to *configure.ac* during successive executions of utilities in the Autotools tool chain.

The result of passing *configure.ac* through autoconf is essentially the same file (now called configure), but with all of the macros fully expanded. You're welcome to take a look at configure, but don't be too surprised if you don't immediately understand what you see. The *configure.ac* file has been transformed, through M4 macro expansions, into a text file containing thousands of lines of complex Bourne shell script.

---

7. You may also pass --verbose --verbose, but this syntax seems a bit . . . verbose to me.

Executing configure

As discussed in "Configuring Your Package" on page 77, the *GNU Coding Standards* indicate that a handwritten configure script should generate another script called config.status, whose job it is to generate files from templates. Unsurprisingly, this is exactly the sort of functionality you'll find in an Autoconf-generated configuration script. This script has two primary tasks:

- Perform requested checks
- Generate and then call config.status

The results of the checks performed by configure are written into config.status in a manner that allows them to be used as replacement text for Autoconf substitution variables in template files (*Makefile.in, config.h.in,* and so on). When you execute ./configure, it tells you that it's creating config.status. It also creates a log file called *config.log* that has several important attributes. Let's run ./configure and then see what's new in our project directory:

```
$ ./configure
configure: creating ./config.status
$
$ ls -1p
autom4te.cache/
config.log
config.status
configure
configure.ac
Makefile
src/
$
```

We see that configure has indeed generated both config.status and *config.log*. The *config.log* file contains the following information:

- The command line that was used to invoke configure (very handy!)
- Information about the platform on which configure was executed
- Information about the core tests configure executed
- The line number in configure at which config.status is generated and then called

At this point in the log file, config.status takes over generating log information and adds the following:

- The command line used to invoke config.status

After `config.status` generates all the files from their templates, it exits, returning control to `configure`, which then appends the following information to the log:

- The cache variables that `config.status` used to perform its tasks
- The list of output variables that may be replaced in templates
- The exit code `configure` returned to the shell

This information is invaluable when you're debugging a `configure` script and its associated *configure.ac* file.

Why doesn't `configure` just execute the code it writes into `config.status` instead of going to all the trouble of generating a second script, only to immediately call it? There are a few good reasons. First, the operations of performing checks and generating files are conceptually different, and the `make` utility works best when conceptually different operations are associated with separate targets. A second reason is that you can execute `config.status` separately to regenerate output files from their corresponding template files, saving the time required to perform those lengthy checks. Finally, `config.status` is written to remember the parameters originally used on the `configure` command line. Thus, when `make` detects that it needs to update the build system, it can call `config.status` to re-execute `configure`, using the command line options that were originally specified.

Executing config.status

Now that you know how `configure` works, you might be tempted to execute `config.status` yourself. This was exactly the intent of the Autoconf designers and the authors of the *GCS*, who originally conceived these design goals. However, a more important reason for separating checks from template processing is that `make` rules can use `config.status` to regenerate makefiles from their templates when `make` determines that a template is newer than its corresponding makefile.

Rather than call `configure` to perform needless checks (your environment hasn't changed—just your template files), makefile rules should be written to indicate that output files depend on their templates. The commands for these rules run `config.status`, passing the rule's target as a parameter. If, for example, you modify one of your *Makefile.in* templates, `make` calls `config.status` to regenerate the corresponding *Makefile*, after which `make` re-executes its own original command line—basically restarting itself.[8]

Listing 4-2 shows the relevant portion of such a *Makefile.in* template, containing the rules needed to regenerate the corresponding *Makefile*.

---

8. This is a built-in feature of GNU `make`. However, for the sake of portability, Automake generates makefiles that carefully reimplement this functionality as much as possible in generic `make` script, rather than relying on the built-in mechanism found in GNU `make`. The Automake solution isn't quite as comprehensive as GNU `make`'s built-in functionality, but it's the best we can do, under the circumstances.

```
Makefile: Makefile.in config.status
        ./config.status $@
```

Listing 4-2: A rule that causes make *to regenerate* Makefile *if its template has changed*

A rule with a target named `Makefile` is the trigger here. This rule allows make to regenerate the source makefile from its template if the template changes. It does this *before* executing either the user's specified targets or the default target, if no specific target was given. This functionality is built into make—if there's a rule whose target is `Makefile`, make always evaluates that rule first.

The rule in Listing 4-2 indicates that *Makefile* is dependent on config `.status` as well as *Makefile.in*, because if configure updates `config.status`, it may generate *Makefile* differently. Perhaps different command line options were provided so that configure can now find libraries and header files it couldn't find previously. In this case, Autoconf substitution variables may have different values. Thus, *Makefile* should be regenerated if either *Makefile.in* or config `.status` is updated.

Since `config.status` is itself a generated file, it stands to reason that you could write such a rule to regenerate this file when needed. Expanding on the previous example, Listing 4-3 adds the required code to rebuild config `.status` if configure changes.

```
Makefile: Makefile.in config.status
        ./config.status $@

config.status: configure
        ./config.status --recheck
```

Listing 4-3: A rule to rebuild config.status *when* configure *changes*

Since `config.status` is a dependency of the `Makefile` target, make will look for a rule whose target is `config.status` and run its commands if needed.

Adding Some Real Functionality

I've suggested before that you should call `config.status` in your makefiles to generate those makefiles from templates. Listing 4-4 shows the code in *configure.ac* that actually makes this happen. It's just a single additional macro call between the two original lines of Listing 4-1.

Git tag 4.1
```
AC_INIT([Jupiter],[1.0])
AC_CONFIG_FILES([Makefile src/Makefile])
AC_OUTPUT
```

Listing 4-4: configure.ac: *Using the* AC_CONFIG_FILES *macro*

This code assumes that templates exist for *Makefile* and *src/Makefile*, called *Makefile.in* and *src/Makefile.in*, respectively. These template files look exactly like their *Makefile* counterparts, with one exception: any text I want Autoconf to replace is marked as an Autoconf substitution variable, using the `@VARIABLE@` syntax.

To create these files, simply rename the existing *Makefile* files to *Makefile.in* in both the top-level and *src* directories. This is a common practice when *autoconfiscating* a project:

```
$ mv Makefile Makefile.in
$ mv src/Makefile src/Makefile.in
$
```

With these changes in place, we are now effectively using our new *configure.ac* file in Jupiter to generate makefiles. To make it useful, let's add a few Autoconf substitution variables to replace the original default values. At the top of these files, I've also added the Autoconf substitution variable, `@configure_input@`, after a comment hash mark. Listing 4-5 shows the comment text that is generated in *Makefile*.

Git tag 4.2
```
# Makefile. Generated from Makefile.in by configure.
--snip--
```

Listing 4-5: Makefile: The text generated from the Autoconf @configure_input@ variable

I've also added the makefile regeneration rules from the previous examples to each of these templates, with slight path differences in each file to account for their different positions relative to `config.status` and configure in the build directory.

Listings 4-6 and 4-7 highlight the required changes to the final recursive versions of *Makefile* and *src/Makefile* from near the end of Chapter 3. We'll consider writing nonrecursive versions of these files later as we cover Automake—the process when using Autoconf with handwritten *Makefile.in* templates is nearly identical to what we did in Chapter 3 with makefiles.[9]

```
# @configure_input@

# Package-specific substitution variables
package = @PACKAGE_NAME@
version = @PACKAGE_VERSION@
tarname = @PACKAGE_TARNAME@
distdir = $(tarname)-$(version)

# Prefix-specific substitution variables
prefix = @prefix@
exec_prefix = @exec_prefix@
bindir = @bindir@

all clean check install uninstall jupiter:
        cd src && $(MAKE) $@
```

---

9. The source repository has a "nonrecursive" branch that contains nonrecursive versions of the Autoconf *Makefile.in* templates. Check out the Git tag 4.8 in the jupiter repository.

```
--snip--
$(distdir): FORCE
        mkdir -p $(distdir)/src
        cp configure.ac $(distdir)
        cp configure $(distdir)
        cp Makefile.in $(distdir)
        cp src/Makefile.in src/main.c $(distdir)/src

distcheck: $(distdir).tar.gz
        gzip -cd $(distdir).tar.gz | tar xvf -
        cd $(distdir) && ./configure
        cd $(distdir) && $(MAKE) all
        cd $(distdir) && $(MAKE) check
        cd $(distdir) && $(MAKE) DESTDIR=$${PWD}/_inst install
        cd $(distdir) && $(MAKE) DESTDIR=$${PWD}/_inst uninstall
        @remaining="`find $${PWD}/$(distdir)/_inst -type f | wc -l`"; \
        if test "$${remaining}" -ne 0; then \
          echo "*** $${remaining} file(s) remaining in stage directory!"; \
          exit 1; \
        fi
        cd $(distdir) && $(MAKE) clean
        rm -rf $(distdir)
        @echo "*** Package $(distdir).tar.gz is ready for distribution."
--snip--
FORCE:
        rm -f $(distdir).tar.gz
        rm -rf $(distdir)

Makefile: Makefile.in config.status
        ./config.status $@

config.status: configure
        ./config.status --recheck

.PHONY: FORCE all clean check dist distcheck install uninstall
```

Listing 4-6: Makefile.in: Required modifications to Makefile from Chapter 3

```
# @configure_input@

# Package-specific substitution variables
package = @PACKAGE_NAME@
version = @PACKAGE_VERSION@
tarname = @PACKAGE_TARNAME@
distdir = $(tarname)-$(version)

# Prefix-specific substitution variables
prefix = @prefix@
exec_prefix = @exec_prefix@
bindir = @bindir@

CFLAGS = -g -O0
--snip--
clean:
        rm -f jupiter
```

```
Makefile: Makefile.in ../config.status
        cd .. && ./config.status src/$@

../config.status: ../configure
        cd .. && ./config.status --recheck

.PHONY: all clean check install uninstall
```

Listing 4-7: src/Makefile.in: Required modifications to src/Makefile from Chapter 3

I've removed the export statements from the top-level *Makefile.in* and added a copy of all the make variables (originally only in the top-level *Makefile*) into *src/Makefile.in*. Since config.status generates both of these files, I can reap excellent benefits by substituting values for these variables directly into both files. The primary advantage of doing this is that I can now run make in any subdirectory without worrying about uninitialized variables that would originally have been passed down by a higher-level makefile.

Since Autoconf generates entire values for these make variables, you may be tempted to clean things up a bit by removing the variables and just substituting @prefix@ where we currently use $(prefix) throughout the files. There are a few good reasons for keeping the make variables. First and foremost, we'll retain the original benefits of the make variables; our end users can continue to substitute their own values on the make command line. (Even though Autoconf places default values in these variables, users may wish to override them.) Second, for variables such as $(distdir), whose values are composed of multiple variable references, it's simply cleaner to build the name in one place and use it everywhere else through a single variable.

I've also changed the commands in the distribution targets a bit. Rather than distribute the makefiles, I now need to distribute the *Makefile.in* templates, as well as the new configure script and the *configure.ac* file.[10]

Finally, I modified the distcheck target's commands to run the configure script before running make.

Generating Files from Templates

Note that you can use AC_CONFIG_FILES to generate *any* text file from a file of the same name with a *.in* extension, found in the same directory. The *.in* extension is the default template-naming pattern for AC_CONFIG_FILES, but you can override this default behavior. I'll get into the details shortly.

Autoconf generates sed or awk expressions into the resulting configure script, which then copies them into config.status. The config.status script uses these expressions to perform string replacement in the input template files.

---

10. Distributing *configure.ac* is not merely an act of kindness—it could also be considered a requirement of GNU source licenses, since *configure.ac* is very literally the source code for configure.

Both sed and awk are text-processing tools that operate on file streams. The advantage of a stream editor (the name *sed* is a contraction of the phrase *stream editor*) is that it replaces text patterns in a byte stream. Thus, both sed and awk can operate on huge files because they don't need to load the entire input file into memory in order to process it. Autoconf builds the expression list that config.status passes to sed or awk from a list of variables defined by various macros, many of which I'll cover in greater detail later in this chapter. It's important to understand that Autoconf substitution variables are the *only* items replaced in a template file while generating output files.

At this point, with very little effort, I've created a basic *configure.ac* file. I can now execute autoreconf, followed by ./configure and then make, in order to build the Jupiter project. This simple, three-line *configure.ac* file generates a configure script that is fully functional, according to the definition of a proper configuration script as specified by the *GCS*.

The resulting configuration script runs various system checks and generates a config.status script that can replace a fair number of substitution variables in a set of specified template files in this build system. That's a lot of functionality in just three lines of code.

Adding VPATH Build Functionality

At the end of Chapter 3, I mentioned that I hadn't yet covered an important concept—that of vpath builds. A *vpath build* is a way of using a make construct (VPATH) to configure and build a project in a directory other than the source directory. This is important if you need to perform any of the following tasks:

- Maintain a separate debug configuration
- Test different configurations side by side
- Keep a clean source directory for patch diffs after local modifications
- Build from a read-only source directory

The VPATH keyword is short for *virtual search path*. A VPATH statement contains a colon-separated list of places to look for relative-path dependencies when they can't be found relative to the current directory. In other words, when make can't find a prerequisite file relative to the current directory, it searches for that file successively in each of the paths in the VPATH statement.

Adding remote build functionality to an existing makefile using VPATH is very simple. Listing 4-8 shows an example of using a VPATH statement in a makefile.

```
VPATH = some/path:some/other/path:yet/another/path

program : src/main.c
        $(CC) ...
```

Listing 4-8: An example of using VPATH in a makefile

In this (contrived) example, if make can't find *src/main.c* in the current directory while processing the rule, it will look for *some/path/src/main.c*, and then for *some/other/path/src/main.c*, and finally for *yet/another/path/src/main.c* before giving up with an error message about not knowing how to make *src/main.c*.

With just a few simple modifications, we can completely support remote builds in Jupiter. Listings 4-9 and 4-10 illustrate the necessary changes to the project's two makefiles.

Git tag 4.3

```
--snip--
# Prefix-specific substitution variables
prefix = @prefix@
exec_prefix = @exec_prefix@
bindir = @bindir@

# VPATH-specific substitution variables
srcdir = @srcdir@
VPATH = @srcdir@
--snip--
$(distdir): FORCE
        mkdir -p $(distdir)/src
        cp $(srcdir)/configure.ac $(distdir)
        cp $(srcdir)/configure $(distdir)
        cp $(srcdir)/Makefile.in $(distdir)
        cp $(srcdir)/src/Makefile.in $(srcdir)/src/main.c $(distdir)/src
--snip--
```

Listing 4-9: Makefile.in: Adding VPATH build capabilities to the top-level makefile

```
--snip--
# Prefix-specific substitution variables
prefix = @prefix@
exec_prefix = @exec_prefix@
bindir = @bindir@

# VPATH-specific substitution variables
srcdir = @srcdir@
VPATH = @srcdir@
--snip--
jupiter: main.c
        $(CC) $(CPPFLAGS) $(CFLAGS) -o $@ $(srcdir)/main.c
--snip--
```

Listing 4-10: src/Makefile.in: Adding VPATH build capabilities to the lower-level makefile

That's it. Really. When config.status generates a file, it replaces an Autoconf substitution variable called @srcdir@ with the relative path to the template's source directory. The value substituted for @srcdir@ in a given *Makefile* within the build directory structure is the relative path to the directory containing the corresponding *Makefile.in* template in the source directory structure. The concept here is that for each *Makefile* in the remote build directory, VPATH provides a relative path to the directory containing the source code for that build directory.

NOTE *Do not expect VPATH to work in commands. VPATH only allows make to find dependencies; therefore, you can only expect VPATH to take effect in target and dependency lists within rules. You may use $(srcdir)/ as a prefix for file system objects in commands, as I've done in Listing 4-10 in the command for the jupiter target rule.*

The changes required for supporting remote builds in your build system are summarized as follows:

- Set a make variable, srcdir, to the @srcdir@ substitution variable.
- Set the VPATH variable to @srcdir@.
- Prefix all file dependencies used *in commands* with $(srcdir)/.

NOTE *Don't use $(srcdir) in the VPATH statement itself, because some older versions of make won't substitute variable references within the VPATH statement.*

If the source directory is the same as the build directory, the @srcdir@ substitution variable degenerates to a dot (.). That means all of these $(srcdir)/ prefixes simply degenerate to ./, which is harmless.[11]

A quick example is the easiest way to show you how this works. Now that Jupiter is fully functional with respect to remote builds, let's give it a try. Start in the Jupiter project directory, create a subdirectory called *build*, and then change into that directory. Execute the configure script using a relative path and then list the current directory contents:

```
$ mkdir build
$ cd build
$ ../configure
configure: creating ./config.status
config.status: creating Makefile
config.status: creating src/Makefile
$
$ ls -1p
config.log
config.status
Makefile
src/
$
$ ls -1p src
Makefile
$
```

---

11. This is not strictly true for non-GNU implementations of make. GNU make is smart enough to know that *file* and *./file* refer to the same filesystem object. However, non-GNU implementations of make aren't always quite so intelligent, so you should be careful to refer to a filesystem object using the same notation for each reference in your *Makefile.in* templates.

The entire build system has been constructed by `configure` and `config` *.status* within the *build* subdirectory. Enter `make` to build the project from within the *build* directory:

```
$ make
cd src && make all
make[1]: Entering directory '.../jupiter/build/src'
cc -g -O0 -o jupiter ../../src/main.c
make[1]: Leaving directory '.../jupiter/build/src'
$
$ ls -1p src
jupiter
Makefile
$
```

No matter where you are, if you can access the project directory using either a relative or an absolute path, you can do a remote build from that location. This is just one more thing that Autoconf does for you in Autoconf-generated configuration scripts. Imagine managing proper relative paths to source directories in your own hand-coded configuration scripts!

Let's Take a Breather

So far, I've shown you a nearly complete build system that includes almost all of the features outlined in the *GCS*. The features of Jupiter's build system are all fairly self-contained and reasonably simple to understand. The most difficult feature to implement by hand is the configuration script. In fact, writing a configuration script by hand is so labor intensive, compared to the simplicity of using Autoconf, that I just skipped the hand-coded version entirely in Chapter 3.

Although using Autoconf as I've used it here is quite easy, most people don't create their build systems in the manner I've shown you. Instead, they try to copy the build system of another project and tweak it to make it work in their own project. Later, when they start a new project, they do the same thing again. This can cause trouble because the code they're copying was never meant to be used the way they're now trying to use it.

I've seen projects in which the *configure.ac* file contained junk that had nothing to do with the project to which it belonged. These leftover bits came from some legacy project, but the maintainer didn't know enough about Autoconf to properly remove all the extraneous text. With the Autotools, it's generally better to start small and add what you need than to start with a copy of *configure.ac* from another full-featured build system and then try to pare it down to size or otherwise modify it to work with a new project.

I'm sure you're feeling like there's a lot more to learn about Autoconf, and you're right. We'll spend the remainder of this chapter examining the most important Autoconf macros and how they're used in the context of the Jupiter project. But first, let's go back and see if we might be able to simplify the Autoconf startup process even more by using another utility that comes with the Autoconf package.

An Even Quicker Start with autoscan

The easiest way to create a (mostly) complete *configure.ac* file is to run the autoscan utility, which is part of the Autoconf package. This utility examines the contents of a project directory and generates the basis for a *configure.ac* file (which autoscan names *configure.scan*) using existing makefiles and source files.

Let's see how well autoscan does on the Jupiter project. First, I'll clean up the droppings from my earlier experiments, and then I'll run autoscan in the *jupiter* directory.

NOTE *If you're using the git repository that accompanies this book, you can simply run* git clean -df *to remove all files and directories not currently under source control by git. Don't forget to switch back into the parent directory if you're still sitting in the build directory.*

Note that I'm *not* deleting my original *configure.ac* file—I'll just let autoscan tell me how to improve it. In less than a second, I have a few new files in the top-level directory:

```
$ cd ..
$ git clean -df
$ autoscan
❶ configure.ac: warning: missing AC_CHECK_HEADERS([stdlib.h]) wanted by:
    src/main.c:2
configure.ac: warning: missing AC_PREREQ wanted by: autoscan
configure.ac: warning: missing AC_PROG_CC wanted by: src/main.c
configure.ac: warning: missing AC_PROG_INSTALL wanted by: Makefile.in:18
$
$ ls -1p
autom4te.cache/
autoscan.log
configure.ac
configure.scan
Makefile.in
src/
$
```

The autoscan utility examines the project directory hierarchy and creates two files called *configure.scan* and *autoscan.log*. The project may or may not already be instrumented for the Autotools—it doesn't really matter, because autoscan is decidedly nondestructive. It will never alter any existing files in a project.

The autoscan utility generates a warning message for each problem it discovers in an existing *configure.ac* file. In this example, autoscan noticed that *configure.ac* should be using the Autoconf-provided AC_CHECK_HEADERS, AC_PREREQ, AC_PROG_CC, and AC_PROG_INSTALL macros. It made these assumptions based on information gleaned from the existing *Makefile.in* templates and from the C-language source files, as you can see by the comments after the warning statements beginning at ❶. You can always see these messages (in even greater detail) by examining the *autoscan.log* file.

NOTE *The notices you receive from autoscan and the contents of your* configure.ac *file may differ slightly from mine, depending on the version of Autoconf you have installed. I have version 2.69 of GNU Autoconf installed on my system (the latest, as of this writing). If your version of autoscan is older (or newer), you may see some minor differences.*

Looking at the generated *configure.scan* file, I note that autoscan has added more text to this file than was in my original *configure.ac* file. After looking it over to ensure that I understand everything, I see that it's probably easiest for me to overwrite *configure.ac* with *configure.scan* and then change the few bits of information that are specific to Jupiter:

```
$ mv configure.scan configure.ac
$ cat configure.ac
#                                                    -*- Autoconf -*-
# Process this file with autoconf to produce a configure script.

AC_PREREQ([2.69])
AC_INIT([FULL-PACKAGE-NAME], [VERSION], [BUG-REPORT-ADDRESS])
AC_CONFIG_SRCDIR([src/main.c])
AC_CONFIG_HEADERS([config.h])

# Checks for programs.
AC_PROG_CC
AC_PROG_INSTALL

# Checks for libraries.

# Checks for header files.
AC_CHECK_HEADERS([stdlib.h])

# Checks for typedefs, structures, and compiler characteristics.

# Checks for library functions.
AC_CONFIG_FILES([Makefile
                 src/Makefile])
AC_OUTPUT
$
```

My first modification involves changing the AC_INIT macro parameters for Jupiter, as illustrated in Listing 4-11.

Git tag 4.4

```
#                                                    -*- Autoconf -*-
# Process this file with autoconf to produce a configure script.

AC_PREREQ([2.69])
AC_INIT([Jupiter], [1.0], [jupiter-bugs@example.org])
```

```
AC_CONFIG_SRCDIR([src/main.c])
AC_CONFIG_HEADERS([config.h])
--snip--
```

Listing 4-11: configure.ac: Tweaking the AC_INIT macro generated by autoscan

The autoscan utility does a lot of the work for you. The *GNU Autoconf Manual*[12] states that you should modify this file to meet the needs of your project before you use it, but there are only a few key issues to worry about (besides those related to AC_INIT). I'll cover each of these issues in turn, but first, let's take care of a few administrative details.

I'd be remiss if I didn't mention autoupdate while discussing autoscan. If you've already got a working *configure.ac* file, and you update to a newer version of Autoconf, you can run autoupdate to update your existing *configure.ac* file with constructs that have changed or been added since the older version of Autoconf.

The Proverbial bootstrap.sh Script

Before autoreconf came along, maintainers passed around a short shell script, often named autogen.sh or bootstrap.sh, which would run all of the Autotools required for their projects in the proper order. The recommended name for this script is bootstrap.sh because Autogen is the name of another GNU project. The bootstrap.sh script can be fairly sophisticated, but to solve the problem of the missing install-sh script (see "Missing Required Files in Autoconf," next), I'll just add a simple temporary bootstrap.sh script to the project root directory, as shown in Listing 4-12.

Git tag 4.5
```
#!/bin/sh
autoreconf --install
❶ automake --add-missing --copy >/dev/null 2>&1
```

Listing 4-12: bootstrap.sh: A temporary bootstrap script that executes the required Autotools

The Automake --add-missing option copies the required missing utility scripts into the project, and the --copy option indicates that true copies should be made (otherwise, symbolic links are created that refer to the files where they're installed with the Automake package).[13]

---

12. See the Free Software Foundation's *GNU Autoconf Manual* at *https://www.gnu.org/software/autoconf/manual/index.html*.

13. This isn't as bad as it sounds, because when make dist generates a distribution archive, it creates true copies in the image directory. Therefore, links work just fine, as long as you (the maintainer) don't move your work area to another host. Note that automake provides a --copy option, but autoreconf provides just the opposite: a --symlink option. Thus, if you execute automake --add-missing and you wish to actually copy the files, you should pass --copy as well. If you execute autoreconf --install, then --copy will be assumed and passed to automake by autoreconf.

NOTE *We don't need to see the warnings from executing automake, so I've redirected the stderr and stdout streams to /dev/null on the automake command line at ❶ in this script. In Chapter 6, we'll remove bootstrap.sh and simply run autoreconf --install, but for now, this solves our missing file problems.*

MISSING REQUIRED FILES IN AUTOCONF

When I first tried to execute autoreconf on the *configure.ac* file in Listing 4-11, I discovered a minor problem related to using Autoconf *without* Automake. When I ran the configure script, it failed with an error: configure: error: cannot find install-sh, install.sh, or shtool in "." "./.." "./../..".

Autoconf is all about portability and, unfortunately, the Unix install utility is not as portable as it could be. From one platform to another, critical bits of installation functionality are just different enough to cause problems, so the Autotools provide a shell script called install-sh (deprecated name: install.sh). This script acts as a wrapper around the system's own install utility, masking important differences between various versions of install.

autoscan noticed that I'd used the install program in my *src/Makefile.in* template, so it generated an expansion of the AC_PROG_INSTALL macro. The problem is that configure couldn't find the install-sh wrapper script anywhere in my project.

I reasoned that the missing file was part of the Autoconf package and it just needed to be installed. I also knew that autoreconf accepts a command line option to install such missing files into a project directory. The --install (-i) option supported by autoreconf is designed to pass tool-specific options down to each of the tools that it calls in order to install missing files. However, when I tried that, I found that the file was still missing, because autoconf doesn't support an option to install missing files.

I could have manually copied install-sh from the Automake installation directory (usually */usr/(local/)share/automake-**), but looking for a more automated solution, I tried manually executing automake --add-missing --copy. This command generated a slew of warnings indicating that the project was not configured for Automake. However, I could now see that install-sh had been copied into my project root directory, and that's all I was after. Executing autoreconf --install didn't run automake because *configure.ac* was not set up for Automake.

Autoconf should ship with install-sh, since it provides a macro that requires it, but then autoconf would have to provide an --add-missing command line option. Nevertheless, there is actually a quite obvious solution to this problem. The install-sh script is not really required by any code Autoconf generates. How could it be? Autoconf doesn't generate any makefile constructs—it only substitutes variables into your *Makefile.in* templates. Thus, there's really no reason for Autoconf to complain about a missing install-sh script.

Updating Makefile.in

Let's make bootstrap.sh executable and then execute it and see what we end up with:

```
$ chmod +x bootstrap.sh
$ ./bootstrap.sh
$ ls -1p
autom4te.cache/
bootstrap.sh
❶ config.h.in
configure
configure.ac
❷ install-sh
Makefile.in
src/
$
```

We know from the file list at ❶ that *config.h.in* has been created, so we know that autoreconf has executed autoheader. We also see the new install-sh script at ❷ that was created when we executed automake in bootstrap.sh. Anything provided or generated by the Autotools should be copied into the archive directory so that it can be shipped with release tarballs. Therefore, we'll add cp commands for these two files to the $(distdir) target in the top-level *Makefile.in* template. Note that we don't need to copy the bootstrap.sh script because it's purely a maintainer tool—users should never need to execute it from a tarball distribution.

Listing 4-13 illustrates the required changes to the $(distdir) target in the top-level *Makefile.in* template.

Git tag 4.6

```
--snip--
$(distdir): FORCE
        mkdir -p $(distdir)/src
        cp $(srcdir)/configure.ac $(distdir)
        cp $(srcdir)/configure $(distdir)
        cp $(srcdir)/config.h.in $(distdir)
        cp $(srcdir)/install-sh $(distdir)
        cp $(srcdir)/Makefile.in $(distdir)
        cp $(srcdir)/src/Makefile.in $(distdir)/src
        cp $(srcdir)/src/main.c $(distdir)/src
--snip--
```

Listing 4-13: Makefile.in: Additional files needed in the distribution archive image directory

If you're beginning to think that this could become a maintenance problem, then you're right. I mentioned earlier that the $(distdir) target was painful to maintain. Luckily, the distcheck target still exists and still works as designed. It would have caught this problem, because attempts to build from the tarball will fail without these additional files—and the distcheck target certainly won't succeed if the build fails. When we discuss Automake in Chapter 6, we will clear up much of this maintenance mess.

Initialization and Package Information

Now let's turn our attention back to the contents of the *configure.ac* file in Listing 4-11 (and the console example immediately preceding that listing). The first section contains Autoconf initialization macros. These are required for all projects. Let's consider each of these macros individually, because they're all important.

AC_PREREQ

The AC_PREREQ macro simply defines the earliest version of Autoconf that may be used to successfully process this *configure.ac* file:

```
AC_PREREQ(version)
```

The *GNU Autoconf Manual* indicates that AC_PREREQ is the only macro that may be used before AC_INIT. This is because it's good to ensure you're using a new enough version of Autoconf before you begin processing any other macros, which may be version dependent.

AC_INIT

The AC_INIT macro, as its name implies, initializes the Autoconf system. Here's its prototype, as defined in the *GNU Autoconf Manual*:[14]

```
AC_INIT(package, version, [bug-report], [tarname], [url])
```

It accepts up to five arguments (autoscan only generates an invocation with the first three): *package*, *version*, and, optionally, *bug-report*, *tarname*, and *url*. The *package* argument is intended to be the name of the package. It will end up (in a canonical form) as the first part of the name of an Automake-generated release distribution tarball when you execute make dist.

NOTE   *Autoconf uses a normalized form of the package name in the tarball name, so you can use uppercase letters in the package name, if you wish. Automake-generated tarballs are named tarname-version.tar.gz by default, but tarname is set to a normalized form of the package name (lowercase, with all punctuation converted to underscores). Bear this in mind when you choose your package name and version string.*

The optional *bug-report* argument is usually set to an email address, but any text string is valid—the URL of a web page that accepts bug reports for the project is a common alternative. An Autoconf substitution variable called @PACKAGE_BUGREPORT@ is created for it, and that variable is also added to the *config.h.in* template as a C-preprocessor definition. The intent here is that you use the variable in your code to present an email address or URL for bug reports at appropriate places—possibly when the user requests help or version information from your application.

---

14. The square brackets used in the macro definition prototypes within this book (as well as the *GNU Autoconf Manual*) indicate optional parameters, not Autoconf quotes.

While the *version* argument can be anything you like, there are a few commonly used OSS conventions that will make things a little easier for you. The most widely used convention is to pass in *major.minor* (for example, 1.2). However, there's nothing that says you can't use *major.minor. revision*, and there's nothing wrong with this approach. None of the resulting VERSION variables (Autoconf, shell, or make) are parsed or analyzed anywhere—they're only used as placeholders for substituted text in various locations.[15] So if you wish, you may even add nonnumerical text into this macro, such as *0.15.alpha1*, which is occasionally useful.[16]

NOTE   *The RPM package manager, on the other hand, does care what you put in the version string. For the sake of RPM, you may wish to limit the version string text to only alphanumeric characters and periods—no dashes or underscores.*

The optional *url* argument should be the URL for your project website. It's shown in the help text displayed by configure --help.

Autoconf generates the substitution variables @PACKAGE_NAME@, @PACKAGE _VERSION@, @PACKAGE_TARNAME@, @PACKAGE_STRING@ (a stylized concatenation of the package name and version information), @PACKAGE_BUGREPORT@, and @PACKAGE _URL@ from the arguments to AC_INIT. You can use any or all of these in your *Makefile.in* template files.

AC_CONFIG_SRCDIR

The AC_CONFIG_SRCDIR macro is a sanity check. Its purpose is to ensure that the generated configure script knows that the directory on which it is being executed is actually the project directory.

More specifically, configure needs to be able to locate itself, because it generates code that executes itself, possibly from a remote directory. There are myriad ways to inadvertently fool configure into finding some other configure script. For example, the user could accidentally provide an incorrect --srcdir argument to configure. The $0 shell script parameter is unreliable, at best—it may contain the name of the shell, rather than that of the script, or it may be that configure was found in the system search path, so no path information was specified on the command line.

The configure script could try looking in the current or parent directories, but it still needs a way to verify that the configure script it locates is actually itself. Thus, AC_CONFIG_SRCDIR gives configure a significant hint that it's looking in the right place. Here's the prototype for AC_CONFIG_SRCDIR:

```
AC_CONFIG_SRCDIR(unique-file-in-source-dir)
```

---

15. As far as M4 is concerned, all data is text; thus, M4 macro arguments, including package and version, are treated simply as strings. M4 doesn't attempt to interpret any of this text as numbers or other data types.

16. A future version of Autoconf will support a public macro that allows lexicographical comparison of version strings, and certain internal constructs in current versions already use such functionality. Therefore, it's good practice to form version strings that increase properly in a lexical fashion from version to version.

The argument can be a path (relative to the project's configure script) to any source file you like. You should choose one that is unique to your project so as to minimize the possibility that configure is fooled into thinking some other project's configuration file is itself. I normally try to choose a file that sort of represents the project, such as a source file named for a feature that defines the project. That way, in case I ever decide to reorganize the source code, I'm not likely to lose it in a file rename. In this case, however, we have only one source file, *main.c*, making it a little difficult to follow this convention. Regardless, both autoconf and configure will tell you and your users if it can't find the file.

The Instantiating Macros

Before we dive into the details of AC_CONFIG_HEADERS, I'd like to spend a little time on the file generation framework Autoconf provides. From a high-level perspective, there are four major things happening in *configure.ac*:

- Initialization
- Check request processing
- File instantiation request processing
- Generation of the configure script

We've covered initialization—there's not much to it, although there are a few more macros you should be aware of. Check out the *GNU Autoconf Manual* for more information—look up AC_COPYRIGHT, for an example. Now let's move on to file instantiation.

There are actually four so-called *instantiating macros*: AC_CONFIG_FILES, AC_CONFIG_HEADERS, AC_CONFIG_COMMANDS, and AC_CONFIG_LINKS. An instantiating macro accepts a list of tags or files; configure will generate these files from templates containing Autoconf substitution variables.

NOTE *You might need to change the name of AC_CONFIG_HEADER (singular) to AC_CONFIG_HEADERS (plural) in your version of* configure.scan. *The singular version is the older name for this macro, and the older macro is less functional than the newer one.[17]*

The four instantiating macros have an interesting common signature. The following prototype can be used to represent each of them, with appropriate text replacing the *XXX* portion of the macro name:

```
AC_CONFIG_XXXS(tag..., [commands], [init-cmds])
```

---

17. This was a defect in autoscan that had not been fixed as of Autoconf version 2.61. However, version 2.62 of autoscan correctly generates a call to the newer, more functional AC_CONFIG_HEADERS. This note is of more historical interest than anything else, as most systems have updated to Autoconf 2.63 or later by now.

For each of these four macros, the tag argument has the form *OUT*[:*INLIST*], where *INLIST* has the form *IN0*[:*IN1*:...:*INn*]. Often, you'll see a call to one of these macros with only a single argument, as in the three examples that follow (note that these examples represent macro *invocations*, not *prototypes*, so the square brackets are actually Autoconf quotes, not indications of optional parameters):

```
AC_CONFIG_HEADERS([config.h])
```

In this example, *config.h* is the *OUT* portion of the preceding specification. The default value for *INLIST* is the *OUT* portion with *.in* appended to it. So, in other words, the preceding call is exactly equivalent to the following:

```
AC_CONFIG_HEADERS([config.h:config.h.in])
```

What this means is that config.status contains shell code that will generate *config.h* from *config.h.in*, substituting all Autoconf variables in the process. You may also provide a list of input files in the *INLIST* portion. In this case, the files in *INLIST* will be concatenated to form the resulting *OUT* file:

```
AC_CONFIG_HEADERS([config.h:cfg0:cfg1:cfg2])
```

Here, config.status will generate *config.h* by concatenating *cfg0*, *cfg1*, and *cfg2* (in that order), after substituting all Autoconf variables. The *GNU Autoconf Manual* refers to this entire *OUT*[:*INLIST*] construct as a *tag*.

Why not just call it a *file*? Well, this parameter's primary purpose is to provide a sort of command line target name—much like makefile targets. It can also be used as a filesystem name if the associated macro generates files, as is the case with AC_CONFIG_HEADERS, AC_CONFIG_FILES, and AC_CONFIG_LINKS.

But AC_CONFIG_COMMANDS is unique in that it doesn't generate any files. Instead, it runs arbitrary shell code, as specified by the user in the macro's arguments. Thus, rather than name this first parameter after a secondary function (the generation of files), the *GNU Autoconf Manual* refers to it more generally, according to its primary purpose—as a command line *tag* that may be specified on the ./config.status command line, in this manner:

```
$ ./config.status config.h
```

This command will regenerate the *config.h* file based on the macro call to AC_CONFIG_HEADERS in *configure.ac*. It will *only* regenerate *config.h*.

Enter ./config.status --help to see the other command line options you can use when executing ./config.status:

```
$ ./config.status --help
`config.status' instantiates files and other configuration actions
from templates according to the current configuration.  Unless the files
and actions are specified as TAGs, all are instantiated by default.
```

❶ Usage: ./config.status [OPTION]... [TAG]...

```
    -h, --help      print this help, then exit
    -V, --version   print version number and configuration settings, then exit
    ❷  --config     print configuration, then exit
    -q, --quiet, --silent
                    do not print progress messages
    -d, --debug     don't remove temporary files
        --recheck   update config.status by reconfiguring in the same
conditions
    ❸ --file=FILE[:TEMPLATE]
                    instantiate the configuration file FILE
      --header=FILE[:TEMPLATE]
                    instantiate the configuration header FILE

❹ Configuration files:
  Makefile src/Makefile

❺ Configuration headers:
  config.h

Report bugs to <jupiter-bugs@example.org>.
$
```

Notice that config.status provides custom help about a project's config
.status file. It lists configuration files ❹ and configuration headers ❺ that
we can use as tags on the command line where the usage specifies [TAG]...
at ❶. In this case, config.status will only instantiate the specified objects. In
the case of commands, it will execute the command set specified by the tag
passed in the associated expansion of the AC_CONFIG_COMMANDS macro.

Each of these macros may be used multiple times in a *configure.ac* file.
The results are cumulative, and we can use AC_CONFIG_FILES as many times
as we need to in *configure.ac*. It is also important to note that config.status
supports the --file= option (at ❸). When you call config.status with tags
on the command line, the only tags you can use are those the help text lists
as available configuration files, headers, links, and commands. When you
execute config.status with the --file= option, you're telling config.status
to generate a new file that's not already associated with any of the calls to
the instantiating macros found in *configure.ac*. This new file is generated
from an associated template using configuration options and check results
determined by the last execution of configure. For example, I could execute
config.status in this manner (using a fictional template called *extra.in*):

```
$ ./config.status --file=extra:extra.in
```

NOTE   *The default template name is the filename with a .in suffix, so this call could have
been made without using the :extra.in portion of the option. I added it here for clarity.*

Finally, I'd like to point out a newer feature of `config.status`—the `--config` option at ❷, added with version 2.65 of Autoconf. Using this option displays the explicit configuration options passed to `configure` on the command line. For instance, assume that we had invoked `./configure` in this manner:

```
$ ./configure --prefix=$HOME
```

When you use the new `--config` option, `./config.status` displays the following:

```
$ ./config.status --config
'--prefix=/home/jcalcote'
```

NOTE   *Older versions of Autoconf generated a `config.status` script that displayed this information when using the `--version` option, but it was part of a larger wall of text. The newer `--config` option makes it easier to find and reuse configuration options originally passed to the `configure` script.*

Let's return now to the instantiating macro signature at the bottom of page 102. I've shown you that the *tag...* argument has a complex format, but the ellipsis indicates that it also represents multiple tags, separated by whitespace. The format you'll see in nearly all *configure.ac* files is shown in Listing 4-14.

```
AC_CONFIG_FILES([Makefile
                src/Makefile
                lib/Makefile
                etc/proj.cfg])
```

Listing 4-14: Specifying multiple tags (files) in `AC_CONFIG_FILES`

Each entry here is one tag specification, which, if fully specified, would look like the call in Listing 4-15.

```
AC_CONFIG_FILES([Makefile:Makefile.in
                src/Makefile:src/Makefile.in
                lib/Makefile:lib/Makefile.in
                etc/proj.cfg:etc/proj.cfg.in])
```

Listing 4-15: Fully specifying multiple tags in `AC_CONFIG_FILES`

Returning to the instantiating macro prototype, there are two optional arguments that you'll rarely see used in these macros: *commands* and *init-cmds*. The *commands* argument may be used to specify some arbitrary shell code that should be executed by `config.status` just before the files associated with the tags are generated. It is unusual for this feature to be used within the file-generating instantiating macros. You will almost always see the *commands* argument used with `AC_CONFIG_COMMANDS`, which generates no files by default, because a call to this macro is basically useless without commands to execute![18] In this case, the *tag* argument becomes a way of telling config .status to execute a specific set of shell commands.

The *init-cmds* argument initializes shell variables at the top of config .status with values available in *configure.ac* and configure. It's important to remember that all calls to instantiating macros share a common namespace along with config.status. Therefore, you should try to choose your shell variable names carefully so they are less likely to conflict with each other and with Autoconf-generated variables.

The old adage about the value of a picture versus an explanation holds true here, so let's try a little experiment. Create a test version of your *configure.ac* file that contains only the contents of Listing 4-16. You should do this in a separate directory, as we're not relying on any of the other files in the Jupiter project directory structure with this experiment.

```
AC_INIT([test], [1.0])
AC_CONFIG_COMMANDS([abc],
                   [echo "Testing $mypkgname"],
                   [mypkgname=$PACKAGE_NAME])
AC_OUTPUT
```

Listing 4-16: Experiment #1—a simple configure.ac *file using* AC_CONFIG_COMMANDS

Now execute autoreconf, ./configure, and ./config.status in various ways to see what happens:

```
$ autoreconf
❶ $ ./configure
configure: creating ./config.status
config.status: executing abc commands
Testing test
$
❷ $ ./config.status
config.status: executing abc commands
Testing test
$
❸ $ ./config.status --help
'config.status' instantiates files from templates according to the current
configuration.
Usage: ./config.status [OPTIONS]... [FILE]...
--snip--
Configuration commands:
  abc

Report bugs to <bug-autoconf@gnu.org>.
$
❹ $ ./config.status abc
config.status: executing abc commands
Testing test
$
```

---

18. The truth is that we don't often use AC_CONFIG_COMMANDS.

As you can see at ❶, executing ./configure caused config.status to be executed with no command line options. There are no checks specified in *configure.ac*, so manually executing ./config.status, as we did at ❷, has nearly the same effect. Querying config.status for help (as we did at ❸) indicates that abc is a valid tag; executing ./config.status with that tag (as we did at ❹) on the command line simply runs the associated commands.

In summary, the important points regarding the instantiating macros are as follows:

- The config.status script generates all files from templates.

- The configure script performs all checks and then executes ./config.status.

- When you execute ./config.status with no command line options, it generates files based on the last set of check results.

- You can call ./config.status to execute file generation or command sets specified by any of the tags given in any of the instantiating macro calls.

- The config.status script may generate files not associated with any tags specified in *configure.ac*, in which case it will substitute variables based on the last set of checks performed.

Generating Header Files from Templates

As you've no doubt concluded by now, the AC_CONFIG_HEADERS macro allows you to specify one or more header files that config.status should generate from template files. The format of a configuration header template is very specific. A short example is given in Listing 4-17.

```
/* Define as 1 if you have unistd.h. */
#undef HAVE_UNISTD_H
```

Listing 4-17: A short example of a header file template

You can place multiple statements like this in your header template, one per line. The comments are optional, of course. Let's try another experiment. Create a new *configure.ac* file like that shown in Listing 4-18. Again, you should do this in an isolated directory.

```
AC_INIT([test], [1.0])
AC_CONFIG_HEADERS([config.h])
AC_CHECK_HEADERS([unistd.h foobar.h])
AC_OUTPUT
```

Listing 4-18: Experiment #2—a simple configure.ac *file*

Create a template header file called *config.h.in* that contains the two lines in Listing 4-19.

```
#undef HAVE_UNISTD_H
#undef HAVE_FOOBAR_H
```

Listing 4-19: Experiment #2 continued—a simple config.h.in *file*

Now execute the following commands:

```
$ autoconf
$ ./configure
checking for gcc... gcc
--snip--
❶ checking for unistd.h... yes
  checking for unistd.h... (cached) yes
  checking foobar.h usability... no
  checking foobar.h presence... no
❷ checking for foobar.h... no
  configure: creating ./config.status
❸ config.status: creating config.h
  $
  $ cat config.h
  /* config.h.  Generated from config.h.in by configure.  */
  #define HAVE_UNISTD_H 1
❹ /* #undef HAVE_FOOBAR_H */
  $
```

You can see at ❸ that config.status generated a *config.h* file from the simple *config.h.in* template we wrote. The contents of this header file are based on the checks executed by configure. Since the shell code generated by AC_CHECK_HEADERS([unistd.h foobar.h]) was able to locate a *unistd.h* header file (❶) in the system include directory, the corresponding #undef statement was converted into a #define statement. Of course, no *foobar.h* header was found in the system include directory, as you can also see by the output of *./configure* at ❷; therefore, its definition was left commented out in the template, as shown at ❹.

Hence, you may add the sort of code shown in Listing 4-20 to appropriate C-language source files in your project.

```
#include "config.h"
#if HAVE_UNISTD_H
# include <unistd.h>
#endif
#if HAVE_FOOBAR_H
# include <foobar.h>
#endif
```

Listing 4-20: Using generated CPP definitions in a C-language source file

NOTE *The* unistd.h *header file is so standard these days that it's not really necessary to check for it in* AC_CONFIG_HEADERS, *but it served here as a file that I was sure existed on my system for this example.*

Using autoheader to Generate an Include File Template

Manually maintaining a *config.h.in* template is more trouble than necessary. The format of *config.h.in* is very strict—for example, you can't have any leading or trailing whitespace on the #undef lines, and the #undef lines you add must use #undef rather than #define, mainly because config.status only knows how to either replace #undef with #define or comment out lines containing #undef.[19]

Most of the information you need from *config.h.in* is available in *configure.ac*. Fortunately, autoheader will generate a properly formatted header file template for you based on the contents of *configure.ac*, so you don't often need to write *config.h.in* templates. Let's return to the command prompt for a final experiment. This one is easy—just delete your *config.h.in* template from experiment #2 and then run autoheader followed by autoconf:

```
$ rm config.h.in
$ autoheader
$ autoconf
$ ./configure
checking for gcc... gcc
--snip--
checking for unistd.h... yes
checking for unistd.h... (cached) yes
checking foobar.h usability... no
checking foobar.h presence... no
checking for foobar.h... no
configure: creating ./config.status
config.status: creating config.h
$
❶ $ cat config.h
/* config.h. Generated from config.h.in by configure. */
/* config.h.in. Generated from configure.ac by autoheader. */
/* Define to 1 if you have the <foobar.h> header file. */
/* #undef HAVE_FOOBAR_H */
--snip--
/* Define to 1 if you have the <unistd.h> header file. */
#define HAVE_UNISTD_H 1
/* Define to the address where bug reports for this package should be sent. */
#define PACKAGE_BUGREPORT ""
/* Define to the full name of this package. */
#define PACKAGE_NAME "test"
/* Define to the full name and version of this package. */
#define PACKAGE_STRING "test 1.0"
/* Define to the one symbol short name of this package. */
#define PACKAGE_TARNAME "test"
/* Define to the version of this package. */
#define PACKAGE_VERSION "1.0"
/* Define to 1 if you have the ANSI C header files. */
#define STDC_HEADERS 1
$
```

---

19. This is an example of config.status using sed or awk to perform token replacement in template files. It looks specifically for #undef token and replaces it with #define token. In this case, token is the name of the header file converted to uppercase, with special characters (such as periods) replaced with underscores, and prepended with HAVE_.

NOTE *Again, I encourage you to use* autoreconf, *which will automatically run* autoheader *if it notices an expansion of* AC_CONFIG_HEADERS *in* configure.ac.

As you can see by the output of the cat command at ❶, an entire set of preprocessor definitions was derived from *configure.ac* by autoheader.

Listing 4-21 shows a much more realistic example of using a generated *config.h* file to increase the portability of your project source code. In this example, the AC_CONFIG_HEADERS macro invocation indicates that *config.h* should be generated, and the invocation of AC_CHECK_HEADERS will cause autoheader to insert a definition into *config.h*.

```
AC_INIT([test], [1.0])
AC_CONFIG_HEADERS([config.h])
AC_CHECK_HEADERS([dlfcn.h])
AC_OUTPUT
```

Listing 4-21: A more realistic example of using AC_CONFIG_HEADERS

The *config.h* file is intended to be included in your source code in locations where you might wish to test a configured option in the code itself using the C preprocessor. This file should be included first in source files so it can influence the inclusion of system header files later in the source.

NOTE *The* config.h.in *template that* autoheader *generates doesn't contain an include-guard construct, so you need to be careful that it's not included more than once in a source file. A good rule of thumb is to always include* config.h *as the very first header in every .c source file and never include it anywhere else. Following this rule will guarantee that it never needs an include guard.*

It's often the case that every *.c* file in a project needs to include *config.h*. In this case, an interesting approach is to use the gcc -include option to include it at the top of every compiled source file from the compiler command line. This can be done within *configure.ac* by appending -include config.h to the DEFS variable (which is currently only used to define HAVE_CONFIG_H—if you're more of a purist, you can use CFLAGS instead). Once done, you may assume *config.h* is part of every translation unit.

Don't make the mistake of including *config.h* in a public header file if your project installs libraries and header files as part of your product set. For more detailed information on this topic, refer to "Item 1: Keeping Private Details out of Public Interfaces" on page 499.

Using the *configure.ac* file from Listing 4-21, the generated configure script will create a *config.h* header file with appropriate definitions for determining, at compile time, whether or not the current system provides the dlfcn interface. To complete the portability check, you can add the code from Listing 4-22 to a source file in your project that uses dynamic loader functionality.

```
   #include "config.h"
❶ #if HAVE_DLFCN_H
```

```
# include <dlfcn.h>
#else
# error Sorry, this code requires dlfcn.h.
#endif
--snip--
❷ #if HAVE_DLFCN_H
    handle = dlopen("/usr/lib/libwhatever.so", RTLD_NOW);
#endif
--snip--
```

Listing 4-22: A sample source file that checks for dynamic loader functionality

If you already had code that included *dlfcn.h*, autoscan would have generated a line in *configure.ac* to call AC_CHECK_HEADERS with an argument list containing *dlfcn.h* as one of the header files to be checked. Your job as maintainer is to add the conditional statements at ❶ and ❷ to your source code around the existing inclusions of the *dlfcn.h* header file and around calls to the *dlfcn* interface functions. This is the crux of Autoconf portability support.

NOTE   *You don't technically need the preprocessor conditionals around the code if you choose to "error out" if the inclusion check fails, but doing so makes it obvious to the reader which portions of the source code are affected by the conditional inclusion.*

Your project might prefer dynamic loader functionality, but could get along without it if necessary. It's also possible that your project requires a dynamic loader, in which case your build should terminate with an error (as this code does) if the key functionality is missing. Often, this is an acceptable stopgap until someone comes along and adds support to the source code for a more system-specific dynamic loader service.

NOTE   *If you have to bail out with an error, it's best to do so at configuration time rather than at compile time. The general rule of thumb is to bail out as early as possible.*

As mentioned earlier, HAVE_CONFIG_H is part of a string of definitions passed on the compiler command line in the Autoconf substitution variable @DEFS@. Before autoheader and AC_CONFIG_HEADERS functionality existed, Automake added all of the compiler configuration macros to the @DEFS@ variable. You can still use this method if you don't use AC_CONFIG_HEADERS in *configure.ac,* but it's not recommended—mainly because a large number of definitions make for very long compiler command lines.

Back to Remote Builds for a Moment

As we wrap up this chapter, you'll notice that we've come full circle. We started out covering some preliminary information before we discussed how to add remote builds to Jupiter. Now we'll return to this topic for a moment, because I haven't yet covered how to get the C preprocessor to properly locate a generated *config.h* file.

Since this file is generated from a template, it will be at the same relative position in the build directory structure as its counterpart template file, *config.h.in*, is in the source directory structure. The template is located in the top-level *source* directory (unless you chose to put it elsewhere), so the generated file will be in the top-level *build* directory. Well, that's easy enough—it's always one level up from the generated *src/Makefile*.

Before we draw any conclusions then about header file locations, let's consider where header files might appear in a project. We might generate them in the current build directory, as part of the build process. We might also add internal header files to the current source directory. We know we have a *config.h* file in the top-level build directory. Finally, we might also create a top-level *include* directory for library interface header files our package provides. What is the order of priority for these various *include* directories?

The order in which we place *include directives* (-I*path* options) on the compiler command line is the order in which they will be searched, so the order should be based on which files are most relevant to the source file currently being compiled. Therefore, the compiler command line should include -I*path* directives for the current build directory (.) first, followed by the source directory [$(srcdir)], then the top-level build directory (..), and, finally, our project's *include* directory, if it has one. We impose this ordering by adding -I*path* options to the compiler command line, as shown in Listing 4-23.

Git tag 4.7

```
--snip--
jupiter: main.c
        $(CC) $(CPPFLAGS) $(CFLAGS) -I. -I$(srcdir) -I.. -o $@ \
            $(srcdir)/main.c
--snip--
```

Listing 4-23: src/Makefile.in: Adding proper compiler include directives

Now that we know this, we need to add another rule of thumb for remote builds to the list we created on page 93:

- Add preprocessor commands for the current build directory, the associated source directory, and the top-level build directory (or other build directory if *config.h.in* is located elsewhere), in that order.

Summary

In this chapter, we covered just about all the major features of a fully functional GNU project build system, including writing a *configure.ac* file, from which Autoconf generates a fully functional `configure` script. We've also covered adding remote build functionality to makefiles with VPATH statements.

So what else is there? Plenty! In the next chapter, I'll continue to show you how you can use Autoconf to test system features and functionality before your users run `make`. We'll also continue enhancing the configuration script so that when we're done, users will have more options and understand exactly how our package will be built on their systems.

5

MORE FUN WITH AUTOCONF: CONFIGURING USER OPTIONS

Hope is not the conviction that something will turn out well,
but the certainty that something makes sense,
regardless of how it turns out.
—*Václav Havel*, Disturbing the Peace

In Chapter 4, we discussed the essentials of Autoconf—how to bootstrap a new or existing project and how to understand some of the basic aspects of *configure.ac* files. In this chapter, we cover some of the more complex Autoconf macros. We'll begin by discussing how to substitute our own variables into template files (for example, *Makefile.in*) and how to define our own preprocessor definitions from within the configuration script. Throughout this chapter, we'll continue to develop functionality in the Jupiter project by adding important checks and tests. We'll cover the all-important AC_OUTPUT macro, and we'll conclude by discussing the application of user-defined project configuration options as specified in the *configure.ac* file.

In addition to all this, I'll present an analysis technique you can use to decipher the inner workings of macros. Using the somewhat complex AC_CHECK_PROG macro as an example, I'll show you some ways to find out what's going on under the hood.

Substitutions and Definitions

We'll begin this chapter by discussing three of the most important macros in the Autoconf suite: AC_SUBST and AC_DEFINE, along with the latter's twin brother, AC_DEFINE_UNQUOTED.

These macros provide the primary mechanisms for communication between the configuration process and the build and execution processes. Values that are *substituted* into generated files provide configuration information to the build process, while values defined in preprocessor variables provide configuration information at build time to the compiler and at runtime to the built programs and libraries. As a result, it's well worth becoming thoroughly familiar with AC_SUBST and AC_DEFINE.

AC_SUBST

You can use AC_SUBST to extend the variable substitution functionality that's such an integral part of Autoconf. Every Autoconf macro that has anything to do with substitution variables ultimately calls this macro to create the substitution variables from existing shell variables. Sometimes the shell variables are inherited from the environment; other times, higher-level macros set the shell variables as part of their functionality before calling AC_SUBST. The signature of this macro is rather trivial (note that the square brackets in this prototype represent optional arguments, not Autoconf quotes):

```
AC_SUBST(shell_var[, value])
```

NOTE *If you choose to omit any trailing optional parameters when invoking M4 macros, you may also omit the trailing commas.[1] However, if you omit any arguments from the middle of the list, you must provide the commas as placeholders for the missing arguments.*

The first argument, *shell_var*, represents a shell variable whose value you wish to substitute into all files generated by config.status from templates. The optional second parameter is the value assigned to the variable. If it isn't specified, the shell variable's current value will be used, whether it's inherited or set by some previous shell code.

The substitution variable will have the same name as the shell variable, except that it will be bracketed with at signs (@) in the template files. Thus, a shell variable named my_var would become the substitution variable reference @my_var@, and you could use it in any template file.

---

1. This is not strictly true. M4 can tell the difference between an omitted trailing variable and an empty trailing variable. This information is available to the macro itself, so it may be written to make use of this difference. Autoconf has a few such macros; the documentation indicates this nuance if it's used. A good example of this is AC_DEFINE, which has two versions: one with a single argument and one with three. Using an empty second argument invokes the three-argument form of the macro.

Calls to AC_SUBST in *configure.ac* should not be made conditionally; that is, they should not be called within conditional shell statements like if-then-else constructs. The reason becomes clear when you carefully consider the purpose of AC_SUBST: you've already hardcoded substitution variable references into your template files, so you'd better use AC_SUBST for each variable unconditionally, or else your output files will retain the variable references rather than the values that should have been substituted.

AC_DEFINE

The AC_DEFINE and AC_DEFINE_UNQUOTED macros define C-preprocessor macros, which can be simple or function-like macros. These are either defined in the *config.h.in* template (if you use AC_CONFIG_HEADERS) or passed on the compiler command line (via the @DEFS@ substitution variable) in *Makefile.in* templates. Recall that if you don't write *config.h.in* yourself, autoheader will write it based on calls to these macros in your *configure.ac* file.

These two macro names actually represent four different Autoconf macros. Here are their prototypes:

```
AC_DEFINE(variable, value[, description])
AC_DEFINE(variable)
AC_DEFINE_UNQUOTED(variable, value[, description])
AC_DEFINE_UNQUOTED(variable)
```

The difference between the normal and the UNQUOTED versions of these macros is that the normal versions use, verbatim, the specified value as the value of the preprocessor macro. The UNQUOTED versions perform shell expansion on the *value* argument, and they use the result as the value of the preprocessor macro. Thus, you should use AC_DEFINE_UNQUOTED if the value contains shell variables that you want configure to expand. (Setting a C-preprocessor macro in a header file to an unexpanded shell variable makes no sense, because neither the C compiler nor the preprocessor will know what to do with it when the source code is compiled.)

The difference between the single- and multi-argument versions lies in the way the preprocessor macros are defined. The single-argument versions simply guarantee that the macro is *defined* in the preprocessor namespace, while the multi-argument versions ensure that the macro is defined with a specific value.

The optional third parameter, *description*, tells autoheader to add a comment for this macro to the *config.h.in* template. (If you don't use autoheader, it makes no sense to pass a description here—hence, its optional status.) If you wish to define a preprocessor macro without a value and provide a *description*, you should use the multi-argument versions of these macros but leave the value argument empty. Another option is to use AH_TEMPLATE—an autoheader-specific macro—which does the same thing as AC_DEFINE when a *description* is given but no *value* is required.

Checking for Compilers

The AC_PROG_CC macro ensures that the user's system has a working C-language compiler. Here's the prototype for this macro:

```
AC_PROG_CC([compiler-search-list])
```

If your code requires a particular flavor or brand of C compiler, you can pass a whitespace-separated list of program names in this argument. For example, if you use AC_PROG_CC([cc cl gcc]), the macro expands into shell code that searches for cc, cl, and gcc, in that order. Usually, the optional argument is omitted, allowing the macro to find the best compiler option available on the user's system.

You'll recall from "An Even Quicker Start with autoscan" on page 95 that when autoscan noticed C source files in the directory tree, it inserted a no-argument call to this macro into Jupiter's *configure.scan* file. Listing 5-1 reproduces the relevant portion of the generated *configure.scan* file.

```
--snip--
# Checks for programs.
AC_PROG_CC
AC_PROG_INSTALL
--snip--
```

Listing 5-1: configure.scan: *Checking for compilers and other programs*

NOTE   *If the source files in Jupiter's directory tree had been suffixed with .cc, .cxx, or .C (all common extensions for C++ source files), autoscan would have instead inserted a call to AC_PROG_CXX.*

The AC_PROG_CC macro looks for gcc and then cc in the system search path. If it doesn't find either, it looks for other C compilers. When it finds a compatible compiler, the macro sets a well-known variable, CC, to the full path of the program, with options for portability as needed, unless the user has already set CC in the environment or on the configure command line.

The AC_PROG_CC macro also defines the following Autoconf substitution variables, some of which you may recognize as *user variables* (listed in Table 3-2 on page 71):

- @CC@ (full path of compiler)
- @CFLAGS@ (for example, -g -02 for gcc)
- @CPPFLAGS@ (empty by default)
- @EXEEXT@ (for example, *.exe*)
- @OBJEXT@ (for example, *o*)[2]

---

2. The value of @OBJEXT@ does not begin with a dot (.) like the value of @EXEEXT@ does because the latter is often empty whereas the former is always present in some form. If the value of @EXEEXT@ did not begin with a dot, a consumer of the replacement reference would have to supply the dot. If @EXEEXT@ resolves to an empty string, the product has a trailing dot.

AC_PROG_CC configures these substitution variables, but unless you use them in your *Makefile.in* templates, you're just wasting time running ./configure. Conveniently, we're already using them in our *Makefile.in* templates, because earlier in the Jupiter project, we added them to our compiler command line and then added a default value for CFLAGS that the user could override on the make command line.

The only thing left to do is ensure that config.status substitutes values for these variable references. Listing 5-2 shows the relevant portions of the *src* directory *Makefile.in* template and the changes necessary to make this happen.

Git tag 5.0

```
--snip--
# VPATH-specific substitution variables
srcdir = @srcdir@
VPATH = @srcdir@

# Tool-specific substitution variables
CC = @CC@
CFLAGS = @CFLAGS@
CPPFLAGS = @CPPFLAGS@

all: jupiter

jupiter: main.c
        $(CC) $(CPPFLAGS) $(CFLAGS) -I. -I$(srcdir) -I.. -o $@ $(srcdir)/main.c
--snip--
```

Listing 5-2: src/Makefile.in: Using Autoconf compiler and flag substitution variables

Checking for Other Programs

Immediately following the call to AC_PROG_CC (refer to Listing 5-1) is a call to AC_PROG_INSTALL. All of the AC_PROG_* macros set (and then substitute, using AC_SUBST) various environment variables that point to the located utilities. AC_PROG_INSTALL does the same thing for the install utility. To use this check, you need to use the associated Autoconf substitution variables in your *Makefile.in* templates, just as we did earlier with @CC@, @CFLAGS@, and @CPPFLAGS@. Listing 5-3 illustrates these changes.

Git tag 5.1

```
--snip--
# Tool-specific substitution variables
CC = @CC@
CFLAGS = @CFLAGS@
CPPFLAGS = @CPPFLAGS@
INSTALL = @INSTALL@
INSTALL_DATA = @INSTALL_DATA@
INSTALL_PROGRAM = @INSTALL_PROGRAM@
INSTALL_SCRIPT = @INSTALL_SCRIPT@
--snip--
install:
        $(INSTALL) -d $(DESTDIR)$(bindir)
```

```
        $(INSTALL_PROGRAM) -m 0755 jupiter $(DESTDIR)$(bindir)
--snip--
```

Listing 5-3: src/Makefile.in: Substituting the `install` utility in your Makefile.in templates

The value of `@INSTALL@` is obviously the path of the located installation program. The value of `@INSTALL_DATA@` is `${INSTALL}` `-m 0644`. Based on this, you might think that the values of `@INSTALL_PROGRAM@` and `@INSTALL_SCRIPT@` would be something like `${INSTALL}` `-m 0755`, but they're not. These values are set simply to `${INSTALL}`.[3]

You might also need to test for other important utility programs, including `lex`, `yacc`, `sed`, and `awk`. If your program requires one or more of these tools, you can add invocations of `AC_PROG_LEX`, `AC_PROG_YACC`, `AC_PROG_SED`, or `AC_PROG_AWK`. If it detects files in your project's directory tree with *.yy* or *.ll* extensions, autoscan will add invocations of `AC_PROG_YACC` and `AC_PROG_LEX` to *configure.scan*.

You can check for about a dozen different programs using these more specialized macros. If a program check fails, the resulting `configure` script will fail with a message indicating that the required utility could not be found and that the build cannot continue until it has been properly installed.

The program and compiler checks cause autoconf to substitute specially named variables into template files. You can find the names of the variables for each macro in the *GNU Autoconf Manual*. You should use these make variables in commands within your *Makefile.in* templates to invoke the tools they represent. The Autoconf macros will set the values of these variables according to the tools they find installed on the user's system, *if the user has not already set them in the environment.*

This is a key aspect of Autoconf-generated `configure` scripts—the user can *always* override anything `configure` will do to the environment by exporting or setting an appropriate variable before executing `configure`.[4]

For example, if the user chooses to build with a specific version of `bison` installed in the home directory, they could enter the following command in order to ensure that `$(YACC)` refers to the correct version of `bison` and that the shell code `AC_PROG_YACC` generates does little more than substitute the existing value of `YACC` for `@YACC@` in your *Makefile.in* templates:

```
$ cd jupiter
$ ./configure YACC="$HOME/bin/bison -y"
--snip--
```

---

3. The `install` program was originally designed to install executables and therefore defaults to applying executable attributes.

4. Since users are not Autoconf experts, it's good practice to add information about variables that affect your project's configuration to your project's *README* or *INSTALL* files. Running `./configure --help` displays many of these variables, but not all users are aware of `configure`'s `--help` option.

NOTE *Passing the variable setting to* configure *as a parameter is functionally similar to setting the variable for the* configure *process on the command line in the shell environment (for example,* YACC="$HOME/bin/bison -y" ./configure*). The advantage of using the syntax given in this example is that* config.status --recheck *can then track the value and properly re-execute* configure *from the makefile with the options that were originally given to it. Thus, you should always use the parameter syntax, rather than the shell environment syntax, to set variables for* configure*. For ways to enforce the use of this syntax, see the documentation for* AC_ARG_VAR *in the Autoconf manual.*

To check for the existence of a program not covered by these more specialized macros, you can use the generic AC_CHECK_PROG macro or write your own special-purpose macro (see Chapter 16).

The key points to take away here are as follows:

- AC_PROG_* macros check for the existence of programs.
- If they find a program, a substitution variable is created.
- You should use these substitution variables in your *Makefile.in* templates to execute associated utilities.

A Common Problem with Autoconf

We should take this opportunity to address a particular problem developers new to the Autotools consistently encounter. Here's the formal definition of AC_CHECK_PROG, as you will find it in the *GNU Autoconf Manual*:

AC_CHECK_PROG(*variable, prog-to-check-for, value-if-found,*
  [*value-if-not-found*], [*path*], [*reject*])

Check whether program *prog-to-check-for* exists in *path*. If it is found, set *variable* to *value-if-found*, otherwise to *value-if-not-found*, if given. Always pass over *reject* (an absolute filename) even if it is the first found in the search path; in that case, set *variable* using the absolute filename of the *prog-to-check-for* found that is not *reject*. If *variable* was already set, do nothing. Calls AC_SUBST for *variable*. The result of this test can be overridden by setting the *variable* variable or the cache variable *ac_cv_prog_variable*.[5]

This is pretty dense language, but after a careful reading, you can extract the following from this description:

- If *prog-to-check-for* is found in the system search path, then *variable* is set to *value-if-found*; otherwise, it's set to *value-if-not-found*.
- If *reject* is specified (as a full path), and it's the same as the program found in the system search path in the previous step, then skip it and continue to the next matching program in the system search path.

---

5. See Section 5.2.2, "Generic Program and File Checks," in version 2.69 (May 1, 2012) of the *GNU Autoconf Manual* (*https://www.gnu.org/software/autoconf/manual/index.html*).

- If *reject* is found first in *path* and then another match (other than *reject*) is found, set *variable* to the absolute path name of the second (non-*reject*) match.

- If the user has already set *variable* in the environment, then *variable* is left untouched (thereby allowing the user to override the check by setting *variable* before running configure).

- AC_SUBST is called on *variable* to make it an Autoconf substitution variable.

Upon first reading this description, there appears to be a conflict: we see in the first item that *variable* will be set to one of two specified values, based on whether or not *prog-to-check-for* is found in the system search path. But then we see in the third item that *variable* will be set to the full path of some program if *reject* is found first and skipped.

Discovering the real functionality of AC_CHECK_PROG is as easy as reading a little shell script. While you could refer to the definition of AC_CHECK_PROG in Autoconf's *programs.m4* macro file, you'll be one level removed from the actual shell code that performs the check. Wouldn't it be better to just look at the shell script that AC_CHECK_PROG generates? We'll use Jupiter's *configure.ac* file to play with this concept. Temporarily modify your *configure.ac* file according to the changes highlighted in Listing 5-4.

```
--snip--
AC_PREREQ(2.69)
AC_INIT([Jupiter], [1.0], [jupiter-bugs@example.org])
AC_CONFIG_SRCDIR([src/main.c])
AC_CONFIG_HEADER([config.h])

# Checks for programs.
AC_PROG_CC
_DEBUG_START_
AC_CHECK_PROG([bash_var], [bash], [yes], [no],, [/usr/sbin/bash])
_DEBUG_END_
AC_PROG_INSTALL
--snip--
```

Listing 5-4: A first attempt at using AC_CHECK_PROG

Now execute autoconf, open the resulting configure script, and search for _DEBUG_START_.

NOTE   *The _DEBUG_START_ and _DEBUG_END_ strings are known as picket fences. I added these to configure.ac for the sole purpose of helping me find the beginning and end of the shell code generated by the AC_CHECK_PROG macro. I chose these names in particular because you're not likely to find them anywhere else in the generated configure script.[6]*

---

6. Don't be tempted to set these "picket fence" tokens to a value in order to keep configure from complaining about them. If you do, configure *won't* complain about them, and you might just forget to remove them.

Listing 5-5 shows the portion of configure this macro generates.

```
--snip--
_DEBUG_START_
```
❶
```
# Extract the first word of "bash" so it can be a program name with args.
set dummy bash; ac_word=$2
{ $as_echo "$as_me:${as_lineno-$LINENO}: checking for $ac_word" >&5
$as_echo_n "checking for $ac_word... " >&6; }
if ${ac_cv_prog_bash_var+:} false; then :
  $as_echo_n "(cached) " >&6
else
  if test -n "$bash_var"; then
  ac_cv_prog_bash_var="$bash_var" # Let the user override the test.
else
  ac_prog_rejected=no
as_save_IFS=$IFS; IFS=$PATH_SEPARATOR
for as_dir in $PATH
do
  IFS=$as_save_IFS
  test -z "$as_dir" && as_dir=.
    for ac_exec_ext in '' $ac_executable_extensions; do
  if as_fn_executable_p "$as_dir/$ac_word$ac_exec_ext"; then
```
❷
```
  if test "$as_dir/$ac_word$ac_exec_ext" = "/usr/sbin/bash"; then
       ac_prog_rejected=yes
       continue
     fi
  ac_cv_prog_bash_var="yes"
  $as_echo "$as_me:${as_lineno-$LINENO}: found $as_dir/$ac_word$ac_exec_ext"
>&5
    break 2
  fi
done
  done
IFS=$as_save_IFS
```
❸
```
if test $ac_prog_rejected = yes; then
    # We found a bogon in the path, so make sure we never use it.
    set dummy $ac_cv_prog_bash_var
    shift
    if test $# != 0; then
      # We chose a different compiler from the bogus one.
      # However, it has the same basename, so the bogon will be chosen
      # first if we set bash_var to just the basename; use the full file name.
      shift
      ac_cv_prog_bash_var="$as_dir/$ac_word${1+' '}$@"
    fi
fi
  test -z "$ac_cv_prog_bash_var" && ac_cv_prog_bash_var="no"
fi
fi
bash_var=$ac_cv_prog_bash_var
if test -n "$bash_var"; then
  { $as_echo "$as_me:${as_lineno-$LINENO}: result: $bash_var" >&5
$as_echo "$bash_var" >&6; }
else
```

```
     { $as_echo "$as_me:${as_lineno-$LINENO}: result: no" >&5
$as_echo "no" >&6; }
fi


_DEBUG_END_
--snip--
```

Listing 5-5: A portion of configure *generated by* AC_CHECK_PROG

The opening comment at ❶ in this shell script is a clue that AC_CHECK_PROG has some undocumented functionality. Apparently, you may pass in arguments along with the program name in the *prog-to-check-for* parameter. Shortly, we'll look at a situation in which you might want to do that.

Farther down in the script at ❷, you can see that the *reject* parameter was added into the mix in order to allow configure to search for a particular version of a tool. From the code at ❸, we can see that our bash_var variable can have three different values: either empty if the requested program is not found in the search path, the program specified if it's found, or the full path of the program specified if *reject* is found first.

Where do you use *reject*? Well, for instance, on Solaris systems with proprietary Sun tools installed, the default C compiler is often the Solaris C compiler. But some software may require the use of the GNU C compiler instead. As maintainers, we don't know which compiler will be found first in a user's search path. AC_CHECK_PROG allows us to ensure that gcc is used with a full path if another C compiler is found first in the search path.

As I mentioned earlier, M4 macros are aware of the fact that arguments are given, empty, or missing, and they do different things based on these conditions. Many of the standard Autoconf macros are written to take full advantage of empty or unspecified optional arguments and generate entirely different shell code in each of these conditions. Autoconf macros may also optimize the generated shell code for these different conditions.

Given what we now know, we probably should have called AC_CHECK_PROG in this manner instead:

```
AC_CHECK_PROG([bash_shell],[bash -x],[bash -x],,,[/usr/sbin/bash])
```

You can see in this example that the manual is technically accurate. If *reject* isn't specified and bash is found in the system path, then bash_shell will be set to bash -x. If bash is *not* found in the system path, then bash_shell will be set to the empty string. If, on the other hand, *reject is* specified and the undesired version of bash is found *first* in the path, then bash_shell will be set to the full path of the *next* version found in the path, along with the originally specified argument (-x). The reason the macro uses the full path in this case is to make sure that configure will avoid executing the version that was found first in the path—*reject*. The rest of the configuration script can now use the bash_shell variable to run the desired Bash shell, as long as it doesn't test out empty.

NOTE   *If you're following along in your own code, don't forget to remove the temporary code from Listing 5-4 from your* configure.ac *file.*

Checks for Libraries and Header Files

The decision of whether or not to use an external library in a project is a tough one. On one hand, you want to reuse existing code to provide required functionality instead of writing it yourself. Reuse is one of the hallmarks of the open source software world. On the other hand, you don't want to depend on functionality that may not exist on all target platforms or that may require significant porting in order to make the libraries you need available where you need them.

Occasionally, library-based functionality can differ in minor ways between platforms. Although the functionality may be essentially equivalent, the libraries may have different package names or different API signatures. The POSIX threads (*pthread*) library, for example, is similar in functionality to many native threading libraries, but the libraries' APIs are usually different in minor ways, and their package and library names are almost always different. Consider what would happen if we tried to build a multithreaded project on a system that didn't support *pthread*; in a case like this, you might want to use the *libthreads* library on Solaris instead.

Autoconf library selection macros allow generated configuration scripts to intelligently select the libraries that provide the necessary functionality, even if those libraries are named differently between platforms. To illustrate the use of the Autoconf library selection macros, we'll add some trivial (and fairly contrived) multithreading capabilities to the Jupiter project that will allow jupiter to print its message using a background thread. We'll use the *pthread* API as our base threading model. In order to accomplish this with our Autoconf-based configuration script, we need to add the *pthread* library to our project build system.

NOTE   *The proper use of multithreading requires the definition of additional substitution variables containing appropriate flags, libraries, and definitions. The* AX_PTHREAD *macro does all of this for you. You can find the documentation for* AX_PTHREAD *at the Autoconf Macro Archive website.[7] See "Doing Threads the Right Way" on page 384 for examples of using* AX_PTHREAD.

First, let's tackle the changes to the source code. We'll modify *main.c* so that the message is printed by a secondary thread, as shown in Listing 5-6.

Git tag 5.2

```
#include <stdio.h>
#include <stdlib.h>
#include <pthread.h>

static void * print_it(void * data)
{
    printf("Hello from %s!\n", (const char *)data);
    return 0;
}

int main(int argc, char * argv[])
```

---

7. See *https://www.gnu.org/software/autoconf-archive/ax_pthread.html.*

```
{
    pthread_t tid;
    pthread_create(&tid, 0, print_it, argv[0]);
    pthread_join(tid, 0);
    return 0;
}
```

Listing 5-6: src/main.c: Adding multithreading to the Jupiter project source code

This is clearly a ridiculous use of a thread; nevertheless, it *is* the prototypical form of thread usage. Consider a hypothetical situation in which the background thread performs some long calculation and main is doing other things while print_it is working. On a multiprocessor machine, using a thread in this manner could literally double a program's throughput.

Now all we need is a way to determine which libraries should be added to the compiler (linker) command line. If we weren't using Autoconf, we'd just add the library to our linker command line in the makefile, as shown in Listing 5-7.

```
program: main.c
        $(CC) ... -lpthread ...
```

Listing 5-7: Manually adding the pthread library to the compiler command line

Instead, we'll use the Autoconf-provided AC_SEARCH_LIBS macro, an enhanced version of the basic AC_CHECK_LIB macro. The AC_SEARCH_LIBS macro allows us to test for required functionality within a list of libraries. If the functionality exists in one of the specified libraries, an appropriate command line option is added to the @LIBS@ substitution variable, which we would then use in a *Makefile.in* template on the compiler (linker) command line. Here is the formal definition of AC_SEARCH_LIBS from the *GNU Autoconf Manual*:

```
AC_SEARCH_LIBS(function, search-libs,
    [action-if-found], [action-if-not-found], [other-libraries])
```

Search for a library defining *function* if it's not already available. This equates to calling AC_LINK_IFELSE([AC_LANG_CALL([], [*function*])]) first with no libraries, then for each library listed in *search-libs*.

Add -l*library* to LIBS for the first library found to contain *function*, and run *action-if-found*. If *function* is not found, run *action-if-not-found*.

If linking with *library* results in unresolved symbols that would be resolved by linking with additional libraries, give those libraries as the *other-libraries* argument, separated by spaces: for example, -lXt -lX11. Otherwise, this macro fails to detect that *function* is present, because linking the test program always fails with unresolved symbols.

The result of this test is cached in the ac_cv_search *function* variable as none required if *function* is already available, as no if no library containing *function* was found, otherwise as the -l*library* option that needs to be prepended to LIBS.[8]

Can you see why the generated configuration script is so large? When you pass a particular function in a call to AC_SEARCH_LIBS, linker command line arguments are added to a substitution variable called @LIBS@. These arguments ensure that you will link with a library that contains the function passed in. If multiple libraries are listed in the second parameter, separated by whitespace, configure will determine which of these libraries are available on your user's system and use the most appropriate one.

Listing 5-8 shows how to use AC_SEARCH_LIBS in Jupiter's *configure.ac* file to find the library that contains the pthread_create function. AC_SEARCH_LIBS won't add anything to the @LIBS@ variable if it doesn't find pthread_create in the *pthread* library.

```
--snip--
# Checks for libraries.
AC_SEARCH_LIBS([pthread_create], [pthread])
--snip--
```

Listing 5-8: configure.ac: Using AC_SEARCH_LIBS to check for the pthread library on the system

As we'll discuss in detail in Chapter 7, naming patterns for libraries differ among systems. For example, some systems name libraries *lib*basename.*so*, while others use *lib*basename.*sa* or *lib*basename.*a*. Cigwin-based systems generate libraries named *cig*basename.*dll*. AC_SEARCH_LIBS addresses this situation (quite elegantly) by using the compiler to calculate the actual name of the library from its *basename*; it does this by attempting to link a small test program with the requested function from the test library. Only -l*basename* is passed on the compiler command line—a near-universal convention among Unix compilers.

We'll have to modify *src/Makefile.in* again in order to properly use the now-populated @LIBS@ variable, as shown in Listing 5-9.

```
--snip--
# Tool-specific substitution variables
CC = @CC@
LIBS = @LIBS@
CFLAGS = @CFLAGS@
CPPFLAGS = @CPPFLAGS@
--snip--
jupiter: main.c
        $(CC) $(CFLAGS) $(CPPFLAGS) -I. -I$(srcdir) -I..\
          -o $@ $(srcdir)/main.c $(LIBS)
--snip--
```

Listing 5-9: src/Makefile.in: Using the @LIBS@ substitution variable

---

8. See Section 5.4, "Library Files," in version 2.69 (May 1, 2012) of the *GNU Autoconf Manual* (*https://www.gnu.org/software/autoconf/manual/index.html*).

NOTE *I added $(LIBS) after the source files on the compiler command line because the linker cares about object file order—it searches files for required functions in the order they are specified on the command line.*

I want *main.c* to be the primary source of object code for jupiter, so I'll continue to add additional objects, including libraries, to the command line after this file.

Is It Right or Just Good Enough?

At this point, we've ensured that our build system will properly use *pthread* on most systems.[9] If our system needs a particular library, that library's name will be added to the @LIBS@ variable and then subsequently used on the compiler command line. But we're not done yet.

This system *usually* works fine, but it still fails in corner cases. Because we want to provide an excellent user experience, we'll take Jupiter's build system to the next level. In doing this, we need to make a design decision: in case configure fails to locate a *pthread* library on a user's system, should we fail the build process or build a jupiter program without multithreading?

If we choose to fail the build, the user will notice, because the build will stop with an error message (though it may not be a very friendly one—either the compile or link process will fail with a cryptic error message about a missing header file or an undefined symbol). On the other hand, if we choose to build a single-threaded version of jupiter, we'll need to display some clear message that the program is being built without multithreading functionality and explain why.

One potential problem is that some users' systems may have a *pthread* shared library installed but not the *pthread.h* header file—most likely because the *pthread* executable (shared-library) package was installed but the developer package wasn't. Shared libraries are often packaged independently of static libraries and header files, and while executables are installed as part of a dependency chain for higher-level applications, developer packages are typically installed directly by a user.[10] For this reason,

---

9. My choice of *pthread* as an example is perhaps unfortunate, because adding multithreading to an application often requires more than simply adding a single library to the command line. Many platforms require additional compiler options (for example, -mthreads, -pthreads, -qthreads, and so on), libraries, and C-preprocessor definitions in order to enable multithreading in an application. Some platforms even require a completely different compiler (for instance, AIX requires the use of the cc_r alias). The examples in this book happen to work fine, even on platforms that require these switches, only because they don't make extensive use of the standard C library.

10. The *pthread* library is so important on most systems that the developer package is often installed by default, even on basic installations of Linux or other modern Unix operating systems.

Autoconf provides macros to test for the existence of both libraries and header files. We can use the AC_CHECK_HEADERS macro to ensure the existence of a particular header file.

Autoconf checks are very thorough. They usually ensure not only that a file exists but also that the file is the correct one, because they allow you to specify assertions about the file that the macro then verifies. The AC_CHECK_HEADERS macro doesn't just scan the filesystem for the requested header. Like AC_SEARCH_LIBS, the AC_CHECK_HEADERS macro builds a short test program in the appropriate language and then compiles it to ensure that the compiler can both find and use the file. In essence, Autoconf macros try to test not just for the existence of specific features but for the functionality required from those features.

The AC_CHECK_HEADERS macro is defined in the *GNU Autoconf Manual* as follows:

AC_CHECK_HEADERS(*header-file...*, [*action-if-found*],
    [*action-if-not-found*], [*includes* = 'AC_INCLUDES_DEFAULT'])

For each given system header file *header-file* in the blank-separated argument list that exists, define HAVE_*header-file* (in all capitals). If *action-if-found* is given, it is additional shell code to execute when one of the header files is found. You can give it a value of break to break out of the loop on the first match. If *action-if-not-found* is given, it is executed when one of the header files is not found.

includes is interpreted as in AC_CHECK_HEADER, in order to choose the set of preprocessor directives supplied before the header under test.[11]

Normally, AC_CHECK_HEADERS is called only with a list of desired header files in the first argument. The remaining arguments are optional and are not often used because the macro works pretty well without them.

We'll add a check for the *pthread.h* header file to *configure.ac* using AC_CHECK_HEADERS. As you may have noticed, *configure.ac* already calls AC_CHECK_HEADERS looking for *stdlib.h*. AC_CHECK_HEADERS accepts a list of filenames, so we'll just add *pthread.h* to the list, using a space to separate the filenames, as shown in Listing 5-10.

Git tag 5.3

```
--snip--
# Checks for header files.
AC_CHECK_HEADERS([stdlib.h pthread.h])
--snip--
```

Listing 5-10: configure.ac: Adding pthread.h *to the* AC_CHECK_HEADERS *macro*

---

11. See Section 5.6.3, "Generic Header Checks," in version 2.69 (May 1, 2012) of the *GNU Autoconf Manual* (*https://www.gnu.org/software/autoconf/manual/index.html*).

In order to make this package available to as many people as possible, we'll use the dual-mode build approach, which will allow us to provide at least *some* form of the jupiter program to users without a *pthread* library. In order to accomplish this, we need to add some conditional preprocessor statements to *src/main.c*, as shown in Listing 5-11.

```
#include "config.h"

#include <stdio.h>
#include <stdlib.h>

#if HAVE_PTHREAD_H
# include <pthread.h>
#endif

static void * print_it(void * data)
{
    printf("Hello from %s!\n", (const char *)data);
    return 0;
}

int main(int argc, char * argv[])
{
#if HAVE_PTHREAD_H
    pthread_t tid;
    pthread_create(&tid, 0, print_it, argv[0]);
    pthread_join(tid, 0);
#else
    print_it(argv[0]);
#endif
    return 0;
}
```

Listing 5-11: src/main.c: Adding conditional code, based on the existence of pthread.h

In this version of *main.c*, we've added a conditional check for the header file. If the shell script generated by AC_CHECK_HEADERS locates the *pthread.h* header file, the HAVE_PTHREAD_H macro will be defined with the value 1 in the user's *config.h* file. If the shell script doesn't find the header file, the original #undef statement will be left commented out in *config.h*. Because we rely on these definitions, we also need to include *config.h* at the top of *main.c*.

If you choose not to use the AC_CONFIG_HEADERS macro in *configure.ac*, then @DEFS@ will contain all the definitions generated by all the macros that call AC_DEFINE. In this example, we've used AC_CONFIG_HEADERS, so *config.h.in* will contain most of these definitions, and @DEFS@ will only contain HAVE_CONFIG_H, which we don't actually use.[12] The *config.h.in* template method significantly

---

12. The use of HAVE_CONFIG_H in *.c* source files around the inclusion of *config.h* is an older pattern that's discouraged today. You may see such code in existing projects, but the Autoconf manual suggests not doing this sort of thing; building the software requires that *config.h* exist, so it's pointless to check for its existence before including it.

shortens the compiler command line (and also makes it simple to take a snapshot of the template and modify it by hand for non-Autotools platforms). Listing 5-12 shows the required changes to the *src/Makefile.in* template.

```
--snip--
# Tool-related substitution variables
CC = @CC@
DEFS = @DEFS@
LIBS = @LIBS@
CFLAGS = @CFLAGS@
CPPFLAGS = @CPPFLAGS@
--snip--
jupiter: main.c
        $(CC) $(CFLAGS) $(DEFS) $(CPPFLAGS) -I. -I$(srcdir) -I..\
          -o $@ $(srcdir)/main.c $(LIBS)
--snip--
```

Listing 5-12: src/Makefile.in: Adding the use of @DEFS@ to the src-level makefile

NOTE *I've added $(DEFS) before $(CPPFLAGS), giving the end user the option to override any of my policy decisions on the command line.*

We now have everything we need to conditionally build the jupiter program. If the user's system has *pthread* functionality installed, the user will automatically build a version of jupiter that uses multiple threads of execution; otherwise, they'll have to settle for serialized execution. The only thing left to do is to add some code to *configure.ac* such that if configure can't find the *pthread* library, it will display a message indicating that it will build a program that uses serialized execution.

Now, consider the unlikely scenario of a user who has the header file installed but doesn't have the library. For example, if the user executes ./configure with CPPFLAGS=-I/usr/local/include but neglects to add LDFLAGS=-L/usr/local/lib, it will seem to configure that the header is available but the library is missing. This condition is easily remedied by simply skipping the header file check entirely if configure can't find the library. Listing 5-13 shows the required changes to *configure.ac*.

Git tag 5.4

```
--snip--
# Checks for libraries.
have_pthreads=no
AC_SEARCH_LIBS([pthread_create], [pthread], [have_pthreads=yes])

# Checks for header files.
AC_CHECK_HEADERS([stdlib.h])

if test "x${have_pthreads}" = xyes; then
    AC_CHECK_HEADERS([pthread.h], [], [have_pthreads=no])
fi

if test "x${have_pthreads}" = xno; then
    AC_MSG_WARN([
```

```
------------------------------------------
Unable to find pthreads on this system.
Building a single-threaded version.
------------------------------------------])
fi
--snip--
```

Listing 5-13: configure.ac: Adding code to indicate that multithreading is not available during configuration

Now, when we run ./bootstrap.sh and ./configure, we'll see some additional output (highlighted here):

```
$ ./bootstrap.sh
$ ./configure
checking for gcc... gcc
--snip--
checking for library containing pthread_create... -lpthread
--snip--
checking pthread.h usability... yes
checking pthread.h presence... yes
checking for pthread.h... yes
configure: creating ./config.status
config.status: creating Makefile
config.status: creating src/Makefile
config.status: creating config.h
$
```

If a user's system is missing the *pthread.h* header file, for instance, they'd see different output. To emulate this for testing purposes, we can use a trick involving Autoconf cache variables. By presetting the cache variable that represents the presence of the *pthread.h* header to no, we can trick configure into not even looking for *pthread.h* because it assumes the search has already been done if the cache variable is already set. Let's try it out:

```
$ ./configure ac_cv_header_pthread_h=no
checking for gcc... gcc
--snip--
checking for library containing pthread_create... -lpthread
--snip--
checking for pthread.h... (cached) no
configure: WARNING:
------------------------------------------
Unable to find pthreads on this system.
Building a single-threaded version.
------------------------------------------
configure: creating ./config.status
config.status: creating Makefile
config.status: creating src/Makefile
config.status: creating config.h
$
```

Had we chosen to fail the build if the *pthread.h* header file or the *pthread* libraries were not found, then the source code would have been simpler; there would have been no need for conditional compilation. In that case, we could change *configure.ac* to look like Listing 5-14.

```
--snip--
# Checks for libraries.
have_pthreads=no
AC_SEARCH_LIBS([pthread_create], [pthread], [have_pthreads=yes])

# Checks for header files.
AC_CHECK_HEADERS([stdlib.h])

if test "x${have_pthreads}" = xyes; then
    AC_CHECK_HEADERS([pthread.h], [], [have_pthreads=no])
fi

if test "x${have_pthreads}" = xno; then
    AC_MSG_ERROR([
    -------------------------------------------
    The pthread library and header files are
    required to build jupiter. Stopping...
    Check 'config.log' for more information.
    ---------------------------------------])
fi
--snip--
```

Listing 5-14: Failing the build if no pthread library is found

NOTE   *Autoconf macros generate shell code that checks for the existence of system features and sets variables based on these tests. However, it's up to you as maintainer to add shell code to* configure.ac *that makes functional decisions based on the contents of the resulting variables.*

Printing Messages

In the preceding examples, we used a few Autoconf macros to display messages to the user during configuration: AC_MSG_WARN and AC_MSG_ERROR. Here are the prototypes for the various AC_MSG_* macros provided by Autoconf:

```
AC_MSG_CHECKING(feature-description)
AC_MSG_RESULT(result-description)
AC_MSG_NOTICE(message)
AC_MSG_ERROR(error-description[, exit-status])
AC_MSG_FAILURE(error-description[, exit-status])
AC_MSG_WARN(problem-description)
```

The AC_MSG_CHECKING and AC_MSG_RESULT macros are designed to be used together. The AC_MSG_CHECKING macro prints a line indicating that it's checking for a particular feature, but it doesn't print a carriage return at the end of this line. Once the feature has been found (or not found) on the user's machine, the AC_MSG_RESULT macro prints the result at the end of the line, followed by a carriage return that completes the line started by AC_MSG_CHECKING. The *result-description* text should make sense in the context of the *feature-description* message. For instance, the message Looking for a C compiler... might be terminated either with the name of the compiler found or with the text not found.

NOTE *You as the* configure.ac *author should strive to not allow additional text to be displayed between these two macro invocations, as it becomes difficult for the user to follow if there is unrelated text between the two sets of output.*

The AC_MSG_NOTICE and AC_MSG_WARN macros simply print a string to the screen. The leading text for AC_MSG_WARN is configure: WARNING:, whereas that of AC_MSG_NOTICE is simply configure:.

The AC_MSG_ERROR and AC_MSG_FAILURE macros generate an error message, stop the configuration process, and return an error code to the shell. The leading text for AC_MSG_ERROR is configure: error:. The AC_MSG_FAILURE macro prints a notice indicating the directory in which the error occurred, the user-specified message, and then the text See 'config.log' for more details. The optional second parameter (*exit-status*) in these macros allows the maintainer to specify a particular status code to be returned to the shell. The default value is 1.

The text messages output by these macros are displayed to stdout and sent to the *config.log* file, so it's important to use these macros instead of simply using shell echo or printf statements.

Supplying multiple lines of text in the first argument of these macros is especially important in the case of warning messages that merely indicate that the build is continuing with limitations. On a fast build machine in a large configuration process, a single-line warning message could zip right past without even being noticed by the user. This is less of a problem in cases where configure terminates with an error, because the user will easily discover the issue at the end of the output.[13]

Supporting Optional Features and Packages

We've discussed the different ways to handle situations when a *pthread* library exists and when it doesn't. But what if a user wants to build a single-threaded version of jupiter when the *pthread* library *is* installed?

---

13. There is a very strong sentiment on the Autoconf mailing list that you should *not* generate multiline messages. The reasons given are many and varied, but they ultimately all boil down to one: many larger projects already generate thousands of lines of configuration output. Much work has gone into making Autoconf-generated configuration scripts as quiet as possible, but they're still not very quiet. My best advice is to use multiline messages in situations where there is simply no other way to effectively notify a user of an important issue, such as building on a platform with unexpected limitations. Many is the time I've finished a 15-minute build only to find that configure notified me in the first minute that the resulting binaries would be missing functionality that I needed.

We certainly don't want to add a note to Jupiter's *README* file telling the user to rename their *pthread* libraries! Neither do we want the user to have to use our Autoconf cache variable trick.

Autoconf provides two macros for working with optional features and external software packages: AC_ARG_ENABLE and AC_ARG_WITH. Their prototypes are as follows:

```
AC_ARG_WITH(package, help-string, [action-if-given], [action-if-not-given])
AC_ARG_ENABLE(feature, help-string, [action-if-given], [action-if-not-given])
```

As with many Autoconf macros, these two are used simply to set some environment variables:

AC_ARG_WITH ${withval} and ${with_*package*}

AC_ARG_ENABLE ${enableval} and ${enable_*feature*}

The macros can also be used in a more complex form, where the environment variables are used by shell script in the macros' optional arguments. In either case, the resulting variable must be used in *configure.ac*, or it will be pointless to perform the check.

The macros are designed to add the options --enable-feature[=yes|no] (or --disable-feature) and --with-package[=arg] (or --without-package) to the generated configuration script's command line interface, along with appropriate help text to the output generated when the user enters ./configure --help. If the user gives these options, the macros set the preceding environment variables within the script. (The values of these variables may be used later in the script to set or clear various preprocessor definitions or substitution variables.)

AC_ARG_WITH controls your project's use of optional external software packages, while AC_ARG_ENABLE controls the inclusion or exclusion of optional software features. The choice to use one or the other is often a matter of perspective on the software you're considering, and sometimes it's simply a matter of preference, as these macros provide somewhat overlapping sets of functionality.

For instance, in the Jupiter project, it could be justifiably argued that Jupiter's use of *pthread* constitutes the use of an external software package, so you'd use AC_ARG_WITH. However, it could also be said that *asynchronous processing* is a software feature that might be enabled via AC_ARG_ENABLE. In fact, both of these statements are true, and which option you use should be dictated by a high-level architectural perspective on the feature or package to which you're providing optional access. The *pthread* library supplies more than just thread creation functions—it also provides mutexes and condition variables, both of which may be used by a library package that doesn't create threads. If a project provides a library that needs to act in a thread-safe manner within a multithreaded process, it will probably use mutex objects from the *pthread* library, but it may never create a thread. Thus, a user may choose to disable asynchronous execution as a feature at configuration time, but the project will still need to link to the *pthread* library in order to access the mutex functionality. In such cases, it makes more sense to specify --enable-async-exec than --with-pthreads.

In general, you should use AC_ARG_WITH when the user needs to choose between implementations of a feature provided by different packages or internally within the project. For instance, if jupiter had some reason to encrypt a file, it might be written to use either an internal encryption algorithm or an external encryption library. The default configuration might use an internal algorithm, but the package might allow the user to override the default with the command line option --with-libcrypto. When it comes to security, the use of a widely understood library can really help your package gain community trust.

Coding Up the Feature Option

Having decided to use AC_ARG_ENABLE, how do we enable or disable the async-exec feature by default? The difference in how these two cases are encoded in *configure.ac* is limited to the help text and the shell script passed in the *action-if-not-given* argument. The help text describes the available options and the default value, and the shell script indicates what we want to happen if the option is *not* specified. (Of course, if it is specified, we don't need to assume anything.)

Say we decide that asynchronous execution is a risky or experimental feature that we want to disable by default. In this situation, we could add the code shown in Listing 5-15 to *configure.ac*.

```
--snip--
AC_ARG_ENABLE([async-exec],
    [  --enable-async-exec     enable async exec],
    [async_exec=${enableval}], [async_exec=no])
--snip--
```

Listing 5-15: Feature disabled by default

On the other hand, if we decide that asynchronous execution is fundamental to Jupiter, we should probably enable it by default, as in Listing 5-16.

```
--snip--
AC_ARG_ENABLE([async-exec],
    [  --disable-async-exec    disable async exec],
    [async_exec=${enableval}], [async_exec=yes])
--snip--
```

Listing 5-16: Feature enabled by default

Now, the question is, do we check for the library and header file regardless of the user's desire for this feature, or do we only check for them if the async-exec feature is enabled? In this case, it's a matter of preference, because we're using the *pthread* library only for this feature. (If we were also using it for non-feature-specific reasons, we'd have to check for it in either case.)

In cases where we need the library even if the feature is disabled, we would add AC_ARG_ENABLE, as in the preceding example, and an additional invocation of AC_DEFINE to create a *config.h* definition specifically for this feature. Since we don't really want to enable the feature if the library or header file is missing—even if the user specifically requested it—we'll also add some shell code to turn the feature off if either is missing, as shown in Listing 5-17.

Git tag 5.5

```
--snip--
# Checks for programs.
AC_PROG_CC
AC_PROG_INSTALL

# Checks for header files.
AC_CHECK_HEADERS([stdlib.h])

# Checks for command line options
AC_ARG_ENABLE([async-exec],
    [  --disable-async-exec    disable async execution feature],
    [async_exec=${enableval}], [async_exec=yes])

have_pthreads=no
AC_SEARCH_LIBS([pthread_create], [pthread], [have_pthreads=yes])

if test "x${have_pthreads}" = xyes; then
    AC_CHECK_HEADERS([pthread.h], [], [have_pthreads=no])
fi

if test "x${have_pthreads}" = xno; then
❶ if test "x${async_exec}" = xyes; then
        AC_MSG_WARN([
-------------------------------------------
Unable to find pthreads on this system.
Building a single-threaded version.
-------------------------------------------])
    fi
    async_exec=no
fi

if test "x${async_exec}" = xyes; then
    AC_DEFINE([ASYNC_EXEC], [1], [async execution enabled])
fi

# Checks for libraries.

# Checks for typedefs, structures, and compiler characteristics.
--snip--
```

Listing 5-17: configure.ac: Properly managing an optional feature during configuration

We're replacing our original library check with a new check for command line arguments, which has the added benefit of checking for the library for the default case that occurs when the user doesn't specify a preference. As you can see, much of the existing code is the same, with some additional script around it to account for user command line choices.

NOTE *There are places in Listing 5-17 that appear to have gratuitous whitespace or arbitrary indentation. This is intentional, as it causes output to be formatted properly when* configure *is being run. We'll fix some of this later as we add additional macros to our toolbox.*

Notice that at ❶, I've also added an additional test for a yes value in the async_exec variable, because this text really belongs to the feature test, not to the *pthread* library test. Remember, we're trying to create a logical separation between testing for *pthread* functionality and testing for the requirements of the async-exec feature itself.

Of course, now we also have to modify *src/main.c* to use the new definition, as shown in Listing 5-18.

```
--snip--
#if HAVE_PTHREAD_H
# include <pthread.h>
#endif

static void * print_it(void * data)
{
    printf("Hello from %s!\n", (const char *)data);
    return 0;
}

int main(int argc, char * argv[])
{
#if ASYNC_EXEC
    pthread_t tid;
    pthread_create(&tid, 0, print_it, argv[0]);
    pthread_join(tid, 0);
#else
    print_it(argv[0]);
#endif
    return 0;
}
```

Listing 5-18: src/main.c: Changing the conditional around async-exec-specific code

Notice that we've left the HAVE_PTHREAD_H check around the inclusion of the header file in order to facilitate the use of *pthread.h* in ways besides those required by this feature.

In order to check for the library and header file only if the feature is enabled, we wrap the original check code in a test of async_exec, as shown in Listing 5-19.

Git tag 5.6

```
--snip--
# Checks for command line options.
AC_ARG_ENABLE([async-exec],
   [ --disable-async-exec    disable async execution feature],
   [async_exec=${enableval}], [async_exec=yes])

if test "x${async_exec}" = xyes; then
    have_pthreads=no
    AC_SEARCH_LIBS([pthread_create], [pthread], [have_pthreads=yes])

    if test "x${have_pthreads}" = xyes; then
        AC_CHECK_HEADERS([pthread.h], [], [have_pthreads=no])
    fi

    if test "x${have_pthreads}" = xno; then
        AC_MSG_WARN([
----------------------------------------
Unable to find pthreads on this system.
Building a single-threaded version.
----------------------------------------])
        async_exec=no
    fi
fi

if test "x${async_exec}" = xyes; then
    AC_DEFINE([ASYNC_EXEC], 1, [async execution enabled])
fi
--snip--
```

Listing 5-19: configure.ac: Checking for the library and header file only if a feature is enabled

This time, we've moved the test for async_exec from being just around the message statement to being around the entire set of header and library checks, which means we won't even look for *pthread* header and libraries if the user has disabled the async_exec feature.

Formatting Help Strings

We'll make one final change to our use of AC_ARG_ENABLE in Listing 5-17. Notice that in the second argument, there are exactly two spaces between the open square bracket and the start of the argument text. You'll also notice that the number of spaces between the argument and the description depends on the length of the argument text, because the description text is supposed to be presented to the user aligned with a particular column. There are four spaces between --disable-async-exec and the description in Listings 5-16 and 5-17, but there are five spaces after --enable-async-exec in Listing 5-15 because the word *enable* is one character shorter than the word *disable*.

But what if the Autoconf project maintainers decide to change the format of the help text for configuration scripts? Or what if you modify your option name but forget to adjust the indentation on your help text?

To solve these potential problems, we'll turn to an Autoconf helper macro called AS_HELP_STRING, whose prototype is as follows:

```
AS_HELP_STRING(left-hand-side, right-hand-side,
    [indent-column = '26'], [wrap-column = '79'])
```

This macro's sole purpose is to abstract away knowledge about the number of spaces that should be embedded in the help text at various places. To use it, replace the second argument in AC_ARG_ENABLE with a call to AS_HELP_STRING, as shown in Listing 5-20.

Git tag 5.7

```
--snip--
AC_ARG_ENABLE([async-exec],
    [AS_HELP_STRING([--disable-async-exec],
        [disable asynchronous execution @<:@default: no@:>@])],
    [async_exec=${enableval}], [async_exec=yes])
--snip--
```

Listing 5-20: configure.ac: Using AS_HELP_STRING

 For details on the funky character sequences around default: no in Listing 5-20, see "Quadrigraphs" on page 143.

Checks for Type and Structure Definitions

Now let's consider how we might test for system- or compiler-provided type and structure definitions. When writing cross-platform networking software, one quickly learns that the data sent between machines needs to be formatted in a way that doesn't depend on a particular CPU or operating system architecture. Some systems' native integer sizes are 32 bits, while others' are 64 bits. Some systems store integer values in memory and on disk from least-significant byte to most-significant byte, while others do the reverse.

Let's consider an example. When using C-language structures to format network messages, one of the first roadblocks you'll encounter is the lack of basic C-language types that have the same size from one platform to another. A CPU with a 32-bit machine word size would likely have a C compiler with 32-bit int and unsigned types. The sizes of the basic integer types in the C language are implementation defined. This is by design, in order to allow implementations to use sizes for char, short, int, and long that are optimal for each platform.

While this language feature is great for optimizing software designed to run on one platform, it's not very helpful when choosing types to move data *between* platforms. In order to address this problem, engineers have tried everything from sending network data as strings (think XML and JSON) to inventing their own sized types.

In an attempt to remedy this shortcoming in the language, the C99 standard provides the sized types int*N*_t and uint*N*_t, where *N* may be 8, 16, 32, or 64. Unfortunately, not all of today's compilers provide these types. (Not surprisingly, GNU C has been at the forefront for some time now, providing C99-sized types with the inclusion of the *stdint.h* header file.)

To alleviate the pain to some extent, Autoconf provides macros for determining whether C99-specific standardized types exist on a user's platform and then defining them if they don't exist. For example, you can add a call to AC_TYPE_UINT16_T to *configure.ac* in order to ensure that uint16_t exists on your users' platforms, either as a system definition in *stdint.h* or the non-standard but more prolific *inttypes.h*, or as an Autoconf definition in *config.h*.

The compiler tests for such integer-based types are typically written by a configuration script as a bit of C code that looks like the code shown in Listing 5-21.

```
int main()
{
❶ static int test_array[1 - 2 * !((uint16_t) -1 >> (16 - 1) == 1)];
    test_array[0] = 0;
    return 0;
}
```

Listing 5-21: A compiler check for a proper implementation of uint16_t

You'll notice that the important line in Listing 5-21 is at ❶, which is where test_array is declared. Autoconf is relying on the fact that all C compilers will generate an error if you attempt to define an array with a negative size. If uint16_t isn't exactly 16 bits of unsigned data on this platform, the array size will be negative.

Notice, too, that the bracketed expression in the listing is a compile-time expression.[14] Whether this could have been done with simpler syntax is anyone's guess, but this code does the trick on all the compilers Autoconf supports. The array is defined with a nonnegative size only if the following three conditions are met:

- uint16_t is defined in one of the included header files.
- The size of uint16_t is exactly 16 bits.
- uint16_t is unsigned on this platform.

Follow the pattern shown in Listing 5-22 to use the definitions provided by this macro. Even on systems where *stdint.h* or *inttypes.h* is not available, Autoconf will add code to *config.h* that defines uint16_t if the system header files don't provide it, so you can use the type in your source code without additional tests.

---

14. It would have to be a compile-time expression anyway, as C-language array sizes must be statically defined.

```
#include "config.h"

#if HAVE_STDINT_H
# include <stdint.h>
#elif HAVE_INTTYPES_H
# include <inttypes.h>
#endif
--snip--
uint16_t x;
--snip--
```

Listing 5-22: Source code that properly uses Autoconf's uint16_t *definitions*

Autoconf offers a few dozen type checks like AC_TYPE_UINT16_T, as detailed in Section 5.9 of the *GNU Autoconf Manual*. In addition, a generic type check macro, AC_CHECK_TYPES, allows you to specify a comma-separated list of questionable types that your project needs.

NOTE *This list is comma separated because some definitions (like* struct fooble*) may have embedded spaces. Since they are comma delimited, you* must *use Autoconf's square-bracket quotes around this parameter if you list more than one type.*

Here is the formal declaration of AC_CHECK_TYPES:

```
AC_CHECK_TYPES(types, [action-if-found], [action-if-not-found],
    [includes = 'AC_INCLUDES_DEFAULT'])
```

If you don't specify a list of header files in the last parameter, the default headers will be used in the compiler test by way of the macro AC_INCLUDES_DEFAULT, which expands to the text shown in Listing 5-23.

```
#include <stdio.h>
#ifdef HAVE_SYS_TYPES_H
# include <sys/types.h>
#endif
#ifdef HAVE_SYS_STAT_H
# include <sys/stat.h>
#endif
#ifdef STDC_HEADERS
# include <stdlib.h>
# include <stddef.h>
#else
# ifdef HAVE_STDLIB_H
#  include <stdlib.h>
# endif
#endif
#ifdef HAVE_STRING_H
# if !defined STDC_HEADERS && defined HAVE_MEMORY_H
#  include <memory.h>
# endif
# include <string.h>
```

```
#endif
#ifdef HAVE_STRINGS_H
# include <strings.h>
#endif
#ifdef HAVE_INTTYPES_H
# include <inttypes.h>
#endif
#ifdef HAVE_STDINT_H
# include <stdint.h>
#endif
#ifdef HAVE_UNISTD_H
# include <unistd.h>
#endif
```

Listing 5-23: The definition of AC_INCLUDES_DEFAULT, as of Autoconf version 2.69

If you know that your type is not defined in one of these header files, you should specify one or more header files to be included in the test, as shown in Listing 5-24. This listing includes the default header files first, followed by the additional header files (which will often need some of the defaults anyway).

```
  AC_CHECK_TYPES([struct doodah], [], [], [
❶ AC_INCLUDES_DEFAULT
#include<doodah.h>
#include<doodahday.h>])
```

Listing 5-24: Using a nondefault set of includes in the check for struct doodah

Notice at ❶ in Listing 5-24 that I've wrapped the last parameter of the macro over three lines in *configure.ac*, without indentation. The text of this argument is included verbatim in the test source file, so you'll want to be sure that whatever you put into this argument is actually valid code in the language you're using.

NOTE   *Test-related problems are often the sorts of things that developers complain about with regard to Autoconf. When you have problems with such syntax, check the* config.log *file for the complete source code for all failed tests, including the compiler output generated during compilation of the test. This information often provides the solution to your problem.*

The AC_OUTPUT Macro

Finally, we come to the AC_OUTPUT macro, which expands, within configure, into shell code that generates the config.status script based on the data specified in the previous macro expansions. All other macros must be used before AC_OUTPUT is expanded, or they will be of little value to your generated configure script. (Additional shell script may be placed in *configure.ac* after AC_OUTPUT, but it will not affect the configuration or file generation performed by config.status.)

Consider adding shell echo or printf statements after AC_OUTPUT to tell the user how the build system is configured based on the specified command line options. You can also use these statements to tell the user about additional useful targets for make. For example, we might add code to Jupiter's *configure.ac* file *after* AC_OUTPUT, as shown in Listing 5-25.

Git tag 5.8

```
--snip--
AC_OUTPUT

cat << EOF
-------------------------------------------------

${PACKAGE_NAME} Version ${PACKAGE_VERSION}

Prefix: '${prefix}'.
Compiler: '${CC} ${CFLAGS} ${CPPFLAGS}'

Package features:
  Async Execution: ${async_exec}

Now type 'make @<:@<target>@:>@'
  where the optional <target> is:
    all                    - build all binaries
    install                - install everything

-------------------------------------------------
EOF
```

Listing 5-25: configure.ac: Adding configuration summary text to the output of configure

Adding such output to the end of *configure.ac* is a handy project feature, because it tells the user, at a glance, exactly what happened during configuration. Since variables such as async_exec are set to yes or no based on configuration, the user can see whether the requested configuration actually took place.

NOTE   *Version 2.62 (and later) of Autoconf does a much better job of deciphering the user's intent with respect to the use of square brackets than earlier versions do. In the past, you might have needed to use a quadrigraph to force Autoconf to display a square bracket, but now you can use the character itself. Most of the problems that occur are a result of not properly quoting arguments. This enhanced functionality comes primarily from enhancements to Autoconf library macros that might accept square bracket characters in arguments. To ensure that square brackets are not misinterpreted in your own* configure.ac *code, you should read up on M4 double quotation in "Quoting Rules" on page 438.*

QUADRIGRAPHS

Those funny character sequences around the word <target> in Listing 5-25 are called *quadrigraph sequences* or simply *quadrigraphs*. They serve the same purpose as escape sequences, but quadrigraphs are a little more reliable than escaped characters or escape sequences because they're never subject to ambiguity.

The sequence @<:@ is the quadrigraph sequence for the open square bracket character, while @:>@ is the quadrigraph for the closed square bracket character. These quadrigraphs will *always* be output by autom4te as literal square bracket characters. This happens after M4 is finished with the file, so it has no opportunity to misinterpret them as Autoconf quote characters.

If you're interested in studying quadrigraphs in more detail, check out Section 8 of the *GNU Autoconf Manual*.

Summary

In this chapter, we covered some of the more advanced constructs found in the *configure.ac* files for many projects. We started with the macros required to generate substitution variables. I refer to these as "advanced" macros because many of the higher-level Autoconf macros use AC_SUBST and AC_DEFINE internally, making them somewhat transparent to you. However, knowing about them helps you to understand how Autoconf works and provides some of the background information necessary for you to learn to write your own macros.

We covered checks for compilers and other tools, as well as checks for less common data types and structures on your users' systems. The examples in this chapter were designed to help you to understand the proper use of the Autoconf type- and structure-definition check macros, as well as others.

We also examined a technique for debugging the use of complex Autoconf macros: using picket fences around a macro invocation in *configure.ac* in order to quickly locate the associated generated text in configure. We looked at checks for libraries and header files, and we examined some of the details involved in the proper use of these Autoconf macros. We went into great detail about building a robust and user-friendly configuration process, including the addition of project-specific command line options to Autoconf-generated configure scripts.

Finally, we discussed the proper placement of the AC_OUTPUT macro in *configure.ac*, as well as the addition of some summary-generation shell code designed to help your users understand what happened during the configuration of your project on their system.

An important Autconf concept to take away from Chapters 4 and 5 was stated at the very start of Chapter 4: Autoconf generates shell scripts from the shell source code you write into *configure.ac*. That means you have *complete* control over what ends up in your configuration script, as long as you understand the proper use of the macros you're invoking. In fact, you can do anything you want in *configure.ac*. Autoconf macros are there simply to make what you choose to do more consistent and simpler for you to write. The less you rely on Autoconf macros to perform configuration tasks, the less consistent your users' configuration experiences will be relative to other open source projects they download and build.

The next chapter takes us away from Autoconf for a while, as we turn our attention to GNU Automake, an Autotools toolchain add-on that abstracts many of the details of creating very functional makefiles for software projects.

6

AUTOMATIC MAKEFILES WITH AUTOMAKE

If you understand, things are just as they are;
if you do not understand, things are just as they are.
—Zen proverb

Shortly after Autoconf began its journey to success, David MacKenzie started working on a new tool for automatically generating makefiles for a GNU project: Automake. During early development of the *GNU Coding Standards (GCS)*, it became apparent to MacKenzie that because the *GCS* is fairly specific about how and where a project's products should be built, tested, and installed, much of a GNU project makefile was boilerplate material. Automake takes advantage of this fact to make maintainers' lives easier and to make the user's experience more consistent.

MacKenzie's work on Automake lasted almost a year, ending around November 1994. A year later, in November 1995, Tom Tromey (of Red Hat and Cygnus fame) took over the Automake project and played a significant role in its development. Although MacKenzie had written the initial version of Automake in Bourne shell script, Tromey completely rewrote the tool in Perl and continued to maintain and enhance Automake over the next five years.

By the end of 2000, Alexandre Duret-Lutz had essentially taken over maintenance of the Automake project. His role as project lead lasted until about mid-2007, at which point Ralf Wildenhues[1] took the wheel, with occasional input from Akim Demaille and Jim Meyering. From 2012 to early 2017, Automake was maintained by Stefano Lattarini while he worked for Google in Switzerland. The current maintainer is Mathieu Lirzin, a computer science master's student at the University of Bordeaux in France.

Most of the complaints I've seen about the Autotools are ultimately associated with Automake. The reasons are simple: Automake provides the highest level of abstraction over the build system and imposes a fairly rigid structure on projects that use it. Automake's syntax is concise—in fact, it's terse, almost to a fault. One Automake statement represents a *lot* of functionality. But once you understand it, you can get a fairly complete, complex, and functionally correct build system up and running in short order—that is, in minutes, not hours or days.

In this chapter, I provide you with some insight into the inner workings of Automake. With such insight, you'll begin to feel comfortable not only with what Automake can do for you but also with extending it in areas where its automation falls short.

Getting Down to Business

Let's face it—getting a makefile right is often difficult. The devil, as they say, is in the details. Consider the following changes to the files in our project directory structure as we continue to improve the project build system for Jupiter. Let's start by cleaning up our work area. You can do this using `make distclean`, or if you're building from a GitHub repository work area, you can use a form of the `git clean` command:[2]

Git tag 6.0

```
$ git clean -xfd
--snip--
❶ $ rm bootstrap.sh Makefile.in src/Makefile.in
❷ $ echo "SUBDIRS = src" > Makefile.am
❸ $ echo "bin_PROGRAMS = jupiter
  > jupiter_SOURCES = main.c" > src/Makefile.am
❹ $ touch NEWS README AUTHORS ChangeLog
$ ls -1
AUTHORS
ChangeLog
configure.ac
Makefile.am
NEWS
```

---

1. I owe many heartfelt thanks to Ralf for kindly answering so many seemingly trivial questions while I worked on the first edition of this book.

2. In the GitHub repository (*https://github.com/NSP-Autotools/jupiter/*), you'll see the *README* file is a symlink to *README.md*. This difference is only for the sake of presentation on the GitHub project site. The important part here is that some form of the *README* file must exist for Automake.

```
README
src
$
```

The `rm` command at ❶ deletes our hand-coded *Makefile.in* templates and
the `bootstrap.sh` script we wrote to ensure that all the support scripts and files
are copied into the root of our project directory. We won't need this script
anymore because we're upgrading Jupiter to Automake proper. (For the sake
of brevity, I used `echo` statements at ❷ and ❸ to write the new *Makefile.am*
files; you can use a text editor if you wish.)

NOTE *There is a hard carriage return at the end of the line at ❸. The shell will continue to
accept input after the carriage return until the quotation is closed.*

I used the `touch` command at ❹ to create new, empty versions of the
NEWS, README, AUTHORS, and *ChangeLog* files in the project root directory. (The *INSTALL* and *COPYING* files are added by `autoreconf -i`.) These
files are required by the *GCS* for all GNU projects. And although they're not
required for non-GNU projects, they've become something of an institution
in the OSS world; users have come to expect them.[3]

NOTE *The* GCS *covers the format and contents of these files. Sections 6.7 and 6.8 cover the*
NEWS *and* ChangeLog *files, respectively, and Section 7.3 covers the* README,
INSTALL, *and* COPYING *files. The* AUTHORS *file is a list of people (names
and optional email addresses) to whom attribution should be given.[4]*

It can be a little painful to maintain a *ChangeLog* file—especially since
you've already done it once as you added commit messages to your repository commits. To simplify the process, consider using a shell script to scrape
your repository log into *ChangeLog* before you make a new release. There are
existing scripts available on the internet; for example, *gnulib* (see Chapter 13)
provides the `gitlog-to-changelog` script, which can be used to import a git
repository's log information into *ChangeLog* prior to release.

Enabling Automake in configure.ac

To enable Automake within the build system, I've added a single line
to *configure.ac*: a call to `AM_INIT_AUTOMAKE` between the calls to `AC_INIT` and
`AC_CONFIG_SRCDIR`, as shown in Listing 6-1.

```
#                                        -*- Autoconf -*-
# Process this file with autoconf to produce a configure script.
```

---

3. The Automake `foreign` option can be used in the `AM_INIT_AUTOMAKE` macro's option list
argument to keep Automake from requiring the standard GNU text files. In Chapter 14,
where I show you how to convert the FLAIM project to use the Autotools, I present the use
of this option.

4. This information is taken from the July 25, 2016, version of the *GNU Coding Standards* at
http://www.gnu.org/prep/standards/.

```
AC_PREREQ([2.69])
AC_INIT([Jupiter], [1.0], [jupiter-bugs@example.org])
AM_INIT_AUTOMAKE
AC_CONFIG_SRCDIR([src/main.c])
--snip--
```

Listing 6-1: Adding Automake functionality to configure.ac

If your project has already been configured with Autoconf, this is the *only* line that's required to enable Automake in a working *configure.ac* file. The AM_INIT_AUTOMAKE macro accepts an optional argument: a whitespace-separated list of option tags, which can be passed into this macro to modify the general behavior of Automake. For a detailed description of each option, see Chapter 17 of the *GNU Automake Manual*.[5] I will, however, point out a few of the most useful options here.

gnits, gnu, foreign

These options set Automake's strictness checks. The default is gnu. The gnits option makes Automake even more pedantic than it already is, and the foreign option loosens things up a bit—with foreign, you aren't required to have the obligatory *INSTALL*, *README*, and *ChangeLog* files normally required for GNU projects.

check-news

The check-news option causes make dist to fail if the project's current version (from *configure.ac*) doesn't show up in the first few lines of the *NEWS* file.

dist-bzip2, dist-lzip, dist-xz, dist-shar, dist-zip, dist-tarZ

You can use the dist-* options to change the default distribution package type. By default, make dist builds a *.tar.gz* file, but developers often want to distribute, for example, *.tar.xz* packages instead. These options make the change quite easy. (Even without the dist-xz option, you can override the current default by using make dist-xz, but using the option is simpler if you always want to build *.xz* packages.)

readme-alpha

The readme-alpha option temporarily alters the behavior of the build and distribution processes during alpha releases of a project. Using this option causes a file named *README-alpha*, found in the project root directory, to be distributed automatically. The use of this option also alters the expected versioning scheme of the project.

---

5. See the Free Software Foundation's *GNU Automake Manual* at *http://www.gnu.org/software/automake/manual/*.

-W *category*, **--warnings=**_category_

The `-W` *category* and `--warnings=`*category* options indicate that the project would like to use Automake with various warning categories enabled. Multiple such options can be used with different category tags. Refer to the *GNU Automake Manual* to find a list of valid categories.

parallel-tests

The parallel-tests feature allows checks to be executed in parallel in order to take advantage of multiprocessor machines during execution of the check target.

subdir-objects

The `subdir-objects` option is required when you intend to reference sources from directories other than the current directory. Using this option causes Automake to generate make commands that cause object and intermediate files to be generated into the same directory as the source file. For more information on this option, see "Nonrecursive Automake" on page 175.

version

The *version* option is actually a placeholder for a version number that represents the lowest version of Automake that is acceptable for this project. For instance, if 1.11 is passed as an option tag, Automake will fail while processing *configure.ac* if its version is earlier than 1.11. This can be useful if you're trying to use features that only exist in later versions of Automake.

With the new *Makefile.am* files in place and Automake enabled in *configure.ac*, let's run autoreconf with the -i option in order to add any new utility files that Automake may require for our project:

```
$ autoreconf -i
configure.ac:11: installing './compile'
configure.ac:6: installing './install-sh'
configure.ac:6: installing './missing'
Makefile.am: installing './INSTALL'
Makefile.am: installing './COPYING' using GNU General Public License v3 file
Makefile.am:    Consider adding the COPYING file to the version control system
Makefile.am:    for your code, to avoid questions about which license your
project uses src/Makefile.am: installing './depcomp'
$
$ ls -1p
aclocal.m4
AUTHORS
autom4te.cache/
ChangeLog
compile
config.h.in
configure
```

```
configure.ac
COPYING
depcomp
INSTALL
install-sh
Makefile.am
Makefile.in
missing
NEWS
README
src/
$
```

Adding the `AM_INIT_AUTOMAKE` macro to *configure.ac* causes `autoreconf -i` to now execute `automake -i`, which includes a few additional utility files: *aclocal.m4, install-sh, compile, missing,* and *depcomp.* Also, Automake now generates *Makefile.in* from *Makefile.am.*

I mentioned *aclocal.m4* in Chapter 2 and `install-sh` in Chapter 4. The `missing` script is a little utility helper script that prints a nicely formatted message when a tool specified on its command line is not available. More detail than this is not really required; if you're curious, execute `./missing --help` in your project directory.

We'll talk about the `depcomp` script shortly, but I'd like to mention the purpose of the `compile` script here. This script is a wrapper around some older compilers that do not understand the concurrent use of the `-c` and `-o` command line options. When you use product-specific flags, which we'll discuss shortly, Automake has to generate code that may compile source files multiple times with different flags for each file. Thus, it has to name the object files differently for each set of flags it uses. The `compile` script facilitates this process.

Automake also adds default *INSTALL* and *COPYING* text files containing boilerplate text that pertains specifically to the GNU project. You can modify these files for your projects as you see fit. I find the default *INSTALL* file text to be useful for general-purpose instructions related to Autotools-built projects, but I like to prepend some project-specific information to the top of this file before committing it to my repository. Automake's `-i` option won't overwrite these text files in a project that already contains them, so feel free to modify the default files as you see fit, once they've been added by `autoreconf -i`.

The *COPYING* file contains the text of the GPL, which may or may not apply to your project. If your project is released under GPL, just leave the text as is. If you're releasing under another license, such as the BSD, MIT, or Apache Commons licenses, replace the default text with text appropriate for that license.[6]

---

6. See the Open Source Initiative website at *http://opensource.org/* for current license text for nearly all known open source licenses.

NOTE *You only need to use the -i option once in a newly checked-out work area or a newly created project. Once the missing utility files have been added, you can drop the -i option in future calls to* autoreconf *unless you add certain macros to* configure.ac, *which may then cause the use of the -i option to add more missing files. We'll see some of this sort of thing in later chapters.*

The preceding commands create an Automake-based build system that contains everything (with the minor exception of check functionality, which we'll get to shortly) that we wrote into our original *Makefile.in* templates, except that this system is more correct and functionally complete according to the *GCS*. A glance at the resulting generated *Makefile.in* template shows that Automake has done a significant amount of work for us. The resulting top-level *Makefile.in* template is nearly 24KB, while the original, hand-coded makefiles were only a few hundred bytes long.

An Automake build system supports the following important make targets (derived from an Automake-generated *Makefile*):

all	check	clean	ctags
dist	dist-bzip2	dist-gzip	dist-lzip
dist-shar	dist-tarZ	dist-xz	dist-zip
distcheck	distclean	distdir	dvi
html	info	install	install-data
install-dvi	install-exec	install-html	install-info
install-pdf	install-ps	install-strip	installcheck
installdirs	maintainer-clean	mostlyclean	pdf
ps	tags	ininstall	

As you can see, this goes far beyond what we could provide in our hand-coded *Makefile.in* templates. Automake writes this base functionality into every project that uses it.

A Hidden Benefit: Automatic Dependency Tracking

In "Dependency Rules" on page 46, we discussed make dependency rules. These are rules we define in makefiles so that make is aware of the hidden relationships between C-language source files and included header files. Automake goes to a lot of trouble to ensure that you don't have to write such dependency rules for languages it understands, like C, C++, and Fortran. This is an important feature for projects containing more than a few source files.

Writing dependency rules by hand for dozens or hundreds of source files is both tedious and error prone. In fact, it's such a problem that compiler writers often provide a mechanism that enables the compiler to write these rules automatically based on its internal knowledge of the source files and the language. The GNU compilers, among others, support a family of -M options (-M, -MM, -MF, -MG, and so on) on the command line. These options tell the compiler to generate a make dependency rule for the specified source file. (Some of these options can be used on the normal compiler command line, so the dependency rule can be generated when the source file is being compiled.)

The simplest of these options is the basic -M option, which causes the compiler to generate a dependency rule for the specified source file on stdout and then terminate. This rule can be captured in a file, which is then included by the makefile so that the dependency information within this rule is incorporated into the directed graph that make builds.

But what happens on systems where the native compilers don't provide dependency generation options, or where they don't work together with the compilation process? In such cases, Automake provides a wrapper script called *depcomp* that executes the compiler twice: once for dependency information and again to compile the source file. When the compiler lacks the options to generate *any* dependency information, another tool may be used to recursively determine which header files affect a given source file. On systems where none of these options is available, automatic dependency generation fails.

NOTE *For a more detailed description of the dependency-generating compiler options, see "Item 10: Using Generated Source Code" on page 529. For more on Automake dependency management, see the relevant sections of the* GNU Automake Manual.

It's time now to bite the bullet and give it a try. As with our build system from the previous chapter, run autoreconf (optional since we ran autoreconf -i earlier, but harmless), followed by ./configure and make.

```
$ autoreconf
$ ./configure
--snip--
$ make
make  all-recursive
make[1]: Entering directory '/.../jupiter'
Making all in src
make[2]: Entering directory '/.../jupiter/src'
gcc -DHAVE_CONFIG_H -I. -I..     -g -O2 -MT main.o -MD -MP -MF .deps/main.Tpo -c -o main.o main.c
mv -f .deps/main.Tpo .deps/main.Po
gcc  -g -O2   -o jupiter main.o  -lpthread
make[2]: Leaving directory '/.../jupiter/src'
make[2]: Entering directory '/.../jupiter'
make[2]: Leaving directory '/.../jupiter'
make[1]: Leaving directory '/.../jupiter'
$
```

You can't truly appreciate what Automake has done here without trying a few of the other make targets we've become familiar with. Try out the install, dist, and distcheck targets on your own to assure yourself that you still have all the functionality you had before you deleted your handwritten *Makefile.in* templates.

NOTE *The check target exists as a do-nothing target at this point, but we need to dive into Automake constructs in a bit more detail before we can add our test back in. When we get to it, you'll see that it's even simpler than the code we originally wrote.*

What's Actually in a Makefile.am File?

In Chapter 4, we discussed how Autoconf accepts as input a shell script sprinkled with M4 macros and then generates the same shell script with those macros fully expanded. Likewise, Automake accepts as input a make-file sprinkled with Automake commands. Just as Autoconf's input files are simply enhanced shell scripts, Automake *Makefile.am* files are nothing more than standard makefiles with additional Automake-specific syntax.

One significant difference between Autoconf and Automake is that the only text Autoconf outputs is the existing shell script in the input file and any additional shell script resulting from the expansion of embedded M4 macros. Automake, on the other hand, assumes that all makefiles should contain a minimal infrastructure designed to support the *GCS*, in addition to any targets and variables that you specify.

To illustrate this point, create a *temp* directory in the root of the Jupiter project and add an empty *Makefile.am* file to it. Next, add this new *Makefile.am* to the project's *configure.ac* file with a text editor and reference it from the top-level *Makefile.am* file, like this:

```
$ mkdir temp
$ touch temp/Makefile.am
❶ $ echo "SUBDIRS = src temp" > Makefile.am
$ vi configure.ac
--snip--
AC_CONFIG_FILES([Makefile
                src/Makefile
        ❷ temp/Makefile])
--snip--
$ autoreconf
$ ./configure
--snip--
$ ls -1sh temp
total 24K
❸ 12K Makefile
0 Makefile.am
❹ 12K Makefile.in
$
```

I used an echo statement at ❶ to rewrite a new top-level *Makefile.am* file that has SUBDIRS reference both *src* and *temp*. I used a text editor to add *temp/ Makefile* to the list of makefiles Autoconf will generate from templates (❷). As you can see, there is a certain amount of support code generated into every makefile that Automake considers indispensable. Even an empty *Makefile.am* file generates a 12KB *Makefile.in* template (❹), from which configure generates a similarly sized *Makefile* (❸).[7]

Since the make utility uses a fairly rigid set of rules for processing makefiles, Automake takes some license with your additional make code. Here are some specifics:

- The make variables defined in *Makefile.am* files are placed at the top of the resulting *Makefile.in* template, immediately following any Automake-generated variable definitions.

- The make rules specified in *Makefile.am* files are placed at the end of the resulting *Makefile.in* template, immediately after any Automake-generated rules.

- Most Autoconf variables substituted by config.status are converted to make variables and initialized to those substitution variables.

The make utility doesn't care where rules are in relation to each other, because it reads every rule into an internal database before processing any of them. Variables are treated similarly, as long as they are defined before the rules that use them. In order to avoid any variable-binding issues, Automake places all variables at the top of the output file in the order in which they're defined in the input file.

Analyzing Our New Build System

Now let's look at what we put into those two simple *Makefile.am* files, beginning with the top-level *Makefile.am* file (shown in Listing 6-2).

```
SUBDIRS = src
```

Listing 6-2: Makefile.am: The top-level Makefile.am *file contains only a subdirectory reference.*

This single line of text tells Automake several things about our project:

- One or more subdirectories contain makefiles to be processed in addition to this file.[8]

- Directories in this space-delimited list should be processed in the order specified.

---

7. It's fairly instructive to examine the contents of this *Makefile.in* template to see the Autoconf substitution variables that are passed in, as well as the framework code that Automake generates.

8. I refer here to actual makefiles, not *Makefile.am* files. Automake determines the list of *Makefile.am* files to process from *configure.ac*'s AC_CONFIG_FILES list. The SUBDIRS list merely exists to tell make which directories to process from the current makefile, and in which order.

- Directories in this list should be recursively processed for all primary targets.

- Directories in this list should be treated as part of the project distribution, unless otherwise specified.

As with most Automake constructs, SUBDIRS is simply a make variable that has special meaning for Automake. The SUBDIRS variable may be used to process *Makefile.am* files within arbitrarily complex directory structures, and the directory list may contain any relative directory references (not just immediate subdirectories). You might say that SUBDIRS is kind of like the glue that holds makefiles together in a project's directory hierarchy, when using a recursive build system.

Automake generates recursive make rules that implicitly process the current directory after those specified in the SUBDIRS list, but it's often necessary to build the current directory before some or all of the other directories in the list. You may change the default ordering by referencing the current directory with a dot anywhere in the SUBDIRS list. For example, to build the top-level directory before the *src* directory, you could change the SUBDIRS variable in Listing 6-2 as follows:

```
SUBDIRS = . src
```

Now let's turn to the *Makefile.am* file in the *src* directory, shown in Listing 6-3.

```
bin_PROGRAMS = jupiter
jupiter_SOURCES = main.c
```

Listing 6-3: src/Makefile.am: The initial version of this Makefile.am file contains only two lines

The first line is a *product list variable* specification, and the second line is a *product source variable* specification.

Product List Variables

Products are specified in a *Makefile.am* file using a *product list variable (PLV)*, which (like SUBDIRS) is a class of make variables that have special meaning to Automake. The following template shows the general format of a PLV:

```
[modifier-list]prefix_PRIMARY = product1 product2 ... productN
```

The PLV name in the first line of Listing 6-3 consists of two components: the *prefix* (*bin*) and the *primary* (PROGRAMS), separated by an underscore (_). The value of the variable is a whitespace-separated list of products generated by this *Makefile.am* file.

Installation Location Prefixes

The *bin* portion of the product list variable shown in Listing 6-3 is an example of an *installation location prefix*. The *GCS* defines many common installation locations, and most are listed in Table 3-1 on page 65. However, any make variable ending in dir, whose value is a filesystem location, is a viable installation location variable and may be used as a prefix in an Automake PLV.

You reference an installation location variable in a PLV prefix by omitting the dir portion of the variable name. For example, in Listing 6-3, the $(bindir) make variable is referred to only as bin when it is used as an installation location prefix.

Automake also recognizes four installation location variables starting with the special pkg prefix: pkglibdir, pkgincludedir, pkgdatadir, and pkglibexecdir. These pkg versions of the standard libdir, includedir, datadir, and libexecdir variables indicate that the listed products should be installed in a subdirectory of these locations named after the package. For example, in the Jupiter project, products listed in a PLV prefixed with lib would be installed into $(libdir), while those listed in a PLV prefixed with pkglib would be installed into $(libdir)/*jupiter*.

Since Automake derives the list of valid installation locations and prefixes from all make variables ending in dir, you may provide your own PLV prefixes that refer to custom installation locations. To install a set of XML files into an *xml* directory within the system data directory, you could use the code in Listing 6-4 in your *Makefile.am* file.

```
xmldir = $(datadir)/xml
xml_DATA = file1.xml file2.xml file3.xml ...
```

Listing 6-4: Specifying a custom installation directory

Installation location variables will contain default values defined either by Automake-generated makefiles or by you in your *Makefile.am* files, but your users can always override these default values on their configure or make command lines. If you don't want certain products to be installed during a particular build, specify an empty value in an installation location variable on the command line; the Automake-generated rules will ensure that products intended for those directories aren't installed. For example, to install only documentation and shared data files for a package, you could enter make bindir='' libdir='' install.[9]

Prefixes Not Associated with Installation

Certain prefixes are not related to installation locations. For example, noinst, check, and EXTRA are used (respectively) to indicate products that are not installed, are used only for testing, or are optionally built. Here's a little more information about these three prefixes:

---

9. Technically, you don't need the empty quotes after the equal sign; a reference to bindir= is the same as bindir='' after shell processing.

noinst

> Indicates that the listed products should be built but not installed. For example, a static so-called *convenience library* might be built as an intermediate product and then used in other stages of the build process to build final products. The noinst prefix tells Automake that the product should not be installed and that only a static library should be built. (After all, it makes no sense to build a shared library that won't be installed.)

check

> Indicates products that are to be built only for testing purposes and will thus not need to be installed. Products listed in PLVs prefixed with check are built only if the user enters make check.

EXTRA

> Used to list programs that are conditionally built. Automake requires that all source files be specified statically within a *Makefile.am* file, as opposed to being calculated or derived during the build process, so that it can generate a *Makefile.in* template that will work for any possible command line. However, a project maintainer may elect to allow some products to be built conditionally based on configuration options given to the configure script. If products are listed in variables generated by the configure script, they should also be listed in a PLV, prefixed with EXTRA, within a *Makefile.am* file. This concept is illustrated in Listings 6-5 and 6-6.

```
AC_INIT(...)
--snip--
optional_programs=
AC_SUBST([optional_programs])
--snip--
if test "x$(build_opt_prog)" = xyes; then
❶ optional_programs=$(optional_programs) optprog
fi
--snip--
```

Listing 6-5: A conditionally built program defined in a shell variable in configure.ac

```
❷ EXTRA_PROGRAMS = optprog
❸ bin_PROGRAMS = myprog $(optional_programs)
```

Listing 6-6: Using the EXTRA prefix to conditionally define products in Makefile.am

At ❶ in Listing 6-5, optprog is appended to an Autoconf substitution variable called optional_programs. The EXTRA_PROGRAMS variable at ❷ in Listing 6-6 lists optprog as a product that may or may not be built, based on end-user configuration choices that determine whether $(optional_programs) at ❸ is empty or contains optprog.

While it may appear redundant to specify optprog in both *configure.ac* and *Makefile.am*, Automake needs the information in EXTRA_PROGRAMS because it cannot attempt to interpret the possible values of $(optional_programs), as defined in *configure.ac*. Hence, adding optprog to EXTRA_PROGRAMS in this example tells Automake to generate rules to build it, even if $(optional _programs) doesn't contain optprog during a particular build.

Primaries

Primaries are like product classes, and they represent types of products that might be generated by a build system. A primary defines the set of steps required to build, test, install, and execute a particular class of products. For example, programs and libraries are built using different compiler and linker commands, Java classes require a virtual machine to execute them, and Python programs require an interpreter. Some product classes, such as scripts, data, and headers, have no build, test, or execution semantics—only installation semantics.

The list of supported primaries defines the set of product classes that can be built automatically by an Automake build system. Automake build systems can still build other product classes, but the maintainer must define the make rules explicitly within the project's *Makefile.am* files.

A thorough understanding the Automake primaries is the key to properly using Automake. The current complete list of supported primaries is as follows.

PROGRAMS

When the PROGRAMS primary is used in a PLV, Automake generates make rules that use compilers and linkers to build binary executable programs for the listed products.

LIBRARIES/LTLIBRARIES

The use of the LIBRARIES primary causes Automake to generate rules that build static archives (libraries) using the system compiler and librarian. The LTLIBRARIES primary does the same thing, but the generated rules also build Libtool shared libraries and execute these tools (as well as the linker) through the libtool script. (I'll discuss the Libtool package in detail in Chapters 7 and 8.) Automake restricts the installation locations for the LIBRARIES and LTLIBRARIES primaries: they can only be installed in $(libdir) and $(pkglibdir).

LISP

The LISP primary was designed mainly to manage the build for Emacs Lisp programs. Hence, it expects to refer to a list of *.el* files. You can find details on the use of this primary in Section 10.1 of the Automake manual.

PYTHON

Python is an interpreted language; the python interpreter converts a Python script, line by line, into Python byte code, executing it as it's converted, so (like shell scripts) Python source files are executable as written. The use of the PYTHON primary tells Automake to generate rules that precompile Python source files (*.py*) into standard (*.pyc*) and optimized (*.pyo*) byte-compiled versions using the py-compile utility. Because of the normally interpreted nature of Python sources, this compilation occurs at install time rather than at build time.

JAVA

Java is a virtual machine platform; the use of the JAVA primary tells Automake to generate rules that convert Java source files (*.java*) into Java class files (*.class*) using the javac compiler. While this process is correct, it's not complete. Java programs (of any consequence) generally contain more than one class file and are usually packaged as *.jar* or *.war* files, both of which may also contain several ancillary text files. The JAVA primary is useful, but only just. (I'll discuss using—and extending—the JAVA primary in "Building Java Sources Using the Autotools" on page 408.)

SCRIPTS

Script, in this context, refers to any interpreted text file—whether it's shell, Perl, Python, Tcl/Tk, JavaScript, Ruby, PHP, Icon, Rexx, or some other. Automake allows a restricted set of installation locations for the SCRIPTS primary, including $(bindir), $(sbindir), $(libexecdir), and $(pkgdatadir). While Automake doesn't generate rules to build scripts, it also doesn't assume that a script is a static file in the project. Scripts are often generated by handwritten rules in *Makefile.am* files, sometimes by processing an input file with the sed or awk utility. For this reason, scripts are not distributed automatically. If you have a static script in your project that you'd like Automake to add to your distribution archive, you should prefix the SCRIPTS primary with the dist modifier as discussed in "PLV and PSV Modifiers" on page 161.

DATA

Arbitrary data files can be installed using the DATA primary in a PLV. Automake allows a restricted set of installation locations for the DATA primary, including $(datadir), $(sysconfdir), $(sharedstatedir), $(localstatedir), and $(pkgdatadir). Data files are not automatically distributed, so if your project contains static data files, use the dist modifier on the DATA primary as discussed in ""PLV and PSV Modifiers" on page 161.

HEADERS

Header files are a form of source file. Were it not for the fact that some header files are installed, they could simply be listed with the product sources. Header files containing the public interface for installed library products are installed into either the $(includedir) or a package-specific subdirectory defined by $(pkgincludedir), so the most common PLVs for such installed headers are the include_HEADERS and pkginclude_HEADERS variables. Like other source files, header files are distributed automatically. If you have a generated header file, use the nodist modifier with the HEADERS primary as discussed in "PLV and PSV Modifiers" on page 161.

MANS

Man pages are UTF-8 text files containing troff markup, which is rendered by man when viewed by a user. Man pages can be installed using the man_MANS or manN_MANS product list variables, where N represents a single-digit section number between 0 and 9 or the letters *l* (for math library topics) or *n* (for Tcl/Tk topics). Files in the man_MANS PLV should have a numeric extension indicating the man section to which they belong and, therefore, their target directory. Files in the manN_MANS PLV may be named with either numeric extensions or a *.man* extension and will be renamed to the associated numeric extensions when they're installed by make install. Project man pages are not distributed by default because man pages are often generated, so you should use the dist modifier as discussed in "PLV and PSV Modifiers" on page 161.

TEXINFOS

When it comes to Linux or Unix documentation, Texinfo[10] is the GNU project format of choice. The makeinfo utility accepts Texinfo source files (*.texinfo, .txi,* or *.texi*) and renders info files (*.info*) containing UTF-8 text annotated with Texinfo markup, which the info utility renders into formatted text for the user. The most common product list variable for use with Texinfo sources is info_TEXINFOS. The use of this PLV causes Automake to generate rules to build *.info, .dvi, .ps,* and *.html* documentation files. However, only the *.info* files are built with make all and installed with make install. In order to build and install the other types of files, you must specify the dvi, ps, pdf, html, install-dvi, install-ps, install-pdf, and install-html targets explicitly on the make command line. Since the makeinfo utility is not installed by default in many Linux distributions, the generated *.info* files are automatically added to distribution archives so your end users won't have to go looking for makeinfo.

Product Source Variables

The second line in Listing 6-3 is an example of an Automake *product source variable* (*PSV*). PSVs conform to the following template:

```
[modifier-list]product_SOURCES = file1 file2 ... fileN
```

Like PLVs, PSVs are composed of multiple parts: the product name (jupiter in this case) and the SOURCES tag. The value of a PSV is a whitespace-separated list of source files from which *product* is built. The value of the

---

10. See the Texinfo project website at *http://www.gnu.org/software/texinfo/*.

PSV in the second line of Listing 6-3 is the list of source files used to build the jupiter program. Ultimately, Automake adds these files to various make rule dependency lists and commands in the generated *Makefile.in* templates.

Only characters that are allowed in make variables (letters, numbers, the at sign, and underscore) are allowed in the product tag of a PSV. As a result, Automake performs a transformation on product names listed in PLVs to render the *product* tags used in the associated PSVs. Automake converts illegal characters into underscores, as shown in Listing 6-7.

```
❶ lib_LIBRARIES = libc++.a
❷ libc___a_SOURCES = ...
```

Listing 6-7: Illegal make variable characters are converted to underscores in product tags.

Here, Automake converts *libc++.a* in the PLV at ❶ into the PSV product tag libc___a (that's three underscores) to find the associated PSV at ❷ in the *Makefile.am* file. You must know the transformation rules so you can write PSVs that match your products.

PLV and PSV Modifiers

The modifier-list portions of the PLV and PSV templates defined previously contain a set of optional modifiers. The following BNF-like rule defines the format of the modifier-list element of these templates:

```
modifier-list = modifier_[modifier-list]
```

Modifiers change the normal behavior of the variable to which they are prepended. The currently defined set of prefix modifiers includes dist, nodist, nobase, and notrans.

The dist modifier indicates a set of files that should be distributed (that is, that should be included in the distribution package that's built when make dist is executed). For example, assuming that some source files for a product should be distributed and some should not, the variables shown in Listing 6-8 might be defined in the product's *Makefile.am* file.

```
dist_myprog_SOURCES = file1.c file2.c
nodist_myprog_SOURCES = file3.c file4.c
```

Listing 6-8: Using the dist and nodist modifiers in a Makefile.am file

Automake normally strips relative path information from the list of header files in a HEADERS PLV. The nobase modifier is used to suppress the removal of path information from installed header files that are obtained from subdirectories by a *Makefile.am* file. For example, take a look at the PLV definition in Listing 6-9.

```
nobase_pkginclude_HEADERS = mylib.h sys/constants.h
```

Listing 6-9: Using the nobase PLV modifier in a Makefile.am file

In this line we can see that *mylib.h* is in the same directory as *Makefile.am*, but *constants.h* is located in a subdirectory called *sys*. Normally, both files would be installed into $(pkgincludedir) by virtue of the pkginclude installation location prefix. However, since we're using the nobase modifier, Automake will retain the *sys/* portion of the second file's path for installation, and *constants.h* will be installed into $(pkgincludedir)/*sys*. This is useful when you want the installation (destination) directory structure to be the same as the project (source) directory structure as files are copied during installation.

The notrans modifier may be used on man page PLVs for man pages whose names should not be transformed during installation. (Normally, Automake will generate rules to rename the extension on man pages from *.man* to *.N*—where *N* is *0, 1, . . . , 9, l, n*—as they're installed.)

You can also use the EXTRA prefix as a modifier. When used with a product source variable (such as jupiter_SOURCES), EXTRA specifies extra source files that are directly associated with the jupiter product, as shown in Listing 6-10.

```
EXTRA_jupiter_SOURCES = possibly.c
```

Listing 6-10: Using the EXTRA prefix with a product SOURCES variable

Here, *possibly.c* may or may not be compiled, based on some condition defined in *configure.ac*.

Unit Tests: Supporting make check

In Chapter 3, we added code to *src/Makefile* that executes the jupiter program and checks for the proper output string when the user makes the check target. We now have enough information to add our check target test back into our new Automake build system. I've duplicated the check target code in Listing 6-11 for reference in the following discussion.

```
--snip--
check: all
        ./jupiter | grep "Hello from .*jupiter!"
        @echo "*** ALL TESTS PASSED ***"
--snip--
```

Listing 6-11: The check target from Chapter 3

Fortunately, Automake has solid support for unit tests. To add our simple grep test back into the new Automake-generated build system, we can add a few lines to the bottom of *src/Makefile.am*, as shown in Listing 6-12.

Git tag 6.1

```
bin_PROGRAMS = jupiter
jupiter_SOURCES = main.c

❶ check_SCRIPTS = greptest.sh
❷ TESTS = $(check_SCRIPTS)

greptest.sh:
```

```
        echo './jupiter | grep "Hello from .*jupiter!"' > greptest.sh
        chmod +x greptest.sh
```

❸ CLEANFILES = greptest.sh

Listing 6-12: src/Makefile.am: Additional code required to support the check target

The check_SCRIPTS line at ❶ is a PLV that refers to a script generated at build time. Since the prefix is check, we know that scripts listed in this line will only be built when the user enters make check. However, we must supply a make rule for building the script as well as a rule for removing the file later, during execution of the clean target. We use the CLEANFILES variable at ❸ to extend the list of files that Automake deletes during make clean.

The TESTS line at ❷ is the important one in Listing 6-12 because it indicates which targets are executed when the user makes the check target. (Since the check_SCRIPTS variable contains a complete list of these targets, I simply referenced it here, as the make variable that it actually is.) Note that in this particular case, check_SCRIPTS is redundant, because Automake generates rules to ensure that all the programs listed in TESTS are built before the tests are executed. However, check_* PLVs become important when additional helper scripts or programs must be built before those listed in TESTS are executed.

It's not necessarily obvious here, but since we added our first test, we need to re-execute autoreconf -i before running make check in order to add a new utility script: *test-driver.* You can find places in the Automake documentation that indicate clearly that you must do this, but it's simpler to just let the build tell you when something is missing and therefore an execution of autoreconf (-i) is required. To give you a flavor for this process, let's try it without running autoreconf first:

```
$ make check
Making check in src
make[1]: Entering directory '/.../jupiter/src'
 cd .. && /bin/bash /.../jupiter/missing automake-1.15 --gnu src/Makefile
parallel-tests: error: required file './test-driver' not found
parallel-tests:   'automake --add-missing' can install 'test-driver'
Makefile:255: recipe for target 'Makefile.in' failed
make[1]: *** [Makefile.in] Error 1
make[1]: Leaving directory '/.../jupiter/src'
Makefile:352: recipe for target 'check-recursive' failed
make: *** [check-recursive] Error 1
$
```

Now let's run autoreconf -i first:

```
$ autoreconf -i
parallel-tests: installing './test-driver'
$
$ make check
/bin/bash ./config.status --recheck
```

```
running CONFIG_SHELL=/bin/bash /bin/bash ./configure --no-create
--no-recursion
checking for a BSD-compatible install... /usr/bin/install -c
checking whether build environment is sane... yes
checking for a thread-safe mkdir -p... /bin/mkdir -p
--snip--
Making check in src
make[1]: Entering directory '/.../jupiter/src'
make  greptest.sh
make[2]: Entering directory '/.../jupiter/src'
echo './jupiter | grep "Hello from .*jupiter!"' > greptest.sh
chmod +x greptest.sh
make[2]: Leaving directory '/.../jupiter/src'
make  check-TESTS
make[2]: Entering directory '/.../jupiter/src'
make[3]: Entering directory '/.../jupiter/src'
PASS: greptest.sh
============================================================================
Testsuite summary for Jupiter 1.0
============================================================================
# TOTAL: 1
# PASS:  1
# SKIP:  0
# XFAIL: 0
# FAIL:  0
# XPASS: 0
# ERROR: 0
============================================================================
make[3]: Leaving directory '/.../jupiter/src'
make[2]: Leaving directory '/.../jupiter/src'
make[1]: Leaving directory '/.../jupiter/src'
make[1]: Entering directory '/.../jupiter'
make[1]: Leaving directory '/.../jupiter'
$
```

After running autoreconf -i (and noting that test-driver was installed into our project), we can see that make check is now successful.

Note that I didn't have to manually invoke configure after running autoreconf -i. The build system is generally smart enough to know when it should re-execute configure for you.

Reducing Complexity with Convenience Libraries

Jupiter is fairly trivial as open source software projects go, so in order to highlight some more of Automake's key features, let's expand it a little. We'll first add a convenience library and then modify jupiter to consume this library.

A *convenience library* is a static library that's only used within the containing project. Such temporary libraries are generally used when multiple binaries in a project need to incorporate the same source code. I'll move the code in *main.c* to a library source file and call the function in the library from jupiter's main routine. Begin by executing the following commands from the top-level project directory:

Git tag 6.2

```
$ mkdir common
$ touch common/jupcommon.h
$ cp src/main.c common/print.c
$ touch common/Makefile.am
$
```

Now add the highlighted text from Listings 6-13 and 6-14 to the *.h* and *.c* files, respectively, in the new *common* directory.

```
int print_routine(const char * name);
```

Listing 6-13: common/jupcommon.h: The initial version of this file

```
#include "config.h"

#include "jupcommon.h"

#include <stdio.h>
#include <stdlib.h>

#if HAVE_PTHREAD_H
# include <pthread.h>
#endif

static void * print_it(void * data)
{
    printf("Hello from %s!\n", (const char *)data);
    return 0;
}

int print_routine(const char * name)
{
#if ASYNC_EXEC
    pthread_t tid;
    pthread_create(&tid, 0, print_it, (void*)name);
    pthread_join(tid, 0);
#else
    print_it(name);
#endif
    return 0;
}
```

Listing 6-14: common/print.c: The initial version of this file

As you can see, *print.c* is merely a copy of *main.c*, with a few small modifications (highlighted in Listing 6-14). First, I renamed `main` to `print_routine`, and then I added the inclusion of the *jupcommon.h* header file after the inclusion of *config.h*. The header file provides `print_routine`'s prototype to *src/main.c*, where it's called from `main`.

Next, we modify *src/main.c*, as shown in Listing 6-15, and then add the text in Listing 6-16 to *common/Makefile.am*.

```
#include "config.h"

#include "jupcommon.h"

int main(int argc, char * argv[])
{
    return print_routine(argv[0]);
}
```

Listing 6-15: src/main.c: Required modifications to have main call into the new library

NOTE *It may seem odd to include* config.h *at the top of* src/main.c *since nothing in that source file appears to use it. The* GCS *recommends following the standard practice of including* config.h *at the top of all source files, before any other inclusions, in case something in one of the other included header files makes use of definitions in* config.h. *I recommend religiously following this practice.*

```
noinst_LIBRARIES = libjupcommon.a
libjupcommon_a_SOURCES = jupcommon.h print.c
```

Listing 6-16: common/Makefile.am: Initial version of this file

Let's examine this new *Makefile.am* file. The first line indicates which products this file should build and install. The `noinst` prefix indicates that this library is designed solely to make using the source code in the *common* directory more convenient.

We're creating a static library called *libjupcommon.a*, also known as an *archive*. Archives are like *.tar* files that only contain object files (*.o*). They can't be executed or loaded into a process address space like shared libraries, but they can be added to a linker command line like object files. Linkers are smart enough to realize that such archives are merely groups of object files.

NOTE *Linkers add to the binary product every object file specified explicitly on the command line, but they only extract from archives those object files that are actually referenced in the code being linked. Therefore, if you link to a static library containing 97 object files, but you only call functions in two of them directly or indirectly, only those two object files are added to your program. In contrast, linking to 97 raw object files adds all 97 of those files to your program, regardless of whether you use any of their functionality.*

The second line in Listing 6-16 is a product source variable that contains the list of source files associated with this library.[11]

Product Option Variables

Now we need to add some additional information to *src/Makefile.am* so that the generated *Makefile* can find the new library and header file we added to the *common* directory. Let's add two more lines to the existing *Makefile.am* file, as shown in Listing 6-17.

```
bin_PROGRAMS = jupiter
jupiter_SOURCES = main.c
❶ jupiter_CPPFLAGS = -I$(top_srcdir)/common
❷ jupiter_LDADD = ../common/libjupcommon.a
--snip--
```

Listing 6-17: src/Makefile.am: Adding compiler and linker directives to Makefile.am files

Like the jupiter_SOURCES variable, these two new variables are derived from the program name. These *product option variables (POVs)* are used to specify product-specific options to tools that are used to build products from source code.

The jupiter_CPPFLAGS variable at ❶ adds product-specific C-preprocessor flags to the compiler command line for all source files that are compiled for the jupiter program. The -I$(top_srcdir)/common directive tells the C preprocessor to add $(top_srcdir)/*common* to its list of locations in which to look for header file references.[12]

The jupiter_LDADD variable at ❷ adds libraries to the jupiter program's linker command line. The file path *../common/libjupcommon.a* merely adds an object to the linker command line so that code in this library can become part of the final program.

NOTE *You can also use $(top_builddir)/ in place of ../ to reference the location of the* common *directory in this path. Using $(top_builddir) has the added advantage of making it simpler to move this* Makefile.am *file to another location without having to modify it.*

---

11. I chose to place both the header file and the source file in this list. I could have used a noinst_HEADERS PLV for the header file, but it isn't necessary because the libjupcommon_a _SOURCES list works just as well. The appropriate time to use noinst_HEADERS is when you have a directory that contains no source files—such as an internal *include* directory. Since header files are associated with compilation only through include references within your source code, the only effect of using noinst_HEADERS is that the listed header files are simply added to the project's distribution file list. (You'd get exactly the same effect by listing such header files in the EXTRA_DIST variable.)

12. The C preprocessor will search for header files referenced with angle brackets in the resulting include search path. It will also search for header files referenced with double quotes within the system include search path, but it will check the current directory first. Therefore, you should use double quotes, rather than angle brackets, to reference header files that can be referenced relative to your project directory structure.

Adding a library to a *program*_LDADD or *library*_LIBADD variable is only necessary for libraries that are built as part of your own package. If you're linking your program with a library that's already installed on the user's system, a call to AC_CHECK_LIB or AC_SEARCH_LIBS in *configure.ac* will cause the generated configure script to add an appropriate reference to the linker command line via the LIBS variable.

The set of POVs supported by Automake are derived mostly from a subset of the standard user variables listed in Table 3-2 on page 71. You'll find a complete list of program and library option variables in the *GNU Autoconf Manual*, but here are some of the important ones.

*product*_CPPFLAGS
> Use *product*_CPPFLAGS to pass flags to the C or C++ preprocessor on the compiler command line.

*product*_CFLAGS
> Use *product*_CFLAGS to pass C-compiler flags on the compiler command line.

*product*_CXXFLAGS
> Use *product*_CXXFLAGS to pass C++-compiler flags on the compiler command line.

*product*_LDFLAGS
> Use *product*_LDFLAGS to pass global and order-independent shared library and program linker configuration flags and options to the linker, including -static, -version-info, -release, and so on.

*program*_LDADD
> Use *program*_LDADD to add Libtool objects (*.lo*) or libraries (*.la*) or non-Libtool objects (*.o*) or archives (*.a*) to the linker command line when linking a program.[13]

*library*_LIBADD
> Use *library*_LIBADD to add non-Libtool linker objects and archives to non-Libtool archives on the ar utility command line. The ar utility will incorporate archives mentioned on the command line into the product archive, so you can use this variable to gather multiple archives together into one.

*ltlibrary*_LIBADD
> Use *ltlibrary*_LIBADD to add Libtool linker objects (*.lo*) and Libtool static or shared libraries (*.la*) to a Libtool static or shared library.

You can use the last three option variables in this list to pass lists of order-dependent static and shared library references to the linker. You can

---

13. The file extensions on non-Libtool objects and archives are not standardized, so my use of *.o* and *.a* here is to provide an example only.

also use these option variables to pass -L and -l options. The following are acceptable formats: -L*libpath*, -l*libname*, [*relpath/*]*archive*.a, [*relpath/*]*objfile*.$(OBJEXT), [*relpath/*]*ltobject*.lo , and [*relpath/*]*ltarchive*.la. (Note that the term *relpath* indicates a relative path within the project, which can be in terms of either relative directory references, using dots, or $(top_builddir).)

Per-Makefile Option Variables

You'll often see the Automake variables AM_CPPFLAGS and AM_LDFLAGS used in a *Makefile.am* file. These per-makefile forms of these flags are used when the maintainer wants to apply the same set of flags to all products specified in the *Makefile.am* file.[14] For example, if you need to set a group of preprocessor flags for all products in a *Makefile.am* file and then add additional flags for a particular product (prog1), you could use the statements shown in Listing 6-18.

```
AM_CFLAGS = ... some flags ...
--snip--
❶ prog1_CFLAGS = $(AM_CFLAGS) ... more flags ...
--snip--
```

Listing 6-18: Using both per-product and per-file flags

The existence of a per-product variable overrides Automake's use of the per-makefile variable, so you need to reference the per-makefile variable in the per-product variable in order to have the per-makefile variable affect that product, as shown in Listing 6-18 at ❶. In order to allow per-product variables to override their per-makefile counterparts, it's a good idea to reference the per-makefile variable first, before adding any product-specific options.

NOTE *User variables, such as CFLAGS, are reserved for the end user and should never be set by configuration scripts or makefiles. Automake will always append them to the appropriate utility command lines, thus allowing the user to override the options specified in the makefile.*

Building the New Library

Next, we need to edit the SUBDIRS variable in the top-level *Makefile.am* file in order to include the new *common* directory we just added. We also need to add the new makefile that was generated in the *common* directory to the list of files generated from templates in the AC_CONFIG_FILES macro invocation in *configure.ac*. These changes are shown in Listings 6-19 and 6-20.

```
SUBDIRS = common src
```

Listing 6-19: Makefile.am: Adding the common *directory to the SUBDIRS variable*

---

14. Using per-makefile flags can generate more compact makefiles, because per-product flags cause Automake to emit per-product rules instead of more general suffix rules. When large file sets are involved, the difference is significant.

```
--snip--
AC_CONFIG_FILES([Makefile
                 common/Makefile
                 src/Makefile])
--snip--
```

Listing 6-20: configure.ac: Adding common/Makefile to the AC_CONFIG_FILES macro

This is the largest set of changes we've made up to this point, but we're reorganizing the entire application, so it seems reasonable. Let's give our updated build system a try. Add the -i option to the autoreconf command line so that it will install any additional missing files that might be required after these enhancements. After so many changes, I like to start with a clean slate, so start with make distclean, or some form of the git clean command if you're running from a git repository work area.

```
$ make distclean
--snip--
$ autoreconf -i
configure.ac:11: installing './compile'
configure.ac:6: installing './install-sh'
configure.ac:6: installing './missing'
Makefile.am: installing './INSTALL'
Makefile.am: installing './COPYING' using GNU General Public License v3 file
Makefile.am:    Consider adding the COPYING file to the version control
system
Makefile.am:    for your code, to avoid questions about which license your
project uses
❶ common/Makefile.am:1: error: library used but 'RANLIB' is undefined
common/Makefile.am:1:   The usual way to define 'RANLIB' is to add 'AC_PROG_
RANLIB'
common/Makefile.am:1:   to 'configure.ac' and run 'autoconf' again.
common/Makefile.am: installing './depcomp'
parallel-tests: installing './test-driver'
autoreconf: automake failed with exit status: 1
$
```

Well, it looks like we're not quite done yet. Since we've added a new type of entity—static libraries—to our build system, automake (via autoreconf) tells us at ❶ that we need to add a new macro, AC_PROG_RANLIB, to the *configure.ac* file.[15]

Add this macro to *configure.ac*, as shown in Listing 6-21.

```
--snip--
# Checks for programs.
AC_PROG_CC
```

---

15. There's a lot of history behind the use of the ranlib utility on archive libraries. I won't get into whether it's still useful with respect to modern development tools, but I will say that whenever you see it used in modern makefiles, there always seems to be a preceding comment about running ranlib "in order to add karma" to the archive, implying that the use of ranlib is somehow unnecessary. You be the judge.

```
AC_PROG_INSTALL
AC_PROG_RANLIB
--snip--
```

Listing 6-21: configure.ac: *Adding* AC_PROG_RANLIB

Finally, enter autoreconf -i once more. Adding -i ensures that, if the new functionality we added to *configure.ac* requires any additional files to be installed, autoreconf will do so.

```
$ autoreconf -i
$
```

No more complaints; all is well.

What Goes into a Distribution?

Automake usually determines automatically what should go into a distribution created with make dist, because it's very aware of every file's role in the build process. To this end, Automake wants to be told about every source file used to build a product and about every file and product installed. This means, of course, that all files must be specified at some point in one or more PLV and PSV variables.[16]

The Automake EXTRA_DIST variable contains a space-delimited list of files and directories that should be added to the distribution package when the dist target is made. For example:

```
EXTRA_DIST = windows
```

You could use the EXTRA_DIST variable to add a source directory to the distribution package that Automake would not automatically add—for example, a Windows-specific directory.

NOTE   *In this case,* windows *is a directory, not a file. Automake will automatically recursively add every file in this directory to the distribution package; this may include some files that you really didn't want there, such as hidden .svn or .CVS status directories. See "Automake -hook and -local Rules" on page 389 for a way around this problem.*

---

16. This bothers some developers—and with good reason. There are cases where dozens of installable files are generated by tools using long, apparently random, and generally unimportant naming conventions. Listing such generated files statically in a variable is painful, to say the least. Regardless, the current requirement is that all files must be specified. Don't bother trying to find a way around it. You'll end up hacking half the Automake source code to get it to work. Do recognize, however, that there is good reason for this requirement: unless every installable file is explicitly made known, the Autotools simply can't generate configure scripts and makefiles that will always work for the user under every possible set of arguments they might supply.

A WORD ABOUT THE UTILITY SCRIPTS

The Autotools have added several files to the root of our project directory structure: compile, depcomp, install-sh, and missing. Because configure or the generated *Makefiles* all execute these scripts at various points during the build process, the end user will need them; however, we can only get them from the Autotools, and we don't want to require the user to have the Autotools installed. For this reason, these scripts are automatically added to the distribution archive.

So, do you check them in to your source code repository or not? The answer is debatable, but generally I recommend that you don't. Any maintainer who will be creating a distribution archive should have the Autotools installed and should be working from a repository work area. As a result, these maintainers will also be running autoreconf -i (possibly in conjunction with the --force option*) to ensure that they have the most up-to-date Autotools-provided utility scripts. If you check them in, it will only make it more probable that they become out-of-date as time goes by. It will also cause unnecessary churn in your repository revision history as contributors ping-pong back and forth between files generated from the different versions of the Autotools they're using.

I extend this sentiment to the configure script as well. Some people argue that checking the utility and configure scripts into the project repository is beneficial, because it ensures that if someone checked out a work area, they could build the project from the work area without having the Autotools installed. However, my personal philosophy is that developers and maintainers should be expected to have these tools installed. Occasionally, an end user will need to build a project from a work area, but this should be the exception rather than the typical case, and in these exceptional cases, the user should be willing to take on the role and requirements of a maintainer.

---

* Use the --force option with caution; it will also overwrite text files such as *INSTALL*, which may have been modified for the project from the default text file that ships with the Autotools.

Maintainer Mode

Occasionally, timestamps on distribution source files will be newer than the current time setting of a user's system clock. Regardless of the cause, this inconsistency confuses make, causing it to think that every source file is out-of-date and needs to be rebuilt. As a result, it will re-execute the Autotools in an attempt to bring configure and the *Makefile.in* templates up-to-date. But as maintainers, we don't really expect our users to have the Autotools installed—or at least not the latest versions that we've installed on our systems.

This is where Automake's *maintainer mode* comes in. By default, Automake adds rules to makefiles that regenerate template files, configuration scripts, and generated sources from maintainer source files such as *Makefile.am* and *configure.ac*, as well as Lex and Yacc input files. However, we can use the Automake AM_MAINTAINER_MODE macro in *configure.ac* to disable the default generation of these maintainer-level make rules.

For maintainers who want these rules in place to keep their build system properly updated after build system changes, the AM_MAINTAINER_MODE macro provides a configure script command line option (--enable-maintainer-mode), which tells configure to generate *Makefile.in* templates that contain rules and commands to execute the Autotools as necessary.

Maintainers must be aware of the use of AM_MAINTAINER_MODE in their projects. They will need to use this command line option when running configure in order to generate full build systems that will properly rebuild Autotools-generated files when their sources are modified.

NOTE   *I also recommend mentioning the use of maintainer mode in the project INSTALL or README files so that end users are not surprised when they modify Autotools sources without effect.*

Although Automake's maintainer mode has its advantages, you should know that there are various arguments against using it. Most focus on the idea that make rules should never be purposely restricted, because doing so generates a build system that will always fail under certain circumstances. I will, however, state that later versions of the Autotools do a much better job of telling you what's happening when a required tool is missing. In fact, this is exactly what the missing script is for. Most tool invocations are wrapped in the missing script, which tells you fairly clearly what's missing and how to install it when it is missing.

Another important consideration when using this macro is that you've now doubled the rows in your test matrix, as every build option has two modes—one that assumes the Autotools are installed and one that assumes the opposite. If you decide to use the macro to disable maintainer mode by default for your end users, keep these points in mind.

Cutting Through the Noise

The amount of noise generated by Autotools-based build systems has been one of the most controversial topics on the Automake mailing list. One camp appreciates quiet builds that just display important information, such as warnings and errors. The other side argues that valuable information is often embedded in this so-called "noise," so all of it is important and should be displayed. Occasionally, a new Autotools developer will post a question about how to reduce the amount of information displayed by make. This almost always spawns a heated debate that lasts for several days over a few dozen email messages. The old-timers just laugh about it and often joke about how "someone has turned on the switch again."

The truth of the matter is that both sides have valid points. The GNU project is all about options, so the Automake maintainers have added the ability to allow you to optionally make silent rules available to your users. *Silent rules* in Automake makefiles are not really silent; they're just somewhat less noisy than traditional Automake-generated rules.

Instead of displaying the entire compiler or linker command line, silent rules display a short line indicating the tool and the name of the file being processed by that tool. Output generated by make is still displayed so the user knows which directory and target are currently being processed. Here is Jupiter's build output, with silent rules enabled (execute make clean first to ensure something actually gets built):

```
$ make clean
--snip--
$ configure --enable-silent-rules
--snip--
$ make
make  all-recursive
make[1]: Entering directory '/.../jupiter'
Making all in common
make[2]: Entering directory '/.../jupiter/common'
  CC      print.o
  AR      libjupcommon.a
ar: `u' modifier ignored since `D' is the default (see `U')
make[2]: Leaving directory '/.../jupiter/common'
Making all in src
make[2]: Entering directory '/.../jupiter/src'
  CC      jupiter-main.o
  CCLD    jupiter
make[2]: Leaving directory '/.../jupiter/src'
make[2]: Entering directory '/.../jupiter'
make[2]: Leaving directory '/.../jupiter'
make[1]: Leaving directory '/.../jupiter'
$
```

As you can see, the use of silent rules doesn't make a lot of difference for Jupiter—Jupiter's build system spends a lot of time moving between directories and very little time actually building things. But in projects with hundreds of source files, you'd see long lists of CC *filename*.o lines, with an occasional indication that make is changing directories or the linker is building a product—compiler warnings tend to jump out at you. For instance, the ar warning in the output would have flown by unnoticed without silent rules.[17]

Silent rules are disabled by default. To enable silent rules by default in Automake-generated *Makefile.am* templates, you may call the AM_SILENT_RULES macro in *configure.ac* with a yes argument.

In any case, the user may always set the default verbosity for a build with --enable-silent-rules or --disable-silent-rules on the configure command

---

17. You may or may not see this ar warning on your system. Some systems default to using the u option; others do not. In any case, for a nifty trick that silences this warning on all systems, check out this commit in the libvirt project: *https://libvirt.org/git/?p=libvirt.git ;a=commitdiff;h=2db6a447*. The m4_divert_text command can be used to modify text as it passes through the M4 utility.

line. The build will then either be "silent" or normal based on the configured default and on whether the user specifies V=0 or V=1 on the make command line.

NOTE   *Neither* configure *option is required—the actual invocation of silent rules is ultimately controlled by the* V *variable in the generated makefile. The* configure *option
merely sets the default value of* V*.*

For smaller projects, I find Automake's silent rules to be less useful than
simply redirecting stdout to */dev/null* on the make command line, in this
manner:

```
$ make >/dev/null
ar: `u' modifier ignored since `D' is the default (see `U')
$
```

As this example shows, warnings and errors are still displayed on stderr,
usually with enough information for you to determine where the problem is
located (though not in this case). Warning-free builds are truly silent in this
case. You should use this technique to clean up compiler warnings in your
source code every so often. Silent rules can help because warnings stand
out in the build output.

Nonrecursive Automake

Now that we've changed our handwritten *Makefile.in* templates over to
Automake *Makefile.am* files, let's take a look at the process of converting this
recursive build system to a nonrecursive build system. In previous chapters,
we saw that using make's include directive can be helpful in dividing makefiles into areas of responsibility relegated to the subdirectories in which
they reside. With Automake, however, it's just simpler to put everything
in a top-level *Makefile.am* file because the content is so short that we can
easily comprehend the entire build system at a glance. If further division
of responsibility is required, a simple comment suffices.

The key here, as in our previous incarnations, is to reference the
content as if make were running from the top-level directory (which—
again—it is).

Listing 6-22 contains the entire contents of the top-level *Makefile.am*
file—the only makefile we'll use in this conversion.

Git tag 6.3
```
noinst_LIBRARIES = common/libjupcommon.a
common_libjupcommon_a_SOURCES = common/jupcommon.h common/print.c

bin_PROGRAMS = src/jupiter
src_jupiter_SOURCES = src/main.c
src_jupiter_CPPFLAGS = -I$(top_srcdir)/common
src_jupiter_LDADD = common/libjupcommon.a

check_SCRIPTS = src/greptest.sh
```

```
TESTS = $(check_SCRIPTS)

src/greptest.sh:
        echo './src/jupiter | grep "Hello from .*jupiter!"' > src/greptest.sh
        chmod +x src/greptest.sh

CLEANFILES = src/greptest.sh
```

Listing 6-22: Makefile.am: A nonrecursive Automake implementation for Jupiter

As you can see here, I've replaced the SUBDIRS variable in the top-level *Makefile.am* file with the full contents of the *Makefile.am* files in each of the directories referenced by this variable. I then added appropriate relative path information to each input object and product reference so that source files are accessed from the top-level directory, where they actually reside in their respective subdirectories, and so that products end up where they belong—with their source input files (or at least in their proper counterpart directories when not building in the source tree). I've highlighted the changes to each of the subdirectory *Makefile.am* files that I pasted into the top-level file.

Note that common_ or src_ was prepended to the product source variables because these prefixes are literally part of the product names now. Ultimately, these names are used to create make targets, which are defined as much by their location as their name. Usually, the location is the current directory, so the directory portions are silently omitted. For our nonrecursive builds, products are now generated into locations other than the current directory, so they must be stated explicitly. As with any other special characters in the product name, the directory-separating slashes become underscores in PSVs.

We also need to add an Automake option and remove the extra *Makefile* references from *configure.ac*, as shown in Listing 6-23.

```
#                                          -*- Autoconf -*-
# Process this file with autoconf to produce a configure script.

AC_PREREQ([2.69])
AC_INIT([Jupiter], [1.0], [jupiter-bugs@example.org])
AM_INIT_AUTOMAKE([subdir-objects])
AC_CONFIG_SRCDIR([src/main.c])
AC_CONFIG_HEADERS([config.h])
--snip--
# Checks for library functions.

AC_CONFIG_FILES([Makefile])
AC_OUTPUT

cat << EOF
--snip--
```

Listing 6-23: configure.ac: Removing extra makefile references for nonrecursive builds

The Automake `subdir-objects` option is necessary to tell Automake that you intend to access source files from directories other than those in which they reside. It's also needed to state that you want the objects and other intermediate products to be generated into the same directory as the source file (or in proper out-of-tree build counterpart directories). This option is not required just for nonrecursive builds but for any situation in which you may need to build one or more source files outside of their own directories. If you omit this option, the build will often still work, but you'll see two effects: warnings will be generated by `autoreconf` (or `automake`) indicating that you should probably use the option, and object files will be left lying in the wrong directories. The latter is only a problem if you happen to have more than one instance of a source file with the same name in different directories, in which case the second object file will overwrite the first, which will most probably result in a linker error when it's not able to find the symbols from the now-overwritten first object.

Finally, we can simply delete the *common* and *src* directories' *Makefile.am* files:

```
$ rm common/Makefile.am src/Makefile.am
$
```

Summary

In this chapter, we've discussed how to instrument a project for Automake using a project that had already been instrumented for Autoconf. (Newer projects are typically instrumented for both Autoconf and Automake at the same time.)

We covered the use of the SUBDIRS variable to tie *Makefile.am* files together, as well as the concepts surrounding product list, product source, and product option variables. Along with product list variables, I discussed Automake primaries—a concept at the very heart of Automake. Finally, I discussed the use of EXTRA_DIST to add additional files to distribution packages, the AM_MAINTAINER_MODE macro to ensure that users don't need to have the Autotools installed, converting to a nonrecursive Automake build system, and the use of Automake silent rules.

Through all of this, we replaced our handwritten *Makefile.in* templates with short, concise *Makefile.am* files that provide significantly more functionality. I hope this exercise has begun to open your eyes to the benefits of using Automake rather than handwritten makefiles.

In Chapters 7 and 8, we'll examine adding Libtool to the Jupiter project. In Chapter 9, we'll finish up our introduction to the Autotools proper by diving into the Autoconf's portable testing framework—autotest. Then, in Chapters 10 through 13, we'll take a short break from the Autotools to tackle some important sideline topics, but we'll return in Chapters 14 and 15, where we'll "Autotool-ize" a real-world project as we explore several other important aspects of Automake.

7

BUILDING LIBRARIES
WITH LIBTOOL

The years teach much which the days never know.
— *Ralph Waldo Emerson, "Experience"*

After too many bad experiences building shared libraries for multiple platforms without the help of GNU Libtool, I have come to two conclusions. First, the person who invented the concept of shared libraries should be given a raise . . . and a bonus. Second, the person who decided that shared library management interfaces and naming conventions should be left to the implementation should be flogged.

The very existence of Libtool stands as a witness to the truth of this sentiment. Libtool exists for only one reason—to provide a standardized, abstract interface for developers who want to create and access shared libraries in a portable manner. It abstracts both the shared-library build process and the programming interfaces used to dynamically load and access shared libraries at runtime.

The Libtool package concept was designed, and initial implementation was done, by Gordon Matzigkeit in March of 1996. Before this time, there was no standard, cross-platform mechanism for building shared libraries. Autoconf and Automake worked great for building portable projects across many platforms—as long as you didn't try to build a shared library. Once you started down this path, however, your code and build system would become littered with conditional constructs for shared-library management. This was a monumental effort because, as we shall see, building shared libraries is significantly different among some platforms.

Thomas Tanner began contributing in November of 1998 with his cross-platform abstraction for shared-library dynamic loading—*ltdl*. Other contributors since that time include Alexandre Oliva, Ossama Othman, Robert Boehne, Scott James Remnant, Peter O'Gorman, and Ralf Wildenhues. Currently, the Libtool package is maintained by Gary V. Vaughn (who has also been contributing to Libtool since 1998) and Bob Friesenhahn (whose excellent suggestions have been incorporated since 1998).

Before I get into a discussion of the proper use of Libtool, I'll spend a few paragraphs on the features and functionality that shared libraries provide so you understand the scope of the material I'm covering here.

The Benefits of Shared Libraries

Shared libraries provide a way to deploy reusable chunks of functionality in a convenient package. You can load shared libraries into a process address space either automatically at program load time, by using the operating system loader, or manually via code in the application itself. The point at which an application binds functionality from a shared library is very flexible, and the developer determines it based on the program's design and the end user's needs.

The interfaces between the program executable and the modules defined as shared libraries must be reasonably well designed because shared-library interfaces must be well specified. This rigorous specification promotes good design practices. When you use shared libraries, the system essentially forces you to be a better programmer.

Shared libraries may be (as the name implies) shared among processes. This sharing is very literal. The code segments for a shared library can be loaded once into physical memory pages. Those same memory pages can then be mapped into the process address spaces of multiple programs at once. The data pages must, of course, be unique for each process, but global data segments are often small compared to the code segments of a shared library. This is true efficiency.

It is easy to update shared libraries during program upgrades. Even if the base program doesn't change between two revisions of a software package, you can replace an old version of a shared library with a new one, as long as the new version's interfaces have not been changed. If interfaces *have* changed, two versions of the same shared library may reside together within the same directory, because the versioning schemes used by shared libraries (and supported by Libtool) on various platforms allow multiple versions of a library to be named differently in the filesystem but treated as the same library by the operating system loader. Older programs will continue to use older versions of the library, while newer programs are free to use the newer versions.

If a software package specifies a well-defined plug-in interface, then shared libraries can be used to implement user-configurable loadable functionality. This means that additional functionality can become available to a program after it has been released, and third-party developers can even add functionality to your program, if you publish a document describing your plug-in interface specification (or if they're smart enough to figure it out on their own).

There are a few widely known examples of these types of systems. Eclipse, for instance, is almost a pure plug-in framework. The base executable supports little more than a well-defined plug-in interface. Most of the functionality in an Eclipse application comes from library functions. Eclipse is written in Java and uses Java class libraries and *.jar* files, but the principle is the same, regardless of the language or platform.

How Shared Libraries Work

The specifics of how POSIX-compliant operating systems implement shared libraries vary from platform to platform, but the general idea is the same. Shared libraries provide chunks of executable code that the operating system can load into a program's address space and execute. The following discussion applies to shared-library references that the linker resolves when a program is built and the operating system loader resolves when the program is loaded.

While the object (*.o*) files produced by compilers do contain executable code, they cannot be executed by themselves from the command line. This is because they're incomplete, containing symbolic references or *links* to external entities (functions and global data items) that must be patched up. This patching is done by using a tool designed to manage such links to combine the complete set of object files containing such references.

Dynamic Linking at Load Time

As a program executable image is being built, the linker (formally called a *link editor*) maintains a table of symbols—function entry points and global data addresses. Each symbol referenced within the accumulating body of object code is added to this table as the linker finds it. As symbol definitions are located, the linker resolves symbol references in the table to their addresses in the code. At the end of the linking process, all object files (or simply *objects*) containing referenced symbol definitions are linked together and become part of the program executable image.

Objects found in static libraries (also called archives) that contain no referenced symbol definitions are discarded, but objects linked explicitly are added to the binary image even if they contain no referenced symbol definitions. If there are outstanding references in the symbol table after all the objects have been analyzed, the linker exits with an error message. On success, the final executable image may be loaded and executed by a user. The image is now entirely self-contained, depending on no external binary code.

Assuming that all undefined references are resolved during the linking process, if the list of objects to be linked contains one or more shared libraries, the linker will build the executable image from all *nonshared* objects specified on the linker command line. This includes all individual object files (*.o*) and all objects contained in static library archives (*.a*). However, the linker will add two tables to the binary image header. The first is the outstanding *external reference table*—a table of references to symbol definitions found only in shared libraries during the linking process. The second is the *shared-library table*, containing the list of shared-library names and versions in which the outstanding undefined references were found.

When the operating system loader attempts to load the program, it must resolve the remaining outstanding references in the external reference table to symbols imported from the shared libraries named in the shared-library table. If the loader can't resolve all of the references, then a load error occurs, and the process is terminated with an operating system error message. Note that these external symbols are not tied to a specific shared library. As long as they're found in any one of the searched libraries in the shared-library table, they're accepted.

NOTE *This process differs slightly from the way a Windows operating system loader resolves symbols in dynamic link libraries (DLLs). On Windows, the linker ties a particular symbol to a specifically named DLL at program build time.[1]*

Using free-floating external references has both pros and cons. On some operating systems, unbound symbols can be satisfied by a library specified by the user. That is, a user can entirely replace a library (or a portion of a library) at runtime by simply preloading one that contains the same symbols. On BSD and Linux-based systems, for example, a user can use the LD_PRELOAD environment variable to inject a shared library into a process address space. Since the loader loads these libraries before any other libraries, the loader will locate symbols in the preloaded libraries when it tries to resolve external references. The program author's intended libraries will not even be checked because the symbols provided by these libraries have already been resolved by the preloaded libraries.

---

1. Windows is not the only system to use hard references in this manner. Modern Windows operating systems are based on the *Common Object File Format (COFF)* system. COFF is also used by other operating systems, such as IBM's AIX. Many Unix (and all Linux) systems today are based on the *Executable and Linking Format (ELF)* system, which promotes the use of soft references. These don't need to be fully resolved until the program is executed.

In the following example, the Linux df utility is executed with an environment containing the LD_PRELOAD variable. This variable has been set to a path referring to a library that presumably contains a heap manager that's compatible with the C *malloc* interface. This technique can be used to debug memory problems in your programs. By preloading your own heap manager, you can capture memory allocations in a log file—in order to debug memory block overruns, for instance. This sort of technique is used by such widely known debugging aids as the *Valgrind* package.[2]

Here, the LD_PRELOAD environment variable is set on the same command line used to execute the df program. This shell code causes only the df child process environment to contain the LD_PRELOAD variable, set to the specified value:

```
$ LD_PRELOAD=$HOME/lib/libmymalloc.so /bin/df
--snip--
$
```

Unfortunately, free-floating symbols can also lead to problems. For instance, two libraries can provide the same symbol name, and the dynamic loader can inadvertently bind an executable to a symbol from the wrong library. At best, this will cause a program crash when the wrong arguments are passed to the mismatched function. At worst, it can present security risks because the mismatched function might be used to capture passwords and security credentials passed by the unsuspecting program.

C-language symbols do not include parameter information, so it's rather likely that symbols will clash in this manner. C++ symbols are a bit safer, in that the entire function signature (minus the return type) is encoded into the symbol name. However, even C++ is not immune to hackers who purposely replace security functions with their own versions of those functions (assuming, of course, that they have access to your run-time shared-library search path).

Automatic Dynamic Linking at Runtime

The operating system loader can also use a very late form of binding, often referred to as *lazy binding*. In this situation, the external reference table entries in the program header are initialized so that they refer to code within the dynamic loader itself.

When a program first calls a *lazy entry*, the call is routed to the loader, which will then (potentially) load the proper shared library, determine the actual address of the function, reset the entry point in the jump table, and, finally, redirect the processor to the shared-library function (which is now available). The next time this happens, the jump table entry will have already been correctly initialized, and the program will jump directly to the called function. This is very efficient both because the overhead for the jump after

---

2. For more information on the Valgrind tool suite, see the Valgrind Developers' website at *http://valgrind.org/*.

fix-up is no more than a normal indirect function call and because the cost of the initial load and link is amortized over many calls to the function during the lifetime of the process.[3]

This lazy binding mechanism makes program startup very fast because shared libraries whose symbols are not bound until they're needed aren't even loaded until the application program first references them. But, consider this—the program may *never* reference them. And that means they may never be loaded, saving both time and space. A good example of this sort of situation might be a word processor with a thesaurus feature implemented in a shared library. How often do you use your thesaurus? If the program is using automatic dynamic linking, chances are that the shared library containing the thesaurus code will never be loaded in most word-processing sessions.

As good as this system appears to be, there can be problems. While using automatic runtime dynamic linking can give you faster load times, better performance, and more efficient use of space, it can also cause your application to terminate abruptly and without warning. In the event that the loader can't find the requested symbol—perhaps the required library is missing—it has no recourse except to abort the process.

Why not ensure that all symbols exist when the program is loaded? Because if the loader resolved all symbols at load time, it might as well populate the jump table entries at that point, too. After all, it had to load all the libraries to ensure that the symbols actually exist, so this would entirely defeat the purpose of using lazy binding. Furthermore, even if the loader did check all external references when the program was first started, there's nothing to stop someone from deleting one or more of these libraries before the program uses them, while the program is still running.[4] Thus, even the pre-check is defeated.

The moral of this story is that there's no free lunch. If you don't want to pay the insurance premium for longer up-front load times and more space consumed (even if you may never really need it), then you may have to take the hit of a missing symbol at runtime, causing a program crash.

Manual Dynamic Linking at Runtime

One possible solution to the aforementioned problem is to take personal responsibility for some of the system loader's work. Then, when things don't go right, you have a little more control over the outcome. In the case of the thesaurus module, was it really necessary to terminate the program if the

---

3. The Spectre security flaw discovered in early 2018 caused the Linux community to make changes to the kernel that make such indirect jumps slightly more expensive. These initial fixes can be removed once processor manufacturers like Intel fix their microarchitecture designs.

4. Unix-like (POSIX) systems will retain deleted files for which outstanding file handles exist within running processes. From the filesystem user's perspective, the file appears to be gone, but the file remains intact until the last file handle is closed. Thus, this argument is not conclusive. As an aside, Windows operating systems simply disallow the delete operation on open files.

thesaurus library could not be loaded or didn't provide the correct symbols? Of course not—but the operating system loader can't know that. Only the software programmer can make such judgment calls.

When a program manages dynamic linking manually at runtime, the linker is left out of the equation entirely, and the program doesn't call any exported shared-library functions directly. Rather, shared-library functions are referenced through function pointers that the program itself populates at runtime.

Here's how it works: A program calls an operating system function (dlopen) to manually load a shared library into its own process address space. This function returns a *handle*, or an opaque value representing the loaded library. The program then calls another loader function (dlsym) to import a symbol from the library to which the handle refers. If all goes well, the operating system returns the address of the requested function or data item from the desired library. The program may then call the function, or access the global data item, through this pointer.

If something goes wrong in this process—the symbol isn't found within the library or the library isn't found—then it becomes the responsibility of the program to define the results, perhaps by displaying an error message indicating that the program was not configured correctly. In the preceding example of the word processor, a simple dialog indicating that the thesaurus is unavailable would be entirely sufficient.

This is a little nicer than the way automatic dynamic runtime linking works; while the loader has no option but to abort, the application has a higher-level perspective and can handle the problem much more gracefully. The drawback, of course, is that you as the programmer have to manage the process of loading libraries and importing symbols within your application code. However, this process is not very difficult, as I'll demonstrate in the next chapter.

Using Libtool

An entire book could be written about the details of shared libraries and how they're implemented on various systems. The short primer you just read should suffice for our immediate needs, so I'll now move on to how you can use Libtool to make a package maintainer's life a little easier.

The Libtool project was designed to extend Automake, but you can use it independently within hand-coded makefiles, as well. As of this writing, the latest version of Libtool, and the one I'm using in the examples here, is version 2.4.6.

Abstracting the Build Process

First, let's look at how Libtool helps during the build process. Libtool provides a script (*ltmain.sh*) that config.status consumes in a Libtool-enabled project. The config.status script converts configure test results and the ltmain.sh script into a custom version of the libtool script, specifically tailored to your project.[5] Your project's makefiles then use this libtool script to build the shared libraries listed in any Automake product list variables defined with the Libtool-specific LTLIBRARIES primary. The libtool script is really just a fancy wrapper around the compiler, linker, and other tools. You should ship the ltmain.sh script in a distribution archive, as part of your build system. Automake-generated rules ensure that this happens properly.

The libtool script insulates the author of the build system from the nuances of building shared libraries on different platforms. This script accepts a well-defined set of options, converting them to appropriate platform- and linker-specific options on the target platform and toolset. Thus, the maintainer doesn't need to worry about the specifics of building shared libraries on each platform—they only need to understand the available libtool script options. These options are well specified in the *GNU Libtool Manual*,[6] and I'll cover many of them in this chapter and the next.

On systems that don't support shared libraries at all, the libtool script uses appropriate commands and options to build and link only static archive libraries. Furthermore, the maintainer doesn't have to worry about the differences between building shared libraries and building static libraries when using Libtool. You can emulate building your package on a static-only system by using the --disable-shared option on the configure command line for your Libtool-enabled project. This option causes Libtool to assume that shared libraries cannot be built on the target system.

Abstraction at Runtime

You can also use Libtool to abstract the programming interfaces the operating system supplies for loading libraries and importing symbols. If you've ever dynamically loaded a library on a Linux system, you're familiar with the standard POSIX shared-library API, including the dlopen, dlsym, and dlclose functions. A system-level shared library, usually called simply *dl*, provides these functions. This translates to a binary image file named *libdl.so* (or something similar on systems that use different library-naming conventions).

Unfortunately, not all Unix systems that support shared libraries provide the *libdl.so* library or functions using these names. To address these differences, Libtool provides a shared library called *ltdl*, which exports a clean, portable, library-management interface that is very similar to the POSIX *dl* interface. The use of this library is optional, of course, but it is highly recommended because it provides more than just a common API across shared-library platforms—it also provides an abstraction for manual dynamic linking between shared-library and non-shared-library platforms.

---

5. Libtool also offers the option of generating the project-specific libtool script when configure is executed. This is done with the LT_OUTPUT macro within *configure.ac*. You may wish to do this if you find you have a need to execute libtool from within configure—for example, to test certain link-related features of your user's environment. In this case, you will need libtool to exist before you execute it for these checks.

6. See the Free Software Foundation's *GNU Libtool Manual*, version 2.4.6 (February 15, 2015) at *https://www.gnu.org/software/libtool/manual/*.

What?! How can that work?! On systems that don't support shared libraries, Libtool actually creates internal symbol tables within the executable that contain all the symbols you would otherwise find within shared libraries (on systems that support shared libraries). By using such symbol tables on these platforms, the lt_dlopen and lt_dlsym functions can make your code appear to be loading libraries and importing symbols, when in fact, the library load function does nothing more than return a handle to the appropriate internal symbol table, and the import function merely returns the address of code that has been statically linked into the program itself. On these systems, a project's shared-library code is linked directly into the programs that would normally load them at runtime.

Installing Libtool

If you want to make use of the latest version of Libtool while developing your packages, you may find that you either have to download, build, and install it manually or look for an updated *libtool* package from your distribution provider.

Downloading, building, and installing Libtool is really trivial, as you'll see here. However, you should check the GNU Libtool website[7] before executing these steps in order to ensure you're getting the most recent package. I've reproduced the basic steps here from Chapter 1:

```
$ wget https://ftp.gnu.org/gnu/libtool/libtool-2.4.6.tar.gz
--snip--
$ tar xzf libtool-2.4.6.tar.gz
$ cd libtool-2.4.6
$ ./configure && make
--snip--
$ sudo make install
--snip--
```

Be aware that the default installation location (as with most of the GNU packages) is */usr/local*. If you wish to install Libtool into the */usr* hierarchy, you'll need to use the --prefix=/usr option on the configure command line. The recommended practice is to install distribution-provided packages into the */usr* hierarchy and user-built packages into the */usr/local* tree, but if you're trying to get a hand-built version of Libtool to interoperate with distribution-provided versions of Autoconf and Automake, you may have to install Libtool into the */usr* hierarchy. The simplest way to avoid problems with package interdependencies is to install hand-built versions of all three packages into */usr/local* or, better still, into a directory within your home directory, which you can then add to your PATH.

---

7. See *https://www.gnu.org/software/libtool/*.

Adding Shared Libraries to Jupiter

Now that I've presented the requisite background information, let's take a look at how we might add a Libtool shared library to the Jupiter project. First, let's consider what functionality we could add to Jupiter using a shared library. Perhaps we want to provide our users with some library functionality that their own applications could use. Or we might have several applications in a package that need to share the same functionality. A shared library is a great tool for both of these scenarios because you get to reuse code and save memory—the cost of the memory used by shared code is amortized across multiple applications, both internal and external to the project.

Let's add a shared library to Jupiter that provides Jupiter's printing functionality. We can do this by having the new shared library call into the *libjupcommon.a* static library. Remember that calling a routine in a static library has the same effect as linking the object code for the called routine right into the calling program. The called routine ultimately becomes an integral part of the calling binary image (program or shared library).[8]

Additionally, we'll provide a public header file from the Jupiter project that will allow external applications to call this same functionality. This allows other applications to display stuff in the same quaint manner that the jupiter program does. (This would be significantly cooler if we were doing something useful in Jupiter, but you get the idea.)

Using the LTLIBRARIES Primary

Automake has built-in support for Libtool; it's the Automake package, rather than the Libtool package, that provides the LTLIBRARIES primary. Libtool doesn't really qualify as a pure Automake extension but rather is more of an add-on package for Automake, where Automake provides the necessary infrastructure for this specific add-on package. You can't access Automake's LTLIBRARIES primary functionality without Libtool because the use of this primary generates make rules that call the libtool script.

Libtool ships separately, rather than as part of Automake, because you can use Libtool quite effectively independently of Automake. If you want to try Libtool by itself, I'll refer you to the *GNU Libtool Manual*; the opening chapters describe the use of the libtool script as a stand-alone product. It's as simple as modifying your makefile commands so that the compiler, linker, and librarian are called through the libtool script, and then modifying some of your command line parameters as required by Libtool.

---

8. Many of you more experienced Autotools (or simply Unix) programmers may be cringing at my engineering choices here. For instance, linking a Libtool library against a traditional static archive is inappropriate for several reasons, which will become clear as we continue. During the process, we'll see that there is a significant difference between a traditional static archive and a Libtool convenience library (on some platforms). Please remember that Jupiter is a learning experience and a work in progress. I promise we'll work out the kinks by the end of the chapter.

Public Include Directories

A project subdirectory named *include* should only contain public header files—those that expose a public interface in your project. We're now going to add just such a header file to the Jupiter project, so we'll create a directory called *include* in the project root directory.

If we had multiple shared libraries, we'd have a choice to make: do we create separate *include* directories, one in each library source directory, or do we add a single, top-level *include* directory? I usually use the following rule of thumb to make my decision: If the libraries are designed to work together as a group, and if consuming applications generally use the libraries together, then I use a single, top-level *include* directory. If, on the other hand, the libraries can be effectively used independently, and if they offer fairly autonomous sets of functionality, then I provide individual *include* directories in the libraries' own directories.

In the end, it doesn't really matter much because the header files for these libraries will be installed in directory structures that are entirely different from the ones where they exist within your project. In fact, you should make sure you don't inadvertently use the same filename for public headers in two different libraries in your project—if you do, you'll have problems installing these files. They generally end up all together in the $(prefix)/*include* directory, although you can override this default by using either the includedir variable or the pkginclude prefix in your *Makefile.am* files.

The includedir variable allows you to specify where you want your header files to be installed by defining the exact value of Automake's $(includedir) variable, the usual value of which is $(prefix)/*include*. The use of the pkginclude prefix indicates to Automake that you want your header files to be in a private, package-specific directory, beneath the directory indicated by $(includedir), called $(includedir)/$(PACKAGE).

We'll also add another root-level directory (*libjup*) for Jupiter's new shared library, *libjupiter*. These changes require you to add references to the new directories to the top-level *Makefile.am* file's SUBDIRS variable and then add corresponding *Makefile* references to the AC_CONFIG_FILES macro in *configure.ac*. Since we're going to make major changes to our project, we'd better clean up the work area before we start. Then we'll create the directories and add a new *Makefile.am* file to the new *include* directory:

Git tag 7.0

```
$ make maintainer-clean
--snip--
$ mkdir libjup include
❶ $ echo "include_HEADERS = libjupiter.h" > include/Makefile.am
$
```

The *include* directory's *Makefile.am* file is trivial—it contains only a single line, in which an Automake HEADERS primary refers to the public header file *libjupiter.h*. Note at ❶ that we're using the include prefix on this primary. You'll recall that this prefix indicates that files specified in this primary are destined to be installed in the $(includedir) directory (for example, */usr/(local/)include*). The HEADERS primary is similar to the DATA primary in that it specifies a set of files that are to be treated simply as data to be installed without modification or preprocessing. The only really tangible difference is that the HEADERS primary restricts the possible installation locations to those that make sense for header files.

The *libjup/Makefile.am* file is a bit more complex, containing four lines as opposed to just one. This file is shown in Listing 7-1.

```
❶ lib_LTLIBRARIES = libjupiter.la
❷ libjupiter_la_SOURCES = jup_print.c
❸ libjupiter_la_CPPFLAGS = -I$(top_srcdir)/include -I$(top_srcdir)/common
❹ libjupiter_la_LIBADD = ../common/libjupcommon.a
```

Listing 7-1: libjup/Makefile.am: *The initial version of this file*

Let's analyze this file, line by line. The line at ❶ is the primary specification, and it contains the usual prefix for libraries: lib. The products this prefix references will be installed in the $(libdir) directory. (We could have also used the pkglib prefix to indicate that we wanted our libraries installed into $(libdir)*/jupiter.*) Here, we're using the LTLIBRARIES primary rather than the original LIBRARIES primary. The use of LTLIBRARIES tells Automake to generate rules that use the libtool script rather than calling the compiler (and possibly the librarian) directly to generate the products.

The line at ❷ lists the sources that are to be used for the first (and only) product.

The line at ❸ indicates a set of C-preprocessor flags that are to be used on the compiler command line for locating the associated shared-library header files. These options indicate that the preprocessor should search the top-level *include* and *common* directories for header files referenced in the source code.

The last line (at ❹) indicates a set of linker options for this product. In this case, we're specifying that the *libjupcommon.a* static library should be linked into (that is, become part of) the *libjupiter.so* shared library.

NOTE   *The more experienced Autotools library developer will notice a subtle flaw in this* Makefile.am *file. Here's a hint: it's related to linking Libtool libraries against non-Libtool libraries. This concept presents a major stumbling block for many newcomers, so I've written the initial version of this file to illustrate the error. Not to worry, however—we'll correct the flaw later in this chapter as we work through this issue in a logical fashion.*

There is an important concept regarding the *_LIBADD variables that you should strive to understand completely: Libraries that are consumed within, and yet built as part of, the same project should be referenced internally using relative paths, via either parent directory references or the $(top_builddir) variable, within the *build* directory hierarchy. Libraries that are external to a project generally don't need to be referenced explicitly at all, because the project's configure script should already have added appropriate -L and -l options for those libraries into the $(LIBS) environment variable when it processed the code generated by the AC_CHECK_LIB or AC_SEARCH_LIBS macro.

Next, we'll hook these new directories into the project's build system. To do so, we need to modify the top-level *Makefile.am* and *configure.ac* files. These changes are shown in Listings 7-2 and 7-3, respectively.

```
SUBDIRS = common include libjup src
```

Listing 7-2: Makefile.am: *Adding* include *and* libjup *to the* SUBDIRS *variable*

```
#                                              -*- Autoconf -*-
# Process this file with autoconf to produce a configure script.

AC_PREREQ([2.69])
AC_INIT([Jupiter],[1.0],[jupiter-bugs@example.org])
AM_INIT_AUTOMAKE
❶ LT_PREREQ([2.4.6])
LT_INIT([dlopen])
AC_CONFIG_SRCDIR([src/main.c])
AC_CONFIG_HEADERS([config.h])

# Checks for programs.
AC_PROG_CC
❷ AC_PROG_INSTALL

# Checks for header files.
AC_CHECK_HEADERS([stdlib.h])
--snip--
AC_CONFIG_FILES([Makefile
                common/Makefile
❸ include/Makefile
                libjup/Makefile
                src/Makefile])
AC_OUTPUT
--snip--
```

Listing 7-3: configure.ac: *Adding the* include *and* libjup *directory makefiles*

Three unrelated changes were required in *configure.ac*. The first is the addition at ❶ of the Libtool setup macros LT_PREREQ and LT_INIT. The LT_PREREQ macro works just like Autoconf's AC_PREREQ macro (used a few lines higher). It indicates the earliest version of Libtool that can correctly process this project. You should choose the lowest reasonable values for the arguments in these macros because higher values needlessly restrict you and your co-maintainers to more recent versions of the Autotools.[9] The LT_INIT macro initializes the Libtool system for this project.

---

9. I don't mean to state that you should only use older functionality provided by the Autotools in order to cater to your users who don't want to upgrade. Remember that those who care what versions of the Autotools you're using are the developers working on your project. This is a significantly smaller audience than the users who will be building from your distribution archives. Choose version numbers that reflect the oldest versions of the Autotools that support the functionality you use in your *configure.ac* file. If you use the latest features, set the version numbers accordingly and don't lose any sleep over it.

The second change is just as interesting. I removed the `AC_PROG_RANLIB` macro invocation after the line at ❷. (And after all we went through to put it there in the first place!) Because Libtool is now building all of the project libraries, and because it understands all aspects of the library build process, we no longer need to instruct Autoconf to make sure `ranlib` is available. In fact, if you leave this macro in, you'll get a warning when you execute `autoreconf -i`.

The last change is found at ❸ in the argument to `AC_CONFIG_FILES`, where we've added references to the two new *Makefile.am* files we added to the *include* and *libjup* directories.

Customizing Libtool with LT_INIT Options

You can specify default values for enabling or disabling static and shared libraries in the argument list passed into `LT_INIT`. The `LT_INIT` macro accepts a single, optional argument: a whitespace-separated list of keywords. The following are the most important keywords allowed in this list, along with an explanation of their proper use.

dlopen

This option enables checking for `dlopen` support. The *GNU Libtool Manual* states that this option should be used if the package makes use of the -dlopen and -dlpreopen flags in `libtool`; otherwise `libtool` will assume that the system does not support *dl-opening*. There's only one reason for using the -dlopen or -dlpreopen flag: you intend to dynamically load and import shared-library functionality at runtime within your project's source code. Additionally, these two options do very little unless you intend to use the *ltdl* library (rather than directly using the *dl* library) to manage your runtime dynamic linking. Thus, you should use this option only if you intend to use the *ltdl* library.

win32-dll

Use this option if your library is properly ported to a Windows DLL using __declspec(dllimport) and __declspec(dllexport). If your library properly uses these keywords to import and export symbols for Windows DLLs, and you don't use this option, then Libtool will only build static libraries on Windows. We'll cover this topic in more detail in Chapter 17.

aix-soname=aix|svr4|both

Adds the flags --with-aix-soname to `configure`'s command line. Prior to version 2.4.4, Libtool always behaved as if `aix-soname` were set to `aix`. If you build shared libraries on AIX often, you'll understand the meaning of this option. If you wish to learn more, read Section 5.4.1 of the *GNU Libtool Manual*.

disable-fast-install

This option changes the default behavior for `LT_INIT` to disable optimization for fast installation on systems where it matters. The concept of fast installation exists because uninstalled programs and libraries may need to be executed from within the build tree (during make check, for example). On some systems, installation location affects the final linked binary image, so Libtool must either relink programs and libraries on these systems when make install is executed or else relink programs and libraries for make check. Libtool chooses to relink for make check by default, allowing the original binaries to be installed quickly without relinking during make install. The user can override this default, depending on platform support, by specifying --enable-fast-install to configure.

shared and disable-shared

These two options change the default behavior for creating shared libraries. The effects of the shared option are default behavior on all systems where Libtool knows how to create shared libraries. The user may override the default shared library-generation behavior by specifying either --disable-shared or --enable-shared on the configure command line.

static and disable-static

These two options change the default behavior for creating static libraries. The effects of the static option are default behavior on all systems where shared libraries have been disabled and on most systems where shared libraries have been enabled. If shared libraries are enabled, the user may override this default by specifying --disable-static on the configure command line. Libtool will always generate static libraries on systems without shared libraries. Hence, you can't (effectively) use the disable-shared and disable-static arguments to LT_INIT or the --disable-shared and --disable-static command line options for configure at the same time. (Note, however, that you may use the shared and static LT_INIT options or the --enable-shared and --enable-static command line options together.)

pic-only and no-pic

These two options change the default behavior for creating and using PIC object code. The user may override the defaults set by these options by specifying --without-pic or --with-pic on the configure command line. I'll discuss the meaning of PIC object code in "So What Is PIC, Anyway?" on page 200.

Now that we've finished setting up the build system for the new library, we can move on to discussing the source code. Listing 7-4 shows the contents of the new *jup_print.c* source file that's referenced in the second line of *libjup/Makefile.am*. Listing 7-5 shows the contents of the new *include/libjupiter.h* library header file.

```
#include "config.h"

#include "libjupiter.h"
#include "jupcommon.h"

int jupiter_print(const char * name)
```

```
{
    return print_routine(name);
}
```

Listing 7-4: libjup/jup_print.c: The initial contents of the shared-library source file

```
#ifndef LIBJUPITER_H_INCLUDED
#define LIBJUPITER_H_INCLUDED

int jupiter_print(const char * name);

#endif /* LIBJUPITER_H_INCLUDED */
```

Listing 7-5: include/libjupiter.h: The initial contents of the shared-library public header file

This leads us to another general software-engineering principle. I've heard it called by many names, but the one I tend to use the most is the *DRY principle*—the acronym stands for *don't repeat yourself*. C function prototypes are very useful because, when used correctly, they enforce the fact that the public's view of a function is identical to the package maintainer's view. All too often I've seen source files that don't include their corresponding header files. It's easy to make a small change in a function or prototype and then not duplicate it in the other location—unless you've included the public header file within the source file. When you do this consistently, the compiler catches any inconsistencies for you.

We also need to include the static library header file (*jupcommon.h*) because we call its function (*print_routine*) from within the public library function. You may have also noticed that I placed *config.h* first, immediately followed by the public header file—there's a good reason for this. I've already stated in Chapter 6 that *config.h* should always come first in every source file. Normally, I'd say the public header file should come first, but public header files should be written so that their functionality is never modified by *config.h*, so, technically, it should not matter if the public header file comes before or after *config.h*. For example, using a compiler-mode dependent type like off_t in a public header file will cause the application binary interface (ABI) to change not only from one platform to another (not necessarily a bad thing) but also on the same platform from one use to another, based on the compilation environment set up by consumer code (not a good thing). The fact is, you should write your public header files in such a way that it doesn't really matter whether you include them before or after *config.h*; they should be purposely designed so they do not depend on anything that *can* be configured by *config.h*. For a more complete treatise on this topic, see "Item 1: Keeping Private Details out of Public Interfaces" on page 499.

By placing the public header file first in the source file (after *config.h*), we ensure that the use of this header file doesn't depend on definitions in any internal header files in the project. For instance, let's say the public header file has a hidden dependency on some construct (such as a type definition, structure, or preprocessor definition) defined in an internal header like *jupcommon.h*. If we include the public header file after *jupcommon.h*, the dependency would be hidden when the compiler begins to process the public header file, because the required construct is already available in the *translation unit* (the source file combined with all of the included header files).

I'd like to make one final point about the contents of Listing 7-5. The preprocessor conditional construct is commonly called an *include guard*. It is a mechanism for preventing your header files from inadvertently being included multiple times within the same translation unit. I use include guards routinely in all my header files, and it's good practice to do so. A good optimizing compiler (gcc, for instance—specifically, its preprocessor) will recognize include guards in header files and skip the file entirely on subsequent inclusions within the same translation unit.[10]

Since a public header file will be consumed by foreign source code, it's even more critical that you use include guards religiously in these header files. While you can control your own code base, you have no say over the code that one of your library consumers writes. What I'm advocating here is that you assume you're the best programmer you know, and everyone else is a little below your skill level. You can do this nicely by not mentioning it to anyone, but you should *act* like it's a fact when you write your public header files.

Next, we'll modify the Jupiter application's `main` function so that it calls into the shared library instead of the common static library. These changes are shown in Listing 7-6.

```
#include "config.h"

#include "libjupiter.h"

int main(int argc, char * argv[])
{
    return jupiter_print(argv[0]);
}
```

Listing 7-6: src/main.c: Changing main to call the shared-library function

Here, we've changed the print function from `print_routine`, found in the static library, to `jupiter_print`, as provided by the new shared library. We've also changed the header file included at the top from *libjupcommon.h* to *libjupiter.h*.

My choices of names for the public function and header file were arbitrary but based on a desire to provide a clean, rational, and informational public interface. The name *libjupiter.h* very clearly indicates that this header file specifies the public interface for *libjupiter.so*. I try to name library interface functions to make it clear that they are part of an interface. How you choose to name your public interface members—files, functions, structures, type definitions, preprocessor definitions, global data, and so on—is up to

---

10. Do not make the mistake of prefixing your include guards with one or two underscores. The problem is that symbols preceded by underscores are reserved by compiler vendors, the standard library, and the C and C++ standards bodies for internal constructs and future enhancements.

you, but you should consider using a similar philosophy. Remember, the goal is to provide a great end-user experience. Intuitive naming should be a significant part of your strategy. For example, it is a good general practice to choose a common prefix for your program and library symbols.[11]

Finally, we must also modify the *src/Makefile.am* file to use our new shared library rather than the *libjupcommon.a* static library. These changes are shown in Listing 7-7.

```
bin_PROGRAMS = jupiter
jupiter_SOURCES = main.c
❶ jupiter_CPPFLAGS = -I$(top_srcdir)/include
❷ jupiter_LDADD = ../libjup/libjupiter.la
--snip--
```

Listing 7-7: src/Makefile.am: Adding shared-library references to the src directory makefile

Here, we've changed the `jupiter_CPPFLAGS` statement at ❶ so that it refers to the new top-level *include* directory rather than the *common* directory. We've also changed the `jupiter_LDADD` statement at ❷ so that it refers to the new Libtool shared-library object rather than the *libjupcommon.a* static library. All else remains the same. The syntax for referring to a Libtool library is identical to that for referring to an older, static library—only the library extension is different. The Libtool library extension *.la* stands for *libtool archive*.

Let's take a step back for a moment. Do we actually need to make this change? No, of course not. The `jupiter` application will continue to work just fine the way we originally wrote it. Linking the code for the static library's `print_routine` directly into the application works just as well as calling the new shared-library routine (which ultimately contains the same code, anyway). In fact, there is slightly more overhead in calling a shared-library routine because of the extra level of indirection when calling through a shared-library jump table.

In a real project, you might actually leave it the way it was. Because both public entry points, `main` and `jupiter_print`, call exactly the same function (`print_routine`) in *libjupcommon.a*, their functionality is identical. Why add even the slight overhead of a call through the public interface? Well, one reason is that you can take advantage of shared code. By using the shared-library function, you're not duplicating code—either on disk or in memory. This is the DRY principle at work.

Another reason is to exercise the interface you're providing for users of your shared library. You'll catch bugs in your public interfaces more quickly if your project code uses your shared libraries exactly the way you expect other programs to use them.

In this situation, you might now consider simply moving the code from the static library into the shared library, thereby removing the need for the static library entirely. However, I'm going to beg your indulgence with my contrived example. In a more complex project, I might very well

---

11. This is especially relevant on ELF systems, where it can be difficult to determine which of your library symbols might conflict with symbols from other libraries.

have a need for this sort of configuration. Common code is often gathered together into static convenience libraries, and more often than not, only a portion of this common code is reused in shared libraries. I'm going to leave it the way it is here for the sake of its educational value.

Reconfigure and Build

Since we've added Libtool—a major new component—to our project build system, we'll add the -i option to the autoreconf command line to ensure that all of the proper auxiliary files are installed into the project root directory:

```
$ autoreconf -i
❶ libtoolize: putting auxiliary files in '.'.
  libtoolize: copying file './ltmain.sh'
  libtoolize: Consider adding 'AC_CONFIG_MACRO_DIRS([m4])' to configure.ac,
  libtoolize: and rerunning libtoolize and aclocal.
  libtoolize: Consider adding '-I m4' to ACLOCAL_AMFLAGS in Makefile.am.
  configure.ac:8: installing './compile'
❷ configure.ac:8: installing './config.guess'
  configure.ac:8: installing './config.sub'
  configure.ac:6: installing './install-sh'
  configure.ac:6: installing './missing'
  Makefile.am: installing './INSTALL'
  Makefile.am: installing './COPYING' using GNU General Public License v3 file
  Makefile.am:    Consider adding the COPYING file to the version control system
  Makefile.am:    for your code, to avoid questions about which license your
  project uses
  common/Makefile.am: installing './depcomp'
  parallel-tests: installing './test-driver'
  $
```

Because we completely removed all generated and copied files from our project directory, most of these notifications have to do with replacing files we've already discussed. However, there are a few noteworthy exceptions.

First, notice the comments from libtoolize at ❶. Most of them are simply suggesting that we move to the new Autotools convention of adding M4 macro files to a directory called *m4* in the project root directory. We're going to ignore these comments for now, but in Chapters 14 and 15, we'll actually do this for a real project.

As you can see at ❷, it appears that the addition of Libtool has caused a few new files to be added to our project—namely, the *config.guess* and *config.sub* files. Another new file was added in the section at ❶ called *ltmain.sh*. The configure script uses *ltmain.sh* to build a project-specific version of libtool for the Jupiter project. I'll describe the config.guess and config.sub scripts later.

Let's go ahead and execute configure and see what happens:

```
$ ./configure
--snip--
checking for ld used by gcc... /usr/bin/ld
checking if the linker (/usr/bin/ld) is GNU ld... yes
checking for BSD- or MS-compatible name lister (nm)... /usr/bin/nm -B
checking the name lister (/usr/bin/nm -B) interface... BSD nm
checking whether ln -s works... yes
checking the maximum length of command line arguments... 1572864
--snip--
checking for shl_load... no
checking for shl_load in -ldld... no
checking for dlopen... no
checking for dlopen in -ldl... yes
checking whether a program can dlopen itself... yes
checking whether a statically linked program can dlopen itself... no
checking whether stripping libraries is possible... yes
checking if libtool supports shared libraries... yes
checking whether to build shared libraries... yes
checking whether to build static libraries... yes
--snip--
configure: creating ./config.status
--snip--
$
```

The first thing to note is that Libtool adds *significant* overhead to the configuration process. I've only shown a few of the output lines here that are new since we added Libtool. All we've added to the *configure.ac* file is the reference to the LT_INIT macro, but we've nearly doubled our configure output. This should give you some idea of the number of system characteristics that must be examined to create portable shared libraries. Fortunately, Libtool does a lot of the work for you.

Now, let's run the make command and see what sort of output we get:

```
$ make
--snip--
Making all in libjup
make[2]: Entering directory '/.../jupiter/libjup'
❶ /bin/bash ../libtool --tag=CC   --mode=compile gcc -DHAVE_CONFIG_H -I. -I..
   -I../include -I../common   -g -O2 -MT libjupiter_la-jup_print.lo -MD -MP -MF
   .deps/libjupiter_la-jup_print.Tpo -c -o libjupiter_la-jup_print.lo `test -f
   'jup_print.c' || echo './'`jup_print.c
❷ libtool: compile:  gcc -DHAVE_CONFIG_H -I. -I.. -I../include -I../common -g
   -O2 -MT libjupiter_la-jup_print.lo -MD -MP -MF .deps/libjupiter_la-jup_print.
   Tpo -c jup_print.c  -fPIC -DPIC -o .libs/libjupiter_la-jup_print.o
❸ libtool: compile:  gcc -DHAVE_CONFIG_H -I. -I.. -I../include -I../common -g
   -O2 -MT libjupiter_la-jup_print.lo -MD -MP -MF .deps/libjupiter_la-jup_print.
   Tpo -c jup_print.c -o libjupiter_la-jup_print.o >/dev/null 2>&1
❹ mv -f .deps/libjupiter_la-jup_print.Tpo .deps/libjupiter_la-jup_print.Plo
❺ /bin/bash ../libtool --tag=CC   --mode=link gcc  -g -O2   -o libjupiter.la
   -rpath /usr/local/lib libjupiter_la-jup_print.lo ../common/libjupcommon.a
   -lpthread
```

```
❻ *** Warning: Linking the shared library libjupiter.la against the
   *** static library ../common/libjupcommon.a is not portable!
   libtool: link: gcc -shared  -fPIC -DPIC  .libs/libjupiter_la-jup_print.o  ../
   common/libjupcommon.a -lpthread  -g -O2   -Wl,-soname -Wl,libjupiter.so.0 -o
   .libs/libjupiter.so.0.0.0
❼ /usr/bin/ld: ../common/libjupcommon.a(print.o): relocation R_X86_64_32 against
   `.rodata.str1.1' can not be used when making a shared object; recompile with
   -fPIC
   ../common/libjupcommon.a: error adding symbols: Bad value
   collect2: error: ld returned 1 exit status
   Makefile:389: recipe for target 'libjupiter.la' failed
   make[2]: *** [libjupiter.la] Error 1
   make[2]: Leaving directory '/.../jupiter/libjup'
   Makefile:391: recipe for target 'all-recursive' failed
   make[1]: *** [all-recursive] Error 1
   make[1]: Leaving directory '/.../jupiter'
   Makefile:323: recipe for target 'all' failed
   make: *** [all] Error 2
   --snip--
   $
```

We seem to have some errors to fix. The first point of interest is that
libtool is being executed at ❶ with a --mode=compile option, which causes
libtool to act as a wrapper script around a somewhat modified version of
a standard gcc command line. You can see the effects of this statement in
the next two compiler command lines at ❷ and ❸. *Two compiler commands?*
That's right. It appears that libtool is running the compiler twice against
our source file.

A careful comparison of these two command lines shows that the first
command is using two additional flags, -fPIC and -DPIC. The first line also
appears to be directing the output file to a *.libs* subdirectory, whereas the
second line is saving it in the current directory. Finally, both the stdout and
stderr output streams are redirected to */dev/null* in the second line.

NOTE *Occasionally, you may run into a situation where a source file compiles fine in the
first compilation but fails in the second due to a PIC-related source code error. These
sorts of problems are rare, but they can be a real pain when they occur because make
halts the build with an error but doesn't give you any error messages to explain the
problem! When you see this situation, simply pass the -no-suppress flag in the CFLAGS
variable on the make command line in order to tell Libtool not to redirect output from
the second compilation to /dev/null.*

This double-compile feature has caused a fair amount of anxiety
on the Libtool mailing list over the years. Mostly, this is due to a lack of
understanding of what Libtool is trying to do and why it's necessary. Using
Libtool's various configure script command line options, you can force a
single compilation, but doing so brings a certain loss of functionality, which
I'll explain here shortly.

The line at ❹ renames the dependency file from **.Tpo* to **.Plo*. You might
recall from Chapters 3 and 6 that dependency files contain make rules that

declare dependencies between source files and referenced header files. The C preprocessor generates these rules when you use the -MT compiler option. However, the overarching concept to understand here is that one Libtool command may (and often does) execute a group of shell commands.

The line at ❺ is another call to the libtool script, this time using the --mode=link option. This option generates a call to execute the compiler in *link mode*, passing all of the libraries and linker options specified in the *Makefile.am* file.

At ❻, we come to the first problem—a portability warning about linking a shared library against a static library. Specifically, this warning is about linking a Libtool shared library against a non-Libtool static library. Notice that this is not an error. Were it not for additional errors we'll encounter later, the library would be built in spite of this warning.

After the portability warning, libtool attempts to link the requested objects together into a shared library named *libjupiter.so.0.0.0* . But here the script runs into the real problem: at ❼, a linker error indicates that somewhere from within *libjupcommon.a*—and more specifically, within *print.o*—an x86_64 object relocation cannot be performed because the original source file (*print.c*) was apparently not compiled correctly. The linker is kind enough to tell us exactly what we need to do to fix the problem (highlighted in the example): we need to compile the source code using a -fPIC compiler option.

Now, if you were to encounter this error and didn't know anything about the -fPIC option, you'd be wise to open the man page for gcc and study it before inserting compiler and linker options willy-nilly until the warning or error disappears (unfortunately, a common practice of inexperienced programmers). Software engineers should understand the meaning and nuances of every command line option used by the tools in their build systems. Otherwise, they don't really know what they have when their build completes. It may work the way it should, but if it does, it's by luck rather than by design. Good engineers know their tools, and the best way to learn is to study error messages and their fixes until the problem is well understood, before moving on.

So What Is PIC, Anyway?

When operating systems create new process address spaces, they typically load program-executable images at the same memory address. This magic address is system specific. Compilers and linkers understand this, and they know what the magic address is on any given system. Therefore, when they generate internal references to function calls or global data, they can generate those references as *absolute* addresses. If you were somehow able to load the executable at a different location in the process virtual address space, it would simply not work properly because the absolute addresses within the code would not be correct. At the very least, the program would crash when the processor jumped to the wrong location during a function call.

Consider Figure 7-1 for a moment. Assume we have a system whose magic executable load address is 0x10000000; this diagram depicts two process address spaces within that system. In the process on the left, an executable image is loaded correctly at address 0x10000000. At some point in the code, a jmp instruction tells the processor to transfer control to the absolute address 0x10001000, where it continues executing instructions in another area of the program.

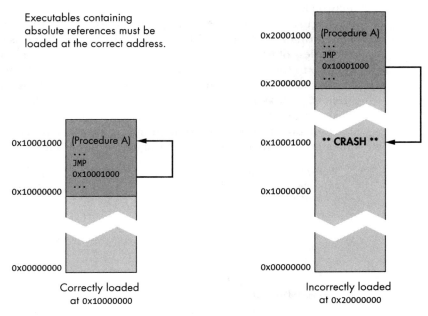

Figure 7-1: Absolute addressing in executable images

In the process on the right, the program is loaded incorrectly at address 0x20000000. When that same branch instruction is encountered, the processor jumps to address 0x10001000 because that address is hardcoded into the program image. This, of course, fails—often spectacularly by crashing, but sometimes with more subtle and dastardly ramifications.

That's how things work for program images. However, when a *shared library* is built for certain types of hardware (AMD64 included), neither the compiler nor the linker knows beforehand where the library will be loaded. This is because many libraries may be loaded into a process and the order in which they are loaded depends on how the *executable* is built, not the library. Furthermore, who's to say which library owns location A and which one owns location B? The fact is, a library may be loaded *anywhere* into the process address space where there is space for it at the time it's loaded. Only the operating system loader knows where it will finally reside—and even then, it only knows just before the library is actually loaded.[12]

As a result, shared libraries can only be built from a special class of object files called PIC objects. *PIC* is an acronym that stands for

---

12. In fact, there's a trend toward operating systems intentionally randomizing the load address for added security. It's harder to exploit a weakness when you're not sure where the weakness will be loaded into the process address space.

position-independent code, and it implies that references within the object code are not absolute but *relative*. When you use the -fPIC option on the compiler command line, the compiler will use somewhat less efficient relative addressing in branching instructions. Such position-independent code may be loaded anywhere.

Figure 7-2 depicts the concept of relative addressing as used when generating PIC objects. With relative addressing, addresses work correctly regardless of where the image is loaded because they're always encoded relative to the current instruction pointer. In Figure 7-2, the diagrams indicate a shared library loaded at the same addresses as those in Figure 7-1 (that is, 0x10000000 and 0x20000000). In both cases, the dollar sign used in the jmp instruction represents the current instruction pointer (IP), so $ + 0xC74 tells the processor that it should jump to the instruction starting 0xC74 bytes ahead of the current position of the instruction pointer.

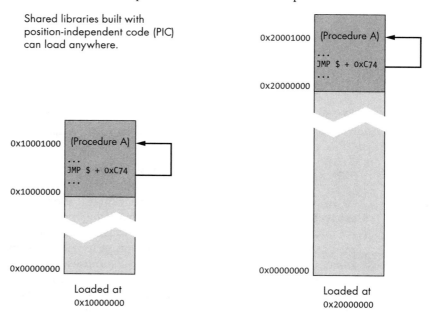

Figure 7-2: Relative addressing in shared-library images

There are various nuances to generating and using position-independent code, and you should become familiar with all of them before using them so you can choose the option that is most appropriate for your situation. For example, the GNU C compiler also supports a -fpic option (lowercase), which uses a slightly quicker but more limited mechanism to generate relocatable object code.[13]

---

13. Wikipedia has a very informative page on position-independent code, although I find its treatment of Windows DLLs to be somewhat outdated. See *https://en.wikipedia.org/wiki/Position-independent_code*.

Fixing the Jupiter PIC Problem

From what we now understand, one way to fix our linker error is to add the -fPIC option to the compiler command line for the source files that comprise the *libjupcommon.a* static library. Listing 7-8 illustrates the changes required to the *common/Makefile.am* file.

```
noinst_LIBRARIES = libjupcommon.a
libjupcommon_a_SOURCES = jupcommon.h print.c
libjupcommon_a_CFLAGS = -fPIC
```

Listing 7-8: common/Makefile.am: *Changes required for generation of PIC objects in a static library*

And now let's retry the build:

```
$ autoreconf
$ ./configure
--snip--
$ make
make  all-recursive
make[1]: Entering directory '/.../jupiter'
Making all in common
make[2]: Entering directory '/.../jupiter/common'
gcc -DHAVE_CONFIG_H -I. -I..    -fPIC -g -O2 -MT libjupcommon_a-print.o -MD
-MP -MF .deps/libjupcommon_a-print.Tpo -c -o libjupcommon_a-print.o `test -f
'print.c' || echo './'`print.c
--snip--
Making all in libjup
make[2]: Entering directory '/.../jupiter/libjup'
/bin/bash ../libtool  --tag=CC   --mode=link gcc -g -O2   -o libjupiter.
la -rpath /usr/local/lib libjupiter_la-jup_print.lo ../common/libjupcommon.a
-lpthread
```

❶ `*** Warning: Linking the shared library libjupiter.la against the`
```
*** static library ../common/libjupcommon.a is not portable!
libtool: link: gcc -shared  -fPIC -DPIC  .libs/libjupiter_la-jup_print.o
../common/libjupcommon.a -lpthread  -g -O2   -Wl,-soname -Wl,libjupiter.so.0
-o .libs/libjupiter.so.0.0.0
libtool: link: (cd ".libs" && rm -f "libjupiter.so.0" && ln -s "libjupiter
.so.0.0.0" "libjupiter.so.0")
libtool: link: (cd ".libs" && rm -f "libjupiter.so" && ln -s "libjupiter
.so.0.0.0" "libjupiter.so")
libtool: link: ar cru .libs/libjupiter.a ../common/libjupcommon.a  libjupiter
_la-jup_print.o
ar: `u' modifier ignored since `D' is the default (see `U')
libtool: link: ranlib .libs/libjupiter.a
libtool: link: ( cd ".libs" && rm -f "libjupiter.la" && ln -s "../libjupiter
.la" "libjupiter.la" )
make[2]: Leaving directory
--snip--
$
```

We now have a shared library built properly with position-independent code, as per system requirements. However, we still have that strange warning at ❶ about the portability of linking a Libtool library against a static library. The problem here is not in *what* we're doing but rather *how* we're doing it. You see, the concept of PIC does not apply to all hardware architectures. Some CPUs don't support any form of absolute addressing in their instruction sets. As a result, native compilers for these platforms don't support a -fPIC option—it has no meaning for them. Unknown options may be silently ignored, but in most cases, compilers stop on unknown options with an error message.

If we tried, for example, to compile this code on an IBM RS/6000 system using the native IBM compiler, it would hiccup when it came to the -fPIC option on the linker command line. This is because it doesn't make sense to support such an option on a system where all code is generated as position-independent code.

One way we could get around this problem would be to make the -fPIC option conditional in *Makefile.am*, based on the target system and the tools we're using. But that's exactly the sort of problem that Libtool was designed to address! We'd have to account for all the different Libtool target system types and tool sets in order to handle the entire set of conditions that Libtool already handles. Additionally, some systems and compilers may require different command line options to accomplish the same goal.

The way around this portability problem, then, is to let Libtool generate the static library, as well. Libtool makes a distinction between static libraries that are installed as part of a developer package and static libraries that are only used internally within a project. It calls such internal static libraries *convenience* libraries, and whether or not a convenience library is generated depends on the prefix used with the LTLIBRARIES primary. If the noinst prefix is used, then Libtool assumes we want a convenience library, because there's no point in generating a shared library that will never be installed. Thus, convenience libraries are always generated as non-installed static archives, which have no value unless they're linked to other code within the project.

The reason for distinguishing between convenience libraries and other forms of static libraries is that convenience libraries are always built, whereas installed static libraries are only built if the --enable-static option is specified on the configure command line—or, conversely, if the --disable-static option is *not* specified and the default library type has been set to static. The conversion from an older static library to a newer Libtool convenience library is simple enough—all we have to do is add LT to the primary name and remove the -fPIC option and the CFLAGS variable (since there were no other options being used in that variable). Note also that I've changed the library extension from *.a* to *.la*. Don't forget to change the prefix on the SOURCES variable to reflect the new name of the library—*libjupcommon.la*. These changes are highlighted in Listings 7-9 and 7-10.

```
noinst_LTLIBRARIES = libjupcommon.la
libjupcommon_la_SOURCES = jupcommon.h print.c
```

Listing 7-9: common/Makefile.am: *Changing from a static library to a Libtool static library*

```
lib_LTLIBRARIES = libjupiter.la
libjupiter_la_SOURCES = jup_print.c
libjupiter_la_CPPFLAGS = -I$(top_srcdir)/include -I$(top_srcdir)/common
libjupiter_la_LIBADD = ../common/libjupcommon.la
```

Listing 7-10: libjup/Makefile.am: *Changing from a static library to a Libtool static library*

Now when we try to build, here's what we get:

```
$ make
--snip--
Making all in libjup
make[2]: Entering directory '/.../jupiter/libjup'
❶ /bin/bash ../libtool --tag=CC   --mode=compile gcc -DHAVE_CONFIG_H -I. -I..
-I../include -I./common   -g -O2 -MT libjupiter_la-jup_print.lo -MD -MP -MF
.deps/libjupiter_la-jup_print.Tpo -c -o libjupiter_la-jup_print.lo `test -f
'jup_print.c' || echo './'`jup_print.c
libtool: compile:  gcc -DHAVE_CONFIG_H -I. -I.. -I../include -I./common -g
-O2 -MT libjupiter_la-jup_print.lo -MD -MP -MF .deps/libjupiter_la-jup_print.
Tpo -c jup_print.c  -fPIC -DPIC -o .libs/libjupiter_la-jup_print.o
libtool: compile:  gcc -DHAVE_CONFIG_H -I. -I.. -I../include -I./common -g
-O2 -MT libjupiter_la-jup_print.lo -MD -MP -MF .deps/libjupiter_la-jup_print.
Tpo -c jup_print.c -o libjupiter_la-jup_print.o >/dev/null 2>&1
mv -f .deps/libjupiter_la-jup_print.Tpo .deps/libjupiter_la-jup_print.Plo
/bin/bash ../libtool  --tag=CC   --mode=link gcc  -g -O2   -o libjupiter.la
-rpath /usr/local/lib libjupiter_la-jup_print.lo ../common/libjupcommon.la
-lpthread
libtool: link: gcc -shared  -fPIC -DPIC  .libs/libjupiter_la-jup_print.o
-Wl,--whole-archive ../common/.libs/libjupcommon.a -Wl,--no-whole-archive
-lpthread  -g -O2   -Wl,-soname -Wl,libjupiter.so.0 -o .libs/libjupiter.
so.0.0.0
libtool: link: (cd ".libs" && rm -f "libjupiter.so.0" && ln -s "libjupiter.
so.0.0.0" "libjupiter.so.0")
libtool: link: (cd ".libs" && rm -f "libjupiter.so" && ln -s "libjupiter.
so.0.0.0" "libjupiter.so")
libtool: link: (cd .libs/libjupiter.lax/libjupcommon.a && ar x "/.../jupiter/
libjup/../common/.libs/libjupcommon.a")
❷ libtool: link: ar cru .libs/libjupiter.a  libjupiter_la-jup_print.o
.libs/libjupiter.lax/libjupcommon.a/print.o
ar: `u' modifier ignored since `D' is the default (see `U')
libtool: link: ranlib .libs/libjupiter.a
libtool: link: rm -fr .libs/libjupiter.lax
libtool: link: ( cd ".libs" && rm -f "libjupiter.la" && ln -s "../libjupiter.
la" "libjupiter.la" )
make[2]: Leaving directory '/.../jupiter/libjup'
--snip--
$
```

You can see at ❷ that the common library is now built as a static convenience library because the ar utility builds *libjupcommon.a*. Libtool also seems to be building files with new and different extensions—a closer look will reveal extensions such as *.la* and *.lo* (check the line at ❶). If you examine these files, you'll find that they're actually descriptive text files containing object and library metadata. Listing 7-11 shows the partial contents of *common/libjupcommon.la*.

```
# libjupcommon.la - a libtool library file
# Generated by libtool (GNU libtool) 2.4.6 Debian-2.4.6-0.1
#
# Please DO NOT delete this file!
# It is necessary for linking the library.

# The name that we can dlopen(3).
dlname=''

# Names of this library.
❶ library_names=''

# The name of the static archive.
❷ old_library='libjupcommon.a'

# Linker flags that cannot go in dependency_libs.
inherited_linker_flags=''

# Libraries that this one depends upon.
❸ dependency_libs=' -lpthread'
--snip--
```

Listing 7-11: common/libjupcommon.la: Textual metadata found in a library archive (.la) file

The various fields in these files help the linker—or rather the libtool wrapper script—to determine certain options that the maintainer would otherwise have to remember and manually pass to the linker on the command line. For instance, the library's shared and static names are documented at ❶ and ❷ here, as well as any library dependencies required by these libraries (at ❸).

NOTE   *This is a convenience library, so the shared library name is empty.*

In this library, we can see that *libjupcommon.a* depends on the *pthreads* library. But, by using Libtool, we don't have to pass a -lpthread option on the libtool command line because libtool can detect from the contents of this metadata file (specifically, the line at ❸) that the linker will need this option, and it passes the option for us.

Making these files human readable was a minor stroke of genius, as they can tell us a lot about Libtool libraries at a glance. These files are designed to be installed on an end user's machine with their associated binaries, and, in fact, the make install rules that Automake generates for Libtool libraries do just this.

Most Linux distros today are leaning toward filtering out *.la* files from official builds of library projects—that is, they don't install them into the */usr* directory structure because *.la* files are only useful during builds where packages are referencing Libtool libraries within a project directory structure. Since the distro provider has already pre-built everything for you and you won't be building those packages yourself, they just take up space (albeit, not very much). When you link against a library (Libtool or otherwise) that's installed on your system in the */usr* directory structure, you're using one of the AC_CHECK/SEARCH macros to find the library and link against the *.a* or *.so* file directly, so the *.la* file isn't used in that case either.

Summary

In this chapter, I outlined the basic rationale for shared libraries. As an exercise, we added a shared library to Jupiter that incorporates functionality from the convenience library we created earlier. We began with a more or less intuitive approach to incorporating a static library into a Libtool shared library, and in the process we discovered a more portable and correct way to do this using Libtool convenience libraries.

As with the other packages in the Autotools toolchain, Libtool gives you a lot of functionality and flexibility. But as you've probably noticed, with this degree of functionality and flexibility comes a price—complexity. The size of Jupiter's configuration script increased dramatically with the addition of Libtool, and the time required to compile and link our project increased accordingly.

In the next chapter, we're going to continue our discussion of Libtool by looking at library-versioning issues and Libtool's solution to the portability problems presented by manual dynamic runtime library management.

8

LIBRARY INTERFACE VERSIONING AND RUNTIME DYNAMIC LINKING

Occasionally he stumbled over the truth, but hastily
picked himself up and hurried on as if nothing had happened.
—*Sir Winston Churchill, quoted in*
The Irrepressible Churchill

In the last chapter, I explained the concepts of dynamically loadable shared libraries. I also showed you how easy it is to add Libtool shared-library functionality and flexibility to your projects, whether your projects provide shared libraries, static libraries, convenience archives, or some mixture of these. There are still two major Libtool topics we need to cover. The first is library versioning, and the second involves using the Libtool *ltdl* library to portably build and consume dynamically loadable modules within your projects.

When I talk about the version of a library, I'm referring specifically to the version of the library's public interface, but I need to clearly define the term *interface* in this context. A *shared-library interface* refers to all aspects of a shared library's connections with the outside world. Besides the function and data signatures that a library exports, these connections include files and file formats, network connections and wire data formats, IPC channels and protocols, and so on. When considering whether to assign a new version to a shared library, you should carefully examine all aspects of the library's interactions with the world to determine if a change will cause the library to act differently from a user's perspective.

Libtool's attempts to hide the differences among shared-library platforms are so well conceived that if you've always used Libtool to build shared libraries, you may not even realize that the way shared libraries are versioned is significantly different between platforms.

System-Specific Versioning

Let's examine how shared-library versioning works on a few different systems to put the Libtool abstraction into context.

Shared-library versioning can be done either internally or externally. *Internal versioning* means that the library name does not reflect its version in any way. Thus, internal versioning implies that some form of executable header information provides the linker with the appropriate function calls for the requested *application binary interface (ABI)*. This also implies that all function calls for all versions of the library are maintained within the same shared-library file. Libtool supports internal versioning where mandated by platform requirements, but it prefers to use external versioning. With *external versioning*, version information is specified in the filename itself.

In addition to library-level versioning, wherein a particular version number or string refers to the entire library interface, many Unix systems support a form of export- or symbol-level versioning, wherein a shared library exports multiple named or numbered versions of the same function or global data item. While Libtool does not hinder the use of such export-level versioning schemes on a per-system basis, it does not provide any specific portability support for them, either. Therefore, I won't go into great detail on this subject.

Linux and Solaris Library Versioning

Modern Linux borrows much of its library versioning system from Oracle's Solaris operating system, version 9.[1] These systems use a form of external library versioning in which version information is encoded in the shared-library filename, following a specific pattern or template. Let's look at a partial directory listing for the */usr/lib/x86_64-linux-gnu* directory on a typical Linux system—specifically, the files associated with a fairly typical library, *libcurl*:

---

1. Note that older Solaris systems and the original Linux shared-library system used the older, so-called *a.out* scheme, in which libraries were managed quite differently. In the *a.out* scheme, all binary code had to be manually mapped into memory using a mapping file that had the same base name as the library and ended in the *.sa* extension. The mapping file had to be manually edited to ensure that the program and all shared libraries were mapped into non-overlapping regions of the process address space. This system was eventually replaced with PIC code, wherein the loader can determine the position of code in memory at runtime.

```
$ ls -lr /usr/lib/x86_64-linux-gnu/libcurl*
❶ -rw-r--r-- ... 947448 ... libcurl.a
   lrwxrwxrwx ...     19 ... libcurl-gnutls.so.3 -> libcurl-gnutls.so.4
   lrwxrwxrwx ...     23 ... libcurl-gnutls.so.4 -> libcurl-gnutls.so.4.4.0
   -rw-r--r-- ... 444800 ... libcurl-gnutls.so.4.4.0
   -rw-r--r-- ...    953 ... libcurl.la
❷ lrwxrwxrwx ...     16 ... libcurl.so -> libcurl.so.4.4.0
   lrwxrwxrwx ...     12 ... libcurl.so.3 -> libcurl.so.4
❸ lrwxrwxrwx ...     16 ... libcurl.so.4 -> libcurl.so.4.4.0
❹ -rw-r--r-- ... 452992 ... libcurl.so.4.4.0
$
```

NOTE *The content in this console directory listing is specific to my system, which is based on a Debian distribution. If your distribution is not based on Debian, you will probably see a somewhat different listing—perhaps even significantly different. In this case, do not try to follow along on your system. Instead, just follow my example here as you read the following description. The concepts, not the filenames, are the important part of this discussion.*

Library names on Linux systems conform to a standard format: *lib*name.*so*.X.Y. The X.Y portion of the format represents the version information, where X is the major version number (always a single number) and Y is the minor version number (which may contain multiple dot-separated parts). The general rule is that changes in X represent non-backward-compatible changes to the library's ABI, while changes in Y represent backward-compatible modifications, including isolated additions to the library's interface and nonintrusive bug fixes.

Often, you'll see what appears to be a third numbered component. The entry at ❹, for example, represents the actual *curl* shared library, *libcurl.so.4.4.0*. In this example, the last two numbers (*4.0*) really just represent a two-part minor version number. Such additional numeric information in the minor version number is sometimes referred to as the library's *patch level*.[2]

The *libcurl.so.4* entry at ❸ is referred to as the library's *shared object name (soname)*[3] and is actually a soft link that points to the binary file. The soname is the format that consuming programs and libraries reference internally—that is, the linker embeds this name in the consuming program or library when it's built. The soft link is created by the ldconfig utility, which (among other things) ensures that an appropriate soname

---

2. According to legend, the entire minor version number can really be any alphanumeric text, though it's usually limited to dot-separated numbers—if only to maintain the sanity of the user. The *GNU Libtool Manual* claims that the ldconfig utility will honor the patch level when it creates the soname link, automatically selecting the highest value found. If this value can be any alphanumeric text, then it's difficult to see how this statement can be true; perhaps the utility uses some heuristic (such as lexicographical value) to attempt to isolate the more "recent" version of the library.

3. *Soname* is pronounced "ess-oh-name."

can locate the latest minor version of an installed library. The `ldconfig` utility is usually executed by post-install scripts and triggers of RPM and Debian packages. Therefore, while the soname is not created or installed by the `make install` target, it is most often created by distro installation packages and, therefore, by Linux packagers.

Notice how this versioning scheme allows multiple sonames for different major versions and multiple binaries with different major and minor versions to all coexist within a single directory.

Development packages for a library (ending in *-dev* or *-devel*) often install a so-called *linker name* entry (at ❷) as well. The linker name is a soft link ending only in *.so* that usually points to the soname, though in some cases (such as this one), it may refer directly to the binary shared library. The linker name is the name by which a library is referred to on the linker command line. The development library allows you to run programs on your system that are linked against the latest version of a library but develop against an older version of that library, or vice versa.

The entry at ❶ refers to the static archive form of the library, which has a *.a* extension on Linux and Solaris systems. The remaining entries represent other forms of the *curl* library set generated for purposes specific to the *curl* package.

The *curl* library has become an important part of modern Linux systems over the years; it's used by many other programs installed on the system, some of which have not been upgraded to the latest major version. The maintainers assert that major version 4 is backward compatible with major version 3. Therefore, sonames referring to version 3 are directed toward version 4 of the libraries on systems where version 4 is installed. This is not necessarily a common practice, but it happens to work in this case.

From here on out, the waters become muddied by a strange array of external and internal shared-library versioning techniques. Each of these less-than-intuitive systems is designed to overcome some of the fundamental problems that have been discovered in the Solaris system over the years.[4] Let's look at a few of them.

IBM AIX Library Versioning

Traditionally, IBM's AIX used a form of internal versioning, storing all library code within a single archive file that follows the pattern *lib*name.*a*. This file may actually contain both static and shared forms of code, as well as 32-bit and 64-bit code. Internally, all shared-library code is stored in a single, logical, shared-object file within the archive file, while static library objects are stored as individual logical object files within the archive.

I say "traditionally" because more recent versions of AIX (including all 64-bit versions) now support the concept of loading shared-library code directly from physical *.so* files.

---

4. In my humble opinion, the solutions provided by these "enhancements" aren't justified due to the additional problems they cause.

Libtool generates shared-library code on AIX using both of these schemes. If the AIX -brtl native linker flag is specified on the command line, Libtool generates shared libraries with *.so* extensions. Otherwise, it generates combined libraries following the older, single-file scheme.[5]

When using the *.so* file scheme on AIX, Libtool generates libraries named in the Linux/Solaris pattern in order to maintain a degree of alliance with these more popular platforms. Regardless of the shared-library extension used, however, version information is still not stored in the filename; it is stored internally, within the library and consuming executables. As far as I can tell, Libtool ensures that the correct internal structures are created to reflect the proper versioning information within the shared-library header. It does this by passing appropriate flags to the native linker with embedded version information derived from the Libtool version string.

Executables on most Unix systems also support the concept of an embedded runtime library search path (called a *LIBPATH* on AIX), which usually specifies a set of colon-separated filesystem paths to be searched for shared-library dependencies. You can use Libtool's -R command line option to specify a library search path for both programs and libraries. Libtool will translate this option to the appropriate GNU or native linker option on any given system.

I say executables *usually* support this option because on AIX, there are a few nuances. If all of the directories specified in the LIBPATH are real directories, everything works as expected—that is, the LIBPATH acts purely as a library search path. However, if the first segment of the LIBPATH is not a real filesystem entry, it acts as a so-called *loader domain*, which is basically a namespace for a particular shared library. Thus, multiple shared libraries of the same name can be stored within the same AIX archive (*.a*) file, each assigned (by linker options) to a different loader domain. The library that matches the loader domain specified in the LIBPATH is loaded from the archive. This can have nasty side effects if you assign a loader domain via the LIBPATH that later becomes (by chance) a real filesystem entry. On the other hand, you could also specify a search directory in the LIBPATH that happens to match a loader domain in a shared library. If that directory is removed later, you'll unintentionally begin to use the loader domain. As you can imagine, strange behavior ensues. Most of these issues have been solved by AIX developers by ensuring that loader domain strings look nothing like filesystem paths.

On AIX systems, all code, whether static or shared, is compiled as position-independent code because AIX has only ever been ported to PowerPC and RS/6000 processors. The architectures of these processors only allow for PIC code, so AIX compilers can't generate non-PIC code.

---

5. The -brtl flag tells the native AIX linker to generate load-time resolved shared objects, wherein external symbol references are resolved at the time the library is loaded, as opposed to the default link-time resolved objects, wherein external symbol references are resolved at link time. Resolving objects at load time is more similar to how objects are treated on Linux and Solaris or, more generally, on ELF systems.

Microsoft DLL Versioning

Consider Microsoft Windows *dynamic link libraries (DLLs)*, which are shared libraries in every sense of the word and provide a proper application programming interface (API). But unfortunately, Microsoft has, in the past, provided no integrated DLL interface versioning scheme. As a result, Windows developers have often referred to DLL versioning issues (tongue-in-cheek, I'm sure) as *DLL hell*.

As a sort of Band-Aid fix to this problem, DLLs on Windows systems can be installed into the same directory as the program that uses them. The Windows operating system loader will always attempt to use the local copy before searching for a copy in the system path. This alleviates a part of the problem because it allows you to install a specific version of the library with the package that requires it. While this is a fair solution, it's not really a good solution, because one of the major benefits of shared libraries is that they can be shared—both on disk and in memory. If every application has its own copy of a different version of the library, then this benefit of shared libraries is lost—both on disk and in memory.

Since the introduction of this partial solution years ago, Microsoft hasn't paid much attention to DLL-sharing efficiency issues. The reasons for this include both a cavalier attitude regarding the cost of disk space and RAM and a technical issue regarding the implementation of Windows DLLs. Instead of generating position-independent code, Microsoft system architects chose to link DLLs with a specific base address and then list all of the absolute address references in a base table in the library image header. When a DLL can't be loaded at the desired base address (because of a conflict with another DLL), the loader *rebases* the DLL by picking a new base address and changing all of the absolute addresses in the code segment that are referred to in the base table. When a DLL is rebased in this manner, it can only be shared with processes that happen to rebase the DLL to the same address. The odds of accidentally encountering such a scenario—especially among applications with many DLL components—are pretty slim.

More recently, Microsoft invented the concept of the *side-by-side cache* (sometimes referred to as *SxS*), which allows developers to associate a unique identification value (a GUID, in fact) with a particular version of a DLL installed in a system location. The location directory name is derived from the DLL name and the version identifier. Applications built against SxS-versioned libraries have metadata stored in their executable headers indicating the specifically versioned DLLs they require. If the right version is found (by newer OS loaders) in the SxS cache, then it is loaded. Based on policy in the EXE header's metadata, the loader can revert to the older scheme of looking for a local copy and then a global copy of the DLL. This is a vast improvement over earlier solutions, and it provides a very flexible versioning system.

The side-by-side cache effectively moves the Windows DLL architecture a step closer to the Unix way of managing shared libraries. Think of the SxS as a system installation location for libraries—much like the */usr/lib* directory on Unix systems. Also similar to Unix, multiple versions of the same DLL may be co-installed in the side-by-side cache.

Regardless of the similarities, since DLLs use the rebasing technique as opposed to PIC code, the side-by-side cache is still a fairly benign efficiency improvement with respect to applications that manage dozens of shared libraries. SxS is really intended for system libraries that many applications are likely to consume. These are generally based at different addresses so that the odds of clashing (and thus rebasing) are decreased but not entirely eliminated.

The entire based approach to shared libraries has the major drawback that the program address space may become fairly fragmented as the system loader honors randomly chosen base addresses throughout a 32-bit address space. Fortunately, 64-bit addressing helps tremendously in this area, so you may find the side-by-side cache to be much more effective with respect to improving memory-use efficiency on 64-bit Windows systems, which are the norm these days anyway.

HP-UX/AT&T SVR4 Library Versioning

Hewlett Packard's version of Unix (since HP-UX version 10.0) adds a form of library-level versioning that's very similar to the versioning used in AT&T UNIX System V Release 4. For our purposes, you can consider these two types of systems to work nearly the same way.

The native linker looks for libraries specified by their base name with a *.sl* extension. However, consuming programs and libraries contain a reference to that library's *internal name*. The internal name is assigned to the library by a linker command line option and should contain the library's interface version number.

The actual library is named with only the major interface version as an extension, and a soft link is created with a *.sl* extension pointing to the library. Thus, a shared library on these systems will follow this pattern:

```
libname.X
libname.sl -> libname.X
```

The only version information we have to work with is a major version number, which should be used to indicate non-backward-compatible changes from one version to the next. Since there's no minor version number, as on Linux or Solaris, we can't keep multiple revisions of a particular interface version around. The only option is to replace version zero of a library with an updated version zero if bug fixes or backward-compatible enhancements (that is to say, non-intrusive additions to the interface) are made.

However, we can still have multiple major versions of the library co-installed, and Libtool takes full advantage of what's available on these systems.

The Libtool Library Versioning Scheme

The authors of Libtool tried hard to provide a versioning scheme that could be mapped to any of the schemes used by any Libtool platform. The Libtool versioning scheme is designed to be flexible enough to be forward compatible with reasonable future changes to existing Libtool platforms and even to new Libtool platforms.

Nevertheless, it's not a panacea. When Libtool has been extended for a new type of shared-library platform, situations have occurred (and continue to occur) that require some serious and careful evaluation. No one can be an expert on all systems, so the Libtool developers rely heavily on outside contributions to create proper mappings from the Libtool versioning scheme to the schemes of new or would-be Libtool platforms.

Library Versioning Is Interface Versioning

You should consciously avoid thinking of library version numbers (either Libtool's or those of a particular platform) as product *major, minor,* and *revision* (also called *patch* or *micro*) values. In fact, these values have very specific meanings to the operating system loader, and they must be updated properly for each new library version in order to keep from confusing the loader. A confused loader could load the wrong version of a library based on incorrect version information assigned to the library.

Several years ago, I was working with my company's corporate versioning committee to come up with a software-versioning policy for the company as a whole. The committee wanted the engineers to ensure that the version numbers incorporated into our shared-library names were in alignment with the corporate software-versioning strategy. It took me the better part of a day to convince them that a shared-library version was not related to a product version in any way, nor should such a relationship be established or enforced by them or by anyone else.

Here's why: the version number on a shared library is not really a library version but an interface version. The *interface* I'm referring to here is the application binary interface presented by a library to the user, another programmer desiring to call functions presented by the interface. An executable program has a single, well-defined, standard entry point (usually called main in the C language). But a shared library has multiple entry points that are generally not standardized in a manner that is widely understood. This makes it much more difficult to determine whether or not a particular version of a library is interface compatible with another version of the same library.

In Libtool's versioning scheme, shared libraries are said to support a range of interface versions, each identified by a unique integer value. If any publicly visible aspect of an interface changes between public releases, it can no longer be considered the same interface; it therefore becomes a new interface, identified by a new integer identifier. Each public release of a library in which the interface has changed simply acquires the next consecutive interface version number. Libraries that change in a backward-compatible manner between releases are said to support both the old and the new interface; thus, a particular release of a library may support interface versions 2 through 5, for example.

Libtool library version information is specified on the `libtool` command line with the `-version-info` option, as shown in Listing 8-1.

```
libname_la_LDFLAGS = -version-info 0:0:0
```

Listing 8-1: Setting shared-library version information in a Makefile.am file

The Libtool developers wisely chose the colon separator over the period in an effort to keep developers from trying to directly associate Libtool version string values with the version numbers appended to the end of shared-library files on various platforms. The three values in the version string are respectively called the interface *current*, *revision*, and *age* values.

The *current* value represents the current interface version number. This is the value that changes when a new interface version must be declared because the interface has changed in some publicly visible way since the last public release of the library. The first interface in a library is given a version number of zero by convention. Consider a shared library in which the developer has added a new function to the set of functions exposed by this library since the last public release. The interface can't be considered the same in this new version because there's one additional function. Thus, its *current* number must be increased from zero to one.

The *age* value represents the number of back versions supported by the shared library. In mathematical terms, the library is said to support the interface range, *current – age* through *current*. In the example I just gave, a new function was added to the library, so the interface presented in this version of the library is not the same as it was in the previous version. However, the previous version is still fully supported, because the previous interface is a proper subset of the current interface. Therefore, the *age* value should also be incremented from zero to one.

The *revision* value merely represents a serial revision of the current interface. That is, if no publicly visible changes are made to a library's interface between releases—perhaps only an internal function was optimized—then the library name should change in some manner, if only to distinguish between the two releases. But both the *current* and *age* values would be the same, because the interface has not changed from the user's perspective. Therefore, the *revision* value is incremented to reflect the fact that this is a new release of the same interface. In the previous example, the *revision* value would be left at zero, because one or both of the other values were incremented.

To simplify the release process for shared libraries, the Libtool versioning algorithm should be followed step-wise for each new version of a library that is about to be publicly released:[6]

1. Start with version information 0:0:0 for each new Libtool library. (This is done automatically if you simply omit the -version-info option from the list of linker flags passed to the libtool script.) For existing libraries, start with the previous public release's Libtool version information.

2. If the library source code has changed at all since the last update, then increment *revision* (*c:r:a* becomes *c:r+1:a*).

3. If any exported functions or data have been added, removed, or changed since the last update, increment *current* and set *revision* to 0.

4. If any exported functions or data have been added since the last public release, increment *age*.

5. If any exported functions or data have been removed since the last public release, set *age* to 0.

Keep in mind that this is an algorithm; as such, it's designed to be followed step-by-step as opposed to jumping directly to the steps that appear to apply to your case. For example, if you removed an API function from your library since the last release, you would not simply jump to the last step and set *age* to zero. Rather, you would follow all of the steps until you reached the last step, and *then* set *age* to zero.

NOTE *Remember to update the version information only immediately before a public release of your software. More frequent updates are unnecessary and only guarantee that the current interface number becomes larger faster.*

Let's look at an example. Assume that this is the second release of a library and the first release used a -version-info string of 0:0:0. One new function was added to the library interface during this development cycle, and one existing function was deleted. The effect on the version information string for this new release of the library would be as follows:

1. Begin with the previous version information: 0:0:0.
2. 0:0:0 becomes 0:1:0 (the library's source was changed).
3. 0:1:0 becomes 1:0:0 (the library's interface was modified).
4. 1:0:0 becomes 1:0:1 (one new function was added).
5. 1:0:1 becomes 1:0:0 (one old function was removed).

It should be clear by now that there is no *direct* correlation between Libtool's *current*, *revision*, and *age* values and Linux's major, minor, and optional patch-level values. Instead, mapping rules are used to transform the values in one scheme to values in the other.

---

6. See the Free Software Foundation's *GNU Libtool Manual* at *https://www.gnu.org/software/libtool/manual/*.

Returning to the preceding example, wherein a second release of a library added one function and removed one function, we ended up with a new Libtool version string of 1:0:0. The version string 1:0:0 indicates that the library is not backward compatible with the previous version (*age* is zero), so the Linux shared-library file would be named *lib*name.*so.1.0.0*. This looks suspiciously like the Libtool version string—but don't be fooled. This fairly common coincidence is perhaps one of the most confusing aspects of the Libtool versioning abstraction.

Let's modify our example just a little to say that we've added a new library interface function but haven't removed anything. Start again with the original version information of 0:0:0 and follow the algorithm:

1. Begin with the previous version information: 0:0:0.
2. 0:0:0 becomes 0:1:0 (the library's source was changed).
3. 0:1:0 becomes 1:0:0 (the library's interface was modified).
4. 1:0:0 becomes 1:0:1 (one new function was added).
5. Not applicable (nothing was removed).

This time, we end up with a Libtool version string of 1:0:1, but the resulting Linux or Solaris shared-library filename is *lib*name.*so.0.1.0*. Consider for a moment what it means, in the face of major, minor, and patch-level values, to have a nonzero *age* value in the Libtool version string. An *age* value of one (as in this case) means that we are effectively still supporting a Linux major value of zero, because this new version of the library is 100 percent backward compatible with the previous version. The minor value in the shared-library filename has been incremented from zero to one to indicate that this is, in fact, an updated version of the soname, *lib*name *.so.0*. The patch-level value remains at zero because this value indicates a bug fix to a particular minor revision of an soname.

Once you fully understand Libtool versioning, you'll find that even this algorithm does not cover all possible interface modification scenarios. Consider, for example, version information of 0:0:0 for a shared library that you maintain. Now assume you add a new function to the interface for the next public release. This second release properly defines version information of 1:0:1 because the library supports both interface versions 0 and 1. However, just before the third release of the library, you realize you didn't really need that new function after all, so you remove it. This is the only publicly visible change made to the library interface in this release. The algorithm would have set the version information string to 2:0:0. But in fact, you've merely removed the second interface and are now presenting the original interface once again. Technically, this library would be properly configured with a version information string of 0:1:0 because it presents a second release of version 0 of the shared-library interface. The moral of this story is that you need to fully understand the way Libtool versioning works and then decide, based on that understanding, what the proper next-version values should be.

I'd also like to point out that the *GNU Libtool Manual* makes little effort to describe the myriad ways an interface can be different from one version of a library to another. An interface version indicates functional semantics as well as API syntax. If you change the way a function works semantically but leave the function signature untouched, you've still changed the function. If you change the network wire format of data sent by a shared library, then it's not really the same shared library from the perspective of the consuming code. All the operating system loader really cares about when attempting to determine which library to load is, *will this library work just as well as that one?* In these cases, the answer would have to be no, because even though the API interface is identical, the publicly visible way the two libraries do things is not the same.

When Library Versioning Just Isn't Enough

These types of changes to a library's interface are so complex that project maintainers will often simply rename the library, thereby skirting library-versioning issues entirely. One excellent way to rename your library is to use Libtool's -release flag. This flag adds a separate class of library versioning information into the base name of the library, effectively making it an entirely new library from the perspective of the operating system loader. The -release flag is used in the manner shown in Listing 8-2.

```
libname_la_LDFLAGS = -release 2.9.0 -version-info 0:0:0
```

Listing 8-2: Setting shared-library release information in a Makefile.am file

In this example, I used -release and -version-info in the same set of Libtool flags, just to show you that they can be used together. You'll note here that the release string is specified as a series of dot-separated values. In this case, the final name of your Linux or Solaris shared library will be *lib*name-*2.9.0.so.0.0.0.*

Another reason developers choose to use release strings is to provide some sort of correlation between library versions across platforms. As demonstrated earlier, a particular Libtool version information string will probably result in different library names across platforms because Libtool maps version information into library names differently from platform to platform. Release information remains stable across platforms, but you should carefully consider how you want to use release strings and version information in your shared libraries, because the way you choose to use them will affect binary compatibility between releases of your libraries. The OS loader will not consider two versions of a library compatible if they have different release strings, regardless of the values of those strings.

Using libltdl

Now let's move on to a discussion of Libtool's *ltdl* library. Once again, we need to add some functionality to the Jupiter project in order to illustrate these concepts. The goal here is to create a plug-in interface that the jupiter program can use to modify output based on end-user policy choices.

Necessary Infrastructure

Currently, jupiter prints *Hello from jupiter!* (Actually, the name printed is more likely, at this point, to be a long, ugly path containing some Libtool directory garbage and some derivation of the name *jupiter*, but just pretend it prints *jupiter* for now.) We're going to add an additional parameter named salutation to the *common* static library method, print_routine. This parameter will also be of type pointer-to-char and will contain the leading word or phrase—the salutation—in jupiter's greeting.

Listings 8-3 and 8-4 indicate the changes we need to make to files in the *common* subdirectory.

Git tag 8.0

```
--snip--
static void * print_it(void * data)
{
    const char ** strings = data;
    printf("%s from %s!\n", strings[0], strings[1]);
    return 0;
}

int print_routine(const char * salutation, const char * name)
{
    const char * strings[] = {salutation, name};
#if ASYNC_EXEC
    pthread_t tid;
    pthread_create(&tid, 0, print_it, strings);
    pthread_join(tid, 0);
#else
    print_it(strings);
#endif
    return 0;
}
```

Listing 8-3: common/print.c: Adding a salutation to the print_routine function

```
int print_routine(const char * salutation, const char * name);
```

Listing 8-4: common/jupcommon.h: Adding a salutation to the print_routine prototype

Listings 8-5 and 8-6 show the changes we need to make to files in the *libjup* and *include* subdirectories.

```
--snip--
int jupiter_print(const char * salutation, const char * name)
{
    print_routine(salutation, name);
}
```

Listing 8-5: libjup/jup_print.c: Adding a salutation to the `jupiter_print` function

```
--snip--
int jupiter_print(const char * salutation, const char * name);
--snip--
```

Listing 8-6: include/libjupiter.h: Adding a salutation to the `jupiter_print` prototype

And finally, Listing 8-7 shows what we need to do to *main.c* in the *src* directory.

```
--snip--
#define DEFAULT_SALUTATION "Hello"

int main(int argc, char * argv[])
{
    const char * salutation = DEFAULT_SALUTATION;
    return jupiter_print(salutation, argv[0]);
}
```

Listing 8-7: src/main.c: Passing a salutation to `jupiter_print`

To be clear, all we've really done here is parameterize the salutation throughout the print routines. That way, we can indicate from main which salutation we'd like to use. I've set the default salutation to *Hello* so that nothing will have changed from the user's perspective. Thus, the overall effect of these changes is benign. Note also that these are all source code changes—we've made no changes to the build system. I wanted to compartmentalize these changes so as to not confuse this necessary refactoring with what we're doing to the build system to add the new module-loading functionality.

After making these changes, should you update the version number of this shared library? That depends on whether you've already shipped this library (that is, posted a tarball) before you made the changes. The point of versioning is to maintain some semblance of control over your public interface—but if you're the only one who has ever seen it, there's no point in changing the version number.

Adding a Plug-In Interface

I'd like to make it possible to change the salutation displayed by simply changing which plug-in module is loaded at runtime. All of the changes we'll need to make to the code and build system to add this functionality will be limited to the *configure.ac* file and to files within the *src* directory and its subdirectories.

First, we need to define the actual plug-in interface. We'll do this by creating a new private header file in the *src* directory called *module.h*. This file is shown in Listing 8-8.

Git tag 8.1

```
#ifndef MODULE_H_INCLUDED
#define MODULE_H_INCLUDED

❶ #define GET_SALUTATION_SYM "get_salutation"

❷ typedef const char * get_salutation_t(void);
❸ const char * get_salutation(void);

#endif /* MODULE_H_INCLUDED */
```

Listing 8-8: src/module.h: *The initial contents of this file*

This header file has a number of interesting aspects. First, let's look at the preprocessor definition, GET_SALUTATION_SYM, at ❶. This string represents the name of the function you need to import from the plug-in module. I like to define these in the header file so all the information that needs to be reconciled exists in one place. In this case, the symbol name, the function type definition, and the function prototype must all be in alignment, and you can use this single definition for all three.

Another interesting item is the type definition[7] at ❷. If we don't provide one, the user is going to have to invent one, or else use a complex typecast on the return value of the dlsym function. Therefore, we'll provide it here for consistency and convenience.

Finally, look at the function prototype at ❸. This isn't so much for the caller as it is for the module itself. Modules providing this function should include this header file so the compiler can catch potential misspellings of the function name.

Doing It the Old-Fashioned Way

For this first attempt, let's use the *dl* interface provided by the Solaris/ Linux *libdl.so* library. In the next section, we'll convert this code over to the Libtool *ltdl* interface for greater portability.

To do this right, we need to add checks to *configure.ac* to look for both the *libdl* library and the *dlfcn.h* header file. These changes to *configure.ac* are highlighted in Listing 8-9.

---

7. Technically, it's bad practice to suffix application-defined types with _t because POSIX reserves this namespace for types defined by the POSIX standard. However, it's not very likely that get_salutation_t is going to conflict with anything POSIX defines in a future version of the standard (though it *is* possible).

```
--snip--
# Checks for header files.
❶ AC_CHECK_HEADERS([stdlib.h dlfcn.h])
--snip--
# Checks for libraries.

# Checks for typedefs, structures, and compiler characteristics.

# Checks for library functions.
❷ AC_SEARCH_LIBS([dlopen], [dl])
--snip--
cat << EOF
-------------------------------------------------

${PACKAGE_NAME} Version ${PACKAGE_VERSION}

Prefix: '${prefix}'.
Compiler: '${CC} ${CFLAGS} ${CPPFLAGS}'
❸ Libraries: '${LIBS}'
--snip--
```

Listing 8-9: configure.ac: Adding checks for the dl library and public header file

At ❶, I added the *dlfcn.h* header file to the list of files passed to the
AC_CHECK_HEADERS macro, and then at ❷, I added a check for the dlopen function in the *dl* library. Note here that the AC_SEARCH_LIBS macro searches a list
of libraries for a function, so this call goes in the "Checks for library functions" section rather than the "Checks for libraries" section. To help us see
which libraries we're actually linking against, I've also added a line to the
cat command at the end of the file. The Libraries: line at ❸ displays the
contents of the LIBS variable, which is modified by the AC_SEARCH_LIBS macro.

NOTE *The LT_INIT macro also checks for the existence of the* dlfcn.h *header file, but I do it
here explicitly so it's obvious to observers that I wish to use this header file. This is a
good rule of thumb to follow, as long as it doesn't negatively affect performance too
much. Since Autoconf caches the results of checks, it's not likely to do so. You can tell
this is happening when you see* (cached) ... *after a check in* configure's *output.*

Adding a module requires several changes, so we'll make them all here,
beginning with the following command:

```
$ mkdir -p src/modules/hithere
$
```

I've created two new subdirectories. The first is *modules*, beneath *src*,
and the second is *hithere*, beneath *modules*. Each new module added to this
project will have its own directory beneath *modules*. The *hithere* module will
provide the salutation *Hi there*.

Listing 8-10 illustrates how to add a SUBDIRS variable to the *src/Makefile.am*
file to ensure that the build system processes the *modules/hithere* directory.

```
❶ SUBDIRS = modules/hithere

   bin_PROGRAMS = jupiter
❷ jupiter_SOURCES = main.c module.h
   --snip--
   greptest.sh:
❸ echo './jupiter | grep ".* from .*jupiter!"' > greptest.sh
   --snip--
```

Listing 8-10: src/Makefile.am: Adding a SUBDIRS *variable to this* Makefile.am *file*

The way I've used SUBDIRS at ❶ presents a new concept. Until now, Jupiter's *Makefile.am* files have only referenced direct descendants of the current directory, but this is not strictly necessary, as you can see. In fact, for Jupiter, the *modules* directory will only contain additional subdirectories, so it makes little sense to provide a *modules/Makefile.am* file just so you can reference its subdirectories.

While you're editing the file, you should add the new *module.h* header file to the SOURCES variable at ❷. If you don't do this, jupiter will still compile and build correctly for you as the maintainer, but the distcheck target will fail because none of the *Makefile.am* files will have mentioned *module.h*.

We also need to change the way the greptest.sh shell script is built so it can test for any type of salutation. A simple modification of the regular expression at ❸ will suffice.

I created a *Makefile.am* file in the new *hithere* subdirectory that contains instructions on how to build the *hithere.c* source file, and then I added the *hithere.c* source file to this directory. These files are shown in Listings 8-11 and 8-12, respectively.

```
   pkglib_LTLIBRARIES = hithere.la
   hithere_la_SOURCES = hithere.c
❶ hithere_la_LDFLAGS = -module -avoid-version
```

Listing 8-11: src/modules/hithere/Makefile.am: The initial version of this file

```
#include "../../module.h"

const char * get_salutation(void)
{
    return "Hi there";
}
```

Listing 8-12: src/modules/hithere/hithere.c: The initial version of this file

The *hithere.c* source file includes the semi-private *module.h* header file using a double-quoted relative path. Since Automake automatically adds -I$(srcdir) to the list of *include* paths used, the C preprocessor will properly sort out the relative path. The file then defines the get_salutation function, whose prototype is in the *module.h* header file. This implementation simply returns a pointer to a static string, and as long as the library is loaded, the caller can access the string. However, callers must be aware of the scope of data references returned by plug-in modules; otherwise, the program may unload a module before a caller is done using it.

The last line of *hithere/Makefile.am* (at ❶ in Listing 8-11) requires some explanation. Here, we're using a -module option on the hithere_la_LDFLAGS variable. This is a Libtool option that tells Libtool you want to call your library *hithere*, not *libhithere*. The *GNU Libtool Manual* makes the statement that modules do not need to be prefixed with *lib*. And since your code will be loading these modules manually, it should not have to be concerned with determining and properly using a platform-specific library name prefix.

If you don't care to use module versioning on your dynamically loadable (dlopen-ed) modules, try using the Libtool -avoid-version option. This option causes Libtool to generate a shared library whose name is *lib*name.*so*, rather than *lib*name.*so.0.0.0*. It also suppresses generation of the *lib*name.*so.0* and *lib*name.*so* soft links that refer to the binary image. Because I'm using both options, my module will simply be named *hithere.so*.

In order to get this module to build, we need to add the new *hithere* module's makefile to the AC_CONFIG_FILES macro in *configure.ac*, as shown in Listing 8-13.

```
--snip--
AC_CONFIG_FILES([Makefile
                common/Makefile
                include/Makefile
                libjup/Makefile
                src/Makefile
                src/modules/hithere/Makefile])
--snip--
```

Listing 8-13: configure.ac: Adding the hithere *directory makefile to* AC_CONFIG_FILES

Finally, in order to use the module, we need to modify *src/main.c* so that it loads the module, imports the symbol, and calls it. These changes to *src/main.c* are highlighted in bold in Listing 8-14.

```
  #include "config.h"

  #include "libjupiter.h"
❶ #include "module.h"

❷ #if HAVE_DLFCN_H
  # include <dlfcn.h>
  #endif

  #define DEFAULT_SALUTATION "Hello"

  int main(int argc, char * argv[])
  {
      int rv;
      const char * salutation = DEFAULT_SALUTATION;
```

```
❸ #if HAVE_DLFCN_H
      void * module;
      get_salutation_t * get_salutation_fp = 0;
   ❹ module = dlopen("./module.so", RTLD_NOW);
      if (module != 0)
      {
          get_salutation_fp = (get_salutation_t *)dlsym(
                  module, GET_SALUTATION_SYM);
          if (get_salutation_fp != 0)
              salutation = get_salutation_fp();
      }
  #endif

      rv = jupiter_print(salutation, argv[0]);

❺ #if HAVE_DLFCN_H
      if (module != 0)
          dlclose(module);
  #endif

      return rv;
  }
```

Listing 8-14: src/main.c: *Using the new plug-in module from the main function*

I'm including the new private *module.h* header file at ❶, and I added a
preprocessor directive to conditionally include *dlfcn.h* at ❷. Finally, I added
two sections of code, one before and one after the original call to jupiter
_print (at ❸ and ❺, respectively). Both are conditionally compiled based on
the existence of a dynamic loader, allowing the code to build and run correctly on systems that do not provide the *libdl* library.

The general philosophy I use when deciding whether or not code
should be conditionally compiled is this: If configure fails because a library
or header file is missing, then I don't need to conditionally compile the
code that uses the item configure checks for. If I check for a library or
header file in configure but allow it to continue if it's missing, then I'd
better use conditional compilation.

There are just a few more minor points to bring up regarding the use
of *dl* interface functions. First, at ❹, dlopen accepts two parameters: a filename or *path* (absolute or relative) and a *flags* word, which is the bitwise
composite of your choice of several flag values defined in *dlfcn.h*. Check the
man page for dlopen to learn more about these flag bits. If you use a path,
then dlopen honors that path verbatim, but if you use a filename, the library
search path is searched in an attempt to locate your module. By prefixing
the name with ./, we're telling dlopen not to search the library path.

We want to be able to configure which module jupiter uses, so we're
loading a generic name, *module.so.* In fact, the built module is located
several directories below the *src* directory in the build tree, so we need to
create a soft link in the current directory called *module.so* that points to
the module we want to load. This is a rather shabby form of configuration

for Jupiter, but it works. In a real application, you would define the desired module to load using policy defined in some sort of configuration file, but in this example, I'm simply ignoring these details for the sake of simplicity.

NOTE *I'm ignoring some error handling in Listing 8-14. In production code, you would probably want to log or display something if the module doesn't load or if the symbol is not exported by the module.*

The following command sequence shows our loadable module in action:

```
$ autoreconf -i
--snip--
$ ./configure && make
--snip--
$ cd src
$ ./jupiter
Hello from ...jupiter!
$
$ ln -s modules/hithere/.libs/hithere.so module.so
$ ./jupiter
Hi there from ...jupiter!
$
```

NOTE *The symlink module.so refers to a file in a hidden .libs directory. Executables and libraries are generated into a .libs directory within the associated source directory by Autotools build systems.*

Converting to Libtool's ltdl Library

Libtool provides a wrapper library called *ltdl* that abstracts and hides some of the portability issues surrounding the use of shared libraries across many different platforms. Most applications ignore the *ltdl* library because of the added complexity involved in using it, but there are really only a few issues to deal with. I'll enumerate them here and then cover them in detail later.

- The *ltdl* functions follow a naming convention based on the *dl* library. The rule of thumb is that *dl* functions in the *ltdl* library have the prefix lt_. For example, dlopen is named lt_dlopen.
- Unlike the *dl* library, the *ltdl* library must be initialized and terminated at appropriate locations within a consuming application.
- Applications should be built using the -dlopen *modulename* option on the linker command line (in the *_LDFLAGS variable). This tells Libtool to link the code for the module into the application when building on platforms without shared libraries or when linking statically.
- The LTDL_SET_PRELOADED_SYMBOLS() macro should be used at an appropriate location within your program source code to ensure that module code can be accessed on non-shared-library platforms or when building static-only configurations.

- Shared-library modules designed to be dlopen-ed using *ltdl* should use the -module option (and, optionally, the -avoid-version option) on the linker command line (in the *_LDFLAGS variable).

- The *ltdl* library provides extensive functionality beyond the *dl* library; this can be intimidating, but know that all of this other functionality is optional.

Let's look at what we need to do to the Jupiter project build system in order to use the *ltdl* library. First, we need to modify *configure.ac* to look for the *ltdl.h* header and search for the lt_dlopen function. This means modifying references to *dlfcn.h* and the *dl* library in the AC_CHECK_HEADERS and AC_SEARCH_LIBS macros, as highlighted in Listing 8-15.

Git tag 8.2

```
--snip--
# Checks for header files.
AC_CHECK_HEADERS([stdlib.h ltdl.h])
--snip--
# Checks for libraries.

# Checks for typedefs, structures, and compiler characteristics.

# Checks for library functions.
AC_SEARCH_LIBS([lt_dlopen], [ltdl])
--snip--
```

Listing 8-15: configure.ac: *Switching from* dl *to* ltdl *in* configure.ac

Even though we're using Libtool, we need to check for *ltdl.h* and *libltdl*, because *ltdl* is a separate library that must be installed on the end user's system. It should be treated the same as any other required third-party library. By searching for these installed resources on the user's system and failing configuration if they're not found, or by properly using preprocessor definitions in your source code, you can provide the same sort of configuration experience with *ltdl* that I've presented throughout this book when using other third-party resources.

I'd like you to recognize that this is the first time we've seen the requirement for the user to install an Autotools package on his system—and this is the very reason most people avoid using *ltdl*. The *GNU Libtool Manual* provides a detailed description of how to package the *ltdl* library with your project so it is built and installed on the user's system when your package is built and installed.[8]

Interestingly, shipping the source code for the *ltdl* library with your package is the only way to get your program to *statically* link with the *ltdl* library. Linking statically with *ltdl* has the side effect of not requiring the user to install the *ltdl* library on their system, since the library becomes part of the

---

8. In fact, the tutorial in the *GNU Libtool Manual* is a great example of adding subprojects to an Autotools build system.

project's executable images. There are a few caveats, however. If your project also uses a third-party library that dynamically links to *ltdl*, you'll have a symbol conflict between the shared and static versions of the *ltdl* libraries.[9]

The next major change we need to make is in the source code—it is limited, in this case, to *src/main.c* and highlighted in Listing 8-16.

```
#include "config.h"

#include "libjupiter.h"
#include "module.h"

#if HAVE_LTDL_H
# include <ltdl.h>
#endif

#define DEFAULT_SALUTATION "Hello"

int main(int argc, char * argv[])
{
    int rv;
    const char * salutation = DEFAULT_SALUTATION;

#if HAVE_LTDL_H
    int ltdl;
❶  lt_dlhandle module;
    get_salutation_t * get_salutation_fp = 0;

❷  LTDL_SET_PRELOADED_SYMBOLS();

❸  ltdl = lt_dlinit();
    if (ltdl == 0)
    {
❹      module = lt_dlopen("modules/hithere/hithere.la");
        if (module != 0)
        {
            get_salutation_fp = (get_salutation_t *)lt_dlsym(
                    module, GET_SALUTATION_SYM);
            if (get_salutation_fp != 0)
                salutation = get_salutation_fp();
        }
    }
#endif

    rv = jupiter_print(salutation, argv[0]);

#if HAVE_LTDL_H
    if (ltdl == 0)
    {
```

---

9. Given how rarely *ltdl* is currently used, this is an unlikely scenario these days, but this could change in the future if more packages begin to use *ltdl*.

```
        if (module != 0)
            lt_dlclose(module);
        lt_dlexit();
    }
#endif

    return rv;
}
```

Listing 8-16: src/main.c: Switching from dl to ltdl in source code

These changes are very symmetrical with respect to the original code. Mostly, items that previously referred to DL or dl now refer to LTDL or lt_dl. For example, #if HAVE_DLFCN_H becomes #if HAVE_LTDL_H, and so forth.

One important change is that the *ltdl* library must be initialized at ❸ with a call to lt_dlinit, whereas the *dl* library did not require initialization. In a larger program, the overhead of calling lt_dlinit and lt_dlexit would be amortized over a much larger code base.

Another important detail is the addition of the LTDL_SET_PRELOADED_SYMBOLS macro invocation at ❷. This macro configures global variables required by the lt_dlopen and lt_dlsym functions on systems that don't support shared libraries or in cases where the end user has specifically requested static libraries. It's benign on systems that use shared libraries.

One last detail is that the return type of dlopen is void *, or a generic pointer, whereas the return type of lt_dlopen is lt_dlhandle (see ❶ and ❹). This abstraction exists so *ltdl* can be ported to systems that use return types that are incompatible with a generic pointer.

When a system doesn't support shared libraries, Libtool actually links all of the modules that might be loaded right into the program. Thus, the jupiter program's linker (libtool) command line must contain some form of reference to these modules. This is done using the -dlopen *modulename* construct, as shown in Listing 8-17.

```
--snip--
jupiter_LDADD = ../libjup/libjupiter.la -dlopen modules/hithere/hithere.la
--snip--
```

Listing 8-17: src/Makefile.am: Adding a -dlopen option to the LDADD line

If you forget this addition to *src/Makefile.am*, you'll get a linker error about an undefined symbol—something like lt__PROGRAM__LTX_preloaded _symbols. If it doesn't detect any modules being linked into the application, Libtool won't clutter your program's global symbol space with symbols that will never be referenced; the symbols required by the *ltdl* library will be missing if the symbol table is empty.

It appears that *ltdl* is not quite as flexible as *dl* regarding the sort of path information you can specify in lt_dlopen to reference a module. In order to fix this problem, I hardwired the proper relative path (*modules/ hithere/hithere.la*) into *main.c*. Additionally, this example is sensitive to the current working directory. If you run jupiter from another directory, it will also fail to find the module. A real program would undoubtedly use a more robust method of configuration, such as a configuration file containing the absolute path to the desired module name.[10]

Preloading Multiple Modules

If Libtool links multiple modules into a program on a system without shared-library support, and if those modules each provide their own version of get_salutation, then there will be a conflict of public symbols within the program's global symbol space. This is because all of these modules' symbols become part of the program's global symbol space and the linker generally won't allow two symbols of the same name to be added to the executable symbol table. Which module's get_salutation function should be honored? Unfortunately, there's no good heuristic to resolve this conflict. The *GNU Libtool Manual* provides for this condition by defining a convention for maintaining symbol-naming uniqueness:

- All exported interface symbols should be prefixed with *modulename*_LTX _ (for example, hithere_LTX_get_salutation).

- All remaining non-static symbols should be reasonably unique. The method Libtool suggests is to prefix them with _*modulename*_ (as in _jupiter_*somefunction*).

- Modules should be named differently even if they're built in different directories.

Although it's not explicitly stated in the manual, the lt_dlsym function first searches for the specified symbol as *modulename*_LTX_*symbolname*, and then, if it can't find a prefixed version of the symbol, it searches for exactly *symbolname*. You can see that this convention is necessary, but only for cases in which Libtool may statically link such loadable modules directly into the application on systems that don't support shared libraries. The price you have to pay for Libtool's illusion of shared libraries on systems that don't support them is pretty high, but it's the going rate for getting the same loadable module functionality on all platforms.

To fix the *hithere* module's source code so that it conforms to this convention, we need to make one change to *hithere.c*, shown in Listing 8-18.

```
❶ #define get_salutation hithere_LTX_get_salutation
❷ #include "../../module.h"

const char * get_salutation(void)
```

---

10. When I tried the same soft-link trick we used earlier to configure the desired module, lt_dlopen failed to find the module. You see, while the filesystem will happily hand lt_dlopen the properly dereferenced filesystem entry (*modules/hithere/hithere.la*), when lt_dlopen parses this text file, it tries to append the relative path it finds there onto the containing directory of the *link* rather than onto the *hithere.la* file to which the filesystem resolved that link.

```
{
    return "Hi there";
}
```

Listing 8-18: src/modules/hithere/hithere.c: Ensuring public symbols are unique when using ltdl

By defining the replacement for get_salutation at ❶ before the inclusion of the *module.h* header file at ❷, we're also able to change the prototype in the header file so that it matches the modified version of the function name. Because of the way the C preprocessor works, this substitution only affects the function prototype in *module.h*, not the quoted symbol string or the type definition. At this point, you may want to go back and examine the way *module.h* is written to prove to yourself that this actually works.

Checking It All Out

You can test your program and modules for both static and dynamic shared-library systems by using the --disable-shared option on the configure command line, like this:

```
  $ make clean
  --snip--
  $ autoreconf
  $ ./configure --disable-shared && make
  --snip--
  $ cd src
  $ ls -1p modules/hithere/.libs
❶ hithere.a
  hithere.la
  hithere.lai
  $
  $ ./jupiter
❷ Hi there, from ./jupiter!
  $
  $ cd ..
  $ make clean
  --snip--
  $ ./configure && make
  $ cd src
  $ ls -1p modules/hithere/.libs
  hithere.a
  hithere.la
  hithere.lai
  hithere.o
  hithere.so
  $
  $ ./jupiter
❸ Hi there, from ...jupiter!
  $
```

As you can see, the output at ❷ and ❸ contains the *hithere* module's salutation in both configurations, yet the file listing at ❶ shows us that, in the --disable-shared version, the shared library doesn't even exist. It appears that *ltdl* is doing its job.

NOTE   *You may have noticed the difference in the example's output for the two executions of jupiter. In the first case, the output shows the name of the program as exactly ./jupiter, while in the second case, it shows ...jupiter. This was my attempt at removing the cruft in the output caused by Libtool's redirection of the shared-library-consuming version referring to the actual program—lt-jupiter—in the jupiter /src/.libs directory. Libtool uses a wrapper around programs linked to built shared libraries in order to make it simpler for uninstalled programs to find the built shared libraries upon which they depend.*

The Jupiter code base has become rather fragile, because I've ignored the issue of where to find shared libraries at runtime. As I've already mentioned, you would ultimately have to fix this problem in a real program. But given that I've finished my task of showing you how to properly use the Libtool *ltdl* library, I'll leave that as an exercise for you.

Summary

The decision to use shared libraries brings with it a whole truckload of issues, and if you're interested in maximum portability, you must deal with each of them. The *ltdl* library is not a solution to every problem. It solves some problems but brings others to the surface. Suffice it to say that using *ltdl* has trade-offs, but if you don't mind the extra maintenance effort, it's a good way to add maximum portability to your loadable-module project.

I hope that by spending some time going through the exercises in this book, you've been able to get your head around the Autotools enough to know how they work and what they're doing for you. At this point, you should be very comfortable *autotool-izing* your own projects—at least at the basic level.

In the coming chapters, I'll discuss additional tools and utilities that are also considered part of the GNU toolbox (and one or two that are not). I'll also show you how to convert a real-world project from a hand-coded build system to a much more concise, and probably more correct, Autotools build system.

9

UNIT AND INTEGRATION TESTING WITH AUTOTEST

. . . to learn is not to know;
there are the learners and the learned.
Memory makes the one, philosophy the other.
—*Alexandre Dumas,* The Count of Monte Cristo

Testing is important. All developers test their software to one degree or another; otherwise, they don't know if the product meets the design criteria. On one end of the spectrum, the author compiles and runs the program. If it presents the general interface they envisioned, they call it done. On the other end, the author writes a suite of tests that attempt to exercise as much of the code as possible under varying conditions, validating that the outputs are correct for the specified inputs. Straining to reach the rightmost point on this line, we find the person who literally writes tests first and then adds and modifies code iteratively until all the tests pass.

In this chapter, I won't attempt to expound on the virtues of testing. I assume every developer agrees that *some* level of testing is important, whether they are a compile-run-and-ship sorta person or a bona-fide test-driven person. I also assume every developer has some level of aversion to testing that lies somewhere along this spectrum. Therefore, our goal here is to let someone else do as much of the work of testing as possible. In this case, "someone else" means the Autotools.

Back in Chapter 3, we added a test to our handcoded makefiles for Jupiter. The output of the test was completely controlled by the make script we put into *src/Makefile*:

```
$ make check
cd src && make check
make[1]: Entering directory '/.../jupiter/src'
cc -g -O0  -o jupiter main.c
./jupiter | grep "Hello from .*jupiter!"
Hello from ./jupiter!
*** All TESTS PASSED
make[1]: Leaving directory '/.../jupiter/src'
$
```

When we moved on to Autoconf in Chapters 4 and 5, not much changed. The output was still controlled by our handwritten make script. We just moved it into *src/Makefile.in*.

In Chapter 6, however, we dropped our handcoded makefiles and templates in favor of Automake's much more terse *Makefile.am* files. Then we had to figure out how to shoehorn our handwritten test into automake script. In doing so, we got a bit of an upgrade on the test output:

```
$ make check
--snip--
PASS: greptest.sh
============================================================================
Testsuite summary for Jupiter 1.0
============================================================================
# TOTAL: 1
# PASS:  1
# SKIP:  0
# XFAIL: 0
# FAIL:  0
# XPASS: 0
# ERROR: 0
============================================================================
$
```

If you're running this code on your terminal with a reasonably late version of Automake, you'll even see colored test output.

Every test added to the TESTS variable in our *src/Makefile.am* template generates a PASS: or FAIL: line, and the summary values account for all of them. This is because Automake has a nice built-in testing framework driven by the TESTS variable. If you need to build any of the files specified in TESTS, you can just create a rule for it as we did for the trivial driver script (*greptest.sh*) that we created. If your test program needs to be built from sources, you can use a check_PROGRAM variable. One minor problem with

Automake's testing framework is that if tests are found in multiple directories, you see multiple such displays during make check, which can be a little annoying, especially when using make -k (continue in the face of errors), because it may not occur to you to scroll up to see the output of earlier, possibly failed tests.

In addition to the TESTS variable, if you set the XFAIL_TESTS variable to a subset of the tests listed in TESTS, you might also see some output in the XFAIL: and XPASS: lines. These are tests that are expected to fail. When such tests pass, they're listed in the XPASS: line as an *unexpected pass*. When they fail, they're listed in the XFAIL: line as an *expected failure*.

A test returning a shell code of 77 increases the count in the SKIP: line, and 99 increases the count in the ERROR: line. I'll provide more detail about special shell codes returned by tests later in this chapter.

As you might have guessed by now, Autoconf also includes a testing framework, called *autotest*, that provides all of the infrastructure required to allow you to simply and easily specify a test that exercises some portion of your code. The results are displayed in a consistent and easy-to-comprehend manner, and failed tests are easy to reproduce in an isolated fashion, complete with a built-in debugging environment. Almost makes you want to write tests, doesn't it? The fact is, a well-designed testing framework, like any other well-designed tool, is a joy to use.

Additionally, autotest is portable—as long as you write your portion of the tests using portable script or code, the entire test suite will be 100 percent portable to any system on which you can run your configure script. That's not as hard as it sounds. Often the shell script you have to write amounts to running a command, and the code behind the command is written using Autotools-provided portability features and is generated using Autotools-provided build processes.

For several years, autotest has been documented as being "experimental." Regardless, its base functionality hasn't changed much during those years, and Autoconf uses it to test itself, as we saw in Chapter 1. So, it's time to stop worrying about whether it's going to change and just start using it for its intended purpose: to make testing less of a chore for software developers, who—let's face it—really just want to write code and let someone else worry about testing.

Being the rational creatures that we are, we can't deny that testing is important. What we can do is make use of good tools that allow us to focus on our code, letting frameworks like autotest worry about ancillary issues like result formatting, success/failure semantics, data gathering for user-submitted bug reports, and portability. As we'll see throughout this chapter, this is the value that autotest provides.

In the spirit of transparency, I'll admit it's difficult to justify using autotest for small test suites like Jupiter's. The testing harness built into Automake is more than adequate for most small project needs. Larger projects—such as Autoconf itself, with its 500-plus unit and integration tests, which test functionality spread out over its entire project directory structure, and even installed components—are a different matter entirely.

Autotest Overview

There are three phases to the files consumed and generated by autotest. The first phase is what the *GNU Autoconf Manual* calls the "prep for distribution" phase. The second phase occurs when `configure` is executed, and the third phase happens during execution of the test suite. Let's take each of these phases in turn.

The first phase, which happens during building of the distribution archive, is essentially the process of generating the executable test program that can be run by users on their systems. It may seem a bit strange that this process must be done during the building of a distribution archive; however, Autoconf is required to be installed on any system that needs to generate this program, so the `testsuite` program must be built when the distribution archive is built so that it may be included in the archive for the user. While the `dist` or `distcheck` targets are being made, `configure` (and `make check`, when using `distcheck`) is executed; `make check` encapsulates rules to rebuild the test program from sources using `autom4te`—the Autoconf caching `m4` driver. The test program is built during execution of the `dist` target by virtue of having it included in the Automake `EXTRA_DIST` variable, which I'll talk about near the end of this chapter.

Figure 9-1 shows the flow of data from maintainer-written source files to the `testsuite` program during `make dist` (or `make distcheck`).

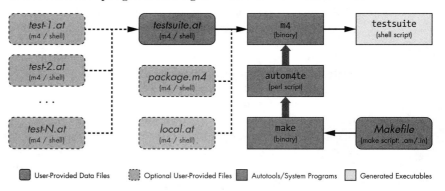

Figure 9-1: Data flow from maintainer-written input files to the testsuite program

The file *testsuite.at*, found at the top of the second column of the diagram, is the main test file written by a project author. This is actually the only maintainer-written file required by autotest. Exactly like *configure.ac*, this file contains shell script sprinkled with M4 macro definitions and invocations. This file is passed through the M4 macro processor, with `autom4te` acting as the driver for `m4`, to generate the `testsuite` program at the top of the last column, which is pure shell script. This process occurs during execution of `autom4te`, which is driven by `make check` reading the makefile generated from a *Makefile.in* or *Makefile.am* file that we write. The prep-for-distribution concept comes from the fact that the `check` target is executed during `make distcheck` (which, of course, builds the distribution archive); the `testsuite` program is added to the distribution archive during this process.

It's built during `make dist`, which does not execute `make check`, because all files listed in `EXTRA_DIST` must be built before they can be included in the distribution archive.

Details like this are normally hidden from us by the Autotools, but as autotest is still considered experimental—meaning, not fully integrated into the Autotools suite—the responsibility for some of this additional infrastructure is relegated to us, the maintainers. We'll cover these details shortly.

The Autoconf manual suggests that test suite authors may put individual sets of related tests, called *test groups*, into separate *.at* files. The *testsuite.at* file, then, contains only a series of `m4_include` directives, including each of these group-specific *.at* files. Therefore, M4 inclusion is the mechanism by which the optional *test-1.at* through *test-N.at* are gathered together into *testsuite.at* for processing by M4.

The *package.m4* and *local.at* files are optional maintainer-written (or generated) input files that are automatically included by `autom4te` when processing *testsuite.at*, if they're found. The former contains basic information about the test suite that's displayed on the console and embedded in bug reports generated by the test suite. The latter, the manual suggests, is an optional mechanism we may choose to use that can help us keep *testsuite.at* uncluttered with global definitions, non-group-related tests, and helper macro definitions and shell functions that may be used by the actual tests. We'll discuss the exact contents of these files later in the chapter.

When a `configure` script is instrumented for autotest, the configuration process generates additional, autotest-related artifacts. Figure 9-2 shows what happens graphically during the configuration process, relative to autotest.

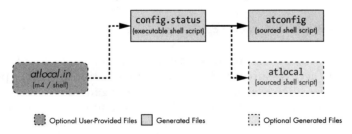

Figure 9-2: The flow of data during `configure` *while generating test-related templates*

Recall from Chapter 2 that `config.status` drives the file-generation portion of the configuration process. When a *configure.ac* file is set up for autotest, `config.status` generates *atconfig*—a shell script that's designed to be sourced by `testsuite` when it's executed.[1] It contains source- and build-tree

---

1. The concept of *shell script sourcing* means that one script reads the contents of another script into its own process space, and then continues as if this data had been originally written directly into the first. This process is very much like when the C preprocessor reads the contents of a header file into the translation unit that the compiler ultimately processes. One script is said to be *sourced* or included by another.

variables such as at_testdir, abs_builddir, abs_srcdir, at_top_srcdir, and so on, in order to facilitate access to files and products in the source and build trees during test suite execution.

The test author may also choose to create a template file called *atlocal.in* that allows them to pass additional Autoconf and project-specific configuration variables through to the test environment, as needed. The product of this template is *atlocal*—also a shell script that's designed to be sourced by testsuite, if it's present. If you choose to write *atlocal.in*, you must add it to the list of tags passed to an invocation of AC_CONFIG_FILES in *configure.ac*. We'll see how this is done later as we export Jupiter's async_exec flag to our test suite.

NOTE *Don't confuse* atlocal *with* local.at *from Figure 9-1. The* atlocal *file in Figure 9-2, sourced by* testsuite *at runtime, is used to pass configuration variables into the test environment from* configure, *while the* local.at *file is written directly by the project maintainer and contains additional test code processed by* autom4te *when* testsuite *is generated.*

Figure 9-3 shows the flow of data during the execution of testsuite.

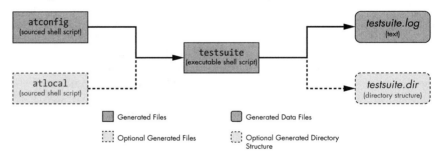

Figure 9-3: The flow of data during execution of the testsuite *script*

As mentioned previously, testsuite sources *atconfig* and *atlocal* (if present) to access source- and build-tree information and other project-related variables, then executes the tests that were generated into it. As it does so, it creates a *testsuite.log* file containing verbose information on the execution of each test. What you see on the screen is a single line of text per test.[2]

The testsuite program generates a directory called *testsuite.dir*. A separate subdirectory is created within this directory for each test. The test suite does not delete test-specific subdirectories for failed tests; we can use the contents of this directory structure to obtain details and to debug the problem. We'll go into detail about what gets added to these directories shortly.

The testsuite program may be executed by hand, of course, but it has to be generated first. The Autoconf manual suggests that the process of generating the testsuite program is best tied directly into the check target so that when make check is executed, testsuite will be generated (if it's missing or out-of-date with respect to its dependencies) and then executed.

_____

2. Unless you use the -v or --verbose option on the testsuite command line.

Since testsuite is added to the distribution archive, end users who run make check will merely execute the existing testsuite program, unless they've touched one of the files that testsuite depends on, in which case make will attempt to regenerate testsuite. Without Autoconf installed, this process would fail. Fortunately, it's not generally in the user's best interest to touch any of testsuite's dependencies in the distribution archive.

Wiring Up Autotest

Since my goal here is to teach you how to use this framework, the approach I chose to take in configuring Jupiter for autotest was to incorporate the entire set of optional files shown in Figures 9-1 through 9-3. This allows us to explore exactly how everything works together. While this approach is probably unwarranted for a project the size of Jupiter, it does work correctly, and it can always be pared down. I'll show you at the end of this chapter just what you can delete to reduce the autotest input file set to the bare minimum.

Before we can write tests, we need to make *configure.ac* aware of our desire to use autotest. This is done by adding two macro invocations to *configure.ac*, as shown in Listing 9-1.

Git tag 9.0

```
#                                          -*- Autoconf -*-
# Process this file with autoconf to produce a configure script.

AC_PREREQ([2.69])
AC_INIT([Jupiter],[1.0],[jupiter-bugs@example.org])
AM_INIT_AUTOMAKE
LT_PREREQ([2.4.6])
LT_INIT([dlopen])
AC_CONFIG_SRCDIR([src/main.c])
AC_CONFIG_HEADERS([config.h])

AC_CONFIG_TESTDIR([tests])
AC_CONFIG_FILES([tests/Makefile
                 tests/atlocal])

# Checks for programs.
AC_PROG_CC
AC_PROG_INSTALL
--snip--
```

Listing 9-1: configure.ac: Wiring autotest into configure.ac

The first of these two macros, AC_CONFIG_TESTDIR, tells Autoconf to enable autotest and specifies that the testing directory will be called *tests*. You may use a dot here to represent the current directory if you wish, but the *GNU Autoconf Manual* recommends that you use a separate directory for ease in managing test output files and directories.

NOTE *The addition of AC_CONFIG_TESTDIR to* configure.ac *is actually the only change required to enable autotest in a project, though changes to makefiles and additional support files are required to make it useful and more automated. Interestingly, this important tidbit is not found anywhere in the manual, though it is implied rather subtly.*

The second line is the standard Autoconf AC_CONFIG_FILES instantiating macro. I'm using a separate instance of it here to generate the test-related files from templates.

Let's look at what goes into each of these files. The first is a makefile for the *tests* directory that's generated from an Autoconf *Makefile.in* template, which itself is generated from the Automake *Makefile.am* file that we need to write. In this makefile, we need to get make check to generate and execute testsuite. Listing 9-2 shows how we might write *tests/Makefile.am* so that Automake and Autoconf will generate such a makefile.

```
❶ TESTSUITE = $(srcdir)/testsuite
❷ TESTSOURCES = $(srcdir)/local.at $(srcdir)/testsuite.at
❸ AUTOM4TE = $(SHELL) $(top_srcdir)/missing --run autom4te
❹ AUTOTEST = $(AUTOM4TE) language=autotest

❺ check-local: atconfig atlocal $(TESTSUITE)
           $(SHELL) '$(TESTSUITE)' $(TESTSUITEFLAGS)

❻ atconfig: $(top_builddir)/config.status
           cd $(top_builddir) && $(SHELL) ./config.status tests/$@

❼ $(srcdir)/package.m4: $(top_srcdir)/configure.ac
           $(AM_V_GEN) :;{ \
           echo '# Signature of the current package.' && \
           echo 'm4_define([AT_PACKAGE_NAME],    [$(PACKAGE_NAME)])' && \
           echo 'm4_define([AT_PACKAGE_TARNAME], [$(PACKAGE_TARNAME)])' && \
           echo 'm4_define([AT_PACKAGE_VERSION], [$(PACKAGE_VERSION)])' && \
           echo 'm4_define([AT_PACKAGE_STRING],  [$(PACKAGE_STRING)])' && \
           echo 'm4_define([AT_PACKAGE_BUGREPORT], [$(PACKAGE_BUGREPORT)])'; \
           echo 'm4_define([AT_PACKAGE_URL],     [$(PACKAGE_URL)])'; \
           } >'$(srcdir)/package.m4'

❽ $(TESTSUITE): $(TESTSOURCES) $(srcdir)/package.m4
           $(AM_V_GEN) $(AUTOTEST) -I '$(srcdir)' -o $@.tmp $@.at; mv $@.tmp $@
```

Listing 9-2: tests/Makefile.am: Getting make check to build and run testsuite

Before we begin dissecting this file, I should mention that these contents were taken from Section 19.4 of the *GNU Autoconf Manual*. I've tweaked them a bit, but essentially these lines comprise a portion of the recommended way to tie autotest into Automake. We'll complete this file as we discuss additional features and requirements of autotest-oriented make script.

NOTE *This lack of more complete integration, along with the fact that Autoconf can be config- ured to use several different test drivers (DejaGNU, for instance), is likely what keeps autotest in experimental mode. While Libtool, for instance, has slowly migrated toward a position of complete integration with Automake, autotest still requires some fiddling to properly integrate into a project build system. Nevertheless, once the requirements are understood, proper integration is pretty simple. Additionally, as we've seen, Automake has its own test framework, which gives Automake maintainers little incentive to fully support autotest.*

The code in Listing 9-2 is pretty straightforward when taken a line at a time—four variables and four rules. The variables are not strictly necessary, but they make for shorter command lines and less duplication in rules and commands. The TESTSUITE variable at ❶ simply keeps us from having to pre- fix testsuite with $(srcdir)/ everywhere we use it.

NOTE *The testsuite program is distributed, so it should be built in the source tree. Files that are destined to end up in the distribution archive should be found in the source directory structure. Additionally, the content of such built and distributed files should be the same, regardless of differences in configuration options used by the original archive creator, or the end user.*

The TESTSOURCES variable at ❷ allows us to easily add additional tests to the makefile. Each *.at* file becomes a dependency of testsuite so that when one of them is changed, testsuite is rebuilt.

The AUTOM4TE variable at ❸ allows us to wrap execution of autom4te with the Automake missing script, which prints a nicer error message if autom4te is not found. This happens when an end user who doesn't have the Autotools installed does something that requires testsuite to be rebuilt—such as modify *testsuite.at.*

NOTE *We can't use Automake's maintainer-rules option to avoid writing these rules into distribution archive Makefiles because we must manually write these rules.*

The AUTOTEST variable at ❹ appends the --language=autotest option to the autom4te command line. *There is actually no program in the Autoconf package called* autotest. If we had to pin down the definition of such a tool, it would be the contents of this AUTOTEST variable.

The check-local rule at ❺ ties execution of testsuite into Automake's check target. Automake standard targets like check have a -local counterpart that you can use to supplement the functionality generated by Automake for the base target. If Automake sees a rule with the target check-local in *Makefile.am*, it generates a command to run $(MAKE) check-local under the generated *Makefile*'s check rule. This gives you a hook into the standard Automake targets. We'll cover such hooks in greater detail in Chapters 14 and 15, where we'll use them extensively in our efforts to convert a real- world project to use an Autotools-based build system.

The check-local target depends on *atconfig*, *atlocal*, and $(TESTSUITE). Recall from Figure 9-3 that *atlocal* is a script sourced by testsuite. It's generated directly by the invocation of AC_CONFIG_FILES that we added to *configure.ac* in Listing 9-1, so we'll cover its contents shortly. The command for this rule executes '$(TESTSUITE)' with $(TESTSUITEFLAGS) as a command line argument. The contents of TESTSUITEFLAGS are user defined, allowing the end user to run make check TESTSUITEFLAGS=-v, for instance, to enable verbose output from testsuite while making targets that invoke testsuite.

You can also use TESTSUITEFLAGS to target specific test groups by number (for instance, TESTSUITEFLAGS=2 testsuite) or, if you've written your tests using the AT_KEYWORDS macro, by tag name. In addition, several command line options are available for the generated testsuite program. You can find complete documentation for testsuite options in Section 19.3 of the *GNU Autoconf Manual*.

NOTE *The single quotes around $(TESTSUITE) allow the path in TESTSUITE to contain spaces, if needed. This technique can and should be used in all makefiles to handle whitespace in paths. I've generally ignored the concept of whitespace in paths within this book in order to reduce the noise in the listings, but you should be aware that makefiles can be written to properly handle whitespace in all filenames and paths—those in targets and dependencies, as well as those in the commands associated with rules.*

I mentioned previously that the *atconfig* script, also sourced by testsuite, is generated automatically beneath the covers by AC_CONFIG_TESTDIR. The problem is, even though config.status understands how to build this file, Automake doesn't know anything about it because it's not listed directly in any of the instantiating macro invocations in *configure.ac*, so we need to add an explicit rule to *Makefile.am* to create or update it. This is where the atconfig rule at ❻ in Listing 9-2 comes in. The check-local rule depends on it, so its commands will be executed if *atconfig* is missing or older than its dependency, $(top_builddir)/config.status, when make check is executed.

The command in the rule for generating $(srcdir)/package.m4 at ❼ (note there is only one command here) merely writes text into the target file if the file is missing or older than *configure.ac*. This is an optional input file (see Figure 9-1), the contents of which are actually required by autotest in some form. Several M4 macros must be defined in the input data that is processed by autotest to create a test suite, including AT_PACKAGE_NAME, AT_PACKAGE_TARNAME, AT_PACKAGE_VERSION, AT_PACKAGE_STRING, AT_PACKAGE_BUGREPORT, and AT_PACKAGE_URL. These variables may be defined directly in *testsuite.at* (or any of the subfiles included by that file), but it makes more sense to generate this information from values already found in *configure.ac* so we don't have to maintain two sets of the same information. This is the very reason why *package.m4* is included automatically by autom4te if it's found while processing *testsuite.at*.

But wait—why not use AC_CONFIG_FILES to have configure generate this file? All we're doing is generating a text file that contains configuration variables, and that sounds like exactly what AC_CONFIG_FILES is for. The problem is, AC_CONFIG_FILES and the other instantiating macros always generate files into the build tree, and *package.m4* must end up in the source tree in order to be added to the distribution archive (not because it's part of any build or execution process the user may instigate, but because it's part of the source code for testsuite). Perhaps the full integration of autotest, at some point in the future, will result in the ability to request the instantiating macros to generate files into the source tree. Until then, this is what we have to work with.

The fourth and final rule, $(TESTSUITE), at ❽, generates $(srcdir)/testsuite using the $(AUTOTEST) command. Because $(TESTSUITE) is a dependency of check-local, it'll get built if it's not up-to-date. The autom4te program, when executed in *autotest mode*, accepts the -I option for specifying include paths for *.at* files that may be included by *testsuite.at* or any of its inclusions. It also accepts the -o option for specifying the output file, testsuite.[3]

NOTE *I've added $(AM_V_GEN) in front of the commands of the last two rules in Listing 9-2 to allow my custom rules to tie into the Autotools' silent build rules system. Any command prefixed with $(AM_V_GEN) will cause the normal command output to be replaced with* GEN target *when building with silent rules enabled. See Section 21.3 of the GNU Automake manual for more details on this and other variables that affect build output when building with silent rules.*

Taken as a whole, all of this allows us to run make check at the command prompt to build (if needed) and execute $(srcdir)/testsuite.

NOTE *There's a bit more we need to do in this* Makefile.am *file to fully integrate autotest functionality into Automake. We'll add some additional administrative rules and variables later in this chapter. For clarity at this point, I limited the content to just what we need to build and run the test suite.*

Well, we've created a new directory and added a new *Makefile.am.* By now, you should be automatically thinking about how this *Makefile.am* file is going to be called if we don't link it into the top-level *Makefile.am* SUBDIRS variable. You're absolutely correct—this must be our next step. Listing 9-3 shows this modification to the top-level *Makefile.am* file.

```
SUBDIRS = common include libjup src tests
```

Listing 9-3: Makefile.am: *Adding the* tests *subdirectory*

NOTE *I added* tests *last. This will almost always be the pattern for a directory such as* tests. *In order to test the system, most, if not all, of the other directories must be built first.*

---

3. It may seem strange that there are two (semicolon-separated) commands in this rule, the first of which generates its output into a temporary file, after which the second moves that temporary file into the final target. This is done because autom4te is generating a script in a piecemeal fashion—the file can be seen and accessed while it's only partially generated. Attempts to execute such a partial script will likely fail but can sometimes cause data loss. This generate-and-move idiom is used to remove risk of the possible data loss scenarios because mv is an atomic filesystem operation.

The second file in Listing 9-1 is the *atlocal* shell script that's automatically sourced by testsuite, if present, which may be used to pass additional configuration variables through to testsuite's runtime environment. We'll use this file in the Jupiter project to pass the async_exec flag through to testsuite so it may know if the program it's testing has been configured with the *async-exec* feature enabled. Listing 9-4 shows how this is done in *atlocal*'s template, *atlocal.in*.

```
async_exec=@async_exec@
```

Listing 9-4: tests/atlocal.in: A template for generating atlocal

Now, this causes a small problem for us because configure is not yet exporting a substitution variable called async_exec. We wrote a shell script that uses a shell variable of this name back in Chapter 5, but recall we only used it to indicate whether we should invoke AC_DEFINE to generate the ASYNC _EXEC preprocessor definition into *config.h.in*. We now need to use AC_SUBST on this variable in order to generate an Autoconf substitution variable of the same name. Listing 9-5 highlights the single-line addition to *configure.ac* required to make this happen.

```
--snip--
  ----------------------------------------
  Unable to find pthreads on this system.
  Building a single-threaded version.
  ----------------------------------------])
        async_exec=no
    fi
fi

AC_SUBST([async_exec])
if test "x${async_exec}" = xyes; then
    AC_DEFINE([ASYNC_EXEC], 1, [async execution enabled])
fi
--snip--
```

Listing 9-5: configure.ac: Making autoconf generate the async_exec substitution variable

One last comment on Listing 9-1: we could have simply added these files to the existing invocation of AC_CONFIG_FILES at the bottom of *configure.ac*, but using a separate invocation here keeps test-related items together. It also serves to illustrate the fact that AC_CONFIG_FILES may indeed be invoked multiple times within *configure.ac*, the results being cumulative.

We now need to create a set of source *.at* files that can be used by autom4te to generate our test program. This set of files can be as simple as a single *testsuite.at* file or as complex as the diagram in Figure 9-1, including *testsuite.at*, a set of test-group-specific *.at* files, and a *local.at* file. These files will contain autotest macro invocations mixed with simple or complex shell script, as required by your testing needs. We'll start with a single line of autotest initialization code in a *tests/local.at* file, as shown in Listing 9-6.

Listing 9-6: tests/local.at: Initialization code for testsuite can be added to a local.at *file*

The `AT_INIT` macro is required by `autom4te` to be found somewhere within the translation unit presented by *testsuite.at* and its inclusions. This single macro invocation expands into several hundred lines of shell script that define the basic testing framework and all of the ancillary boilerplate functionality associated with it.

We also need to create an empty *testsuite.at* file in the *tests* directory. We'll add items to it as we progress:

```
$ touch tests/testsuite.at
```

We now have the basis for generation and execution of the autotest framework in Jupiter. Every project that uses autotest will have to be configured in the manner we've shown so far. For smaller projects, some of the optional pieces may be omitted, the contents of which would then be combined directly into *testsuite.at*. We'll discuss how to simplify when we've completed our exploration of autotest. For now, let's give it a try:

```
$ autoreconf -i
--snip--
$ ./configure
--snip--
config.status: creating tests/Makefile
config.status: creating tests/atlocal
--snip--
config.status: executing tests/atconfig commands
------------------------------------------------

Jupiter Version 1.0
--snip--
$ make check
--snip--
Making check in tests
make[1]: Entering directory '/.../jupiter/tests'
make  check-local
make[2]: Entering directory '/.../jupiter/tests'
:;{ \
  echo '# Signature of the current package.' && \
  echo 'm4_define([AT_PACKAGE_NAME], [Jupiter])' && \
  echo 'm4_define([AT_PACKAGE_TARNAME], [jupiter])' && \
  echo 'm4_define([AT_PACKAGE_VERSION], [1.0])' && \
  echo 'm4_define([AT_PACKAGE_STRING], [Jupiter 1.0])' && \
  echo 'm4_define([AT_PACKAGE_BUGREPORT], [jupiter-bugs@example.org])'; \
  echo 'm4_define([AT_PACKAGE_URL], [])'; \
} >'tests/package.m4'
/bin/bash ../missing --run autom4te --language=autotest \
    -I '.' -o testsuite.tmp testsuite.at
mv testsuite.tmp testsuite
/bin/bash './testsuite'
```

```
## ---------------------- ##
## Jupiter 1.0 test suite. ##
## ---------------------- ##

## ------------ ##
## Test results. ##
## ------------ ##

0 tests were successful.
--snip--
$
```

We can see from the output of configure that our generated files were created in the *tests* directory, as expected. It also appears that the code generated by AC_CONFIG_TESTDIR has wired in the generation of the *tests/atconfig* file as a *command* tag, rather than as a simple template file, using AC_CONFIG _COMMANDS internally.

We then see from the output of make check that testsuite was both built and executed. We can't yet incorporate testsuite into a distribution archive from the dist or distcheck targets because we haven't wired our autotest functionality into Automake. However, when we complete our changes at the end of this chapter, you'll find that running make check against the contents of a distribution archive will not build testsuite, as it will have shipped with the archive (assuming we haven't touched any of testsuite's dependencies).

NOTE   *One interesting item of note near the top of the make check output is highlighted by the lines starting with make[1]: and make[2]:, where make indicates it's entering the jupiter/tests directory twice. This happens because of the check-local hook we added, where the check target recursively invokes $(MAKE) check-local as a command within the same directory.*

Great, it works—with make check anyway. But it doesn't do anything yet except print a few extra lines of text to the console. To make it do something useful, we need to add some tests. Therefore, our first task will be to move the original Automake-based jupiter execution test from *src/Makefile.am* into our autotest test suite.

Adding a Test

Autotest tests are bundled into sets called *test groups*. The purpose of a test group is to allow tests within a group to interact with each other. For example, the first test in a group may generate some data files used by subsequent tests within the same group.

Tests that interact with each other are harder to debug, and broken tests are harder to reproduce if they require other tests to run first. Multi-test groups are hard to avoid when striving for full coverage; the ideal is to have only one test per test group as much as possible. Where you just can't do it, test groups exist to facilitate the required interaction. The crux of this facility is that *tests within the same test group are executed within the same temporary directory*, allowing initial tests to generate files that subsequent tests can then see and access.

Our single test will not suffer from these problems—mainly because we haven't yet put much effort into testing Jupiter (and, if we're honest with ourselves, there isn't much actual code to test). Right now, when you execute make check, you see two sets of test output on the screen:

```
$ make check
--snip--
❶ make   check-TESTS
make[2]: Entering directory '/.../jupiter/src'
make[3]: Entering directory '/.../jupiter/src'
PASS: greptest.sh
=========================================================================
Testsuite summary for Jupiter 1.0
=========================================================================
# TOTAL: 1
# PASS:  1
# SKIP:  0
# XFAIL: 0
# FAIL:  0
# XPASS: 0
# ERROR: 0
--snip--
❷ Making check in tests
make[1]: Entering directory '/.../jupiter/tests'
make   check-local
make[2]: Entering directory '/.../jupiter/tests'
/bin/bash './testsuite'
## ---------------------- ##
## Jupiter 1.0 test suite. ##
## ---------------------- ##

## ------------ ##
## Test results. ##
## ------------ ##

0 tests were successful.
--snip--
$
```

The first set of tests (beginning at ❶) are executed in the *jupiter/src* directory. This is our original grep-based test where we check jupiter's output against a pattern. As you can see, the basic test framework built into Automake is not bad. We're hoping to improve on that framework with autotest. The second set of tests (beginning at ❷) are executed in the *jupiter/tests* directory and involve autotest.

Defining Tests with AT_CHECK

The grep-based test we've been using in *src/Makefile.am* is a perfect example for use in the AT_CHECK macro provided by the autotest framework. Here are the prototypes for the AT_CHECK family of macros:

AT_CHECK(*commands*, [*status* = '0'], [*stdout*], [*stderr*], [*run-if-fail*], [*run-if-pass*])
AT_CHECK_UNQUOTED(*commands*, [*status* = '0'], [*stdout*], [*stderr*], [*run-if-fail*], [*run-if-pass*])

AT_CHECK executes *commands*, checks the returned status against *status*, and compares the output on stdout and stderr with the contents of the *stdout* and *stderr* macro arguments. If *status* is omitted, autotest assumes a successful status code of zero. If *commands* returns a status code to the shell that does not match the expected status code specified in *status*, the test fails. In order to ignore the status code of *commands*, you should use the special command ignore in the *status* parameter.

Regardless, there are a couple of status codes that even ignore will not ignore: a status code of 77 (skip) returned by *commands* will cause autotest to skip the rest of the tests in the current test group, while 99 (hard failure) will cause autotest to fail the entire test group immediately.

Like *status*, the *stdout* and *stderr* parameters appear to be optional, but looks can be deceiving. If you pass nothing in these arguments, this merely tells Autoconf that the test's stdout and stderr output streams are expected to be empty. Anything else will fail the test. So how do we tell autotest we don't want to check the output? As with *status*, we can use special commands in *stdout* or *stderr*, including those shown in Table 9-1:

Table 9-1: Special Commands Allowed in *stdout* and *stderr* Arguments.

Command	Description
ignore	Do not check this output stream, but do log it to the test group's log file.
ignore-no-log	Do not check or log this output stream.
stdout	Log and capture the test's stdout output to the file stdout.
stderr	Log and capture the test's stderr output to the file stderr.
stdout-nolog	Capture the test's stdout output to the file stdout, but do not log.
stderr-nolog	Capture the test's stderr output to the file stderr, but do not log.
expout	Compare the test's stdout output to the file expout, created earlier; log the differences.
experr	Compare the test's stderr output to the file experr, created earlier; log the differences.

The *run-if-fail* and *run-if-pass* arguments allow you to optionally specify shell code that should be executed upon test failure or success, respectively.

AT_CHECK_UNQUOTED does exactly the same thing as AT_CHECK, except that it performs shell expansion on *stdout* and *stderr* first, before making the comparison with the output of *commands*. Since AT_CHECK doesn't do shell expansion on *stdout* and *stderr*, it stands to reason that you need to use AT_CHECK_UNQUOTED if you reference any shell variables in the text of these parameters.

Defining Test Groups with AT_SETUP and AT_CLEANUP

The AT_CHECK macro must be invoked between invocations of AT_SETUP and AT_CLEANUP, the pair of which define a test group and, therefore, the temporary directory from which the tests in the group are executed. The prototypes for these macros are defined as follows:

```
AT_SETUP(test-group-name)
AT_CLEANUP
```

If you've got any experience with the *xUnit* family of unit test frameworks (JUnit, NUnit, CPPUnit, and so on), you've probably got a pretty strong notion of what the setup and cleanup (or teardown) functions should be used for. Usually a *setup* function runs some common code before each test in a test set, and a *cleanup* or *teardown* function executes some common code at the end of each test in the set.

Autotest is a bit different—there is no formal setup or teardown functionality shared by tests belonging to the same group (although this sort of functionality can be emulated with shell functions defined within the test group in *testsuite.at*, or in its included subfiles). As with *xUnit* frameworks, Autotest runs every test in total isolation, because every test runs within its own subshell. The only way a test can affect a subsequent test is by sharing the same test group and leaving filesystem droppings around for subsequent tests to examine and act upon.

AT_SETUP accepts only one argument, *test-group-name*, which is the name of the test group that we're starting, and this argument is required. AT_CLEANUP accepts no arguments.

We'll add the group setup and cleanup macro invocations, wrapping a call to AT_CHECK, to a new file, *tests/jupiter.at*, as shown in Listing 9-7.

Git tag 9.1
```
AT_SETUP([jupiter-execution])
AT_CHECK([../src/jupiter],,[Hello from ../src/.libs/lt-jupiter!])
AT_CLEANUP
```

Listing 9-7: tests/jupiter.at: Adding our first test group—attempt #1

Libtool adds a wrapper script in the *src* directory for any executables that use Libtool shared libraries. This wrapper script allows jupiter to find the uninstalled Libtool libraries it's trying to use. As mentioned in Chapter 7, it's a convenience mechanism that Libtool provides so we don't have to jump through hoops to test programs using Libtool shared libraries before they're installed.

The end result is that the *src/jupiter* script is executing the real jupiter program from *src/.libs/lt-jupiter*. Because jupiter displays its own location, based on its argv[0] contents, we need to expect it to print this path.

We then need to add an m4_include statement to our currently empty *testsuite.at* file in order to include *jupiter.at*, as shown in Listing 9-8.

```
m4_include([jupiter.at])
```

Listing 9-8: tests/testsuite.at: Including jupiter.at in testsuite.at

We'll also want to add this new source file to our *tests/Makefile.am* file's TESTSOURCES variable so it becomes a prerequisite of testsuite, as shown in Listing 9-9.

```
--snip--
TESTSUITE = $(srcdir)/testsuite
TESTSOURCES = $(srcdir)/local.at $(srcdir)/testsuite.at \
  $(srcdir)/jupiter.at
AUTOM4TE = $(SHELL) $(top_srcdir)/missing --run autom4te
AUTOTEST = $(AUTOM4TE) --language=autotest
--snip--
```

Listing 9-9: tests/Makefile.am: Adding additional sources to TESTSOURCES

We'll follow this practice for every test we add to our test suite. In the end, the only thing in *testsuite.at* will be several invocations of m4_include, one for each test group. Executing this code renders the following output:

```
$ autoreconf -i
--snip--
$ ./configure
--snip--
$ make check
--snip--
/bin/bash './testsuite'
## --------------------- ##
## Jupiter 1.0 test suite. ##
## --------------------- ##
  1: jupiter-execution                                      FAILED (jupiter.at:2)

## ------------- ##
## Test results. ##
## ------------- ##

ERROR: 1 test was run,
1 failed unexpectedly.
## ------------------------ ##
## testsuite.log was created. ##
## ------------------------ ##

Please send `tests/testsuite.log' and all information you think might help:

   To: <jupiter-bugs@example.org>
   Subject: [Jupiter 1.0] testsuite: 1 failed

You may investigate any problem if you feel able to do so, in which
case the test suite provides a good starting point. Its output may
be found below `tests/testsuite.dir'.
--snip--
$
```

NOTE *Running autoreconf and configure was required only because we updated the* tests/ Makefile.am *file. If we'd just touched an existing .at file, which is rebuilt by the check target in* tests/Makefile, *then neither* autoreconf *nor* configure *would have been necessary.*

I'll admit here that our single test failed because I deliberately coded the test incorrectly in order to show you what a failed test looks like.

Although not obvious from the output, there is more than one *testsuite.log* file created by testsuite when tests fail. The first is a master *testsuite.log* file in the *tests* directory, which is always created, even when all tests pass, and is designed to be sent in bug reports to the project maintainer. There is also a log file of the same name in a separate numbered directory within the *tests/ testsuite.dir* directory for failed tests. The name of each of these directories is the number of the test group that failed. The test group number can be seen in the output. While you only need the master *testsuite.log* file, since it contains the entire contents of all of the individual tests' *testsuite.log* files, this file also contains a lot of other information about your project and the test environment that the maintainer would want to see but just gets in the way for our purposes here.

To see exactly how our test failed, let's examine the contents of the *testsuite.log* file left in the *tests/testsuite.dir/1* directory:

```
$ cat tests/testsuite.dir/1/testsuite.log
#                          -*- compilation -*-
1. jupiter.at:1: testing jupiter-execution ...
./jupiter.at:2: ../src/jupiter ❶
--- /dev/null    2018-04-21 17:27:23.475548806 -0600 ❷
+++ /.../jupiter/tests/testsuite.dir/at-groups/1/stderr 2018-06-01 16:08:04.391926296 -0600
@@ -0,0 +1 @@
+/.../jupiter/tests/testsuite.dir/at-groups/1/test-source: line 11: ../src/jupiter: ❸
  No such file or directory
--- -    2018-06-01 16:08:04.399436755 -0600 ❹
+++ /.../jupiter/tests/testsuite.dir/at-groups/1/stdout 2018-06-01 16:08:04.395926314 -0600
@@ -1 +1 @@
-Hello from ../src/.libs/lt-jupiter!
+
./jupiter.at:2: exit code was 127, expected 0 ❺
1. jupiter.at:1: 1. jupiter-execution (jupiter.at:1): FAILED (jupiter.at:2)
$
```

First note that autotest writes, as often as possible, the related source line into the *testsuite.log* file. This isn't a big win for us at this point, but if *testsuite.at* or its included files were long and complicated, you can see how this information could be very helpful.

At ❶, we see the argument we passed to the *commands* parameter of AT_CHECK, along with the number of the line at which this argument was passed to the macro in *jupiter.at*.

However, now things start to get a bit muddy. The entire point of the *stdout* and *stderr* arguments in AT_CHECK is to provide some comparison text for what is actually sent by the *commands* to these output streams. In accordance with the general Unix philosophy of not duplicating existing functionality, the autotest authors chose to use the diff utility to make these comparisons. The log lines from ❷ to ❸ (inclusive) show the *unified*[4] output of the diff utility when comparing the *original* file (*/dev/null* since we passed no value in the *stderr* argument) to the *modified* file— the text sent to the stderr output stream during the attempt to execute *../src/jupiter*.

If you're not familiar with unified diff output, a brief explanation is in order. The two lines starting at ❷ indicate the objects being compared. The original, or minus (---), line indicates the left side of the comparison, while the modified, or plus (+++), line indicates the right side of the comparison. Here, we're comparing */dev/null* with a temporary file called */.../jupiter/tests/testsuite.dir/at-groups/1/stderr* that was used by autotest to capture the stderr stream during the attempt to execute *../src/jupiter*.

The next line, starting and ending with @@, is a *chunk* marker—diff's way of telling us about a portion of the two files that does not match. There can be more than one chunk in the output displayed by diff. In this case, the entire output text is so short that only one chunk was required to show us the differences.

The numbers in the chunk marker represent two ranges, separated by a space. The first range starts with a minus (-) sign, indicating the range associated with the *original* file, and the second range starts with a plus (+) sign, indicating the range associated with the *modified* file. Here's the line we're currently discussing:

```
@@ -0,0 +1 @@
```

A range is made up of two integer values separated by a comma (,) or a single value with a default second value of 1. In this example, the range being compared starts at zero in the original file and is zero lines long, while the comparison range in the second file starts at line 1 and is one line long. These ranges are 1-based, meaning line 1 is the first line in the file. Therefore, the first range specification, -0,0, is a special range that means there's no content in the file.

The lines following the range specification contain the full text of these ranges, showing us the actual differences. The original file lines are printed first, each prefixed with a minus sign, and then the modified file lines are printed afterward, each prefixed with a plus sign. When there is enough content around the modified lines to do so, additional unprefixed lines are added before and after these lines, showing some context around the changes. In this case, the entire content of this section is:

```
+/.../jupiter/tests/testsuite.dir/at-groups/1/test-source: line 11: ../src/jupiter:
  No such file or directory
```

---

4. The diff utility has a couple different output styles that may be selected using command line arguments. The -u option usually selects *unified* output format.

Since the original file was empty, as indicated by the -0,0 range in the chunk marker, there are no lines starting with minus. All we see is the one modified file line starting with a plus.

Well, clearly these files are not the same—we expected nothing on the stderr stream, but we got some error text instead. The shell experienced an error attempting to execute *../src/jupiter*—it could not be found. If you try this at the shell prompt, you'll see the following output:

```
$ ../src/jupiter
bash: ../src/jupiter: No such file or directory
$
```

NOTE
Obviously you should not do this from the tests *directory, or any other directory that's a sibling to the* src *directory, or it'll actually find* jupiter *(if it's been built) rather than print this error.*

If you put this line into a shell script called (arbitrarily) *abc.sh* and execute the script on the bash command line, you'll see output that matches the format shown in *testsuite.log*:

```
$ bash abc.sh
abc.sh: line 2: ../src/jupiter: No such file or directory
$
```

We can see at ❺ in *testsuite.log* that the shell returned a 127 status code, indicating an error of some sort. The value 127 is used by the shell to indicate execution errors—file not found or file not executable.

To be complete, let's also consider the lines between ❹ and ❺ for a moment. This is the unified diff output seen when comparing the text specified in AT_CHECK's *stdout* argument with what was actually written to stdout by jupiter (actually the shell, since we know *../src/jupiter* was not found). In this case, we see that the minus text is the original comparison text we specified and the plus text is a single newline character, as this is what the shell sent to stdout. The chunk marker range specification, fully expanded, would be:

```
@@ -1,1 +1,1 @@
```

There was one line of text in each of the original and modified sources to be compared, but, as we can see by the output, the text in these sources was completely different.

So What Happened?

This first attempt assumed that the jupiter program (or, rather, the Libtool wrapper script) is found at *../src/jupiter*, relative to the *tests* directory. While this assumption is true, I've already alluded to the fact that each test group is executed in its own temporary directory, so it makes perfect sense that this relative path is not going to work from another directory. Even if we figured out, by trial and error, how many parent directory references to use, it would be quite fragile; if we ran testsuite from a different directory, it would fail because it depends so intimately on running from a specific position relative to the jupiter program.

Let's try a different tack. We'll make use of the variables generated by configure into *atconfig*. One of them, abs_top_builddir, contains the absolute path to the top build directory. Therefore, we should be able to successfully reference jupiter from anywhere using ${abs_top_builddir}/*src/jupiter*.

But now we have another problem: jupiter prints its own path and we've just decided to obtain that path using a shell variable, so we'll also need to change the comparison text to use this variable, as well. This change, however, causes yet another issue—we'll need to change AT_CHECK to AT_CHECK_UNQUOTED if we expect that shell variable in AT_CHECK's *stdout* parameter to be expanded before the macro makes the comparison. Let's make these modifications by changing *jupiter.at* as shown in Listing 9-10.

Git tag 9.2

```
AT_SETUP([jupiter-execution])
AT_CHECK_UNQUOTED(["${abs_top_builddir}"/src/jupiter],,
                  [Hello from ${abs_top_builddir}/src/.libs/lt-jupiter!
])
AT_CLEANUP
```

Listing 9-10: tests/jupiter.at: Adding our first test group—attempt #2

Here, we've switched to using AC_CHECK_UNQUOTED, and we've changed both the jupiter program path in the first argument and the comparison text in the third argument to use the abs_top_builddir variable we inherit from *atconfig*.

NOTE *The newline at the end of the* stdout *argument is intentional and explained shortly.*

Let's try it out:

```
$ make check
--snip--
/bin/bash './testsuite'
## ---------------------- ##
## Jupiter 1.0 test suite. ##
## ---------------------- ##
  1: jupiter-execution                              ok

## ------------- ##
## Test results. ##
## ------------- ##

1 test was successful.
--snip--
$
```

Once again, I only had to run make. Even though our changes were dramatic, affecting even the macros we called in the test suite, remember that the entire test suite is generated from make check. The only time we need to execute autoreconf and configure is if we make changes to *configure.ac* or any of the templates from which it generates files used by make check, or if we make any changes to *Makefile.am* files.

This attempt had much better results, but what's with that extra newline at the end of our comparison text in Listing 9-10? Well, remember what it is that we're sending to stdout from jupiter. Listing 9-11 provides a reminder.

```
--snip--
    printf("Hello from %s!\n", (const char *)data);
--snip--
```

Listing 9-11: common/print.c: What jupiter sends to stdout

The comparison text is an exact duplicate of what we expect to find on jupiter's stdout, so we'd better be sure to include every character we write; the trailing newline is part of that data stream.

It's probably a good idea at this point to remove the *src/Makefile.am* code that builds and runs the Automake version of the test. Change *src/Makefile.am* as shown in Listing 9-12.

```
SUBDIRS = modules/hithere

bin_PROGRAMS = jupiter
jupiter_SOURCES = main.c module.h
jupiter_CPPFLAGS = -I$(top_srcdir)/include
jupiter_LDADD = ../libjup/libjupiter.la -dlopen modules/hithere/hithere.la
```

Listing 9-12: src/Makefile.am: The updated full contents of this file after removing tests

NOTE *All we did here was remove the test-related lines from the bottom half of the file.*

Unit Testing vs. Integration Testing

On the whole, the autotest version is not much better than what we had when our test was being executed by Automake's test framework in *src/Makefile.am*. Adding new tests is, however, a bit simpler than what we'd have to do in *src/Makefile.am*'s TESTS variable. In fact, the only way the Automake version becomes simpler is if we actually write test programs and build them in check primaries. We might still have to build test programs in check primaries, but calling them and validating their output is trivial when using autotest.

If you're thinking that it feels like autotest is more attuned to system and integration testing than unit testing, you're pretty close to the mark. Autotest is designed to test your project from the outside, but it's not limited to such tests. Anything you can call from the command line can be a test, from autotest's perspective. What autotest actually offers you is a framework for generating uniform test output, regardless of the kind of tests you're using.

One approach to unit testing that I've used for years involves writing test programs where the main source module of my test program literally #includes the *.c* file I'm testing. This gives me the option of calling static methods within the module under test and provides direct access to internal structures defined within that module.[5] This approach is pretty C oriented, but other languages have their own ways of performing the same sort of tricks. The idea is to create a test program that can reach into the private parts of a module and exercise functionality in small chunks. When you put those chunks together, you can feel confident that the individual chunks are working as designed; if there's a problem, it's probably in the way you glued them together.

Let's add some unit testing to Jupiter by creating a test module that tests the functions in the *common/print.c* module. Create a file called *test _print.c* in the *common* directory that contains the content in Listing 9-13.

Git tag 9.3

```
#define printf mock_printf
#include "print.c"

#include <stdarg.h>
#include <string.h>

static char printf_buf[512];

int mock_printf(const char * format, ... )
{
    int rc;
    va_list ap;
    va_start(ap, format);
    rc = vsnprintf(printf_buf, sizeof printf_buf, format, ap);
    va_end(ap);
    return rc;
}

int main(void)
{
    const char *args[] = { "Hello", "test" };
    int rc = print_it(args);
    return rc != 0 || strcmp(printf_buf, "Hello from test!\n") != 0;
}
```

Listing 9-13: common/test_print.c: A unit test program for the print.c *module*

---

5. I used to do this by adding a main function bracketed with #ifdef UNIT_TEST and #endif at the bottom of the file, but I found that this approach cluttered my source file—often the test code ended up being longer than the code I was testing—and it was more difficult to maintain a logical separation between test code and the code under test. The #include method allows the compiler to assure you that the code under test compiles and links without the test code.

The first line uses the preprocessor to rename any calls to `printf` to `mock_printf`. The second line then uses the preprocessor to #include *print.c* directly into *test_print.c*. Now, any calls to `printf` inside of *print.c* will be redefined to call `mock_printf` instead—including any prototypes defined in system header files like *stdio.h*.[6]

The idea here is to verify that the `print_it` function actually prints *Hello from* argument*!\n* and returns zero to the caller. We don't need any output— a shell return code is sufficient for this test to indicate to the user that `print_it` is working as designed.

Neither do we need this module's `main` routine to accept any command line arguments. If we had several tests in here, however, it might be convenient to accept some sort of argument that allows us to tell the code which test we want to run.

All we're really doing here is directly calling `print_it` with a short string and then attempting to verify that `print_it` returned zero and actually passed what we expected to `printf`. Note that `print_it` is a static function, which should make it inaccessible to other modules, but because we're including *print.c* at the top of *test_print.c*, we're effectively combining both source files into a single translation unit.

Now, let's write the build code for this test program. First, we need to add some lines to *common/Makefile.am* so that a test_print program gets built when we run make check. Modify *common/Makefile.am* as shown in Listing 9-14.

```
noinst_LIBRARIES = libjupcommon.la
libjupcommon_la_SOURCES = jupcommon.h print.c

check_PROGRAMS = test_print
test_print_SOURCES = test_print.c
```

Listing 9-14: common/Makefile.am: Adding `test_print` as a `check_PROGRAM`

When we make the check target, we'll now get a new program, test_print, in the *common* directory. Now we need to add a call to this program to our test suite. Create a new file in *tests* called *print.at*, as shown in Listing 9-15.

```
AT_SETUP([print])
AT_CHECK(["${abs_top_builddir}/common/test_print"])
AT_CLEANUP
```

Listing 9-15: tests/print.at: Adding the `print` test

We also need to add an `m4_include` statement for this test to *testsuite.at*, as in Listing 9-16.

```
m4_include([jupiter.at])
m4_include([print.at])
```

Listing 9-16: tests/testsuite.at: Adding print.at to testsuite.at

---

6. This can occasionally have an undesirable side effect when, for example, the system header file uses implementation-specific tricks to prototype functions in a nonportable (and non-standard) manner for the sake of performance, but it usually works just fine.

And finally, we need to add this new source file to the TESTSOURCES variable in *tests/Makefile.am*, as shown in Listing 9-17.

```
--snip--
TESTSUITE = $(srcdir)/testsuite
TESTSOURCES = $(srcdir)/local.at $(srcdir)/testsuite.at \
  $(srcdir)/jupiter.at $(srcdir)/print.at
AUTOM4TE = $(SHELL) $(top_srcdir)/missing --run autom4te
AUTOTEST = $(AUTOM4TE) -language=autotest
--snip--
```

Listing 9-17: tests/Makefile.am: Adding print.at to TESTSOURCES

Since the test_print program only uses the shell status code to indicate an error, using it in AT_CHECK is as simple as it gets. You only need the first argument—the name of the program itself. If test_print had more than one test, you might accept a command line argument (within the same parameter) that indicates which test you want to run and then add several invocations of AC_CHECK, each running test_print with a different argument.

Notice that we've started a new test group—as I mentioned earlier, you should try hard to limit your test groups to a single test unless the nature of the tests are such that they work together on the same file-based data set.

Let's give it a shot. Note that in order to run the new test, we really only need to make the check target to update and execute testsuite. However, since we added the *print.at* dependency to *tests/Makefile.am*, we should probably also run autoreconf and configure. Had we enabled maintainer mode, the extra maintainer-mode rules would have done this for us:

```
$ autoreconf -i
--snip--
$ ./configure
--snip--
$ make check
--snip--
/bin/bash './testsuite'
## ---------------------- ##
## Jupiter 1.0 test suite. ##
## ---------------------- ##
  1: jupiter-execution                              ok
  2: print                                          ok

## ------------- ##
## Test results. ##
## ------------- ##

All 2 tests were successful.
--snip--
$
```

Administrative Details

I mentioned while describing the contents of *tests/Makefile.am* back near Listing 9-2 that we needed to add some additional infrastructure to that file in order to complete the tie-in with Automake. Let's take care of those details now.

We've seen that make (all) and make check work just fine, building our products and building and executing our test suite. But we've neglected some of the other targets that Automake wires up for us—specifically, installcheck, clean, and distribution-related targets like dist and distcheck. There's a general lesson to be considered here: whenever we add custom rules to *Makefile.am*, we need to consider the impact on the standard targets generated by Automake.

Distributing Test Files

There are several generated files that need to be distributed. These files are not inherently known by Automake, and, therefore, Automake needs to be told explicitly about them. This is done with the Automake-recognized EXTRA_DIST variable, which we'll add to the top of *tests/Makefile.am*, as shown in Listing 9-18.

Git tag 9.4

```
EXTRA_DIST = testsuite.at local.at jupiter.at print.at \
  $(TESTSUITE) atconfig package.m4

TESTSUITE = $(srcdir)/testsuite
TESTSOURCES = $(srcdir)/local.at $(srcdir)/testsuite.at \
  $(srcdir)/jupiter.at $(srcdir)/print.at
--snip--
```

Listing 9-18: tests/Makefile.am: Ensuring test files get distributed with EXTRA_DIST

Here, I've added all the test suite source files, including *testsuite.at*, *local.at*, *jupiter.at*, and *print.at*. I've also added the testsuite program and any input files we generated using a non-Automake mechanism. These include *atconfig*, which is generated by code provided by the AC_CONFIG_TESTDIR macro internally, and *package.m4*, which is generated by a custom rule we added earlier to this *Makefile.am* file. It's important to understand here that adding files to EXTRA_DIST causes them to be built, if needed, when make dist is executed.

NOTE   *I would have like to have just used $(TESTSOURCES) in EXTRA_DIST, but the sources in that variable were formatted for rules and commands. EXTRA_DIST, as interpreted by Automake, is designed to refer to a list of files relative to the current directory within the source tree.*

As a reminder, we distribute *.at* files because the GNU General Public License says we must distribute the source code for our project and these files are the source code for testsuite, just as *configure.ac* is the source code for configure. However, even if you're not using the GPL, you should still consider shipping the preferred editing format of all files in your project; it is an open source project, after all.

Checking Installed Products

We wrote a check target; it's probably a good idea to support the *GCS* installcheck target, which Automake also supports. This is done by adding the installcheck-local target to this *Makefile.am* file, as shown in Listing 9-19.

```
--snip--
check-local: atconfig atlocal $(TESTSUITE)
        $(SHELL) '$(TESTSUITE)' $(TESTSUITEFLAGS)

installcheck-local: atconfig atlocal $(TESTSUITE)
        $(SHELL) '$(TESTSUITE)' AUTOTEST_PATH='$(DESTDIR)$(bindir)' $(TESTSUITEFLAGS)
--snip--
```

Listing 9-19: tests/Makefile.am: Supporting installed-product testing

The only difference between check-local and installcheck-local is the addition of the AUTOTEST_PATH command line option to testsuite, pointing testsuite to the copy of jupiter found in $(DESTDIR)$(bindir), where it was installed. AUTOTEST_PATH is prepended to the shell PATH variable before invoking *commands* in AT_CHECK_UNQUOTED; therefore, you could write test code that assumes PATH contains the path to an installed copy of jupiter. However, tests are designed to be executed on either installed or uninstalled programs, so it's a good idea to continue deriving and using a full path to programs within your test commands.

Now we'll need to make a decision. When the user types make check, they clearly mean to test the copy of jupiter in the build tree. But when they type make installcheck, certainly they want to check the installed version of the program, either in the default install location or wherever the user indicates by using command line make variables like DESTDIR, prefix, and bindir.

This brings up a new issue: when we run tests for uninstalled jupiter, we're relying on Libtool's wrapper script to ensure jupiter can find *libjupiter.so*. Once we start testing installed jupiter, we'll become responsible for showing jupiter where *libjupiter.so* is located. If jupiter is installed in standard places (such as */usr/lib*), the system will naturally find *libjupiter.so*. Otherwise, we'll have to set the LD_LIBRARY_PATH environment variable to point to it.

So, how do we write our tests to work correctly in both situations? One interesting (but broken) approach is shown in Listing 9-20.

```
AT_SETUP([jupiter-execution])

set -x
find_jupiter()
{
```

```
        jupiter="$(type -P jupiter)"
        LD_LIBRARY_PATH="$(dirname "${jupiter}")/../lib" export LD_LIBRARY_PATH
        compare="${jupiter}"
        if test "x${jupiter}" == x; then
          jupiter="${abs_top_builddir}/src/jupiter"
          compare="$(dirname "${jupiter}")/.libs/lt-jupiter"
        fi
      }

      find_jupiter
      AT_CHECK_UNQUOTED(["${jupiter}"],,
                        [Hello from ${compare}!
      ])
      AT_CLEANUP
```

Listing 9-20: tests/jupiter.at: Testing execution for both installed and uninstalled jupiter—attempt#1

The find_jupiter shell function attempts to locate jupiter in the PATH by using the shell's type command. If the first result is empty, we revert to using the uninstalled version of jupiter.

The function sets two shell variables, jupiter and compare. The jupiter variable is the full path to jupiter. The compare variable is derived from jupiter and contains either the value of ${jupiter} or the Libtool location and name for uninstalled versions. We can set LD_LIBRARY_PATH in both cases to the *../lib* directory, relative to where jupiter is found because that's probably[7] where it's installed.

The problems with this approach are numerous. First, it doesn't handle, very well, the situation where jupiter *should* be installed but isn't found in the specified or implied install path. In this case, the code quietly reverts to testing the uninstalled version—likely not what you wanted. Another issue is that find_jupiter will locate jupiter anywhere in the PATH, even if the instance is not the one you intended to test. But there's an even more nefarious bug: if you execute make check, intending to test the uninstalled version, and an installed version of jupiter happens to be somewhere in the PATH, that's the version that will be tested.

It's unfortunate that AUTOTEST_PATH defaults to a non-empty value when it's not specified on the command line, as this would be a good way to differentiate the use of make check from make installcheck. However, AUTOTEST_PATH does default to the name of the directory specified in AC_CONFIG_TESTDIR, which also happens to be the value of ${at_testdir}—one of the variables generated by AC_CONFIG_TESTDIR in *atconfig*. We can use this fact to differentiate between make check and make installcheck by comparing ${AUTOTEST_PATH} to ${at_testdir}. Change *tests/jupiter.at* as shown in Listing 9-21.

---

7. It's possible that the user installed the software such that the *lib* directory is not a direct sibling of *bin*, in which case this test would fail. A better way of dealing with this issue is to figure out where the library was installed by querying the value of ${libdir}, which would have to be passed into the test environment through *atlocal.in*.

```
--snip--
set -x
find_jupiter()
{
  if test "x${AUTOTEST_PATH}" == "x${at_testdir}"; then
    jupiter="${abs_top_builddir}/src/jupiter"
    compare="$(dirname "${jupiter}")/.libs/lt-jupiter"
  else
    jupiter="${AUTOTEST_PATH}/jupiter"
    LD_LIBRARY_PATH="${AUTOTEST_PATH}/../lib" export LD_LIBRARY_PATH
    compare="${jupiter}"
  fi
}
jupiter=$(find_jupiter)
--snip--
```

Listing 9-21: tests/jupiter.at: A better way to use AUTOTEST_PATH

Now, when make check is executed, the jupiter variable will always be set directly to the uninstalled version in the build tree (and compare will be set to .../.libs/lt-jupiter), but when make installcheck is entered, it will be set to ${AUTOTEST_PATH}/jupiter (and compare will be set to the same value). Additionally, since we're able to fully distinguish between installed and uninstalled testing, we can set the LD_LIBRARY_PATH only for installed versions of jupiter.

If AUTOTEST_PATH has been set incorrectly, which can happen (for example, when the user sets DESTDIR or prefix incorrectly on the make command line), the test will fail because ${jupiter} will not be found.

If I were to add additional tests that needed to run the jupiter program, these lines would be a perfect candidate for *local.at*. The problem is that a shell script designed to run within tests *must* be defined and executed between calls to AT_SETUP and AT_CLEANUP; otherwise, it's simply omitted from the autom4te output stream while generating testsuite. So, how exactly is *local.at* useful to us? Well, you can't write shell code directly in *local.at*, but you can define M4 macros that can be invoked from within your test modules. Let's move the find_jupiter functionality into a macro definition in *local.at*, as shown in Listing 9-22.

Git tag 9.5

```
AT_INIT

m4_define([FIND_JUPITER], [[set -x
if test "x${AUTOTEST_PATH}" == "x${at_testdir}"; then
  jupiter="${abs_top_builddir}/src/jupiter"
  compare="$(dirname ${jupiter})/.libs/lt-jupiter"
else
  LD_LIBRARY_PATH="${AUTOTEST_PATH}/../lib" export LD_LIBRARY_PATH
  jupiter="${AUTOTEST_PATH}/jupiter"
  compare="${jupiter}"
fi]])
```

Listing 9-22: tests/local.at: Moving find_jupiter to an M4 macro

Using a shell function was, perhaps, a good idea when we started, but it's become a bit extraneous at this point, so I modified the code to just set the jupiter variable directly. Notice the second (*value*) argument of the call to m4_define is set verbatim to the shell script we want to have generated when the macro is invoked.

The set -x command in the first line of the *value* argument enables shell diagnostic output so you can see the contents of this macro executing, but only if you set TESTSUITEFLAGS=-v on the make command line. This is the default setting for the output generated into *testsuite.log*, so you'll be able to see what the code generated by the macro invocation is actually doing.

NOTE *You may have noticed the code in the second argument of* m4_define *has two sets of square brackets around it. This is not strictly necessary in this example because there are no special characters in the embedded code snippet. However, if there had been, double quoting would have allowed the special characters—embedded square brackets, or even a previously defined M4 macro name—to be generated exactly as is, without interference from the* m4 *utility.*

Now change *tests/jupiter.at* as shown in Listing 9-23.

```
AT_SETUP([jupiter-execution])
FIND_JUPITER
AT_CHECK_UNQUOTED(["${jupiter}"],,
                  [Hello from ${compare}!
])
AT_CLEANUP
```

Listing 9-23: tests/jupiter.at: Calling the FIND_JUPITER *macro*

With this change in place, all tests that needed to run the jupiter program may do so merely by invoking the FIND_JUPITER macro and then executing ${jupiter} within the *commands* argument. As you can see, the options available to you are endless—they're limited only by your mastery of the shell. Because each test is run in a separate subshell, you can feel free to set any environment variables you want without affecting subsequent tests.

Let's try it out. Note we haven't changed anything except for *.at* files, so we need only run make to see the effects of our changes. First, we'll install into a local directory using DESTDIR so that we can see how make installcheck works:

```
$ make install DESTDIR=$PWD/inst
--snip--
$ make check TESTSUITEFLAGS=-v
--snip--
Making check in tests
make[1]: Entering directory '/.../jupiter/tests'
make  check-local
make[2]: Entering directory
  '/home/jcalcote/dev/book/autotools2e/book/sandbox/tests'
/bin/bash './testsuite' -v
## ---------------------- ##
## Jupiter 1.0 test suite. ##
```

```
## ---------------------- ##
1. jupiter.at:1: testing jupiter-execution ...
++ test xtests == xtests
❶ ++ jupiter=/.../jupiter/src/jupiter
+++ dirname /.../jupiter/src/jupiter
++ compare=/.../jupiter/src/.libs/lt-jupiter
++ set +x
./jupiter.at:3: "${jupiter}"
1. jupiter.at:1:  ok
--snip--
$ make installcheck DESTDIR=$PWD/inst TESTSUITEFLAGS=-v
--snip--
Making installcheck in tests
make[1]: Entering directory '/.../jupiter/tests'
/bin/bash './testsuite' AUTOTEST_PATH='/.../jupiter/inst/usr/local/bin' -v
## ---------------------- ##
## Jupiter 1.0 test suite. ##
## ---------------------- ##
1. jupiter.at:1: testing jupiter-execution ...
++ test x/.../jupiter/inst/usr/local/bin == xtests
++ LD_LIBRARY_PATH=/.../jupiter/inst/usr/local/bin/../lib
++ export LD_LIBRARY_PATH
❷ ++ jupiter=/.../jupiter/inst/usr/local/bin/jupiter
++ compare=/.../jupiter/inst/usr/local/bin/jupiter
++ set +x
./jupiter.at:3: "${jupiter}"
1. jupiter.at:1:  ok
--snip--
$
```

Here, the execution of make check TESTSUITEFLAGS=-v shows us at ❶
that jupiter is being picked up from the build tree and compare is set
to the path of the Libtool binary, while make installcheck DESTDIR=$PWD/
inst TESTSUITEFLAGS=-v indicates at ❷ that jupiter being picked up from
the installation path we specified and compare is set to the same location.
LD_LIBRARY_PATH is also being set in this code path.

Cleaning Up

The testsuite program has a --clean command line option that cleans up
the *tests* directory of all test droppings. To wire that into the clean target, we
add a clean-local rule, as shown in Listing 9-24.

<table>
<tr><td>Git tag 9.6</td><td>

```
installcheck-local: atconfig atlocal $(TESTSUITE)
        $(SHELL) '$(TESTSUITE)' AUTOTEST_PATH='$(bindir)' $(TESTSUITEFLAGS)

clean-local:
        test ! -f '$(TESTSUITE)' || $(SHELL) '$(TESTSUITE)' --clean
        rm -rf atconfig

atconfig: $(top_builddir)/config.status
        cd $(top_builddir) && $(SHELL) ./config.status tests/$@
```

</td></tr>
</table>

Listing 9-24: tests/Makefile.am: *Adding support for make clean in tests*

If testsuite exists, it's asked to clean up after itself. I've also added a command to remove the generated *atconfig* script, as `make distcheck` fails if this file is not removed during execution of the `clean` target while checking the distribution directory from which the package is built, and *atconfig* is not generated by Automake or by any Autoconf code that Automake monitors. At this point, you could try out `make dist` or `make distcheck` to see whether it now works as it should.

NOTE *You might think the* `clean-local` *target is optional, but it's required so that* `make` *distcheck won't fail when building a distribution archive due to extra files being left around after* `distcheck` *runs* `make clean`*.*

Note also that we don't need to, nor should we, attempt to clean up files generated into the source tree, such as *package.m4* and testsuite itself. Why not? Much like `configure`, autotest products like testsuite sit somewhere between source files we write by hand and product files found in the build tree. They're built from sources, but stored in the source tree and ultimately distributed in the archive.

Niceties

One more thing I like to do is add a call to `AT_COLOR_TESTS` to my *local.at* file, right after the invocation of `AT_INIT`. Users can always specify colored test output using a command line argument to testsuite (`--color`), but using this macro allows you to enable colored test output by default. Change your *local.at* file as shown in Listing 9-25.

Git tag 9.7

```
AT_INIT
AT_COLOR_TESTS

m4_define([FIND_JUPITER], [set -x
if test "x${AUTOTEST_PATH}" == "x${at_testdir}"; then
--snip--
```

Listing 9-25: tests/local.at: Making colored test output the default

You should notice that the ok text after each successful test, as well as the summary line `All 2 tests were successful.`, is now green. If you had experienced any failed tests, those tests would have shown `FAILED` (testsuite.at:*N*) in red after the failed tests, with summary lines in red as follows:

```
ERROR: All 2 tests were run,
1 failed unexpectedly.
```

A Minimal Approach

I mentioned at the outset that we'd take a look at cutting this system down to the bare necessities for smaller projects. Here's what we can do without:

tests/local.at   Copy the contents of this file to the top of *testsuite.at* and delete the file.

tests/atlocal.in   Assuming you don't need to pass any configuration variables into your testing environment, remove the reference to the product file, *atlocal*, from the call to AC_CONFIG_FILES in *configure.ac* and delete this template file.

tests/*.at (subtest files included by *testsuite.at*)   Copy the content of these files serially into *testsuite.at* and delete the files; remove the m4_include macro invocations that were originally used to include these files.

Edit *tests/Makefile.am* and remove all references to the preceding files (templates and generated products) from the EXTRA_DIST and TESTSOURCES variables, as well as from the check-local and installcheck-local rules.

I would not remove the *package.m4* rule, although you may do so if you wish by copying the m4_define macro invocations generated into this file directly to the top of *testsuite.at*. Since make generates this file using variables defined by Autoconf, a generated instance of *package.m4* already contains the values that replace the variable references in the command in *tests/Makefile.am*. In my opinion, the value of not having to edit this information in two places far outweighs the overhead of maintaining the file generation rule.

Summary

We've covered the basics of autotest, but there are a dozen more or less useful AT_* macros you can use in your *testsuite.at* file. Section 19.2 of the *GNU Autoconf Manual* documents them in detail. I've shown you how to wire autotest into the Automake infrastructure. If you've chosen to use autotest without Automake, there will be some differences between the Automake's *tests/Makefile.am* and Autoconf's *tests/Makefile.in*, as you can no longer rely on Automake to do some things for you. However, if you're writing your own *Makefile.in* templates, these modifications will quickly become obvious to you.

I've also shown you a technique for creating unit tests in C that allows you full access to the private implementation details of a source module. I've attained nearly 100 percent code coverage using this technique in past projects, but I'll warn you of one caveat now: writing unit tests at this level makes it much more difficult to change the functionality of your application. Fixing a bug is not so bad, but making design changes will generally require you to disable or entirely rewrite the unit tests associated with the code affected by your changes. It's a good idea to be pretty sure of your design before committing yourself to this level of unit testing. It has been said that one of the primary values of writing unit tests is that you can *set them and forget them*—that is, you can write the tests once, wire them into your build system, and no longer pay attention to whether the code under test is working. Well, this is mostly true, but if you ever have to modify a well-tested feature in your project, unless you comment your test code well, you'll find yourself wondering what you were thinking when you wrote that test code.

Nevertheless, I sleep better at night knowing that code I just committed to my company repository is fully tested. I've also been more confident in discussions with colleagues regarding bugs surrounding my well-tested code. Autotest has helped reduce the effort involved in these projects.

With the end of this chapter, we've also come to the end of the Jupiter project—and it's a good thing, because I've taken the *Hello, world!* concept *much* further than anyone has a right to. From here on out, we'll be focusing on more isolated topics and real-world examples.

10

FINDING BUILD DEPENDENCIES
WITH PKG-CONFIG

*A common mistake that people make when trying
to design something completely foolproof is to
underestimate the ingenuity of complete fools.*
—*Douglas Adams,* The Hitchhiker's
Guide to the Galaxy

Let's say your project depends on a library
called *stpl*—some third-party library. How
can your build system determine where
libstpl.so is installed on an end user's system?
Where are *stpl*'s header files to be found? Do you simply assume */usr/lib* and */usr/include*? This is effectively
what Autoconf does if you don't tell it to look elsewhere, and for many packages perhaps that's fine—it's a common convention to install libraries and
headers into these directories.

But what if they aren't installed in these locations? Or perhaps they were
built locally and installed into the */usr/local* tree. What compiler and linker
options should be used when using *stpl* in your project? These are a few issues
that have plagued developers from the beginning, and the Autotools don't
really do much to manage this problem. Autoconf expects any libraries you
use to be installed into "standard places," meaning into directories where the
preprocessor and linker automatically look for header files and libraries. If

a user has the library but it's installed in a different location, the Autotools expect end users to know how to interpret the configuration failure. As it happens, there are several good reasons why many libraries are not installed in these standard places.

Additionally, Autoconf can't easily find libraries that are not installed at all. For example, you may have just built a library package and you want another package to pick up headers and libraries from the first package's build directory structure. This is possible with Autoconf, but it involves the end user setting variables like CPPFLAGS and LDFLAGS on the configure command line. The project maintainer can make things a little easier by providing user-specified configuration options with the AC_ARG_ENABLE and AC_ARG_WITH macros for libraries they anticipate might not be easy to find in standard places, but if your project uses a lot of third-party libraries, it's just guesswork trying to determine which of these will be particularly problematic for users. And then, users are not often programmers; we cannot rely on them to have enough background to know what to use for option values, even if we do supply command line options for problematic dependencies. Throughout this chapter, I'll refer to these as *build dependency issues.*

There is a tool that provides a solution for these types of problems in a more elegant fashion using a very common tactic in software design—by providing another layer of indirection. And it's not part of the GNU Autotools. Regardless, its use has become so prolific over the past 20 years that it would be an oversight to exclude a description of *pkg-config* in any book that discusses Linux build systems. Pkg-config is as useful as it is today because a lot of projects have begun to use it—especially library projects, and most especially library projects that install into nonstandard locations.

In this chapter, we'll look at pkg-config's components and functionality and how to use it in your projects. For your own library projects, we'll also discuss how to update the pkg-config database as your package is installed onto users' (or as I like to call them, potential contributors') systems.

Before we get started, allow me to state up front that this chapter makes a lot of references to filesystem objects that may or may not exist on your flavor of Linux, or perhaps exist in a different form or location. It's the nature of what we're doing here—we're discussing packages with libraries and header files that probably exist in different locations on different Linux flavors. I'll try to point out such potential differences as we come to them so you'll not be too surprised when things don't line up exactly on our two systems.

A pkg-config Overview

Pkg-config is an open source software utility maintained by the *freedesktop.org* project. The pkg-config website is a part of the freedesktop.org project website. The pkg-config project is the result of an effort to turn the gnome-config script—a part of the gnome build system—into a more general-purpose tool, and it seems to have caught on. Some inspiration for pkg-config also came from the gtk-config program.

A NOTE ABOUT PKG-CONFIG CLONES

The freedesktop.org pkg-config project has been around for a long time and has garnered a somewhat loyal following. As a direct result, other projects now exist that offer functionality that is similar—often identical—to the pkg-config project. One that comes to mind is the pkgconf project (which Red Hat's Fedora Linux seems to prefer when you ask the yum package manager to install pkg-config for you).

Do not confuse these two. The pkgconf project is a modern clone of the original pkg-config project that comes with claims of higher efficiency—and as far as I'm concerned, it may very well live up to these claims. Regardless, this chapter is about pkg-config. If you've found a project that provides functionality similar to pkg-config and you like it, then by all means use it. My goal here is to teach you about pkg-config. If this material helps you understand how to use pkgconf, or another pkg-config clone, then you're getting from this book exactly what I hoped to convey.

That said, however, I cannot cover all of the nuanced differences among the different pkg-config clones. If you want to follow along with my examples and your flavor of Linux won't install the original pkg-config package for you, you can always navigate in your browser to *https://www.freedesktop.org/wiki/Software/pkg-config* and download and install version 0.29.2 of the original pkg-config project.

Pkg-config is used simply by invoking the `pkg-config` command line utility with options that display the desired data. When you're looking for libraries and header files, the "desired data" includes package version information as well as compiler and linker options wherein library and header file locations are specified. For instance, to obtain C-preprocessor flags required to access a library's header files, one need only specify the `--cflags` option on the `pkg-config` command line, and compiler options appropriate for the package are displayed to `stdout`. This display can be captured and appended to compiler command lines in your configuration scripts and makefiles as needed.

Perhaps you're wondering why we even need a tool like `pkg-config` when Autoconf provides the `AC_CHECK_LIB` and `AC_SEARCH_LIBS` macros. In the first place, as I mentioned previously, the Autoconf macros are designed to only look in "standard locations" for libraries. You can trick the macros into looking in other places by preloading search paths into `CPPFLAGS` (using `-I` options) and `LDFLAGS` (using `-L` options). However, pkg-config is designed to help you find libraries that may be installed in places only your users' pkg-config installations know about; the best thing about pkg-config is that it knows how to find libraries and headers on end users' systems that these users don't even know about. Pkg-config can also tell your build system about additional dependencies required when your users are trying to statically link to your libraries. Therefore, it effectively hides such details from users, and that's the sort of user experience we're looking for.

There are, however, some caveats to using pkg-config with the Autotools. In the first edition of this book, I suggested that the PKG_CHECK_MODULES add-on M4 macro that's shipped with pkg-config was a good approach to using it with Autoconf. I've since amended my thoughts on this issue as I've discovered over the years that the use of this macro can cause more problems than it solves under some rather common conditions. Additionally, pkg-config is so simple to use directly in shell script that it makes little sense to wrap it with less-than-transparent M4 macros. We'll discuss this topic in much more detail in this chapter, but I wanted to set the stage for the pattern of use you'll see in the examples that follow.

Diving In

Let's start by looking at the output of the --help option for the pkg-config command:

```
$ pkg-config --help
Usage:
  pkg-config [OPTION...]
--snip--
Application Options:
--snip--
  --modversion                  output version for package
--snip--
  --libs                        output all linker flags
  --static                      output linker flags for static linking
--snip--
  --libs-only-l                 output -l flags
  --libs-only-other             output other libs (e.g. -pthread)
  --libs-only-L                 output -L flags
  --cflags                      output all pre-processor and compiler flags
  --cflags-only-I               output -I flags
  --cflags-only-other           output cflags not covered by the cflags-only-I option
  --variable=NAME               get the value of variable named NAME
  --define-variable=NAME=VALUE  set variable NAME to VALUE
  --exists                      return 0 if the module(s) exist
  --print-variables             output list of variables defined by the module
  --uninstalled                 return 0 if the uninstalled version of one or more
                                module(s) or their dependencies will be used
  --atleast-version=VERSION     return 0 if the module is at least version VERSION
  --exact-version=VERSION       return 0 if the module is at exactly version VERSION
  --max-version=VERSION         return 0 if the module is at no newer than version
                                VERSION
  --list-all                    list all known packages
  --debug                       show verbose debug information
  --print-errors                show verbose information about missing or conflicting
                                packages (default unless --exists or
                                --atleast/exact/max-version given on the command line)
--snip--
$
```

I've only shown what I consider the most useful options here. There are another dozen or so, but these are the ones we'll use all the time in our *configure.ac* files. (I've taken the liberty of wrapping long description lines as pkg-config seems to think everyone has a 300-column monitor.)

Let's start by listing all of the modules pkg-config is aware of on the system. Here's a sampling of the ones on my system:

```
$ pkg-config --list-all
--snip--
systemd                 systemd - systemd System and Service Manager
fontutil                FontUtil - Font utilities dirs
usbutils                usbutils - USB device database
bash-completion         bash-completion - programmable completion for the bash shell
libcurl                 libcurl - Library to transfer files with ftp, http, etc.
--snip--
notify-python           notify-python - Python bindings for libnotify
nemo-python             Nemo-Python - Nemo-Python Components
libcrypto               OpenSSL-libcrypto - OpenSSL cryptography library
libgdiplus              libgdiplus - GDI+ implementation
shared-mime-info        shared-mime-info - Freedesktop common MIME database
libssl                  OpenSSL-libssl - Secure Sockets Layer and cryptography libraries
xbitmaps                X bitmaps - Bitmaps that are shared between X applications
--snip--
xkbcomp                 xkbcomp - XKB keymap compiler
dbus-python             dbus-python - Python bindings for D-Bus
$
```

Pkg-config becomes *aware* of a package by having the package installation process update the pkg-config database, which is nothing more than a well-known directory that pkg-config examines to resolve queries. The database entries are simply plaintext files ending in a *.pc* extension. Therefore, making pkg-config aware of your library project during installation is nothing more difficult than generating and installing a text file, and Autoconf can help us generate this file, as we'll see later.

The pkg-config utility looks in several directories to find these files. We can discover what directories it looks in, and the order of the search, by calling it with the --debug option and piping the output (sent to stderr) through grep in this manner:

```
$ pkg-config --debug |& grep directory
Cannot open directory #1 '/usr/local/lib/x86_64-linux-gnu/pkgconfig' in package search path: No
such file or directory
Cannot open directory #2 '/usr/local/lib/pkgconfig' in package search path: No such file or
directory
Cannot open directory #3 '/usr/local/share/pkgconfig' in package search path: No such file or
directory
Scanning directory #4 '/usr/lib/x86_64-linux-gnu/pkgconfig'
Scanning directory #5 '/usr/lib/pkgconfig'
Scanning directory #6 '/usr/share/pkgconfig'
$
```

The first three directories that `pkg-config` tried to look in on my system did not exist. These are all in the */usr/local* tree. I haven't built and installed many packages on this system; as a result, I haven't installed any *.pc* files into the */usr/local* tree.

From the output, it's clear that *.pc* files must be installed into one of these six directories: */usr(/local)/lib/x86_64-linux-gnu/pkgconfig, /usr(/local) /lib/pkgconfig,* or */usr(/local)/share/pkgconfig.* When you think about it, you'll recognize these paths as what amounts to pkg-config's ${`libdir`}*/pkgconfig* and ${`datadir`}*/pkgconfig* directories if, that is, pkg-config didn't need to choose between */usr* and */usr/local* during its installation. In the early days, these were indeed just pkg-config's library and data installation paths, but it didn't take long for the project developers to realize that where pkg-config was installed was not really germane to where pkg-config should search on users' systems for *.pc* files—they could be found in many locations, depending not on where pkg-config was installed but on where the users had installed packages on their systems—all over the place.

But what about packages installed into custom locations or packages not yet installed? Pkg-config has solutions for these cases as well. The `PKG_CONFIG_PATH` environment variable can prepend user-specified paths to the default search path `pkg-config` uses to search for its data files. We'll discover how to use this functionality as we cover more details about using the `pkg-config` command in *configure.ac.*

Writing pkg-config Metadata Files

As stated earlier, pkg-config's *.pc* files are simply short text files that describe critical aspects of the build-and-link process to consumer build processes that use components of these dependent packages.

Let's take a look at a sample *.pc* file on my system—the one for the *libssl* library, a part of the OpenSSL package. First we'll need to find it:

```
$ pkg-config --variable pcfiledir libssl
/usr/lib/x86_64-linux-gnu/pkgconfig
$
```

The --variable option allows you to query the value of a variable, and pcfiledir is a pkg-config-defined variable that exists for every *.pc* file. I cover the complete list of predefined variables later in this chapter. The pcfiledir variable shows you the current location of the file as discovered by `pkg-config`. The nice thing about this variable is that it can also be used within your *.pc* file to provide a sort of relocation mechanism. If your library and include file paths are all relative to ${pcfiledir} within your *.pc* file, you can move it anywhere you like (as long as you move the libraries and header files it locates to the same relative locations).

I've provided the full contents of my *libssl.pc* file in Listing 10-1.

```
❶ prefix=/usr
exec_prefix=${prefix}
libdir=${exec_prefix}/lib/x86_64-linux-gnu
includedir=${prefix}/include

❷ Name: OpenSSL-libssl
Description: Secure Sockets Layer and cryptography libraries
Version: 1.0.2g
Requires.private: libcrypto
Libs: -L${libdir} -lssl
Libs.private: -ldl
Cflags: -I${includedir}
```

Listing 10-1: libssl.pc: A sample .pc file

A *.pc* file contains two types of entities: variable definitions (starting at ❶), which may reference other variables using Bourne shell–like syntax, and key-value tags (starting at ❷), which define the types of data that pkg-config can return about an installed package. These files can contain as little or as much of the pkg-config specification as is required by the package. Besides these entities, *.pc* files may also contain comments—any text preceded by a hash (#) mark. While these types of entities may be intermixed, it's common convention to put variable definitions at the top, followed by the key-value pairs.

Variables look and act like shell variables; definitions are formatted as a variable name, followed by an equal (=) sign, followed by a value. You do not need to quote the value portion, even if it contains whitespace.

Pkg-config makes some predefined variables available for use within the *.pc* file and (as we've already seen) from the command line. Table 10-1 shows these variables.

Table 10-1: Pre-Defined Variables Recognized by pkg-config

Variable	Description
pc_path	The default search path used by pkg-config to find the *.pc* file
pcfiledir	The installed location of the *.pc* file
pc_sysrootdir	The system root directory set by the user, or / by default
pc_top_builddir	The location of the user's top build directory when executing pkg-config

After you've looked at enough *.pc* files, you may begin to wonder if variables like prefix, exec_prefix, includedir, libdir, and datadir have any special meaning to pkg-config. They don't; it's just nice to define these paths relative to each other to reduce duplication.

Key-value pairs are formatted as a well-known keyword, followed by a colon (:) character, followed by some text that makes up the value portion. Values may reference variables; referencing an undefined variable merely expands to nothing. Quotes are not needed in these values either.

The keys of key-value pairs are well-known and documented, although putting unknown keys in the file has no effect on pkg-config's ability to use the rest of the data in the file. The keys shown in Table 10-2 are well-known:

Table 10-2: Well-Known Keys in Key-Value Pairs Recognized by pkg-config

Key	Description
Name	A human-readable name for the library or package.
Description	A brief human-readable description of the package.
URL	A URL associated with the package—perhaps the package download site.
Version	A version string for the package.
Requires	A list of packages required by this package; specific versions may be specified.
Requires.private	A list of private packages required by this package.
Conflicts	An optional field describing packages that this package conflicts with.
Cflags	Compiler flags that should be used with this package.
Libs	Linker flags that should be used with this package.
Libs.private	Linker flags for private libraries required by this package.

Informational Fields

To get the pkg-config --exists command to return zero to the shell, you need to specify Name, Description, and Version, at the very least. To be complete, consider also providing a URL if your project has one.

If you're not sure why a particular pkg-config command is not working as expected, use the --print-errors option. Where pkg-config would normally silently return a shell code, --print-errors will display a reason for a nonzero shell code:

```
$ cat test.pc
prefix=/usr
libdir=${prefix}/lib dir

Name: test
#Description: a test pc file
Version: 1.0.0
$
$ pkg-config --exists --print-errors test.pc
Package 'test' has no Description: field
$ echo $?
1
$
```

NOTE *The --validate option will also provide this information for both installed and uninstalled .pc files.*

One obvious oversight is the lack of options for displaying the name and description information belonging to a package. The description is displayed when using the --list-all option; however, even the package name that shows up in that listing is actually the base name of the *.pc* file, not the value of the Name field from within the file. In spite of this, as mentioned previously, these three fields—Name, Description, and Version—are required; otherwise, as far as pkg-config is concerned, the package does not exist.

Functional Fields

The category of the Version field crosses over from informational to functional, as there are some pkg-config command line options that can make use of the value of this field to provide data about the package to configuration scripts. The Requires, Requires.private, Cflags, Libs, and Libs.private fields also provide machine-readable information to configuration scripts and makefiles. Cflags, Libs, and Libs.private directly provide command line options for the C compiler and linker. The options to be added to these tools' command lines are accessed by using various of the pkg-config command line options.

While pkg-config is conceptually simple, some of the details are a bit elusive if you haven't played with it long enough to glean a proper understanding. Let's cover each of these fields in more detail.

The informational fields are designed to be read by people. The package version, for instance, can be displayed using the --modversion option:

```
$ pkg-config --modversion libssl
1.0.2g
$
```

NOTE *Do not confuse the --version option with the --modversion option. If you do, you'll quietly get pkg-config's version, regardless of what module you specify after --version.*

However, the Version field can also be used to indicate to configuration scripts if a package's version meets requirements:

```
$ pkg-config --atleast-version 1.0.2 libssl && echo "libssl is good enough"
libssl is good enough
$ pkg-config --exists "libssl >= 2.0" || echo "nope - too old :("
nope - too old :(
$
```

NOTE *Library version checks go against Autoconf's general philosophy of checking for required functionality, rather than checking for a specific version of a library, because some distro providers backport functionality to older versions of libraries so they can use that functionality without upgrading the library on a target version of their distro (mostly for convenience, as newer versions of libraries sometimes come with newer dependency requirements that can propagate for several levels). These examples are provided merely to show you what's possible with the functionality provided by pkg-config.*

Of the functional fields, some are more obvious than others. We'll cover each of them, starting with the more trivial ones. The Cflags field is probably the simplest to comprehend. It merely provides include path additions and other options to the C preprocessor and compiler. All options for both of these tools are combined into this one field, but pkg-config provides command line options for returning portions of the field value:

```
$ pkg-config --cflags xorg-wacom
-I/usr/include/xorg
$ pkg-config --cflags-only-other xorg-wacom
$
```

NOTE *The important thing to notice here is that the Cflags field contains compiler command line options, not portions of compiler command line options. For example, to define an include path for your library, ensure the value in Cflags contains both the -I flag and the path, just as you would on the compiler command line.*

The other options that affect the output of the Cflags field are --cflags-only-I and --cflags-only-other. As you can see, pkg-config is aware of the difference between -I options and other options; if you specify --cflags-only-I, you'll only see the -I options in the *.pc* file.

The Libs field provides a place to set -L, -l, and any other options destined for the linker. For instance, if your package provides the *stpl* library, *libstpl.so*, you would add the -L/*installed/lib/path* and -lstpl options to the Libs field. Pkg-config's --libs option returns the entire value, and, as with Cflags, there are separate options (--libs-only-l, --libs-only-L, and --libs-only-other) that separate and return subsets of the Libs options.

Somewhat harder to grasp is the use of the Libs.private field. This field is documented as being for libraries "required by this package but not exposed to applications." In reality, however, while these are libraries required to build the library published by the package, they're also libraries required by the consumer of the package if they're linking statically to the package's library.[1] In fact, the use of pkg-config's --static command line option, in conjunction with the --libs (or variations thereof) option, will display a combination of the Libs and Libs.private field options. This is because linking to a static library requires, at link time, all of the symbols required by all of the code pulled in from the static library to which you are directly linking.

This is an important concept, and understanding how it works is the key to properly writing *.pc* files for your projects. Think about it from the end user's perspective: they want to compile some project, and they want it to be linked statically with your library (we must also assume your project builds and installs a static version of your library, of course). In order to do

---

1. Which, in my opinion, would have been a better way of documenting the feature in pkg-config's man page because to the user (or even to the maintainer using a third-party library), it doesn't matter that some other library is required to the build a dependency of your project. All that matters to you, the consumer, is what's required for your project to use that library.

this, what options and libraries will be required on the compiler and linker command lines, *in addition to those already required when linking to your dynamic library*, in order to successfully perform this task? The answer to this question will tell you what should go into the Libs.private field in the *.pc* file for your project.

Now that those topics are behind us, we can properly discuss the Requires and Requires.private fields. The values in these fields are other pkg-config package names, with optional version specifications. If your package requires a particular version of another package that's also managed by pkg-config, you need only specify that package in the Requires field if its Cflags and Libs field values are required by your users' build processes in order to consume your package's shared library, or in the Requires.private field if its Cflags and Libs.private field values are required in order to consume your package's static library.

With this understanding of Requires and Requires.private, we can now see that additional options required by pkg-config packages that you'd normally put into your Cflags and Libs or Libs.private fields are not needed in those fields because you can simply reference the package by name (and version or version range) in Requires or Requires.private. Pkg-config will recursively find and combine options from all the packages' fields as needed.

If the package required by your package is not managed by pkg-config, you must add the options you'd normally find in such *.pc* files into your own Cflags, Libs, and Libs.private fields.

The version specification used in the Requires and Requires.private fields matches that of the RPM version specification. You may use >, >=, =, <=, or <. Sadly, these fields only allow one instance of a given library, which means you can't apply both upper and lower bounds on the versions of required packages. Listing 10-2 provides a contrived example of using version ranges.

```
--snip--
Name: music
Description: A library for managing music files
Version: 1.0.0
Requires: chooser >= 1.0.1, player < 3.0
--snip--
```

Listing 10-2: Specifying version and version ranges in Requires

The Requires field indicates that two libraries are required here: *chooser* and *player*. The version of *chooser* must be 1.0.1 or higher, and the version of *player* must be less than 3.0.[2]

Finally, the Conflicts field merely allows you as the author of a package to define packages that conflict with your package, and the format of the

---

2. It's actually rare to see a situation where a library requires a version of another library *less* than a given value, as this scenario implies that *player* changed its interface at 3.0 and the author of music chose not to upgrade to that new interface.

field is identical to that of `Requires` and `Requires.private`. For this field you may provide the same package more than once in order to define specific ranges of versions that conflict with your package.

When you've completed writing your *.pc* file, you can validate it using the `--validate` option:

```
$ pkg-config --validate test.pc
$
```

 You can use any `pkg-config` options that provide field information on either an installed .pc file, by using just the base name of the file, or on an uninstalled .pc file, by specifying the full name of the file, as is done in this example.

If you have any errors that `pkg-config` can detect, they'll be displayed. If you see nothing, then you know that `pkg-config` can at least parse your file properly and that a few basic checks pass.

Generating .pc Files with Autoconf

Now that you understand the basic structure of a *.pc* file, let's consider how we might use configuration data generated by our configuration scripts to generate a *.pc* file. Consider the types of information provided by pkg-config. Much of it is path information, and configuration scripts are designed to manage all these paths, including install locations for built products.

For example, the user may specify an installation prefix on the `configure` command line. This prefix determines where the package's include files and libraries will end up on their system when they install the package. The *.pc* file had better know these locations, and it would be nice of us to provide a build system that automatically updates this file to reflect the prefix path the user specified on their `configure` command line.

Generating pc Files from pc.in Templates

To accomplish this, we won't write *.pc* files. Instead, we'll write *.pc.in* template files for Autoconf and set the value of the `prefix` variable to `@prefix@` in these templates. That way, `configure` will replace this reference with the actual configured prefix when it converts the *.pc.in* template into the installable *.pc* file.

We can also set the value of the `Version` field to `@PACKAGE_VERSION@`, which is defined by the value you pass to the Autoconf `AC_INIT` macro in *configure.ac*. To facilitate an experiment, create a *configure.ac* file in an empty directory, as shown in Listing 10-3.

```
AC_INIT([test],[3.1])
AC_OUTPUT([test.pc])
```

Listing 10-3: configure.ac: Generating test.pc from test.pc.in

Now create a *test.pc.in* file in the same directory, like the one shown in Listing 10-4.

```
❶ prefix=@prefix@
   libdir=${prefix}/lib/test
   includedir=${prefix}/include/test

   Name: test
   Description: A test .pc file
❷ Version: @PACKAGE_VERSION@

   CFlags: -I${includedir} -std=c11
   Libs: -L${libdir} -ltest
```

Listing 10-4: test.pc.in: A .pc template file

Here we've specified the prefix and Version field values at ❶ and ❷ as Autoconf substitution variable references.

Generate the file and check the result:

```
$ autoreconf -i
$ ./configure --prefix=$HOME/test
configure: creating ./config.status
config.status: creating test.pc
$
$ cat test.pc
❶ prefix=/home/jcalcote/test
   libdir=${prefix}/lib/test
   includedir=${prefix}/include/test

   Name: test
   Description: A test .pc file
❷ Version: 3.1

   CFlags: -I${includedir} -std=c11
   Libs: -L${libdir} -ltest
$
$ pkg-config --cflags test.pc
-std=c11 -I/home/jcalcote/test/include/test
$
```

As you can see at ❶ and ❷ in the console output, the Autoconf variable references were replaced in the generated *test.pc* file with the values of those Autoconf variables.

Generating .pc Files with make

Generating *.pc* files from templates using Autoconf has the disadvantage of inhibiting the user's ability to change their prefix choices when they run make. This minor issue can be overcome by writing *Makefile.am* rules to generate the *.pc* files. Change the *configure.ac* file from the previous experiment, as shown in Listing 10-5.

```
AC_INIT([test],[3.1])
AM_INIT_AUTOMAKE([foreign])
AC_OUTPUT([Makefile])
```

Listing 10-5: configure.ac: Changes required to Listing 10-3 to generate test.pc *using make*

Now add a *Makefile.am* file, as shown in Listing 10-6.

```
EXTRA_DIST = test.pc
%.pc : %.pc.in
        sed -e 's|[@]prefix@|$(prefix)|g'\
            -e 's|[@]PACKAGE_VERSION@|$(PACKAGE_VERSION)|' $< >$@
```

Listing 10-6: Makefile.am: Adding make rules to generate test.pc

The key functionality in Listing 10-6 is encapsulated in the pattern rule that converts *.pc.in* files into *.pc* files using a simple sed command. The only odd bit in this sed command is the use of square brackets around the leading at (@) sign on the variables to be replaced. Those brackets are treated by sed as extraneous regular expression syntax, but the effect they have on Autoconf is to inhibit it from interpreting the sequence as the opening character of a replacement variable. We don't want Autoconf replacing this variable. Rather, we want sed to look for the sequence in *test.pc.in*. Another solution is to come up with your own format for variables to be replaced, but do note that this syntax is fairly common in the Autotools community for this very purpose.

NOTE *Pattern rules are specific to GNU make and are therefore not portable. There has been some chatter recently on the Automake mailing list of relaxing the restriction requiring the generation of portable make syntax and simply requiring GNU make because GNU make has been ported so widely these days.*

For this example, I've added *test.pc* to the Automake EXTRA_DIST variable so it will be built when make dist or distclean is executed, but you can add *test.pc* as a prerequisite to any target in your *Makefile.am* files to make it available to that stage of the build if required. Let's try it out:

```
$ autoreconf -i
configure.ac:2: installing './install-sh'
configure.ac:2: installing './missing'
$
$ ./configure
checking for a BSD-compatible install... /usr/bin/install -c
checking whether build environment is sane... yes
checking for a thread-safe mkdir -p... /bin/mkdir -p
checking for gawk... gawk
checking whether make sets $(MAKE)... yes
checking whether make supports nested variables... yes
checking that generated files are newer than configure... done
configure: creating ./config.status
config.status: creating Makefile
$
$ make prefix=/usr dist
```

```
make  dist-gzip am__post_remove_distdir='@:'
make[1]: Entering directory '/home/jcalcote/dev/book/autotools2e/book/test'
sed -e 's|[@]prefix@|/usr|g'\
     -e 's|[@]PACKAGE_VERSION@|3.1|' test.pc.in >test.pc
--snip--
$
$ cat test.pc
prefix=/usr
--snip--
```

Note that I added prefix=/usr to the make command line; thus, *test.pc* was
generated with that value in the prefix variable.

Uninstalled .pc Files

I mentioned at the outset of this chapter that pkg-config had the ability to
handle resolving references to uninstalled libraries and header files also. By
uninstalled, I mean products that have been built but not installed; they're
still sitting in another project's build output directory. Let's now consider
how this is done.

To use it, a user would set PKG_CONFIG_PATH to point to a directory containing a *-uninstalled* variant of a required package's *.pc* file. By "*-uninstalled*
variant," I mean that a *.pc* file named *test.pc* would have a *-uninstalled* variant named *test-uninstalled.pc*. The *-uninstalled* variant is not installed in
a pkg-config database directory but, rather, is still sitting in the project
source directory for the third-party dependency that the user has built.
Here's an example:

```
$ ./configure PKG_CONFIG_PATH=$HOME/required/pkg
--snip--
$
```

NOTE   *I'm following the Autoconf recommended procedure here of passing environment vari-*
ables as parameters to configure. Setting the variable in the environment or setting
it on the same command line before configure *works also, but is not recommended*
because configure *is less aware of variables set in these other ways.*

Assuming *$HOME/required/pkg* was where the required package was
unpacked and built, and assuming the same directory held the (possibly generated) *.pc* files for that package and that there was a *-uninstalled*
variant in that directory, that file would be accessed by executions of the
pkg-config utility that reference the required package's name from within
our configure script.

Obviously, you would not want to install the *-uninstalled* variant of any of
your *.pc* files—they're designed to be used only in this fashion, from within
a build directory. Perhaps not quite as obvious is the fact that the *-uninstalled*
variants of your *.pc* files don't contain all the same options as their installed
counterparts. The difference, simply stated, is in the path options. The
-uninstalled variants should contain absolute paths relative to the source

location of your header files and the build location of your libraries so that when the options are passed to consumers' tools, they'll be able to find the products (header files and libraries) in those paths.

Let's try it out. Edit the *configure.ac* file you created from Listing 10-3 to be like that shown in Listing 10-7.

```
AC_INIT([test],[3.1])
AC_OUTPUT([test.pc test-uninstalled.pc])
```

Listing 10-7: configure.ac: *Generating the* -uninstalled *variant of* test.pc

Absolute paths can be derived by using appropriate Autoconf substitution variables, like @abs_top_srcdir@ and @abs_top_builddir@, in the manner shown in Listing 10-8.

```
❶ libdir=@abs_top_builddir@/lib/test
❷ includedir=@abs_top_srcdir@/include/test

  Name: test
  Description: A test .pc file
❸ Version: @PACKAGE_VERSION@

  CFlags: -I${includedir} -std=c11
  Libs: -L${libdir} -ltest
```

Listing 10-8: test-uninstalled.pc: *A* -uninstalled *variant of* test.pc.in

This is the *-uninstalled* variant of the *.pc* file from Listing 10-4. I've removed the prefix variable, as it no longer makes sense in this context. I've replaced the ${prefix} references with @abs_top_builddir@ in the libdir pkg-config variable at ❶ and @abs_top_srcdir@ in the includedir pkg-config variable at ❷. Let's try it out:

```
$ autoreconf
$ ./configure
configure: creating ./config.status
config.status: creating test.pc
config.status: creating test-uninstalled.pc
$ pkg-config --cflags test.pc
-std=c11 -I/home/jcalcote/dev/book/autotools2e/book/temp/include/test
$
```

You may be asking yourself at this point why this is supposed to be so much easier than just setting CFLAGS (or CPPFLAGS) and LDFLAGS on the configure command line. Well, for one thing, it's easier to remember PKG_CONFIG_PATH than all of the potentially required individual tool variables. Another reason is the options are encapsulated where they're best understood—within *.pc* files written by the required package's author. Finally, if these options change, you'll have to change your use of individual variables accordingly, but the PKG_CONFIG_PATH will remain the same. The extra level of indirection afforded by pkg-config hides all the details from both you and your power users and contributors.

Using pkg-config in configure.ac

We've seen the way *.pc* files are put together. Now let's take a look at how to consume this functionality in *configure.ac*. As mentioned in the previous section, the --cflags option provides access to the Cflags fields your compiler needs in order to compile this package. Let's try this out with the *libssl.pc* file we saw previously. I've reproduced the relevant portion of Listing 10-1 here in Listing 10-9.

```
prefix=/usr
--snip--
includedir=${prefix}/include
--snip--
Cflags: -I${includedir}
```

Listing 10-9: libssl.pc: Relevant portion of this .pc file

When we use the --cflags option against this *.pc* file, we now understand that we should see a -I compiler command line option.

```
$ pkg-config --cflags libssl

$
```

And . . . nothing is printed. Huh, did we do something wrong? The *libssl.pc* file shows us that if we mentally expand the variables, we should see something like -I/usr/include, right? Actually, pkg-config is doing exactly what it should do—it's printing *the additional command line options* necessary to find the *libssl* header files. We don't need to tell the compiler about the */usr/include* directory, as this is a standard location and pkg-config knows this and omits such options automatically.[3]

Let's try a *.pc* file whose Cflags value includes something other than standard include locations. Note here that I'm using pkg-config itself to find the location of its database directory for the *xorg-wacom.pc* file because it's different on different Linux distributions:

```
$ cat $(pkg-config --variable pcfiledir xorg-wacom)/xorg-wacom.pc
sdkdir=/usr/include/xorg

Name: xorg-wacom
Description: X.Org Wacom Tablet driver.
Version: 0.32.0
Cflags: -I${sdkdir}
$
$ pkg-config --cflags xorg-wacom
-I/usr/include/xorg
$
```

---

3. You can override this default functionality by setting the PKG_CONFIG_ALLOW_SYSTEM_CFLAGS environment variable to any value. The PKG_CONFIG_ALL_SYSTEM_LIBS variable works the same way for linker options. Nevertheless, the use of these environment variables is rarely necessary.

Since */usr/include/xorg* is not a standard include path, a -I option for that path is displayed.[4] All of this means that you can be complete in documenting your package's requirements in your *.pc* files without worrying about cluttering consumer compiler and linker command lines with pointless redundant definitions.

So, how do we use this output? Nothing more difficult than a little shell script, as shown in Listing 10-10.

```
--snip--
LIBSSL_CFLAGS=$(pkg-config --cflags libssl)
--snip--
```

Listing 10-10: Using pkg-config to populate CFLAGS in configure.ac

The dollar-parens notation captures the output of this pkg-config command in the LIBSSL_CFLAGS environment variable.

NOTE *You may, of course, use backticks rather than the dollar-parens notation I used in Listing 10-10 to accomplish the same goal. The backtick format is older and slightly more portable, but it has the drawback of not being easily nestable. For example, you cannot do something like $(pkg-config --cflags $(cat libssl-pc-file.txt)) with backticks without a lot of escape magic.*

The linker options are accessed in a similar manner:

```
$ pkg-config --libs libssl
-lssl
$
```

Referring back to the *libssl.pc* file in Listing 10-1, we can indeed see that the Libs line contained -lssl. Also, as we just discovered, the -L option, referring a standard linker location, */usr/lib/x86_64-linux-gnu*, was automatically omitted. We can add this to our *configure.ac* file in the manner shown in Listing 10-11.

```
--snip--
LIBSSL_LIBS=$(pkg-config --libs libssl)
--snip--
```

Listing 10-11: Using pkg-config to populate LIBS in configure.ac

Let's tie it all together by populating all required variables for compiling *libssl* header files and linking with *libssl*. Listing 10-12 shows how this might be done.

---

4. It's exactly because xorg project consumers were having so much trouble finding X libraries and headers that pkg-config was originally created. Most of the X libraries and headers are installed into nonstandard locations.

```
--snip--
if pkg-config --atleast-version=1.0.2 libssl; then
  LIBSSL_CFLAGS=$(pkg-config --cflags libssl)
  LIBSSL_LIBS=$(pkg-config --libs libssl)
else
  m4_fatal([Requires libssl v1.0.2 or higher])
fi
--snip--
CFLAGS="${CFLAGS} ${LIBSSL_CFLAGS}"
LIBS="${LIBS} ${LIBSSL_LIBS}"
--snip--
```

Listing 10-12: Using pkg-config to access libssl *in* configure.ac

Could it be any simpler or any more readable? I doubt it. Let's look at
one more example—that of linking statically to *libssl*, which also requires
(privately) *libcrytpo*:

```
$ cat $(pkg-config --variable pcfiledir libssl)/libssl.pc
prefix=/usr
exec_prefix=${prefix}
libdir=${exec_prefix}/lib/x86_64-linux-gnu
includedir=${prefix}/include

Name: OpenSSL-libssl
Description: Secure Sockets Layer and cryptography libraries
Version: 1.0.2g
```
❶ `Requires.private: libcrypto`
```
Libs: -L${libdir} -lssl
```
❷ `Libs.private: -ldl`
```
Cflags: -I${includedir}
$
$ cat $(pkg-config --variable pcfiledir libcrypto)/libcrypto.pc
prefix=/usr
exec_prefix=${prefix}
libdir=${exec_prefix}/lib/x86_64-linux-gnu
includedir=${prefix}/include

Name: OpenSSL-libcrypto
Description: OpenSSL cryptography library
Version: 1.0.2g
Requires:
Libs: -L${libdir} -lcrypto
```
❸ `Libs.private: -ldl`
```
Cflags: -I${includedir}
$
$ pkg-config --static --libs libssl
```
❹ `-lssl -ldl -lcrypto -ldl`
```
$
```

As you can see in this console example at ❶, *libssl* privately requires the pkg-config-managed *libcrypto* package, meaning that linking to the *libssl* shared library does not require the addition of -lcrypto on the linker command line, but linking to it statically does require this additional library option. We can also see at ❷ that *libssl* privately requires a library that's not maintained by pkg-config, *libdl.so*.

NOTE *You may find the contents of your* libssl.pc *and* libcrypto.pc *files are somewhat different from mine, depending on your Linux distribution and the version of openssl you have installed. Don't worry about the differences—things will work fine on your system with your* .pc *files. The important part of this example is to understand the concepts I'm explaining.*

Moving down to the *libcrypto.pc* file, we see at ❸ that *libcrypto* also privately requires *libdl.so*.

The noteworthy item at ❹ is that pkg-config is "smart enough" to understand the linker's library ordering requirements and set -ldl on the output line after both -lssl and -lcrypto.[5] We humans have a hard enough time doing this stuff manually at times. It's nice when a tool comes along that just manages everything the way it should without making us worry about how it's done. Ultimately, the point I'm trying to make is that pkg-config puts control of the options squarely in the hands of the person most likely to understand how all these options should be specified and ordered—the maintainer of our dependencies.

pkg-config Autoconf Macros

As I mentioned at the outset of this chapter, pkg-config also ships a set of Autoconf extension macros in a file called *pkg.m4* that's installed into the */usr(/local)/share/aclocal* directory, which is where autoconf looks for *.m4* files containing the Autoconf standard macros that you can use in your *configure.ac* files.

Why didn't I use these in my examples? Well, there are a couple of reasons why I tend to avoid these macros, one obvious and the other more subtle—nefarious even. The obvious reason is how easy it is to use the pkg-config utility directly in shell script in *configure.ac*. Why would you try to wrap that in an M4 macro?

As for the second reason, recall from previous discussions that the input to autoconf is a data stream containing the contents of your *configure.ac* file, along with all of the macro definitions required to allow M4 to expand all macro invocations into shell script. These macro definitions become part of the input stream because autoconf reads all of the *.m4* files in the

---

5. In reality, pkg-config merely outputs both sets of private libraries, as specified in *libssl.pc* and *libcrytpo.pc*, without trying to sort them or reduce them. This is exactly what the linker needs, but it tends to make pkg-config look a bit smarter than it really is. The actual intelligence here is that pkg-config *doesn't* reorder -l (and -L) options but, rather, allows the package maintainer to specify them in the order required by the linker and then honors that ordering.

/usr(/local)/share/aclocal directory first, before reading your *configure.ac* file. In other words, there is no indication to autoconf that a required *.m4* file is missing. It simply expects all macro definitions required by *configure.ac* to be found in the *.m4* files in the installation paths' *aclocal* directory. As a result, autoconf cannot tell you if a macro definition in the input stream is not present. It simply fails to realize that PKG_CHECK_MODULES is a macro and, therefore, does not expand it to valid shell script. All of this happens when you run autoconf (or autoreconf). When you then try to run configure, it fails with messages that are so far removed from the actual problem that you couldn't possibly know just from reading them what they mean.

A picture is worth a thousand words, so let's try a quick experiment. Create a *configure.ac* file in an empty directory, as shown in Listing 10-13.

```
AC_INIT([test],[1.0])
AN_UNDEFINED_MACRO()
```

Listing 10-13: A configure.ac file with an unknown macro expansion

Now execute autoconf, followed by ./configure:

```
$ autoconf
$ ./configure
./configure: line 1675: syntax error: unexpected end of file
$
```

Notice how you get no error while autoconf is converting *configure.ac* into *configure*. This makes complete sense because m4, being a text-based macro processor, doesn't try to interpret anything in the data stream except for known macros. Everything else is passed directly though to the output stream, as if it were actual shell script.

When we ran configure, we got a cryptic error about unexpected end-of-file (at line 1675. . . from a two-line *configure.ac* file). What's really happening here is that you unintentionally started defining a shell function called AN_UNDEFINED_MACRO but didn't supply a body in curly braces for the function. The shell thought this was not cool and told you about it in its usual succinct manner.

Had we left the parentheses off of AN_UNDEFINED_MACRO, the shell would have been a bit more informational:

```
$ cat configure.ac
AC_INIT([test],[1.0])
AN_UNDEFINED_MACRO
$ ./configure
./configure: line 1674: AN_UNDEFINED_MACRO: command not found
$
```

At least this time, the shell told us the name of the problematic item, giving us the opportunity to go looking for it in *configure.ac* and perhaps figure out what's wrong.[6]

The point is, this is exactly what happens when you *think* you're using a pkg-config macro but *pkg.m4* was not found by autoconf while it was looking through the usual macro directories. Not very enlightening. In my humble opinion, you're much better off just skipping the hundreds of lines of nontransparent macro code and using pkg-config directly in your *configure.ac* file.

The reasons why autoconf might not find your installed *pkg.m4* file are enlightening, however. One common reason is that you installed the *pkg-config* package (or it was automatically installed when the OS was installed) from your distro's package repository, using yum or apt. But you downloaded, built, and installed Autoconf from the GNU website because your distro's version of Autoconf is four revisions behind and you need the latest. Where did pkg-config's installation process install *pkg.m4*? (Hint: */usr/share/aclocal*.) Where is autoconf getting its macro files from? (Hint: */usr/local/share/aclocal*.) You can, of course, easily remedy this problem by copying */usr/share/aclocal/pkg.m4* into */usr/local/share/aclocal*, and once you've hit this problem one or two times, you'll never be caught by it again. But your power users and contributors will have to go through the same process—or you could just tell them all to buy this book and read Chapter 10.

Summary

In this chapter, we've discussed the benefits of using Autoconf with pkg-config, how to generate *.pc* files from Autoconf templates, how to use pkg-config from *configure.ac* files, and various nuances of pkg-config features.

You can read *somewhat* more about the proper use of the *pkg-config* package on the official pkg-config website at *https://www.freedesktop.org/wiki/Software/pkg-config*. Dan Nicholson has written a concise and easy-to-follow tutorial on using pkg-config on his personal page at freedesktop.org (*http://people.freedesktop.org/~dbn/pkg-config-guide.html*). This page can also be accessed via links on the pkg-config website.

The *pkg-config* man page has a bit more information about the proper use of pkg-config, but, honestly, there isn't much more out there, other than blog entries written by enterprising individuals. Thankfully, there really isn't very much more to figure out about pkg-config. It's well written and well documented (as far as software goes), with a few minor exceptions, which I've tried to cover here.

---

6. The Autoconf project maintainers recognized this potential source of errors long ago and added code to Autoconf that at least tells you if you're attempting to invoke a misspelled Autoconf macro. The standard prefixes, AC_, AH_, and so on, are watched for by Autoconf. If it sees one it doesn't recognize, it will tell you about it when you execute autoconf (or autoreconf). You can take advantage of this mechanism by adding a line like m4_pattern_forbid([^PKG_]) to your *configure.ac* files if you choose to use the pkg-config macros. This construct tells Autoconf that anything in the input stream that starts with PKG_ is supposed to be a macro and should therefore generate an error if it's not a macro. You may recognize the argument as a regular expression.

11

INTERNATIONALIZATION

*Like all great travelers
I have seen more than I remember,
and remember more than I have seen.*
—*Benjamin Disraeli*, Vivian Grey

When it comes to making software available in other languages, native English speakers are a little arrogant—but who can blame us? We've been taught almost from childhood, through every experience we've had in this industry, that English is the only important language—so much so that we don't even think about it anymore. All computer science–oriented research and academic discussions in communities of any consequence are in English. Even our programming languages have English keywords.

Some might counter that this is because most programming languages are invented by English speakers. Not even close! Take for instance Niklaus Wirth of Switzerland, a native German speaker, who invented or had a hand in the invention of several important programming languages, including Euler, Pascal, and Modula. Not popular enough for you? Bjarne Stroustrup, Danish by birth, invented C++. Guido van Rossum, born and raised in the Netherlands, invented Python. Rasmus Lerdorf, who was born in Denmark and later moved to Canada, wrote PHP. Ruby was written by Yukihiro Matsumoto in Japan.

My point is, it never occurs to developers—even non-native English speakers—to invent programming languages that use, for example, German keywords. Why not? Probably because no one would use them if they did—not even Germans. New programming languages are often conceived in academic or corporate research environments, and the industry journals, forums, and standardization organizations that facilitate the discussion of the pros and cons of these inventions are written or managed almost exclusively in English—for pragmatic reasons, of course. No one is really saying that English is the best language. Rather, we need a common medium in which to publish, and English, being one of the most widely spoken languages on Earth, just sort of fell into that role.

What we miss because of this English-only attitude is that there are entire communities of non-English-speaking software users out there who struggle to understand applications written entirely in English. It's as uncomfortable for them to use these applications as it is for English-only speakers to look at a web page in Chinese or Russian.

Larger companies often provide language packs to allow these communities to use software offerings in their native languages. Some of these commercial native-language offerings are extensive, providing support even for the more difficult Arabic and Asian languages.[1] However, most smaller commercial and open source software package authors don't even try because, they say, it's too expensive, too difficult, or just not important enough to their communities or markets. The first of these arguments *may* have some merit in the corporate world. Let's talk about our options for solving these problems and, hence, for expanding our communities.

Obligatory Disclaimer

Before I dive into this topic, I'll state up front that multivolume works could (and should) be written on software internationalization and localization. The topic is simply huge. I will not even come close to covering everything in a couple of chapters. My goal here is to give an introduction to a topic that may seem daunting from the outside. If you're already familiar with these concepts, you'll probably be disgusted by the amount of material I don't cover. Please understand that these chapters are not for you, although you may find some ideas of value in them. Rather, these chapters are for the beginner with little experience in this area.

In this chapter, I'll cover what's in the C standard and what works with the UTF-8 codeset, and I'll go a bit beyond. I'll also cover major portions of the *GNU gettext* library because integration of *gettext* into Autotools projects is, in fact, the point of this chapter, but I won't cover third-party

---

1. I'm not trying to imply that Arabic and Asian languages are more difficult to learn or understand—only that GUI layouts, conceived in English, or perhaps other European languages, are usually not designed with right-to-left or vertical presentations in mind. This is why it's important to consider these topics up front, rather than (as is more common) at the very end, just before publishing your software. At least make a conscious decision about what languages you'll support.

libraries and solutions, though I do mention them where appropriate. Neither will I cover wide-character string manipulation and multibyte-to-wide-character (and vice versa) transformations; there are plenty of resources out there that cover these topics in detail.

I just stated that I'll cover *major portions* of *gettext*, meaning that there are parts I'll leave out because they're used only under special conditions. Once you have the basics down, you can easily pick up the rest, as needed, from the manual.

Speaking of the manual, like many software manuals, the *gettext* manual is more a reference than a tutorial for beginners. You may have tried to read the *gettext* manual, intending to become familiar with internationalization and localization though this channel, and walked away thinking, "Either this is a terrible manual or I'm just way out of my depth here." I know I did at one point. If so, you'd be somewhat correct on both counts. First, it's pretty apparent that the manual was written by non-English speakers. Is that such a surprise? We've already decided that native English speakers don't really care that much, in general, about this topic. Some of the idioms used in the manual are simply not familiar to English speakers, and some of the phraseology is clearly foreign. But, provenance aside, the manual is also not organized in a manner helpful to someone trying to become familiar with the topic. As I was doing research for this chapter, I found several online tutorials that would be much more helpful than the manual for programmers just trying to figure out where to start.

So, let's begin by first covering some definitions.

Internationalization (I18n)

Internationalization, sometimes referred to as *i18n* in the literature because it's easier to write,[2] is the process of preparing a software package to be published in other languages or for other cultures. This preparation includes writing (or refactoring) the software in such a way as to be easily configured to display human-readable text in other languages, or according to other cultural customs and standards. The text I'm referring to here includes strings, numbers, dates and times, currency values, postal addresses, salutations and greetings, paper sizes, measurements, and any other aspect of human communications you can think of that may be done differently in different languages and cultures.

Internationalization is specifically *not* about converting embedded text from one language to another. Rather, it's about preparing your software so that static and generated text can easily be displayed in a target language or in formats that conform to target cultural norms. People of British culture, for instance, expect to see dates, decimal numbers, and local currency

---

2. *Internationalization* and *localization* are often abbreviated *i18n* and *l10n* in literature, respectively, because they're such long words. The abbreviations are derived from the first and last letters of the words, with the number of intervening letters in between.

displayed differently than do Americans, even though both speak English natively. So internationalization encompasses not only language support but also general culture support.

To be clear, this preparation is not about building a version of your software specifically for Spanish speakers, for example. That topic is reserved for Chapter 12, where I'll discuss the concept of *localization*. Rather, internationalization is about designing or modifying your software such that it *can* be easily used by Spanish speakers. This means first locating and tagging the strings in your software that should be translated and finding the places in your code where times and dates, currency, numbers, and other locale-specific content is formatted for display. Then you need to make those bits of static text and text-generation code configurable based on a global or specified locale. Of course, it also means configuring your software to be aware of the current system locale and switch into it automatically.

There are two areas of software internationalization that are different enough that we should discuss them separately:

- Dynamic, runtime-generated text messages
- Static text messages that are hardcoded into your application

Let's cover generated messages first, since we often get some help in this area from programming language standard libraries. Most such libraries provide some form of support for locale management, and C is no exception. C++ provides the same sort of functionality in an object-oriented manner.[3] Once you understand what's available in C, the C++ version is pretty easy to pick up on your own, so we'll cover the functionality provided by the C standard library here.

I'll also introduce you to the extended interfaces provided by the POSIX 2008 and X/Open standards because, as we'll see, the functionality provided by the standard C library, while usable, is a bit weak, and the POSIX and X/Open standard functionality is pretty widely available. Finally, GNU extensions to the C standard can make your application shine in other cultures, as long as you're willing to break away from the standards a bit.

Instrumenting Source Code for Dynamic Messages

The standard C library offers the setlocale and localeconv functions exposed by the *locale.h* header file, as shown in the synopsis in Listing 11-1.

```
#include <locale.h>

char *setlocale(int category, const char *locale);
struct lconv *localeconv(void);
```

Listing 11-1: A synopsis of the standard C library setlocale and localeconv functions

---

3. The C++ standardization process seems always to be a step ahead of that of the C language, so C++ internationalization and localization features are a little more advanced. I'm not talking about accessing the features in an object-oriented manner but rather about how the actual locale category information available to the software is usually a little broader in C++ than it is in C.

The task of setlocale is to tell the standard C library which locale to use for a given class of library functionality. This function accepts a *category*—an enumeration value representing a segment of locale-specific functionality in the library that should be changed from the current locale to a new target *locale*. The available standard category enumeration values are as follows.

LC_ALL

LC_ALL represents all categories. Changing the value of this category sets all of the available categories to the specified locale. This is the most common and recommended value to use, unless you have a very specific reason for not setting all categories to the same locale.

LC_COLLATE

Changing LC_COLLATE affects the way collation functions like strcoll and strxfrm work. Different languages and cultures collate and sort based on different character- or glyph-ordering rules. Setting the collation locale changes the rules used by the library's text collation functions.

LC_CTYPE

Changing LC_CTYPE affects the way the character attribute functions defined in *ctype.h* operate (except for isdigit and isxdigit). It also affects the multibyte and wide-character versions of these functions.

LC_MONETARY

Changing LC_MONETARY affects the monetary-formatting information returned by localeconv (which we'll discuss later in this section), as well as the resulting strings returned by the X/Open standard and POSIX extension strfmon.

LC_NUMERIC

Changing LC_NUMERIC affects the decimal point character in formatted input and output operations performed by functions like printf and scanf and the values related to decimal formatting returned by localeconv, as well as the resulting strings returned by the X/Open standard and POSIX extension strfmon.

LC_TIME

Changing LC_TIME affects the way time and date strings are formatted by strftime.

The return value of setlocale is a string representing the previous locale, or set of locales if all categories are not set the same. If you're only interested in determining the current locale, you can pass NULL in the *locale* parameter and setlocale will not change anything. If you have set any of the individual categories independently to different locale values, the returned string's format, when passing LC_ALL, is implementation defined and therefore not quite as useful as it might otherwise be. Nevertheless, most implementations will allow you to pass this string back into setlocale, using LC_ALL, to reset category-specific locales to those of a previously retrieved state.

Once the desired locale has been set up, `localeconv` may be called to return a pointer to a structure containing *some* of the attributes of the current locale. Why not all of them? Because the designers of this API—otherwise intelligent people—were on pain medication or something when they created it. Seriously, the *GNU C Library* manual has something to say about it:

> Together with the `setlocale` function the ISO C people invented the `localeconv` function. It is a masterpiece of poor design. It is expensive to use, not extensible, and not generally usable as it provides access to only `LC_MONETARY` and `LC_NUMERIC` related information. Nevertheless, if it is applicable to a given situation it should be used since it is very portable.[4]

In addition to these criticisms, I'll add that it's not thread safe; the contents of the structure are subject to modification (by another thread calling `setlocale`) while you're accessing it. Nevertheless, the rules are clear about how it can get modified—only by calls to `setlocale` with a non-`NULL` *locale* parameter value—so it is usable, but it's neither elegant nor complete. As the preceding excerpt indicates, you should try to use `localeconv` if you don't need additional information for your application, because it's part of the C standard and is, therefore, extremely portable.

To be completely fair, the fields in the structure returned by `localeconv` are those that presumably require some direct programmer intervention to use correctly, given the functionality provided by the C standard library. For example, the `printf` family of functions provides no special format specifiers for locale-specific number and currency values, so information related to the `LC_NUMERIC` and `LC_MONETARY` categories must be made available to the developer in some fashion in order to make proper use of these categories in a program designed to print numbers and currency amounts in locale-specific formats. It also means, of course, that without third-party libraries or extensions to the C standard, you'll be writing some tedious text-formatting functions that vary their output based on the rules returned by `localeconv`.

On the other hand, the `LC_COLLATE`, `LC_TIME`, and `LC_CTYPE` categories all directly affect various existing standard library functionality, making it presumably unnecessary for the program author to have direct access to the locale attributes used by these library functions.[5]

Setting and Using Locales

The C and C++ standards require that all implementations of the standard library be initialized in every process with the default "C" locale so that

---

4. Taken from *The GNU C Library* manual, last updated February 1, 2018, for glibc 2.27. See *https://www.gnu.org/software/libc/manual/html_node/The-Lame-Way-to-Locale-Data .html#The-Lame-Way-to-Locale-Data*.

5. When you think about it, you'll realize that the locale attributes missing from `struct lconv` must be maintained by an implementation of the library somewhere or else the library functions that need to act differently based on these attributes could not be implemented to do so. The GNU C library implementation actually exposes these fields in `struct lconv` as internal-only fields, named with a leading underscore.

all programs not explicitly selecting a locale will act in a predictable and consistent manner. Therefore, the first thing you must do to internationalize your software is to change the locale. The easiest and most consistent way to change the locale within your application is to call setlocale with a *category* value of LC_ALL somewhere near the start of the program. But what string should we pass as a *locale* argument? Well, that's the beauty of this function—you don't need to pass any specific locale string at all. Passing an empty string disables the *default locale*, allowing the library to select the *environment locale* that's in effect on the host. This allows your users to determine how your program will display times and dates, decimal numbers, and currency values and how collation and character set management will work.

Listing 11-2 shows the code for a program that configures the standard C library to use the host environment locale and then displays the standard locale attributes available from localeconv to the console.

NOTE *The example programs in this chapter can be found in the online GitHub repository named* NSP-Autotools/gettext, *found at* https://github.com/NSP-Autotools /gettext/. *The small utility programs presented in this chapter are found in the* small-utils *directory in that repository, and a makefile is provided that will build them all by default. Use a command like* make lc, *for instance, to build just the* lc *program presented in Listing 11-2.*

Git tag 11.0

```c
#include <stdio.h>
#include <stdlib.h>
#include <stdbool.h>
#include <string.h>
#include <limits.h>
#include <locale.h>

static void print_grouping(const char *prefix, const char *grouping)
{
    const char *cg;
    printf("%s", prefix);
    for (cg = grouping; *cg && *cg != CHAR_MAX; cg++)
        printf("%c %d", cg == grouping ? ':' : ',', *cg);
    printf("%s\n", *cg == 0 ? " (repeated)" : "");
}

static void print_monetary(bool p_cs_precedes, bool p_sep_by_space,
        bool n_cs_precedes, bool n_sep_by_space,
        int p_sign_posn, int n_sign_posn)
{
    static const char * const sp_str[] =
    {
        "surround symbol and quantity with parentheses",
        "before quantity and symbol",
        "after quantity and symbol",
        "right before symbol",
        "right after symbol"
    };
```

```c
    printf("    Symbol comes %s a positive (or zero) amount\n",
            p_cs_precedes ? "BEFORE" : "AFTER");
    printf("    Symbol %s separated from a positive (or zero) amount by a space\n",
            p_sep_by_space ? "IS" : "is NOT");
    printf("    Symbol comes %s a negative amount\n",
            n_cs_precedes ? "BEFORE" : "AFTER");
    printf("    Symbol %s separated from a negative amount by a space\n",
            n_sep_by_space ? "IS" : "is NOT");
    printf("    Positive (or zero) amount sign position: %s\n",
            sp_str[p_sign_posn == CHAR_MAX? 4: p_sign_posn]);
    printf("    Negative amount sign position: %s\n",
            sp_str[n_sign_posn == CHAR_MAX? 4: n_sign_posn]);
}

int main(void)
{
    struct lconv *lc;
    char *isym;

    setlocale(LC_ALL, "");  // enable environment locale
    lc = localeconv();       // obtain locale attributes

    printf("Numeric:\n");
    printf("  Decimal point: [%s]\n", lc->decimal_point);
    printf("  Thousands separator: [%s]\n", lc->thousands_sep);

    print_grouping("  Grouping", lc->grouping);

    printf("\nMonetary:\n");
    printf("  Decimal point: [%s]\n", lc->mon_decimal_point);
    printf("  Thousands separator: [%s]\n", lc->mon_thousands_sep);

    print_grouping("  Grouping", lc->mon_grouping);

    printf("  Positive amount sign: [%s]\n", lc->positive_sign);
    printf("  Negative amount sign: [%s]\n", lc->negative_sign);
    printf("  Local:\n");
    printf("    Symbol: [%s]\n", lc->currency_symbol);
    printf("    Fractional digits: %d\n", (int)lc->frac_digits);

    print_monetary(lc->p_cs_precedes, lc->p_sep_by_space,
            lc->n_cs_precedes, lc->n_sep_by_space,
            lc->p_sign_posn, lc->n_sign_posn);

    printf("  International:\n");
    isym = lc->int_curr_symbol;
    printf("    Symbol (ISO 4217): [%3.3s], separator: [%s]\n",
            isym, strlen(isym) > 3 ? isym + 3 : "");
    printf("    Fractional digits: %d\n", (int)lc->int_frac_digits);

#ifdef __USE_ISOC99
    print_monetary(lc->int_p_cs_precedes, lc->int_p_sep_by_space,
            lc->int_n_cs_precedes, lc->int_n_sep_by_space,
            lc->int_p_sign_posn, lc->int_n_sign_posn);
```

```
#endif

    return 0;
}
```

The struct lconv structure contains both char * and char fields. The char * fields mostly refer to strings whose values are determined according to the current locale. Some of the char fields are intended to be taken as Boolean values, while the rest are designed to be read as small integer values. The code shown in Listing 11-2 should indicate pretty clearly which are Boolean and which are small integers. The documentation for your compiler's standard library should also make it clear.

The only weird ones are the grouping and mon_grouping fields, which indicate how digits in numbers and currency values (respectively) should be grouped, with groups separated by the corresponding *thousands separator* string. The grouping and mon_grouping fields are char * fields designed to be read not as strings but as arrays of small integers. They're terminated with either a zero or the value CHAR_MAX (defined in *limits.h*). If they're terminated with zero, the final grouping value is repeated forever; otherwise, the final grouping includes the remaining digits in the value.

Finally, note the call to the internal print_monetary routine that's wrapped in a check for __USE_ISOC99 (near the bottom of the listing). The international forms of these currency attributes were added with the C99 standard. Everyone should be up to C99 by now, so this is not generally an issue. I added the conditional compilation check because, for this utility program, it's possible and appropriate. For an application trying to use these fields, you should probably just insist that C99 be required to build the application.

Building and executing this program from a US English Linux system generates the following console output:

```
$ gcc lc.c -o lc
$ ./lc
Numeric:
  Decimal point: [.]
  Thousands separator: [,]
  Grouping: 3, 3 (repeated)

Monetary:
  Decimal point: [.]
  Thousands separator: [,]
  Grouping: 3, 3 (repeated)
  Positive amount sign: []
  Negative amount sign: [-]
  Local:
    Symbol: [$]
    Fractional digits: 2
    Symbol comes BEFORE a positive (or zero) amount
    Symbol is NOT separated from a positive (or zero) amount by a space
    Symbol comes BEFORE a negative amount
```

```
         Symbol is NOT separated from a negative amount by a space
         Positive (or zero) amount sign position: before quantity and symbol
         Negative amount sign position: before quantity and symbol
    International:
         Symbol (ISO 4217): [USD], separator: [ ]
         Fractional digits: 2
         Symbol comes BEFORE a positive (or zero) amount
         Symbol IS separated from a positive amount by a space
         Symbol comes BEFORE a negative amount
         Symbol IS separated from a negative amount by a space
         Positive (or zero) amount sign position: before quantity and symbol
         Negative amount sign position: before quantity and symbol
$
```

To change the environment locale, set the LC_ALL environment variable to the name of the locale you want to use. The values you can use are the locales that are generated and installed on your system.

NOTE   *You can also set individual locale categories using environment variables with the same names as the category names. For example, to change the locale to Spanish (in Spain), but only for the LC_TIME category, you could set the LC_TIME environment variable to es_ES.utf8. This works for all the standard categories defined earlier.[6]*

To find out which locales are available, run the locale utility with the -a option, in this manner:

```
$ locale -a
C
C.UTF-8
en_AG
en_AG.utf8
en_AU.utf8
en_BW.utf8
en_CA.utf8
en_DK.utf8
en_GB.utf8
en_HK.utf8
en_US.utf8
en_ZA.utf8
en_ZM
en_ZM.utf8
en_ZW.utf8
ja_JP.utf8
POSIX
sv_SE.utf8
$
```

---

6. Well, it mostly works. For instance, you should not mix LC_CTYPE from a single-byte locale such as ISO-8859-15 with LC_COLLATE from another local that uses, say, UTF-8. Essentially, you should ensure that the same character encoding is used in all of the category changes you make.

NOTE *My example console listings are performed on a Debian-based system. If you're using a Fedora-based distribution, for example, you should expect to see different results, as Fedora has significantly different default functionality with respect to installed language packs and how the* locale *utility works. I'll discuss Red Hat specifics later on in the chapter where it really matters.*

Normally, a US English installation of Linux will have several locales configured that begin with the string en. I've generated Swedish (sv_SE.utf8) and Japanese (ja_JP.utf8) locales on my Debian-based system, as well, in order to show examples of output when the environment is configured for non-English languages and cultures.

NOTE *I also use the French (*fr_FR.utf8*) locale later in the chapter. You may wish to pre-build or preinstall all of these locales using whatever mechanism is provided by your distribution to make it easier to follow along with my examples on your system. Of course, if you are not a native English speaker, you're probably already using a different locale by default. In this case, you might also want to build or install the* en_US.utf8 *locale—though, not surprisingly, this locale generally comes preinstalled even on systems not built or sold in the United States.*

You may have noticed the C, C.UTF-8, and POSIX locales in the preceding list. The C locale, as already mentioned, is the default locale for programs that do not set the locale explicitly. The POSIX locale is currently defined as an alias for the C locale.

Generating and Installing Locales

The process of generating and installing a locale is pretty specific to a distribution, but there are a few common implementations. On a Debian- or Ubuntu-based system, for instance, you can look at the */usr/share/i18n/ SUPPORTED* file to see which locales can be generated and installed from sources on your system:

```
$ cat /usr/share/i18n/SUPPORTED
aa_DJ.UTF-8 UTF-8
aa_DJ ISO-8859-1
aa_ER UTF-8
--snip--
zh_TW BIG5
zu_ZA.UTF-8 UTF-8
zu_ZA ISO-8859-1
$
```

There are 480 locale names in this file on my Linux Mint system. The general format of a locale name, as defined by the X/Open standard, is as follows:

language[*_territory*][*.codeset*][*@modifier*]

There are up to four parts of a locale name. The first part, *language*, is required. The remaining parts, *territory*, *codeset*, and *modifier*, are optional. For example, the locale name for US English using the UTF-8 character set is en_US.utf8. The *language* is represented in the form of a two-letter ISO 639 language code.[7] For instance, en refers to the English language, which could be American, Canadian, British, or some other dialect of English.

The *territory* portion indicates the location of the language and takes the form of a two-letter ISO 3166 country code.[8] For example, US is for the United States, CA is for Canada, and GB is for Great Britain.

The portion after the dot (.) indicates the *codeset* or character encoding, formatted as a standard ISO character-encoding name like UTF-8 or ISO-8859-1.[9] The most common character encoding is UTF-8 (represented in the locale name as utf8) since it can represent all characters in the world. It doesn't represent all of them efficiently, however; some languages don't use utf8 because they require several bytes per character in this encoding.

The *modifier* portion is not often used.[10] One possible use is to generate a locale that differs only in case sensitivity, or in some other attribute that is not a normal locale attribute. For instance, when setting LC_MESSAGES=en@boldquot, you get an English message set that differs from the normal English message set only in that quoted text is bolded. Another historically common case is where the en_IE@euro locale is distinguished only by a difference in the currency symbol used. Suffice it to say that the differences applied by using a locale with a particular modifier are designed for very special use cases.

To generate and install a particular locale on a Debian- or Ubuntu-based system, you can add a file to the */var/lib/locales/supported.d* directory containing the line from *SUPPORTED* representing the locale you want to add. The name of the file added to the *supported.d* directory is not particularly important, although I advise not using filenames that are too far from something reasonably similar to what you find already in this directory structure. It's only important that a file exists in that directory and that it contains the exact contents of the desired line from *SUPPORTED*.

For instance, to add sv_SE.utf8, I'd find the line in *SUPPORTED* that represents this language, add a file to *supported.d* containing this line, and then run the locale-gen program, in this manner:

```
$ cat /usr/share/i18n/SUPPORTED | grep sv_SE
sv_SE.UTF-8 UTF-8
```

---

7. See *https://en.wikipedia.org/wiki/List_of_ISO_639-1_codes* for a complete table of ISO 639 language codes.

8. See *https://en.wikipedia.org/wiki/ISO_3166-1#Current_codes* for a complete list of ISO 3166 country codes.

9. See *https://en.wikipedia.org/wiki/Character_encoding* for a comprehensive list of widely used character encodings.

10. Very few even know what it's used for. For instance, I found a description in the *Oracle International Language Environment Guide* at *https://docs.oracle.com/cd/E23824_01/html/E26033 /glmbx.html* that defines modifier as "the name of the characteristics that differentiate the locale from the locale without the modifier." That seems like an excellent example of a recursive definition.

```
sv_SE ISO-8859-1
sv_SE.ISO-8859-15 ISO-8859-15
$
$ echo "sv_SE.UTF-8 UTF-8" | sudo tee -a /var/lib/locales/supported.d/sv
[sudo] password for jcalcote: *****
sv_SE.UTF-8 UTF-8
$ sudo locale-gen
Generating locales (this might take a while)...
  en_AG.UTF-8... done
--snip--
  en_ZW.UTF-8... done
  ja_JP.UTF-8... done
  sv_SE.UTF-8... done
Generation complete.
$
$ locale -a
C
C.UTF-8
en_AG
--snip--
ja_JP.utf8
POSIX
sv_SE.utf8
$
```

Each line in *SUPPORTED* contains a locale database entry name,
followed by a codeset name. For Swedish, the entry we're interested in is
sv_SE.UTF-8, with the UTF-8 codeset. I chose to add a file called *sv* to */var/lib
/locales/supported.d*. You may add as many lines as you want to the file; each
line will be processed as a separate locale. Because the files in */var/lib/locale*
are owned by root, you'll need to have root-level permissions to create or
write to them. I used a common trick involving the tee and echo commands
to add the line I wanted to *supported.d/sv* as root.[11] You could also just use a
text editor started with sudo, of course.

To generate a locale on a Red Hat– or CentOS-based system, you can
use the localedef utility in this manner:

```
$ localedef --list-archive
aa_DJ
aa_DJ.iso88591
aa_DJ.utf8
--snip--
sv_SE.utf8
--snip--
zu_ZA
zu_ZA.iso88591
zu_ZA.utf8
$
$ sudo localedef -i sv_SE -f UTF-8 sv_SE.UTF-8
$
```

---

11. This trick is needed because directly using sudo echo >>/var/lib/locale/supported.d/sv will
attempt to apply the redirection before sudo.

```
$ locale -a | grep sv_SE.utf8
sv_SE.utf8
$
```

The -i option on the localedef command line signifies the input file, which is taken from the output of the localedef --list-archive command. The -f option indicates the codeset to use.

> **NOTE** *I've found that recent Red Hat (and therefore CentOS) systems generally come prein-stalled with many locales. You may find, upon using locale -a, that you do not need to generate any locales. Anything that shows up in locale -a is immediately usable as a locale in the LANG and LC_* environment variables. Fedora systems, on the other hand, require the installation of language-specific langpacks, even if the locale shows up in the list displayed by locale -a. Swedish, for instance, requires the installation of glibc-langpack-sv. Additionally, the language sources do not seem to be installed on Fedora. Therefore, the localedef command will not work on that platform, but installation of the langpack will provide a precompiled version of the locale.*

Now that we have a Swedish locale available to us, let's see what's displayed when we execute the lc program built from the code in Listing 11-2 when using that locale:

```
$ LC_ALL=sv_SE.utf8 ./lc
Numeric:
  Decimal point: [,]
  Thousands separator: [ ]
  Grouping: 3, 3 (repeated)

Monetary:
  Decimal point: [,]
  Thousands separator: [ ]
  Grouping: 3, 3 (repeated)
  Positive amount sign: []
  Negative amount sign: [-]
  Local:
    Symbol: [kr]
    Fractional digits: 2
    Symbol comes AFTER a positive (or zero) amount
    Symbol IS separated from positive (or zero) amount by a space
    Symbol comes AFTER a negative value
    Symbol IS separated from negative value by a space
    Positive (or zero) amount sign position: before quantity and symbol
    Negative amount sign position: before quantity and symbol
  International:
    Symbol (ISO 4217): [SEK], separator: [ ]
    Fractional digits: 2
    Symbol comes AFTER a positive value
    Symbol IS separated from positive value by a space
    Symbol comes AFTER a negative value
    Symbol IS separated from negative value by a space
    Positive (or zero) amount sign position: before quantity and symbol
    Negative amount sign position: before quantity and symbol
$
```

Unfortunately, as I mentioned earlier, localeconv only returns information on the numeric (LC_NUMERIC) and monetary (LC_MONETARY) categories, which isn't quite as bad as it sounds because the remaining ones are handled nearly automatically for you by the library. Regardless, there are other options for accessing the complete set of locale attributes, which we'll discuss later in this chapter.

Formatting Time and Date for Display

The standard C library quietly handles time and date behind the scenes, depending on which format specifiers you use in the format string passed to strftime. Here's the prototype for strftime:

```
#include <time.h>

size_t strftime(char *s, size_t max, const char *format, const struct tm *tm);
```

Briefly, the strftime function places up to *max* bytes in the buffer pointed to by *s*. The content is determined by the text and format specifiers in *format*. Only a single time value format can be specified in *format*, and its value is obtained from *tm*. Since this is a standard library function, you can refer to any standard C library manual for details on the way format specifiers work in this function.

Listing 11-3 provides the source code for a small program that prints the current time and date in a general format supported in some form by all languages and territories.[12]

```c
#include <stdio.h>
#include <locale.h>
#include <time.h>

int main(void)
{
    time_t t = time(0);
    char buf[128];

    setlocale(LC_ALL, "");  // enable environmental locale

    strftime(buf, sizeof buf, "%c", gmtime(&t));
    printf("Calendar time: %s\n", buf);
    return 0;
}
```

Listing 11-3: td.c: A small program to print the calendar date and time in the environment locale

---

12. It's possible for gmtime to return NULL, though highly unlikely in this example because of how we obtained the value passed to it. Regardless, a robust program would check for errors before assuming it returned a proper structure for strftime to use.

Building and executing this program displays something like the following output on the console; your times and dates will very likely not match mine:

```
$ gcc td.c -o td
$ LC_ALL=C ./td
Calendar time: Tue Jul  2 03:57:56 2019
$ ./td
Calendar time: Tue 02 Jul 2019 03:57:58 AM GMT
$ LC_ALL=sv_SE.utf8 ./td
Calendar time: tis  2 jul 2019 03:57:59
$
```

I set LC_ALL=C on the first execution to show how you can execute your localized programs using the default C locale. This can be a handy debugging aid for testing your internationalized software.

NOTE *The C locale is not the "American" locale. Rather, it's referred to as the minimal locale. If you execute the lc program with LC_ALL=C, you'll find that many of the options are blank. The standard library expects and handles such blank options in an appropriate manner.*

Compare the English and Swedish outputs. The day and month names are in the locale language. For July, the month name happens to be the same in English and Swedish. However, notice the case difference in both day names and month names. In English, the names are capitalized, while in Swedish, they are not. Another difference is the 12-hour AM/PM time format in English and the 24-hour time format in Swedish. Swedish and C omit the leading zero on the day, whereas the US locale does not. Finally, the US time is followed by the Greenwich mean time zone name, GMT. There is only one time zone in Sweden—Central European Time, CET—and this fact is reflected in the simplicity of Sweden's standard general time and date format.

All of these differences are defined by the environment locale, but a quick glance at the code in Listing 11-3 shows that I'm merely using the %c format specifier in the call to strftime in all cases. The effective locale is causing this format specifier to output general time and date information in a format specific to the locale.

Not all of the format specifiers accepted by strftime are as helpful, however. For example, while using a format string like "%X %D" may seem like a good approach, it will not yield correct results in all locales. The %X specifier formats the time in a locale-specific manner, but %D formats the date in a very US-English way. Additionally, full time-date strings are formatted in different locales with the time and date portions in different orders. Later in the chapter, I'll show you how to work around these issues with nl_langinfo.

Collation and Character Classes

Now let's consider the less obvious categories—those whose information is not returned in struct lconv: LC_COLLATE and LC_CTYPE.

LC_COLLATE affects the way the functions strcoll and strxfrm work. It's more difficult for an English speaker to comprehend these functions' inner workings because, in the English language, locale-specific comparisons of characters just happen to collate in the same order as their lexicographical orderings in the *ASCII* table.

NOTE   *The original* American Standard Code for Information Interchange (ASCII) *was invented in 1963 by the* American Standards Association (ASA). *At first, it included only US English capital letters and numbers. In 1967, it was amended to include control characters and lowercase US English letters. Since the standard limited code length to 7 bits, it included only 128 characters, using the codes 0 through 127. This 7-bit limitation was imposed because the eighth bit in each byte was commonly used for error correction during data transmission. In 1981, IBM incorporated the ASCII code into the lower half of an 8-bit, 256-character code it named* code page 437 *and incorporated this code into the firmware of its IBM PC line of personal computers. In this chapter, when I mention the* ASCII table, *I'm actually referring to* code page 437. *Technically, ASCII is still limited to 128 characters.*

This is not the case in many other languages. For instance, in English and Spanish, the accented vowels sort properly immediately after their unaccented counterparts, while in Japanese, neither vowels nor accented vowels exist in the alphabet, so they sort according to their ordinal values in the ASCII table. Since all the accented vowels are in the upper half of the ASCII table and all non-accented vowels are in the lower half, it should be clear that the sort order of a list of Spanish words will be different when using an English or Spanish language locale than it will for any locales based on languages that don't have Latin characters in their alphabet.

Listing 11-4 contains a short program that uses the C qsort function to sort a list of Spanish words using different comparison routines.

```
#include <stdio.h>
#include <stdlib.h>
#include <locale.h>
#include <string.h>

#define ECOUNT(x) (sizeof(x)/sizeof(*(x)))

int lex_count = 0;
int loc_count = 0;

static int compare_lex(const void *a, const void *b)
{
    lex_count++;
    return strcmp(*(const char **)a, *(const char **)b);
}
```

```
static int compare_loc(const void *a, const void *b)
{
    loc_count++;
    return strcoll(*(const char **)a, *(const char **)b);
}

static void print_list(const char * const *list, size_t sz)
{
    for (int i = 0; i < sz; i++)
        printf("%s%s", i ? ", " : "", list[i]);
    printf("\n");
}

int main()
{
    const char *words[] = {"rana", "rastrillo", "radio", "rápido", "ráfaga"};

    setlocale(LC_ALL, "");  // enable environment locale

    printf("Unsorted        : ");
    print_list(words, ECOUNT(words));

    qsort(words, ECOUNT(words), sizeof *words, &compare_lex);

    printf("Lex (strcmp)    : ");
    print_list(words, ECOUNT(words));

    qsort(words, ECOUNT(words), sizeof *words, &compare_loc);

    printf("Locale (strcoll): ");
    print_list(words, ECOUNT(words));

    return 0;
}
```

Listing 11-4: sc.c: A short program that illustrates sort order differences between locales

First, the unsorted words list is printed to the console; then, the pointers in the words list are sorted with qsort using the compare_lex function, which uses strcmp to determine the collation order of the letters in each pair of words compared. The strcmp function doesn't know anything about locales. It simply uses the order of the words' letters in the ASCII table. Then the sorted list is printed to the console.

Next, qsort is called once again on words—this time using compare_loc, which uses strcoll to determine the sort order of the word pairs. The strcoll function uses the current locale to determine the relative order of the letters in the words being compared. The re-sorted list is then printed to the console.

Building and executing this program with different locales displays the following output:

```
$ gcc sc.c -o sc
$ ./sc
Unsorted        : rana, rastrillo, radio, rápido, ráfaga,
Lex (strcmp)    : radio, rana, rastrillo, ráfaga, rápido,
Locale (strcoll): radio, ráfaga, rana, rápido, rastrillo,
$ LC_ALL=es_ES.utf8 ./sc
Unsorted        : rana, rastrillo, radio, rápido, ráfaga,
Lex (strcmp)    : radio, rana, rastrillo, ráfaga, rápido,
Locale (strcoll): radio, ráfaga, rana, rápido, rastrillo,
$ LC_ALL=ja_JP.utf8 ./sc
Unsorted        : rana, rastrillo, radio, rápido, ráfaga,
Lex (strcmp)    : radio, rana, rastrillo, ráfaga, rápido,
Locale (strcoll): radio, rana, rastrillo, ráfaga, rápido,
$
```

English and Spanish sort accented vowels the same way. The C locale, represented by the results obtained using strcmp, always sorts strictly according to the ASCII table. Japanese, however, sorts differently than the Latin languages because Japanese makes no assumptions about how characters (accented or otherwise) not found in its alphabet should be ordered.

Internally, strcoll uses an algorithm to transform the characters in the comparison strings into numeric values that order naturally in the current locale; then it compares these byte arrays using the strcmp function. The algorithm used by strcoll can be pretty heavyweight because, for each set of two strings it compares, it transforms the locale-specific multibyte character sequences of these string pairs into sequences of bytes that can be compared lexicographically, by codeset ordinal values, and then internally compares those byte sequences using strcmp.

If you know you're going to be comparing the same string or set of strings, it can be much more efficient to use the strxfrm function first, which exposes the transformation algorithm that strcoll uses internally. You can then simply use strcmp against these transformed strings to obtain the same collation you'd get from strcoll against untransformed strings.

Listing 11-5 illustrates this process by converting the contents of Listing 11-4 to use strxfrm on the words in the words array, writing the transformed words into a two-dimensional array large enough to hold the transformed strings.

```
#include <stdio.h>
#include <stdlib.h>
#include <locale.h>
#include <string.h>

#define ECOUNT(x) (sizeof(x)/sizeof(*(x)))

❶ typedef struct element
{
    const char *input;
```

```
        const char *xfrmd;
} element;

static int compare(const void *a, const void *b)
{
    const element *e1 = a;
    const element *e2 = b;
  ❷ return strcmp(e1->xfrmd, e2->xfrmd);
}

static void print_list(const element *list, size_t sz)
{
    for (int i = 0; i < sz; i++)
      ❸ printf("%s, ", list[i].input);
    printf("\n");
}

int main()
{
    element words[] =
    {
        {"rana"}, {"rastrillo"}, {"radio"}, {"rápido"}, {"ráfaga"}
    };

    setlocale(LC_ALL, "");  // enable environment locale

    // point each xfrmd field at corresponding input field
    for (int i = 0; i < ECOUNT(words); i++)
      ❹ words[i].xfrmd = words[i].input;

    printf("Unsorted          : ");
    print_list(words, ECOUNT(words));

    qsort(words, ECOUNT(words), sizeof *words, &compare);

    printf("Lex (strcmp)      : ");
    print_list(words, ECOUNT(words));

    for (int i = 0; i < ECOUNT(words); i++)
    {
        char buf[128];
        strxfrm(buf, words[i].input, sizeof buf);
      ❺ words[i].xfrmd = strdup(buf);
    }

    qsort(words, ECOUNT(words), sizeof *words, &compare);

    printf("Locale (strxfrm/cmp): ");
    print_list(words, ECOUNT(words));

    return 0;
}
```

Listing 11-5: sx.c: The sc program, rewritten to use strxfrm

There are several items of note here. The strxfrm function returns a zero-terminated byte buffer that looks and acts like an ordinary C string. There are no internal null characters; it can be acted upon by other string functions in the standard C library, but it's not necessarily intelligible from a human-readability standpoint. Because of this weird characteristic, the transform buffer contents can only be used for comparison purposes during sorting. The original input value must be used for display. Therefore, we need to keep track of, and sort as pairs, the input buffer and the transform buffer for each word in our list. The element structure at ❶ manages this for us.

Since we no longer need to use strcoll, I've removed the compare_loc function and renamed compare_lex to compare, and I've changed the code to compare the xfrmd fields of the element structures passed in (at ❷). Note, however, that the print_list function still prints the input field of the elements (at ❸). This works because the words array has been converted to an array of pairs, where each element of the array contains both the original and the transformed words.

In order to make this code work with the original flow of main in *sc.c*, immediately after setting the locale, *sx.c* iterates over words (at ❹), setting each element's xfrmd pointer to the same value as its input pointer. This allows us to see what happens when using strcmp on untransformed strings during the first call to qsort.

At ❺, after printing the results of that first sort operation, the program iterates over words again, this time calling strxfrm on each input string and pointing the corresponding xfrmd field at a strdup copy of the transform buffer, buf.[13]

Building and executing the code in Listing 11-5 should show us the same output we got when we ran the code from Listing 11-4:

```
$ gcc sx.c -o sx
$ ./sx
Unsorted              : rana, rastrillo, radio, rápido, ráfaga,
Lex (strcmp)          : radio, rana, rastrillo, ráfaga, rápido,
Locale (strxfrm/cmp): radio, ráfaga, rana, rápido, rastrillo,
$ LC_ALL=es_ES.utf8 ./sx
Unsorted              : rana, rastrillo, radio, rápido, ráfaga,
Lex (strcmp)          : radio, rana, rastrillo, ráfaga, rápido,
Locale (strxfrm/cmp): radio, ráfaga, rana, rápido, rastrillo,
$ LC_ALL=ja_JP.utf8 ./sx
Unsorted              : rana, rastrillo, radio, rápido, ráfaga,
Lex (strcmp)          : radio, rana, rastrillo, ráfaga, rápido,
Locale (strxfrm/cmp): radio, rana, rastrillo, ráfaga, rápido,
$
```

It's a bit more complicated—the value of this version is not immediately apparent when sorting five words, but the time savings over transforming the strings within strcoll is significant when sorting hundreds of strings, even with the overhead of allocating and freeing the transform buffers.

---

13. Please forgive me for not freeing the allocated buffers—I didn't want to confuse the important code with cleanup details.

NOTE *This sample takes shortcuts in order to highlight the important points of* strxfrm. *A real program would check the result of* strxfrm, *which returns the number of bytes required by the transformation (minus the terminating null character). If the value is larger than the buffer size specified, the program should reallocate and call* strxfrm *again. There is no reasonable way to predetermine the required buffer size for any given locale and codeset. I made my buffer large enough to handle just about any possibility, so I skipped this check for the sake of code readability, but this is* not *a recommended practice.*

Now let's turn our attention to the LC_CTYPE locale category. Changing this locale category affects the way most of the character classification functions in *ctype.h* work, including isalnum, isalpha, isctrl, isgraph, islower, isprint, ispunct, isspace, and isupper (but specifically not isdigit or isxdigit). It also affects the way toupper and tolower work—sort of. The fact is, the functions in *ctype.h* are broken in many ways with respect to internationalization. The problem is they rely on algorithmic mechanisms to convert character case, which work fine as long as you stick with the ASCII table. As soon as you leave this familiar playing field, however, all bets are off. Sometimes they work, and sometimes they don't. The most consistent way to make them work is to use wide characters, because the wide-character versions of these functions are newer in the C and C++ standards and the UTF-16 and UTF-32 codesets allow for similar algorithmic conversion for an expanded set of characters. However, even when wide characters are used, there are still cases where the algorithmic approach fails to convert properly, as some languages have digraphs that come in three forms: lowercase, uppercase, and title case. There's just no algorithmic way to deal properly with these types of situations.

The source code in Listing 11-6 shows one way to properly convert a Spanish word from uppercase to lowercase.

```
#include <stdio.h>
#include <locale.h>
#include <wctype.h>
#include <wchar.h>

int main()
{
    const wchar_t *orig = L"BAÑO";
    wchar_t xfrm[64];

    setlocale(LC_ALL, "");  // enable environment locale

    int i = 0;
    while (i < wcslen(orig))
    {
        xfrm[i] = towlower(orig[i]);
        i++;
    }
```

```
        xfrm[i] = 0;
        printf("orig: %ls, xfrm: %ls\n", orig, xfrm);

        return 0;
}
```

Listing 11-6: ct.c: Converting a Spanish word using wide characters

The output is as follows:

```
$ gcc ct.c -o ct
$ ./ct
orig: BAÑO, xfrm: baño
$
```

This program doesn't work if you change to char buffers and use UTF-8. It barely works using wide characters. If you set LC_ALL=C, it prints only orig: because, had we checked the return value of printf in Listing 11-6 (as we should do—especially when dealing with character set conversions like this), we'd have seen it return a -1, which is what it returns when it fails to convert a wide-character string to a multibyte string using %ls.

Rather than cover all the nuances of what does and doesn't work in the LC_CTYPE category, I'll just say that if you have to do a lot of this sort of conversion and character classification, I'd highly recommend using a third-party library like IBM's *International Components for Unicode (ICU)*[14] or GNU libunistring[15] (both of which, to put it succinctly, just do the right thing in all cases). ICU is a large library, and there's a bit of a learning curve, but it's worth the effort if you need it. GNU libunistring is a little easier to get your head around, but it still presents a lot of new functionality. There are also wrapper libraries like *Boost::locale*,[16] if you're using C++, that make accessing ICU a bit simpler, although *Boost::locale*, itself, is pretty complex.

X/Open and POSIX Standard Extensions

It's a shame there is not a standard C library function to format numeric and currency amounts by locale in the same manner that strftime formats time and date by locale. There is, however, an extension provided by the X/Open and POSIX standards and implemented in the GNU C library—the strfmon function, whose prototype is as follows:

```
#include <monetary.h>

ssize_t strfmon(char *s, size_t max, const char *format, ...);
```

---

14. See *http://site.icu-project.org/home/*.

15. See *https://www.gnu.org/software/libunistring/*.

16. See *https://www.boost.org/doc/libs/1_67_0/libs/locale/doc/html/index.html*.

It works very much like strftime, placing the formatted value string in the *max*-sized buffer pointed to by s. The *format* string works like the format strings in the printf family of functions and like that of strftime. The format specifiers are specific to this function but, like those of the other functions, begin with percent sign (%) and end with a format character. Several supported modifier characters may be used between the percent and the format character. The two valid format characters are i for international and n for local.

This function is designed to format currency amounts and follows all the localeconv-provided LC_CURRENCY rules, but it can also be used to format decimal numbers according to localeconv-provided LC_NUMERIC rules. Listing 11-7 provides example code for formatting a currency value in local and international formats without any special modifiers and for formatting a decimal number. Unlike strftime, strfmon can format multiple values.

```c
#include <stdio.h>
#include <locale.h>
#include <monetary.h>

int main()
{
    double amount = 12654.376;
    char buf[256];

    setlocale(LC_ALL, "");  // enable environment locale

    strfmon(buf, sizeof buf, "Local: %n, Int'l: %i, Decimal: %!6.2n",
            amount, amount, amount);
    printf("%s\n", buf);
    return 0;
}
```

Listing 11-7: amount.c: An example of calling strfmon to format currency and decimal values

Let's build and execute this program to see what's displayed when using different locales:

```
$ gcc amount.c -o amount
$ LC_ALL=C ./amount
Local: 12654.38, Int'l: 12654.38, Decimal: 12654.38
$ ./amount
Local: $12,654.38, Int'l: USD 12,654.38, Decimal: 12,654.38
$ LC_ALL=sv_SE.utf8 ./amount
Local: 12 654,38 kr, Int'l: 12 654,38 SEK, Decimal: 12 654,38
$ LC_ALL=ja_JP.utf8 ./amount
Local: ¥12,654, Int'l: JPY 12,654, Decimal: 12,654.38
$
```

All the characteristics displayed by the lc program in Listing 11-2 for monetary and numeric categories are taken into account by strfmon in the same manner the standard strftime function does for time and date characteristics. For instance, in both English and Japanese, the currency symbols are displayed before the values, while the Swedish currency symbols, kr and SEK, follow the value. The decimal separator is a comma in Sweden (and many other European locales), and Japanese yen values do not display a fractional part.

The exclamation mark (!) modifier in the decimal format specifier is used to suppress display of the currency symbol. By explicitly specifying a format precision, we can override the default Japanese locale characteristic that indicates that monetary values should not have a fractional part. The strfmon function was obviously designed for formatting currency values but, as we can see here, it can just as well be used to format plain old numeric decimal and integer values.

Overcoming localeconv's Shortcomings

The X/Open and POSIX standards also provide a better and more functional version of localeconv called nl_langinfo. Here is the prototype for this function:

```
#include <langinfo.h>

char *nl_langinfo(nl_item item);
```

The advantages of this interface over the standard library interface are numerous. First, it's more efficient, only acquiring and returning the field you request on an as-needed basis, rather than filling and returning the entire locale attribute structure for each request. The nl_langinfo function is used to acquire a single attribute, specified by item, of the global environment locale.

If your application is required to manage multiple locales simultaneously, check out the POSIX interface for managing multiple discrete locales within the same application. I won't cover them in detail here because they manage the same set of locale categories as the interfaces I've already shown you. Instead, see the POSIX 2008 standard for information on the newlocale, duplocale, uselocale, and freelocale functions, in connection with the nl_langinfo_l function, which accepts a second argument of type locale_t returned by newlocale. I will mention that the uselocale function can be used to set the locale of the current thread. All of the functions I've mentioned so far are implemented by the GNU C library.

The GNU C library also provides support for additional classes of locale information, including LC_MESSAGES, LC_PAPER, LC_NAME, LC_ADDRESS, LC_TELEPHONE, LC_MEASUREMENT, and LC_IDENTIFICATION. The LC_MESSAGES category has been standardized by POSIX and is the basis for *gettext*, which I'll discuss shortly. The others are not standardized in C or POSIX, but they've been incorporated for many years into so many aspects of Linux, including Linux ports of the X Window System, that it's hard to conceive of them being replaced or removed in the foreseeable future. Hence, I recommend their use if you do not intend to port your software outside of GNU tools.

These additional categories are not accessible though localeconv and the struct lconv structure. Rather, you'll need to use nl_langinfo to access the values in the locale that are associated with these categories.

Listing 11-8 is the same program found in Listing 11-2, except that this version uses nl_langinfo to display the locale information available through that interface. It's intentionally organized to display the content that's common to both interfaces in exactly the same format.

```c
#include <stdio.h>
#include <stdlib.h>
#include <string.h>
#include <stdbool.h>
#include <limits.h>
#include <stdint.h>
#include <locale.h>
#include <langinfo.h>

static void print_grouping(const char *prefix, const char *grouping)
{
    const char *cg;
    printf("%s", prefix);
    for (cg = grouping; *cg && *cg != CHAR_MAX; cg++)
        printf("%c %d", cg == grouping ? ':' : ',', *cg);
    printf("%s\n", *cg == 0 ? " (repeated)" : "");
}

static void print_monetary(bool p_cs_precedes, bool p_sep_by_space,
        bool n_cs_precedes, bool n_sep_by_space,
        int p_sign_posn, int n_sign_posn)
{
    static const char * const sp_str[] =
    {
        "surround symbol and quantity with parentheses",
        "before quantity and symbol",
        "after quantity and symbol",
        "right before symbol",
        "right after symbol"
    };

    printf("    Symbol comes %s a positive (or zero) amount\n",
            p_cs_precedes ? "BEFORE" : "AFTER");
    printf("    Symbol %s separated from a positive (or zero) amount by a space\n",
            p_sep_by_space ? "IS" : "is NOT");
    printf("    Symbol comes %s a negative amount\n",
            n_cs_precedes ? "BEFORE" : "AFTER");
    printf("    Symbol %s separated from a negative amount by a space\n",
            n_sep_by_space ? "IS" : "is NOT");
    printf("    Positive (or zero) amount sign position: %s\n",
            sp_str[p_sign_posn == CHAR_MAX? 4: p_sign_posn]);
    printf("    Negative amount sign position: %s\n",
            sp_str[n_sign_posn == CHAR_MAX? 4: n_sign_posn]);
}
```

```c
#ifdef OUTER_LIMITS

#define ECOUNT(x) (sizeof(x)/sizeof(*(x)))

static const char *_get_measurement_system(int system_id)
{
    static const char * const measurement_systems[] = { "Metric", "English" };
    int idx = system_id - 1;
    return idx < ECOUNT(measurement_systems)
            ? measurement_systems[idx] : "unknown";
}

#endif

int main(void)
{
    char *isym;

    setlocale(LC_ALL, "");

    printf("Numeric\n");
    printf("  Decimal: [%s]\n", nl_langinfo(DECIMAL_POINT));
    printf("  Thousands separator: [%s]\n", nl_langinfo(THOUSANDS_SEP));

    print_grouping("  Grouping", nl_langinfo(GROUPING));

    printf("\nMonetary\n");
    printf("  Decimal point: [%s]\n", nl_langinfo(MON_DECIMAL_POINT));
    printf("  Thousands separator: [%s]\n", nl_langinfo(MON_THOUSANDS_SEP));
    printf("  Grouping");

    print_grouping("  Grouping", nl_langinfo(MON_GROUPING));

    printf("  Positive amount sign: [%s]\n", nl_langinfo(POSITIVE_SIGN));
    printf("  Negative amount sign: [%s]\n", nl_langinfo(NEGATIVE_SIGN));
    printf("  Local:\n");
    printf("    Symbol: [%s]\n", nl_langinfo(CURRENCY_SYMBOL));
    printf("    Fractional digits: %d\n", *nl_langinfo(FRAC_DIGITS));

    print_monetary(*nl_langinfo(P_CS_PRECEDES), *nl_langinfo(P_SEP_BY_SPACE),
            *nl_langinfo(N_CS_PRECEDES), *nl_langinfo(N_SEP_BY_SPACE),
            *nl_langinfo(P_SIGN_POSN), *nl_langinfo(N_SIGN_POSN));

    printf("  International:\n");
    isym = nl_langinfo(INT_CURR_SYMBOL);
    printf("    Symbol (ISO 4217): [%3.3s], separator: [%s]\n",
        isym, strlen(isym) > 3 ? isym + 3 : "");
    printf("    Fractional digits: %d\n", *nl_langinfo(INT_FRAC_DIGITS));

    print_monetary(*nl_langinfo(INT_P_CS_PRECEDES), *nl_langinfo(INT_P_SEP_BY_SPACE),
            *nl_langinfo(INT_N_CS_PRECEDES), *nl_langinfo(INT_N_SEP_BY_SPACE),
            *nl_langinfo(INT_P_SIGN_POSN), *nl_langinfo(INT_N_SIGN_POSN));
```

```
    printf("\nTime\n");
    printf("  AM: [%s]\n", nl_langinfo(AM_STR));
    printf("  PM: [%s]\n", nl_langinfo(PM_STR));
    printf("  Date & time format: [%s]\n", nl_langinfo(D_T_FMT));
    printf("  Date format: [%s]\n", nl_langinfo(D_FMT));
    printf("  Time format: [%s]\n", nl_langinfo(T_FMT));
    printf("  Time format (AM/PM): [%s]\n", nl_langinfo(T_FMT_AMPM));
    printf("  Era: [%s]\n", nl_langinfo(ERA));
    printf("  Year (era): [%s]\n", nl_langinfo(ERA_YEAR));
    printf("  Date & time format (era): [%s]\n", nl_langinfo(ERA_D_T_FMT));
    printf("  Date format (era): [%s]\n", nl_langinfo(ERA_D_FMT));
    printf("  Time format (era): [%s]\n", nl_langinfo(ERA_T_FMT));
    printf("  Alt digits: [%s]\n", nl_langinfo(ALT_DIGITS));

    printf("  Days (abbr)");
    for (int i = 0; i < 7; i++)
        printf("%c %s", i == 0 ? ':' : ',', nl_langinfo(ABDAY_1 + i));
    printf("\n");

    printf("  Days (full)");
    for (int i = 0; i < 7; i++)
        printf("%c %s", i == 0 ? ':' : ',', nl_langinfo(DAY_1 + i));
    printf("\n");

    printf("  Months (abbr)");
    for (int i = 0; i < 12; i++)
        printf("%c %s", i == 0 ? ':' : ',', nl_langinfo(ABMON_1 + i));
    printf("\n");

    printf("  Months (full)");
    for (int i = 0; i < 12; i++)
        printf("%c %s", i == 0 ? ':' : ',', nl_langinfo(MON_1 + i));
    printf("\n");

    printf("\nMessages\n");
    printf("  Codeset: %s\n", nl_langinfo(CODESET));

#ifdef OUTER_LIMITS

    printf("\nQueries\n");
    printf("  YES expression: %s\n", nl_langinfo(YESEXPR));
    printf("  NO expression:  %s\n", nl_langinfo(NOEXPR));

    printf("\nPaper\n");
    printf("  Height: %dmm\n", (int)(intptr_t)nl_langinfo(_NL_PAPER_HEIGHT));
    printf("  Width:  %dmm\n", (int)(intptr_t)nl_langinfo(_NL_PAPER_WIDTH));
    printf("  Codeset: %s\n", nl_langinfo(_NL_PAPER_CODESET));

    printf("\nName\n");
    printf("  Format: %s\n", nl_langinfo(_NL_NAME_NAME_FMT));
    printf("  Gen:    %s\n", nl_langinfo(_NL_NAME_NAME_GEN));
    printf("  Mr:     %s\n", nl_langinfo(_NL_NAME_NAME_MR));
    printf("  Mrs:    %s\n", nl_langinfo(_NL_NAME_NAME_MRS));
    printf("  Miss:   %s\n", nl_langinfo(_NL_NAME_NAME_MISS));
    printf("  Ms:     %s\n", nl_langinfo(_NL_NAME_NAME_MS));
```

```
    printf("\nAddress\n");
    printf("  Country name:  %s\n", nl_langinfo(_NL_ADDRESS_COUNTRY_NAME));
    printf("  Country post:  %s\n", nl_langinfo(_NL_ADDRESS_COUNTRY_POST));
    printf("  Country abbr2: %s\n", nl_langinfo(_NL_ADDRESS_COUNTRY_AB2));
    printf("  Country abbr3: %s\n", nl_langinfo(_NL_ADDRESS_COUNTRY_AB3));
    printf("  Country num:   %d\n",
            (int)(intptr_t)nl_langinfo(_NL_ADDRESS_COUNTRY_NUM));
    printf("  Country ISBN:  %s\n", nl_langinfo(_NL_ADDRESS_COUNTRY_ISBN));
    printf("  Language name: %s\n", nl_langinfo(_NL_ADDRESS_LANG_NAME));
    printf("  Language abbr: %s\n", nl_langinfo(_NL_ADDRESS_LANG_AB));
    printf("  Language term: %s\n", nl_langinfo(_NL_ADDRESS_LANG_TERM));
    printf("  Language lib:  %s\n", nl_langinfo(_NL_ADDRESS_LANG_LIB));
    printf("  Codeset:       %s\n", nl_langinfo(_NL_ADDRESS_CODESET));

    printf("\nTelephone\n");
    printf("  Int'l format:    %s\n", nl_langinfo(_NL_TELEPHONE_TEL_INT_FMT));
    printf("  Domestic format: %s\n", nl_langinfo(_NL_TELEPHONE_TEL_DOM_FMT));
    printf("  Int'l select:    %s\n", nl_langinfo(_NL_TELEPHONE_INT_SELECT));
    printf("  Int'l prefix:    %s\n", nl_langinfo(_NL_TELEPHONE_INT_PREFIX));
    printf("  Codeset:         %s\n", nl_langinfo(_NL_TELEPHONE_CODESET));

    printf("\nMeasurement\n");
    printf("  System: %s\n",_get_measurement_system(
            (int)*nl_langinfo(_NL_MEASUREMENT_MEASUREMENT)));
    printf("  Codeset: %s\n", nl_langinfo(_NL_MEASUREMENT_CODESET));

    printf("\nIdentification\n");
    printf("  Title:       %s\n", nl_langinfo(_NL_IDENTIFICATION_TITLE));
    printf("  Source:      %s\n", nl_langinfo(_NL_IDENTIFICATION_SOURCE));
    printf("  Address:     %s\n", nl_langinfo(_NL_IDENTIFICATION_ADDRESS));
    printf("  Contact:     %s\n", nl_langinfo(_NL_IDENTIFICATION_CONTACT));
    printf("  Email:       %s\n", nl_langinfo(_NL_IDENTIFICATION_EMAIL));
    printf("  Telephone:   %s\n", nl_langinfo(_NL_IDENTIFICATION_TEL));
    printf("  Language:    %s\n", nl_langinfo(_NL_IDENTIFICATION_LANGUAGE));
    printf("  Territory:   %s\n", nl_langinfo(_NL_IDENTIFICATION_TERRITORY));
    printf("  Audience:    %s\n", nl_langinfo(_NL_IDENTIFICATION_AUDIENCE));
    printf("  Application: %s\n", nl_langinfo(_NL_IDENTIFICATION_APPLICATION));
    printf("  Abbr:        %s\n", nl_langinfo(_NL_IDENTIFICATION_ABBREVIATION));
    printf("  Revision:    %s\n", nl_langinfo(_NL_IDENTIFICATION_REVISION));
    printf("  Date:        %s\n", nl_langinfo(_NL_IDENTIFICATION_DATE));
    printf("  Category:    %s\n", nl_langinfo(_NL_IDENTIFICATION_CATEGORY));
    printf("  Codeset:     %s\n", nl_langinfo(_NL_IDENTIFICATION_CODESET));

#endif // OUTER_LIMITS

    return 0;
}
```

Listing 11-8: nl.c: *Using* nl_langinfo *to display available locale information*

To build this code, you need to add a couple of definitions to the command line: _GNU_SOURCE and OUTER_LIMITS. The first definition belongs to the GNU C library and allows *nl.c* to access the extended international monetary fields that were not part of the C standard until C99. The second is my own invention that allows you to build the program without the extended categories provided by the GNU C library:

```
$ gcc -D_GNU_SOURCE -DOUTER_LIMITS nl.c -o nl
$ ./nl
Numeric
  Decimal: [.]
  Thousands separator: [,]
  Grouping: 3, 3 (repeated)

Monetary
  Decimal point: [.]
  Thousands separator: [,]
  Grouping  Grouping: 3, 3 (repeated)
  Positive amount sign: []
  Negative amount sign: [-]
  Local:
    Symbol: [$]
    Fractional digits: 2
    Symbol comes BEFORE a positive (or zero) amount
    Symbol is NOT separated from a positive (or zero) amount by a space
    Symbol comes BEFORE a negative amount
    Symbol is NOT separated from a negative amount by a space
    Positive (or zero) amount sign position: before quantity and symbol
    Negative amount sign position: before quantity and symbol
  International:
    Symbol (ISO 4217): [USD], separator: [ ]
    Fractional digits: 2
    Symbol comes BEFORE a positive (or zero) amount
    Symbol IS separated from a positive (or zero) amount by a space
    Symbol comes BEFORE a negative amount
    Symbol IS separated from a negative amount by a space
    Positive (or zero) amount sign position: before quantity and symbol
    Negative amount sign position: before quantity and symbol

Time
  AM: [AM]
  PM: [PM]
  Date & time format: [%a %d %b %Y %r %Z]
  Date format: [%m/%d/%Y]
  Time format: [%r]
  Time format (AM/PM): [%I:%M:%S %p]
  Era: []
  Year (era): []
  Date & time format (era): []
  Date format (era): []
  Time format (era): []
  Alt digits: []
  Days (abbr): Sun, Mon, Tue, Wed, Thu, Fri, Sat
  Days (full): Sunday, Monday, Tuesday, Wednesday, Thursday, Friday, Saturday
  Months (abbr): Jan, Feb, Mar, Apr, May, Jun, Jul, Aug, Sep, Oct, Nov, Dec
  Months (full): January, February, March, April, May, June, July, August,
September, October, November, December
```

```
Messages
  Codeset: UTF-8

Queries
  YES expression: ^[yY].*
  NO expression:  ^[nN].*

Paper
  Height:  279mm
  Width:   216mm
  Codeset: UTF-8

Name
  Format: %d%t%g%t%m%t%f
  Gen:
  Mr:     Mr.
  Mrs:    Mrs.
  Miss:   Miss.
  Ms:     Ms.

Address
  Country name:  USA
  Country post:  USA
  Country abbr2: US
  Country abbr3: USA
  Country num:   840
  Country ISBN:  0
  Language name: English
  Language abbr: en
  Language term: eng
  Language lib:  eng
  Codeset:       UTF-8

Telephone
  Int'l format:    +%c (%a) %l
  Domestic format: (%a) %l
  Int'l select:    11
  Int'l prefix:    1
  Codeset:         UTF-8

Measurement
  System: English
  Codeset: UTF-8

Identification
  Title:      English locale for the USA
  Source:     Free Software Foundation, Inc.
  Address:    http://www.gnu.org/software/libc/
  Contact:
  Email:      bug-glibc-locales@gnu.org
  Telephone:
  Language:   English
  Territory:  USA
  Audience:
  Application:
```

```
Abbr:
Revision:     1.0
Date:         2000-06-24
Category:     en_US:2000
Codeset:      UTF-8
$
```

The highlighted section of the preceding output shows the portion of nl's output that goes beyond the lc program from Listing 11-2. The additional locale categories are defined as follows.

LC_MESSAGES

This category provides one additional item value, CODESET, which defines the codeset used by this locale. This item is categorized under "Messages" because it's intended to be helpful when translating static text messages in application code. The value can also be used as an environment variable on Linux systems in order to help select the static message catalog to be used.

LC_PAPER

The paper category provides two items, _NL_PAPER_HEIGHT and _NL_PAPER _WIDTH, which return paper dimensional values in millimeters for the most commonly used printer paper in the locale. This can be very helpful when formatting print output or when auto-selecting paper sizes—*letter* and *A04*, for example. Be aware that the pointer values returned from these item enumeration values should be treated like native-word-sized integer values, rather than as actual pointers. See the *nl.c* code in Listing 11-8 for details.

LC_NAME

The name category provides information on formatting salutations such as Mr., Mrs., Miss, and Ms. in the locale. The items in this category allow your software to automatically select how to state such salutations in the current language and territory.

LC_ADDRESS

The address category provides items that return geographical information for the locale, such as country name, postal code, and two- and three-letter country name abbreviations. It also returns the language name and library used by the locale.

LC_TELEPHONE

The telephone category provides format-specifier strings usable within the *printf* family of functions to display telephone numbers in a style that's common in the current locale.

LC_MEASUREMENT

The measurement category provides a single item for returning the system of measurement used in the current locale. The _NL_MEASUREMENT _MEASUREMENT item returns a string whose first character is a short integer value: 0 for Metric or 1 for English.

LC_IDENTIFICATION

The identification category is actually locale metadata. That is, the fields of this category return information about the territory, author, and process used to create the current locale (for example, the locale author's name, email address, phone number, and so on). It also returns versioning information about the locale. Be aware that the pointer value returned from _NL_ADDRESS_COUNTRY_NUM should be treated like a native-word-sized integer value rather than a pointer. See the *nl.c* code in Listing 11-8 for details.

You can access the same information using the -k option with the locale command line program that comes preinstalled on your Linux distro, as follows:

```
$ locale -k LC_PAPER
height=279
width=216
paper-codeset="UTF-8"
$
```

You can query the GNU C library nl_langinfo function for individual time- and date-formatting attributes such as AM and PM strings, various more granular format-specifier strings, and full and abbreviated days of the week and months of the year in the current locale.

The GNU C library nl_langinfo implementation even returns regular expressions intended to be used for matching query responses. The regular expressions returned from the YESEXPR and NOEXPR item enumeration values can be used to match *yes* or *no* answers to questions prompted by software.

Instrumenting Source Code for Static Messages

Instrumenting access to locale-specific static text messages in your source code is also part of the process of internationalizing software, so we'll cover instrumentation of static text display messages here. Then we'll move on to how to generate and consume language packs in Chapter 12, where I'll discuss localization.

It should be clear by now that something needs to be done with the static portion of the "*greeting*, from *progname*!" text we printed from Jupiter (for example). I'm not going to take Jupiter any further, but it does provide a concise example of something that needs to change in our programs when the locale changes. The process of instrumenting source code for translating static display messages involves scanning your source code for all string literals that can be displayed to a user during the execution of a program and then doing something that makes it possible for the program to use a version of that string that specifically targets the current locale.

There are a few open source (and several third-party commercial) libraries that can be used to accomplish this task, but we're going to focus on the GNU *gettext* library. The *gettext* library is very simple from a software perspective. In its simplest form, there's one function for tagging a message to be translated and two functions for selecting the message catalog to be used for display. The tagging function is named gettext, and its prototype is shown in Listing 11-9.

```
#include <libintl.h>

char *gettext(const char *msgid);
```

Listing 11-9: The prototype for the gettext function

This function accepts a message identifier in the *msgid* parameter and returns the display message to the user. The message identifier can be any string but is usually the display message itself, in US-ASCII. The reason for this is that if the message catalog cannot be found, gettext returns the *msgid* value itself, which will then be used by the program in the same way the translated message would have been used, had it been found. Thus, the gettext function cannot fail in a manner that will cause the program to not work in some reasonable fashion under any conceivable set of conditions.

This convention makes it very simple to both instrument existing programs and write new programs that use locale-based message catalogs. You simply need to find all of the static text messages within the program source files and wrap them in calls to gettext.

Occasionally, it's necessary to provide the translator with more contextual information than just the string. For a common example, when providing message IDs for menu items such as the Open submenu option in the File menu, the programmer may have communicated to the translator that the programmer has provided the entire menu hierarchy in a format such as |File|Open. When the translator sees this, they know that only the portion following the last vertical bar symbol should be translated. But if there is no translation for the current locale, the message ID will be the full string. In this case, the programmer must write the code to check for a leading vertical bar. If it's found, only the portion following the last vertical bar should actually be displayed.

The code in Listing 11-10 shows a very short (and somehow familiar) example program that uses gettext.

```
#include <stdio.h>
#include <libintl.h>

#define _(x) gettext(x)

int main()
```

```
{
    printf(_("Hello, world!\n"));
    return 0;
}
```

Listing 11-10: gt.c: A short program that illustrates the use of the gettext library

The printf function sends the return value of gettext to stdout. The gettext function is exported by the GNU C library, so no additional libraries are required to use it. When using gettext without GNU C, just link the *intl* library (shared object or static archive).

We could call gettext directly in printf, but the underscore (_) macro is a common idiom used when internationalizing software for two reasons: First, it decreases the visual impact of instrumenting an existing code base for *gettext*. Second, it allows us a single point of replacement if we choose to wrap gettext with additional functionality or if we decide to use a more functional variant of gettext (for example, dgettext and dcgettext). I haven't discussed these variants here, but you can find out more about them in the *GNU C Library* manual.[17]

Message Catalog Selection

Selection of the message catalog is done in two phases: the programmer phase and the user phase. The programmer phase is handled by the functions textdomain and bindtextdomain. The prototypes for these functions (also exported by the GNU C library) are shown in Listing 11-11.

```
#include <libintl.h>

char *textdomain(const char *domainname);
char *bindtextdomain(const char *domainname, const char *dirname);
```

Listing 11-11: The prototypes for textdomain and bindtextdomain

The textdomain function allows the software author to determine the message catalog domain that is in use at any given point within the program. The domain represents a given message catalog containing some portion of the messages in a program. All strings extracted from the source code belonging to a specific domain end up in the message catalog for that domain.

A package may have several domains. The typical boundary between domains, and therefore between message catalogs, is an executable module—either a program or a library. For example, the *curl* package installs the command line curl program and the *libcurl.so* shared library. The *curl* library is designed to be used by both the curl program and by other third-party programs and libraries. If the *curl* package were internationalized, the package author might decide to use the *curl* domain for the curl program and the *libcurl* domain for the library so that third-party applications that use *libcurl* aren't required to have the curl message catalog installed.

---

17. See *https://www.gnu.org/software/libc/manual/html_node/Translation-with-gettext
.html#Translation-with-gettext*.

The example used by the *GNU C Library* manual[18] is one where *libc* itself uses libc as the domain name, but programs using *libc* would use their own domain. Simply put, the *domainname* parameter in these functions directly corresponds to a message catalog filename.

The dirname parameter in bindtextdomain is used to specify a base directory in which to search for the well-defined message catalog directory structure, which I'll discuss shortly. Normally, the value passed in this parameter is the absolute path in the Automake datadir variable, suffixed with */locale*. Recall that datadir contains, by default, $(prefix)/*share* and prefix contains */usr/local*, so the full path used here would be */usr/local/share/locale*. For distribution-provided packages, prefix is more often simply */usr*, so the full path would then become */usr/share/locale*. It's therefore up to the maintainer to ensure that datadir is available within the software (using techniques discussed in Chapter 3) and referenced in the argument passed to this parameter.

Listing 11-12 shows how to add the code necessary to select the proper message catalog based on the current locale. Of course, we must first make the program aware of the current locale in the usual manner by calling setlocale.

```c
#include <stdio.h>
#include <locale.h>
#include <libintl.h>

#ifndef LOCALE_DIR
# define LOCALE_DIR "/usr/local/share/locale"
#endif

#ifdef TEST_L10N
# include <stdlib.h>
# undef LOCALE_DIR
# define LOCALE_DIR getenv("PWD")
#endif

#define _(x) gettext(x)

int main()
{
    const char *localedir = LOCALE_DIR;

    setlocale(LC_ALL, "");
    bindtextdomain("gt", localedir);
    textdomain("gt");

    printf(_("Hello, world!\n"));

    return 0;
}
```

Listing 11-12: gt.c: Enhancements to enable the current locale and select the message catalog

---

18. See *https://www.gnu.org/software/libc/manual/html_node/Locating-gettext-catalog
.html#Locating-gettext-catalog*.

I'm using *gt* as the domain name here because that's the name of the program. If this program were part of a package wherein all the components used the same domain, then the package name might be a better choice.

The directory name passed into `bindtextdomain`'s second parameter is derived from a future *config.h* inclusion. We'll add that later when we incorporate this program into an Autotools build system. If we define `TEST_L10N` on the compiler command line, the directory name resolves to the value of the `PWD` environment variable, allowing us to test our program in any location containing the locale directory structure. (We'll replace this hack later with a more Autotool-ish mechanism in Chapter 12.)

That's really all there is to instrumenting your code for message catalog lookup. In the next section, I'll discuss how to generate and build message catalogs, which is part of the process of localizing a software package. I'll also talk about the internal workings of the *gettext* library, which allows the user to select (during the user phase) the message catalog that should be used by their choice of environment variable settings.

Summary

In this chapter, my goal was to give you enough background that you could easily continue learning about internationalizing your software projects. I've covered the C standard library functionality that's designed to help you internationalize your software.

In the next chapter, we'll continue our exploration of this topic by diving into localization. We'll also discover how to tie all of this into the Autotools so that language packs get built and installed with make commands generated by Automake.

12

LOCALIZATION

When I'm working on a problem, I never think about beauty.
I think only how to solve the problem. But when I have finished,
if the solution is not beautiful, I know it is wrong.
—R. Buckminster Fuller

Once you've accomplished the significant work involved in internationalizing your software, you can begin to consider building message catalogs for other languages and cultures. Building a language-specific message catalog is known as *localization*. You are localizing your software for a target locale. We will spend this chapter discussing the way GNU projects approach these topics, including how to hook message catalog management into an Autotools build system.

Getting Started

Message catalogs must be located where applications can find them. It could have been decided that applications should just store their language-specific message catalogs in a location selected by each project, but Linux (and Unix, in general) has long practiced the subtle art of quietly guiding application developers by convention. Not only do such conventions keep developers from having to make the same decisions over and over, but they also maximize the potential for reuse wherever possible. To these ends, the established convention for message catalogs is to place them in a common directory under the system data directory—what the *GNU Coding Standards* refers to as the datadir—most often defined as $(prefix)/*share*.[1] A special directory, $(datadir)/*locale*, houses all application message catalogs in a format that provides some nice features for the user.

Language Selection

I mentioned in Chapter 11 that application selection of the current language, and hence the message catalog used by the application, is done in two phases. I've discussed the programmer phase already. Now let's turn to the user phase, which allows the user some choice over which message catalog is selected. As with the selection of locale, the selection of the message catalog can be directed through the use of environment variables. The following environment variables are used to select the message catalog an application will use:

- LANGUAGE
- LC_ALL
- LC_*xxx*
- LANG

Up to this point, we've only focused on the LC_ALL variable, but, in actuality, the application's global locale is selected by first examining LC_ALL, then a category-specific variable (LC_TIME, for example), and finally LANG, in that order. In other words, if LC_ALL is not set or is set to the empty string, setlocale will look for LC_*xxx* variables (specifically, LC_COLLATE, LC_CTYPE, LC_MESSAGES, LC_MONETARY, LC_NUMERIC, and LC_TIME) and use their values to determine which locales are used for the associated areas of library functionality. Finally, if none of those are set, LANG is examined. If LANG is not set, you get an implementation-defined default, which is not always the same as the C locale.

On top of these variables used by setlocale, the *gettext* functions look at LANGUAGE first, which, if set, will override all the others for message catalog selection. Additionally, the *value* set in the LANGUAGE variable has some impact on the selection criteria. Before we get into value formats, let's take a look at the directory structure beneath $(datadir)/*locale*. If you look at this directory on your own system, you'd see something like this:

```
$ ls -1p /usr/share/locale
aa/
ab/
ace/
ach/
af/
```

---

1. See *https://www.gnu.org/prep/standards/html_node/Directory-Variables.html*.

```
all_languages
--snip--
locale.alias
--snip--
sr@ijekavian/
sr@ijekavianlatin/
sr@latin/
sr@Latn/
--snip--
zh_CN/
zh_HK/
zh_TW/
zu/
$
```

The format of these directory names should look somewhat familiar—it's the same format used by locale names defined in "Generating and Installing Locales" on page 303. Under each locale directory containing message catalogs (it's rather sparse—application localization is not as prevalent as you might imagine), you'll find a directory named *LC_MESSAGES*, containing one or more *message object* (*.mo*) files, which are compiled message catalogs. Here's Spanish in Spain, for instance:

```
$ tree --charset=ASCII /usr/share/locale/es_ES/LC_MESSAGES
/usr/share/locale/es_ES/LC_MESSAGES
`-- libmateweather.mo

0 directories, 1 file
$
```

If you examine the region-independent Spanish locale directory, */usr/share/locale/es*, you'll see a lot more message catalogs. Most programs don't bother differentiating regional locales when translating:

```
$ tree --charset=ASCII /usr/share/locale/es/LC_MESSAGES
/usr/share/locale/es/LC_MESSAGES/
|-- apt.mo
|-- apturl.mo
|-- bash.mo
|-- blueberry.mo
|-- brasero.mo
--snip--
|-- xreader.mo
|-- xviewer.mo
|-- xviewer-plugins.mo
|-- yelp.mo
`-- zvbi.mo
$
```

As I mentioned earlier, the base name of a message object file is the domain of the owning application. When you call textdomain and bindtextdomain, the *domain* you specify selects a message object file by name. In this directory listing, *blueberry* is the message catalog domain of the application that uses the *blueberry.mo* message catalog.

Building Message Catalogs

The *gettext* library provides a set of utilities that help you build message catalogs from source code that's been internationalized for message catalog selection. Figure 11-1 depicts the flow of data through the *gettext* utilities, from source code to binary message object.

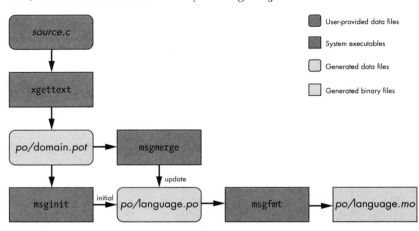

Figure 12-1: The flow of data from source file to message object file

The xgettext utility extracts messages from programming language source files and builds a *portable object template* (*.pot*) file. This is done each time the message strings in source files are changed or updated in some way. Perhaps existing messages are modified or removed or new messages are added. In any case, the *.pot* file must be updated. The *.pot* file is usually named after the message catalog domain used by the package or program.

Assuming we created a message catalog in a project for a French locale, this process generates files in the current directory, but we're going to follow a common convention by generating all of our message artifacts into a directory off the project root called *po*:

```
project/
    po/
        fr.mo
        fr.po
        project.pot
```

The internal layout of the *po* directory in my examples is arbitrary—you can tell the *gettext* tools how to name output files, and you can put them anywhere you like. I chose this structure because it's what we're going to use when we integrate *gettext* with an Autotools project later in this chapter.

Let's generate a *.pot* file for the source code of the gt program in Listing 11-12 on page 328:

```
$ mkdir po
$ cd po
$ xgettext -k_ -c -s -o gt.pot ../gt.c
$ cat gt.pot
# SOME DESCRIPTIVE TITLE.
# Copyright (C) YEAR THE PACKAGE'S COPYRIGHT HOLDER
# This file is distributed under the same license as the PACKAGE package.
# FIRST AUTHOR <EMAIL@ADDRESS>, YEAR.
#
#, fuzzy
msgid ""
msgstr ""
"Project-Id-Version: PACKAGE VERSION\n"
"Report-Msgid-Bugs-To: \n"
"POT-Creation-Date: 2018-07-12 17:22-0600\n"
"PO-Revision-Date: YEAR-MO-DA HO:MI+ZONE\n"
"Last-Translator: FULL NAME <EMAIL@ADDRESS>\n"
"Language-Team: LANGUAGE <LL@li.org>\n"
"Language: \n"
"MIME-Version: 1.0\n"
"Content-Type: text/plain; charset=CHARSET\n"
"Content-Transfer-Encoding: 8bit\n"

#: ../gt.c:25
#, c-format
msgid "Hello, world!\n"
msgstr ""
$
```

The xgettext utility is designed to build *.pot* files from the source files of many different programming languages, so it accepts --language or -L command line options as hints. However, it will also guess the language, based on the file extension, if no such option is given.

Because xgettext is designed to parse many different types of source file, it can sometimes require help locating the messages we want it to extract. It assumes text to be extracted is somehow associated with the gettext function in the C language. For other language source files, it looks for appropriate variations of this function name. Unfortunately, we threw a monkey wrench into the works when we replaced gettext with the underscore (_) macro name. This is where the --keyword (-k) option can be used to tell xgettext where to look for message text to be extracted. Our use of -k_ causes xgettext to look for _ instead of gettext. Without this option, xgettext won't find any messages to extract and, therefore, won't generate a *.pot* file.[2]

---

2. It's also very quiet about the entire process, so it can look like something just refused to work until you understand that it can't find any occurrences of static text to be extracted that are near calls to gettext.

I'm also telling it to add comments (-c) and to sort the output messages (-s) as they're added to the *.pot* file. If you don't tell it otherwise (with the -o option), it'll create a file called *messages.po*. Files not associated with a command line option are considered input files by xgettext.

At this point, though not strictly required, and depending on the work-flow you choose to use, you may want to hand-edit *gt.pot* to update place-holder values that xgettext adds. For example, you may want to replace the PACKAGE and VERSION placeholder strings in the Project-Id-Version field and perhaps add an email address to the Report-Msgid-Bugs-To field. These can be added during generation by using the --package-name, --package-version, and --msgid-bugs-address command line options. There are a few others; you can look them up in the manual.

From this template, we can now generate *portable object* (*.po*) files for different locales. The msginit utility is used to create an initial version of a locale-specific *.po* file, while msgmerge is used to update an existing *.po* file that was previously generated with msginit.

Let's create a French *fr.po* file from our template, *gt.pot*:

```
$ msginit --no-translator --locale=fr
Created fr.po.
$ cat fr.po
# French translations for PACKAGE package.
# Copyright (C) 2019 THE PACKAGE'S COPYRIGHT HOLDER
# This file is distributed under the same license as the PACKAGE package.
# Automatically generated, 2019.
#
msgid ""
msgstr ""
"Project-Id-Version: PACKAGE VERSION\n"
"Report-Msgid-Bugs-To: \n"
"POT-Creation-Date: 2019-07-02 23:19-0600\n"
"PO-Revision-Date: 2019-07-02 23:19-0600\n"
"Last-Translator: Automatically generated\n"
"Language-Team: none\n"
"Language: fr\n"
"MIME-Version: 1.0\n"
"Content-Type: text/plain; charset=ASCII\n"
"Content-Transfer-Encoding: 8bit\n"
"Plural-Forms: nplurals=2; plural=(n > 1);\n"

#: ../gt.c:21
#, c-format
msgid "Hello, world!\n"
msgstr ""
$
```

If you don't specify any input or output files, it looks in the current directory for a *.pot* file and derives the output filename from the locale you specify with the --locale (-l) option. I've also added the --no-translator option to suppress an interactive aspect of this utility. If you leave it off, msginit attempts to find your email address on the local host and use it. If it gets confused, it stops and asks you which address to use.

In addition to the specified or implied *.pot* file, it also examines make-files and other build files within the near vicinity of the source files you specify to see what the project might be called. The project name, PROJECT, that you see in this output is the default it uses when it can't find a project name, but it may surprise you how thorough msginit can be when searching for a project name.

Now, there's nothing very French about this *.po* file yet—would that it were so simple! No, you still have to translate the strings from English to French manually. So what's different about this *.po* file from its source template file? Essentially, everything related to a locale-specific implementation in the template has been filled in, including the title and copyright year in the comments at the top as well as the PO-Revision-Date, Last-Translator, Language-Team, Language, and Plural-Forms fields.

The next step in the process is to actually translate the file. Normally, I'd go find a native French speaker with a good grasp of English and ask them to fill in the blanks for me. Since there's little chance of misusing an internet translator with gt's one simple message, I'll just look it up myself and set the msgstr field at the bottom of the file to "Bonjour le monde!\n".

Once translated, the *.po* file is passed through the msgfmt utility to create the locale-specific *message object* (*.mo*) file. Let's do this for *fr.po*:

```
$ msgfmt -o fr.mo fr.po
$ ls -1p
fr.mo
fr.po
gt.pot
$
```

There are lots of options you can use with msgfmt. For our example, the default functionality is quite sufficient. Still, I specified the output file (with -o) because the default output file is *messages.mo* and I wanted to be clear that this is the French language message file.

To test our French message catalog, we could copy *fr.mo* over to */usr/local/share/locale/fr/LC_MESSAGES/gt.mo* as root and then execute gt with the LANGUAGE variable set to fr, but a simpler way is to use that hack I added to gt that lets us build a version that treats the current directory as the localedir.

NOTE *I named the output file* fr.mo, *but the installed message file must be named after the project's or program's message domain—*gt *in this case—so during installation* fr.mo *should be renamed to* gt.mo. *It's installed into a language-specific subdirectory of* localedir, *so the French nature of the* .mo *file is maintained after installation by virtue of its location in the filesystem.*

First, let's install our *fr.mo* file locally and then rebuild gt so that it looks in the current directory rather than the system data directory. Then we'll run gt with English and French locales, as follows:

```
$ cd ..
$ mkdir -p fr/LC_MESSAGES
$ cp po/fr.mo fr/LC_MESSAGES/gt.mo
$ gcc -DTEST_L10N gt.c -o gt
$ ./gt
Hello, world!
$ LANGUAGE=french ./gt
Bonjour le monde!
$
```

This console example should raise a few concerns: Why did I use LANGUAGE rather than LC_ALL? How was I able to use french instead of fr as the value of LANGUAGE without causing gt heartache while searching for the French version of *gt.mo*?

To answer the first question, I cannot use the LC_* or LANG variables here, because I don't have any French locales installed on my system and these variables merely set the locale, leaving textdomain and bindtextdomain to determine the locale based on queries to the structure returned by localeconv (or, rather, a more extensive internal form of that structure) in the C library. Because I don't have any French locales installed, setlocale will not be able to set a locale based on the values of the LANG or LC_* variables, so it will simply leave the current global locale set to the system default—English, on my host. Therefore, the language used will continue to be English.

The answer to the second question brings us back to an as yet unproven statement I made in "Language Selection" on page 332, where I said that the *value* the user sets in the LANGUAGE variable has some impact on the selection criteria used by textdomain and bindtextdomain. The *gettext* library allows the user to select *fallback* message catalogs when a requested locale is not available on the system. This is done by being *less specific* in the LANGUAGE variable (which is specifically used by textdomain and bindtextdomain) than in the other variables, which are examined by setlocale. The value format supported by LANGUAGE can exactly duplicate the strict format required in LC_* and LANG, but it also supports locale names with missing components and language aliases.

First, let's consider what I mean by missing components. Recall the components of a locale name:

```
language[_territory][.codeset][@modifier]
```

The bindtextdomain function attempts to find message catalogs in the specified locale directories that match this entire format, as specified either in the LANGUAGE variable or in the current locale string, as provided by localeconv. But then it backs off by dropping first the *codeset*, then a normalized form of the *codeset*,[3] then the *territory*, and finally the *modifier*.

---

3. See *https://www.gnu.org/software/libc/manual/html_node/Using-gettextized-software.html#Using-gettextized-software*.

If all components are dropped, we're left with just the *language* portion of the locale name (or whatever other random text was specified in LANGUAGE). If a match still cannot be found, bindtextdomain then looks at the */usr/share/locale/locale.alias* file for an alias matching the value in LANGUAGE (french is an alias on my system for fr_FR.ISO-8859-1). This algorithm allows users to be rather vague about which message catalog they want to use and still obtain one that's reasonably close to their native language, if an exact match for the current locale is not available.

Integrating gettext with the Autotools

Up to this point in this chapter, I've been building little utilities and programs like gt by just using gcc from the command line. Now it's time to turn gt into an Autotools project so we can add *Native Language Support (NLS)* functionality in the manner the GNU project recommends. It's really best to go this route, because it allows translators out there—people who love to do this sort of thing, and who like your program—to more easily add a message catalog for their language.

The information in this section was mostly taken from Section 13, "The Maintainer's View," of the *GNU gettext Utilities Manual*.[4] The *gettext* manual is a little out-of-date with respect to the Autotools and even the *gettext* package itself, but it's otherwise well organized and very detailed on the topics of internationalization and localization. In fact, it's so complete that it's hard to get your head around it until you have some of the basics behind you. My goal in this chapter is to give you the background you need to dig into the *gettext* manual without fear. In fact, this chapter only lightly brushes over many topics that the manual covers in great detail.

Let's move *gt.c* into a project *src* directory and create *configure.ac*, *Makefile.am*, and the other GNU-mandated text files. Assuming you're in the directory where our original *gt.c* file was created, do the following:

Git tag 12.0
```
$ mkdir -p gettext/src
$ mv gt.c gettext/src
$ cd gettext
$ autoscan
$ mv configure.scan configure.ac
$ touch NEWS README AUTHORS ChangeLog
$
```

The *Makefile.am* file should look like the one shown in Listing 12-1.

```
bin_PROGRAMS = src/gt
src_gt_SOURCES = src/gt.c
src_gt_CPPFLAGS = -DLOCALE_DIR=\"$(datadir)/locale\"
```

Listing 12-1: Makefile.am: The initial contents of this Automake input file

---

4. See *https://www.gnu.org/software/gettext/manual/gettext.html#Maintainers*.

I've added target-specific CPPFLAGS to allow me to pass the LOCALE_DIR on the compiler command line. We should also edit our *src/gt.c* file and add the *config.h* header file to it so we'll have access to the LOCALE_DIR variable we're defining in there. Listing 12-2 shows the changes we need to make. You can also remove the TEST_L10N hack; we will no longer need this because we can test Autotools-built gt using a local installation.

```
#include "config.h"

#include <stdio.h>
#include <locale.h>
#include <libintl.h>

#ifndef LOCALE_DIR
# define LOCALE_DIR "/usr/local/share/locale"
#endif

#define _(x) gettext(x)
--snip--
```

Listing 12-2: src/gt.c: Changes required to configure the LOCALE_DIR

Now edit the new *configure.ac* file and make the changes shown in Listing 12-3.

```
#                                          -*- Autoconf -*-
# Process this file with autoconf to produce a configure script.

AC_PREREQ([2.69])
AC_INIT([gt], [1.0], [gt-bugs@example.org])
AM_INIT_AUTOMAKE([subdir-objects])
AC_CONFIG_SRCDIR([src/gt.c])
AC_CONFIG_HEADERS([config.h])
AC_CONFIG_MACRO_DIRS([m4])

# Checks for programs.
AC_PROG_CC

# Checks for libraries.

# Checks for header files.
AC_CHECK_HEADERS([locale.h])

# Checks for typedefs, structures, and compiler characteristics.

# Checks for library functions.
AC_CHECK_FUNCS([setlocale])

AC_CONFIG_FILES([Makefile])

AC_OUTPUT
```

Listing 12-3: configure.ac: Changes necessary to the autoscan-generated .scan file

Note that some header file references were removed in the AC_CHECK
_HEADERS line in Listing 12-3.

At this point, you should be able to execute autoreconf -i, followed by
configure and make to build gt:

```
$ mkdir m4
$ autoreconf -i
configure.ac:12: installing './compile'
configure.ac:6: installing './install-sh'
configure.ac:6: installing './missing'
Makefile.am: installing './INSTALL'
Makefile.am: installing './COPYING' using GNU General Public License v3 file
Makefile.am:     Consider adding the COPYING file to the version control
system
Makefile.am:     for your code, to avoid questions about which license your
project uses
Makefile.am: installing './depcomp'
$ ./configure && make
--snip--
configure: creating ./config.status
config.status: creating Makefile
config.status: creating config.h
config.status: config.h is unchanged
config.status: executing depfiles commands
make  all-am
make[1]: Entering directory '/.../gettext'
depbase=`echo src/gt.o | sed 's|[^/]*$|.deps/&|;s|\.o$||'`;\
gcc -DHAVE_CONFIG_H -I.     -g -O2 -MT src/gt.o -MD -MP -MF $depbase.Tpo -c -o
src/gt.o src/gt.c &&\
mv -f $depbase.Tpo $depbase.Po
gcc  -g -O2    -o src/gt src/gt.o
make[1]: Leaving directory '/.../gettext'
$
```

NOTE   *I created the* m4 *directory before running* autoreconf *because* autoreconf *complains
about* m4 *not being present when it finds* AC_CONFIG_MACRO_DIRS *in* configure.ac.
*It still works, but warns you that the directory is missing. Creating it in advance
just reduces noise.*

The first step in enhancing an existing Autotools project for NLS support with *gettext* is to add a bunch of *gettext*-specific files to your project. It's
actually kind of tedious, so the *gettext* people have created a little utility
called gettextize that works pretty well. When you run gettextize, it does a
small amount of analysis, dumps a bunch of files into your project's *po* directory (it creates one if it's not there yet), and then displays a six- or seven-step
procedure on your console. To ensure you don't ignore this output, it waits
until you press ENTER to terminate the program, obtaining from you in
the process a promise that you'll read and perform those steps. Sadly, the
instructions are a little out-of-date—not all of them are actually necessary,
and some of them don't apply if you're using the full Autotools suite. Like
many programs that integrate with the Autotools, *gettext* was written to be

usable by packages that use Autoconf alone and by programs that use the full Autotools suite. I'll explain which ones are important as we go.

Let's start by running gettextize on our gt project directory:

```
$ gettextize
❶ Creating po/ subdirectory
Copying file ABOUT-NLS
Copying file config.rpath
Not copying intl/ directory.
Copying file po/Makefile.in.in
Copying file po/Makevars.template
Copying file po/Rules-quot
Copying file po/boldquot.sed
Copying file po/en@boldquot.header
Copying file po/en@quot.header
Copying file po/insert-header.sin
Copying file po/quot.sed
Copying file po/remove-potcdate.sin
Creating initial po/POTFILES.in
Creating po/ChangeLog
Copying file m4/gettext.m4
Copying file m4/iconv.m4
Copying file m4/lib-ld.m4
Copying file m4/lib-link.m4
Copying file m4/lib-prefix.m4
Copying file m4/nls.m4
Copying file m4/po.m4
Copying file m4/progtest.m4
❷ Updating Makefile.am (backup is in Makefile.am~)
Updating configure.ac (backup is in configure.ac~)
Adding an entry to ChangeLog (backup is in ChangeLog~)

❸ Please use AM_GNU_GETTEXT([external]) in order to cause autoconfiguration
to look for an external libintl.

❹ Please create po/Makevars from the template in po/Makevars.template.
You can then remove po/Makevars.template.

❺ Please fill po/POTFILES.in as described in the documentation.

❻ Please run 'aclocal' to regenerate the aclocal.m4 file.
You need aclocal from GNU automake 1.9 (or newer) to do this.
Then run 'autoconf' to regenerate the configure file.

❼ You will also need config.guess and config.sub, which you can get from the
CVS of the 'config' project at http://savannah.gnu.org/. The commands to fetch
them are
$ wget 'http://savannah.gnu.org/cgi-bin/viewcvs/*checkout*/config/config/
config.guess'
$ wget 'http://savannah.gnu.org/cgi-bin/viewcvs/*checkout*/config/config/
config.sub'

❽ You might also want to copy the convenience header file gettext.h
from the /usr/share/gettext directory into your package.
```

```
It is a wrapper around <libintl.h> that implements the configure --disable-nls
option.

Press Return to acknowledge the previous 6 paragraphs.
[ENTER]
$
```

The first thing gettextize does is create a *po* subdirectory (at ❶) in
the root of our project directory, if needed. This will be where all the
NLS-related files are kept and managed by an NLS-specific makefile,
which gettextize also provides, as you can see from the third `Copying file`
message found in the first few lines of the output.

After copying files from your system's *gettext* installation folder to the
po directory, it then updates the root-level *Makefile.am* file and *configure.ac*
(at ❷). Listings 12-4 and 12-5 show the changes it makes to these files.

```
bin_PROGRAMS = src/gt
src_gt_SOURCES = src/gt.c src/gettext.h
src_gt_CPPFLAGS = -DLOCALE_DIR=\"$(localedir)\"

SUBDIRS = po

ACLOCAL_AMFLAGS = -I m4

EXTRA_DIST = config.rpath
```

Listing 12-4: Makefile.am: Changes to this file made by gettextize

A `SUBDIRS` variable is added (or updated, if one exists) to the top-level
Makefile.am file so that *po/Makefile* will be processed by make, and `AC_LOCAL`
`_AMFLAGS` is added to support the *m4* directory, which gettextize would have
added had we not done so first. Finally, gettextize adds an `EXTRA_DIST` variable to ensure that *config.rpath* gets distributed.

NOTE *Adding* `AC_LOCAL_AMFLAGS = -I m4` *is no longer necessary with later versions of
Automake, because it provides the* `AC_CONFIG_MACRO_DIRS` *macro, which handles
this include directive for* aclocal *transparently.*

I manually changed $(datadir)/*locale* to the Autoconf-provided
$(localedir) in the src_gt_CPPFLAGS line.

```
#                                              -*- Autoconf -*-
# Process this file with autoconf to produce a configure script.
--snip--
# Checks for library functions.
AC_CHECK_FUNCS([setlocale])

AC_CONFIG_FILES([Makefile po/Makefile.in])

AC_OUTPUT
```

Listing 12-5: configure.ac: Changes to this file made by gettextize

The only change made to *configure.ac* by gettextize is to add the *po/Makefile.in* file to the `AC_CONFIG_FILES` file list. An astute reader would notice the *.in* on the end of this reference and perhaps believe that gettextize had made a mistake. Looking back at the list of files copied by the utility shows us, however, that the file copied into the *po* directory really is called *Makefile.in.in*. Autoconf processes the file first, and then *gettext* utilities process it again later to remove the second *.in* extension.

Referring back to the output of gettextize, we see at ❸ that gettextize is asking us to add a macro invocation, `AM_GNU_GETTEXT([external])`, to *configure.ac*. This may perhaps seem strange, given that it just finished editing *configure.ac* for us. The displayed text isn't clear on this point, but the fact is, an entire copy of the *gettext* runtime used to be added to projects on demand. This line is simply telling us that if we do not intend to use such an internal version of the *gettext* library, we should indicate so by using the external option in a call to this macro so that configure will know to look outside the project for the *gettext* utilities and libraries. As it happens, using an internal version of the *gettext* library is no longer generally promoted—mainly because *gettext* is now integrated into *libc* (at least on Linux systems), so everyone has ready access to an external version of *gettext*. If you're using another type of system with GNU tools, you should install the *gettext* package so you can use that external version.

I noticed also when I added this macro that autoreconf complained that I was using `AM_GNU_GETTEXT` but not `AM_GETTEXT_VERSION`, which indicates to the build system the lowest allowable version of *gettext* that may be used with this project. I added this macro as well, with a version value corresponding to the output of gettext --version on my system. I might have used a lower version value to allow my project to build on other, perhaps older systems, but I'd have had to do a bit of research to ensure that all the options I used were valid back to the version I chose to use.

Listing 12-6 shows this addition to *configure.ac*.

```
--snip--
# Checks for programs.
AC_PROG_CC
AM_GNU_GETTEXT_VERSION([0.19.7])
AM_GNU_GETTEXT([external])

# Checks for libraries.
--snip--
```

Listing 12-6: configure.ac: Adding `AM_GNU_GETTEXT`

The next step, at ❹, indicates that we should copy *po/Makevars.template* to *po/Makevars* and edit it to ensure the values are correct. I say "copy" rather than "move" because removing the template will just cause it to be replaced the next time you run autoreconf -i anyway, so there's no point in being pedantic about it.

Listing 12-7 shows a pared-down version of this file—I've removed the comments so we can more easily see the functional content, but the comments are extensive and really quite useful, so please do examine the full file.

```
DOMAIN = $(PACKAGE)
subdir = po
top_builddir = ..
XGETTEXT_OPTIONS = --keyword=_ --keyword=N_
COPYRIGHT_HOLDER = John Calcote
PACKAGE_GNU = no
MSGID_BUGS_ADDRESS = gt-bugs@example.org
EXTRA_LOCALE_CATEGORIES =
USE_MSGCTXT = no
MSGMERGE_OPTIONS =
MSGINIT_OPTIONS =
PO_DEPENDS_ON_POT = no
DIST_DEPENDS_ON_UPDATE_PO = yes
```

Listing 12-7: po/Makevars.template: A list of variables that control the NLS build

I've highlighted the changes I made to gt's version of this file. As you can see, the defaults are mostly just fine. I changed the copyright holder from the default, Free Software Foundation. I've also indicated that gt is not a GNU package—the default here was blank, which tells *gettext* to attempt to figure it out at runtime.

I've specified a value for MSGID_BUGS_ADDRESS, which is a value in the generated *.pot* file. The value generate by the *po* directory's makefile will be the email address (or web link) you specify here. Finally, I've set PO_DEPENDS_ON_POT to no because otherwise, anytime the *gt.pot* file changes in insignificant ways, the locale-specific *.po* files all get regenerated, and I'd rather just generate the *.po* files in my project when a distribution is created. This is an arbitrary decision based on personal preference; you can choose to leave it at its default value of yes, if you want.

At ❺, we see a request to add some text to *po/POTFILES.in*. This is a result of Automake's requirement that all source files be specified in makefiles. We're being asked to add all of the source files that must be processed by xgettext to extract messages. Files may be added one per line, and comments starting with a hash (#) mark may be used in this file if desired. Listing 12-8 highlights what I've added to gt's version of *po/POTFILES.in*.

```
# List of source files that contain translatable strings.
src/gt.c
```

Listing 12-8: po/POTFILES.in: Changes made to the generated version of this file

The files listed in *po/POTFILES.in* should be relative to the project directory root.

The steps listed at ❻ and ❼ are no longer necessary with late versions of the Autotools. The configure script will automatically run aclocal and rebuild itself for you when you execute it, if necessary. The *config.sub* and *config.guess* files are now automatically installed by autoreconf based on the use of the *gettext* macros. Unfortunately, autoreconf installs the versions of these files that ship with Autoconf; they're likely out-of-date, so the advice to find and install the latest versions is still valid. If you need to, you can pull the latest versions of these files from the GNU Savannah *config* repository using the supplied wget commands. You'll know if you need to if *gettext* has problems figuring out your platform using the ones installed by autoreconf. Be sure to run autoreconf -i at least once more after these steps are completed.

The request to copy and consume *gettext.h* at ❽ is optional but helpful, in my opinion, because it enables a configure script option added by *gettext* Autoconf macros that allows the user to disable NLS processing while building from a distribution archive. I copied */usr/share/gettext/gettext.h* into gt's *src* directory and added it to the list of source files for the gt program in *Makefile.am*, as shown in Listing 12-9.

```
bin_PROGRAMS = src/gt
src_gt_SOURCES = src/gt.c src/gettext.h
--snip--
```

Listing 12-9: Makefile.am: Adding src/gettext.h to src_gt_SOURCES

Let's try building after all these changes:

```
$ autoreconf -i
Copying file ABOUT-NLS
Copying file config.rpath
Creating directory m4
Copying file m4/codeset.m4
Copying file m4/extern-inline.m4
Copying file m4/fcntl-o.m4
--snip--
Copying file po/insert-header.sin
Copying file po/quot.sed
Copying file po/remove-potcdate.sin
configure.ac:12: installing './compile'
configure.ac:13: installing './config.guess'
configure.ac:13: installing './config.sub'
configure.ac:6: installing './install-sh'
configure.ac:6: installing './missing'
Makefile.am: installing './depcomp'
$
$ ./configure
checking for a BSD-compatible install... /usr/bin/install -c
checking whether build environment is sane... yes
checking for a thread-safe mkdir -p... /bin/mkdir -p
--snip--
checking for locale.h... yes
checking for setlocale... yes
checking that generated files are newer than configure... done
configure: creating ./config.status
config.status: creating Makefile
config.status: creating po/Makefile.in
config.status: creating config.h
config.status: executing depfiles commands
config.status: executing po-directories commands
config.status: creating po/POTFILES
config.status: creating po/Makefile
```

```
$
$ make
make  all-recursive
make[1]: Entering directory '/.../gettext'
Making all in po
make[2]: Entering directory '/.../gettext/po'
make gt.pot-update
make[3]: Entering directory '/.../gettext/po'
--snip--
case `/usr/bin/xgettext --version | sed 1q | sed -e 's,^[^0-9]*,,'` in \
  '' | 0.[0-9] | 0.[0-9].* | 0.1[0-5] | 0.1[0-5].* | 0.16 | 0.16.[0-1]*) \
    /usr/bin/xgettext --default-domain=gt --directory=.. \
      --add-comments=TRANSLATORS: --keyword=_ --keyword=N_  \
      --files-from=./POTFILES.in \
      --copyright-holder='John Calcote' \
      --msgid-bugs-address="$msgid_bugs_address" \
    ;; \
  *) \
    /usr/bin/xgettext --default-domain=gt --directory=.. \
      --add-comments=TRANSLATORS: --keyword=_ --keyword=N_  \
      --files-from=./POTFILES.in \
      --copyright-holder='John Calcote' \
      --package-name="${package_prefix}gt" \
      --package-version='1.0' \
      --msgid-bugs-address="$msgid_bugs_address" \
    ;; \
esac
--snip--
make[3]: Leaving directory '/.../gettext/po'
test ! -f ./gt.pot || \
  test -z "" || make
touch stamp-po
make[2]: Leaving directory '/.../gettext/po'
make[2]: Entering directory '/.../gettext'
gcc -DHAVE_CONFIG_H -I.  -DLOCALE_DIR=\"/usr/local/share/locale\"  -g -O2
-MT src/src_gt-gt.o -MD -MP -MF src/.deps/src_gt-gt.Tpo -c -o src/src_gt-gt.o
`test -f 'src/gt.c' || echo './'`src/gt.c
mv -f src/.deps/src_gt-gt.Tpo src/.deps/src_gt-gt.Po
gcc  -g -O2   -o src/gt src/src_gt-gt.o
make[2]: Leaving directory '/.../gettext'
make[1]: Leaving directory '/.../gettext'
$
```

NOTE   *Some corner-case conditions may cause files written by autoreconf to be considered "modified locally," which would generate errors without the -f or --force flag. I recommend you try it first using only -i. If you get errors about files like ABOUT-NLS being modified locally, then re-execute it with the -f flag also. Just be aware that -f will overwrite some files you may have intentionally modified.*

As you can see from this output, xgettext is run against our source code—specifically, the files we mentioned in *po/POTFILES.in*—whenever we build, if any of the files are missing at build time. If a file is found, it won't be rebuilt automatically, but there is a manual make target I'll mention shortly.

What Should Be Committed?

We've added a lot of new files to the gt project. In "A Word About the Utility Scripts" on page 172, I gave you my philosophy on what should be committed to a source repository, which is that people who check out your project from its repository should be willing to take on the role of maintainer or developer, rather than user. Users build from distribution archives, but maintainers and developers use a different set of tools. Therefore, people who check out source from repositories should be willing to use the Autotools.

It's now time to consider which of these new files you should commit to gt's repository. Following my philosophy, I would only commit those files that are actually assets of the project. Anything that can be easily regenerated or recopied from other sources during the Autotools bootstrap process (autoreconf -i) should be left out.

The gettextize utility runs a program called autopoint, which acts for NLS-enabled projects as autoreconf -i does for Autotools projects, copying files into the project directory structure as needed. The AM_GETTEXT_* macros we added to *configure.ac* earlier ensure that the appropriate *.m4* files are added to the *m4* directory, the appropriate NLS files are added to the *po* directory, and (if you were using an internal version of the *gettext* library) the *gettext* source and build files are added to the *intl* directory. In fact, autopoint is a sort of contraction of the phrase *auto-po-intl-m4*.[5] More recent versions of autoreconf are aware of autopoint and will execute it for you if they notice you have an NLS-enabled project, but only if you provide the -i option to autoreconf, because autopoint only installs missing files and file installation is a function of the -i option.

Because autopoint installs all required non-asset files in your *po* directory, the only thing you need to commit in that directory are the files you modified, including *POTFILES.in, Makevars, ChangeLog*,[6] and, of course, your *.po* files. You don't need to commit your *.pot* file because the *po/Makefile* will regenerate that from your source code if it's missing. You don't need to commit your *.mo* files, as those get generated directly from *.po* files at install time. You don't need the *ABOUT-NLS* file unless you've modified it. You don't need anything in the *m4* directory except macro files you wrote and added yourself. You will need to commit the *src/gettext.h* file since you manually copied that file from your system *gettext* install directory.[7]

---

5. I had to install the *autopoint* package on my system separately. It doesn't come with *gettext*, and it must be installed in order to use gettextize. On my Debian-based system, I used the command sudo apt install autopoint.

6. There are tools now that will scrape a git log for *ChangeLog* data. If you already do this in your projects, then don't bother committing your *ChangeLog* either; there's no point in writing that information twice.

7. This is true unless you want to provide a bootstrap.sh script that copies that file over and runs autoreconf -i. However, you don't necessarily know where it will be located on someone else's system, so I'd recommend not doing this.

This leaves us with the following files in gt's directory structure:

```
$ tree --charset=ascii
.
|-- AUTHORS
|-- ChangeLog
|-- configure.ac
|-- COPYING
|-- INSTALL
|-- Makefile.am
|-- NEWS
|-- po
|   |-- ChangeLog
|   |-- Makevars
|   `-- POTFILES.in
|-- README
`-- src
    |-- gettext.h
    `-- gt.c

2 directories, 13 files
$
```

Adding a Language

Let's add our French language *.po* file:

Git tag 12.2

```
$ cd po
$ msginit --locale fr_FR.utf8
--snip--
$ echo "
fr" >>LINGUAS
$
```

While we do run msginit, we don't need to specify input and output files. Rather, msginit automatically discovers and uses all *.pot* files in the current directory as input files, and it automatically names the output *.po* file after the language specified. The only option we need to use is the --locale option to specify the target locale for which a *.po* file should be generated.

NOTE *I didn't use the --no-translator option this time because when I run* msginit, *I'm acting in the role of the translator for the target language. That is to say, the person who runs* msginit *for a given locale or language should be the translator for that language. Therefore, that person should also be willing to provide contact information for the translation, which they can input at the interactive prompt for their email address when they run* msginit *in this manner.*

We also need to add all supported languages to a file named *LINGUAS* in the *po* directory (and this new file should also be committed). This tells the build system which languages to support. We may actually have more languages in the *po* directory than we currently support. The languages in the *LINGUAS* file are those for which *.mo* files will be generated and installed when we run make install. The format of *LINGUAS* is fairly loose; you only need some sort of whitespace between languages. You may also use hash-preceded comments, if you want.

You'll find a file named *fr.po* in the *po* directory now. Of course, it still has to be translated by someone who speaks both languages fairly well. The contents should look something like that of Listing 12-10 after translation. I've updated mine, filling in all the blanks, so to speak.

```
# French translations for gt package.
# Copyright (C) 2019 John Calcote
# This file is distributed under the same license as the gt package.
# John Calcote <john.calcote@gmail.com>, 2019.
#
msgid ""
msgstr ""
"Project-Id-Version: gt 1.0\n"
"Report-Msgid-Bugs-To: gt-bugs@example.org\n"
"POT-Creation-Date: 2019-07-02 01:17-0600\n"
"PO-Revision-Date: 2019-07-02 01:35-0600\n"
"Last-Translator: John Calcote <john.calcote@gmail.com>\n"
"Language-Team: French\n"
"Language: fr\n"
"MIME-Version: 1.0\n"
"Content-Type: text/plain; charset=UTF-8\n"
"Content-Transfer-Encoding: 8bit\n"
"Plural-Forms: nplurals=2; plural=(n > 1);\n"

#: src/gt.c:21
#, c-format
msgid "Hello, world!\n"
msgstr "Bonjour le monde!\n"
```

Listing 12-10: po/fr.po: The translated French portable object file

Installing Language Files

Installation of language files is no harder than running make with the usual Automake-provided install target:

```
$ sudo make install
--snip--
Making install in po
--snip--
$
```

You may also use a DESTDIR variable on the make command line to test your installation in a local staging directory in order to see what gets installed. Of course you don't need sudo when you do this, as long as you have write privileges in your DESTDIR location.

Testing is not quite as simple as executing your program from the *src* directory after building, but neither is it that difficult. The problem is the entire Linux NLS system is designed to work with installed language files. You'll need to install into a local prefix directory, such as $PWD/*root*, for instance:

```
$ ./configure --prefix=$PWD/root
--snip--
$ make
--snip--
$ make install
--snip--
$ cd root/bin
$ ./gt
Hello, world!
$ LANGUAGE=french ./gt
Bonjour le monde!
$
```

Why does this work? Because we're passing the locale directory, based on prefix, into *gt.c* in the makefile on the gcc command line. Therefore, the prefix you use tells gt where the NLS files will be installed.

NOTE   *Don't try this with the* DESTDIR *variable. The* prefix *will still be set to* /usr/local, *but the* install *target will put everything into* $(DESTDIR)/$(prefix). *The locale directory is based only on* prefix, *which tricks built software into thinking it's being installed into* $(prefix), *while allowing packagers to stage the installation locally.*

Manual make Targets

The *gettext* makefile provides a couple of targets that can be used manually from the *po* directory (in fact, they'll only work from the *po* directory). If you want to manually update one of your *.pot* files, you can run make *domain* .update-pot, where *domain* is the name of the NLS domain you specified when you called textdomain and bindtextdomain in your source code.

If you want to update the translated language files using msgmerge, which will merge new messages from the *.pot* files into the locale-specific *.po* files, you can run make update-po. This will update all of the *.po* files whose locales are specified in *LINGUAS*.

Note that *.mo* files are not created at build time but only at installation time. The reason for this is that they're useless before they're installed. If you really need to have the *.mo* files without installing your package, you can install into a local prefix or into a DESTDIR staging directory, in the manner outlined earlier.

Summary

In this chapter, I barely grazed the surface of the topic of adding NLS support to projects.

What did I skip? Well, for instance, there are dozens of options in the *gettext* tools that help localizers build language files for programs that cause the software to display messages sensibly.

For another example, in a typical `printf` statement in C, you might provide a format string in English such as `"There are %d files in the '%s' directory."` In this example, `%d` and `%s` are placeholders for a count and a directory name, of course, but in German, the translated string would become something like `"Im verzeichnis '%s' befinden sich %d dateien."` Even a non-German-speaking programmer can see what's wrong here—the order of the format specifiers has changed. One solution, of course, is to use `printf`'s newer positional format specifiers.

There are dozens of other issues you will want to consider; the *GNU gettext Utilities Manual* is a great place to start.

13

MAXIMUM PORTABILITY
WITH GNULIB

*Nothing was ever created by two men. There are no good collaborations,
whether in art, in music, in poetry, in mathematics, in philosophy.
Once the miracle of creation has taken place, the group can build and
extend it, but the group never invents anything.*
—*John Steinbeck,* East of Eden

You know those cool scripting languages you've been using for the last 10 years or so—Python, PHP, Perl, JavaScript, Ruby, and so on? One of the coolest features of these languages, and even some compiled languages like Java, is the ability to access community-provided library functionality through the use of tools like pip and maven, from repositories like PEAR, RubyGems, CPAN, and Maven Central.

Don't you wish you could do that sort of thing with C and C++? You can have that experience in C with the *GNU Portability Library (Gnulib)*[1], with its companion command line tool `gnulib-tool`. Gnulib is a library of source code designed to be widely portable, even to platforms like Windows, using

---

1. See *https://www.gnu.org/software/gnulib/*.

both native- and Cygwin-based compilation (though Gnulib is tested on Cygwin a little more than it is with native Windows builds).

There are literally hundreds of portable utility functions in Gnulib that are designed with one goal in mind—portability to many different platforms. This chapter is about how to get started with Gnulib and how to use it to your best advantage.

License Caveat

Before I continue, I should mention that much of the Gnulib source code is licensed under GPLv3+ or LGPLv3+. Some of the Gnulib source code is, however, licensed under LGPLv2+, which may make that functionality a bit more palatable. The Gnulib functions that can reasonably be used in libraries are licensed under either LGPLv2 or LGPLv3+; all else is licensed either under GPLv3+ or under a sort of hybrid mix of "LGPLv3+ and GPLv2" (which is ultimately more compatible with GPLv2 than LGPLv2). If this bothers you, then you may want to skip this chapter, but before discarding Gnulib entirely, consider checking the license on the functionality you wish to use to see if your project can accommodate it.

Since Gnulib is distributed in source format, and designed to be incorporated into applications and libraries in that format, the use of Gnulib implies the incorporation of GPL and LGPL source code directly into your source base. At the very least, this means you'll need to license portions of your code using GPL and LGPL licenses. This may explain why Gnulib is not extremely popular, except with maintainers of other GNU packages.

If, on the other hand, you're writing an open source program already licensed under the GPL, or an open source library already using the LGPL, then your project is a perfect fit for Gnulib. Read on.

Getting Started

As mentioned, Gnulib is distributed in source format. While you can always go to the Savannah git repository and browse and download individual files online, it's much simpler to just clone the Gnulib repository to a work area on your local host. The Gnulib repository provides the gnulib-tool utility in the repository's root directory, which you can use to copy desired source modules, with companion Autoconf macros and build scripts, directly into your projects.

The gnulib-tool utility runs as is right from the root of the repository. To make it easy to access, create a soft link somewhere in your PATH to this program; then you can run gnulib-tool from your project directory to add Gnulib modules to your Autotools-based project:

```
$ git clone https://git.savannah.gnu.org/git/gnulib.git
--snip--
$ ln -s $PWD/gnulib/gnulib-tool $HOME/bin/gnulib-tool
$
```

That's all you need to make Gnulib usable in the most effective manner on your system.

NOTE *The upstream Gnulib project doesn't do releases but rather simply incorporates changes and bug fixes directly into the master branch. The programming examples in this chapter were written to use Gnulib source code from commit f876e0946c730 fbd7848cf185fc0dcc712e13e69 in the Savannah Gnulib git repository. If you're having trouble getting the code in this chapter to build correctly, it could be because something has changed in the Gnulib source since this book was written. Try backing off to this commit of Gnulib.*

Adding Gnulib Modules to a Project

To help you understand how to use Gnulib, let's create a project that does something useful. We'll write a program that converts data to and from base64 strings, which are widely used today, and Gnulib has a portable library of base64 conversion functionality. We'll start by creating a small program containing only a main function that will act as a driver for the Gnulib base64 conversion functionality we'll add later.

NOTE *The source code for this project is in the NSP-Autotools GitHub repository called* b64 *at* https://github.com/NSP-Autotools/b64/.

Git tag: 13.0

```
$ mkdir -p b64/src
$ cd b64
$
```

Edit *src/b64.c* and add the contents shown in Listing 13-1.

```
#include "config.h"
#include <stdio.h>

int main(void)
{
    printf("b64 - convert data to and from base64 strings.\n");
    return 0;
}
```

Listing 13-1: src/b64.c: The initial contents of the driver program main source file

Now let's run autoscan to provide a base *configure.ac* file, rename the new *configure.scan* file to *configure.ac*, and then create a *Makefile.am* file for our project. Note that I'm creating a nonrecursive Automake project here, adding the single source file, *src/b64.c*, directly to the top-level *Makefile.am* file.

Since we're not creating a "foreign" project, we also need to add the standard GNU text files (but you may certainly add foreign to the AM_INIT_AUTOMAKE macro argument list in *configure.ac* to avoid having to do this if you wish):

```
$ autoscan
$ mv configure.scan configure.ac
$ echo "bin_PROGRAMS = src/b64
src_b64_SOURCES = src/b64.c" >Makefile.am
$ touch NEWS README AUTHORS ChangeLog
$
```

Edit the new *configure.ac* file and make the changes shown in Listing 13-2.

```
#                                          -*- Autoconf -*-
# Process this file with autoconf to produce a configure script.

AC_PREREQ([2.69])
AC_INIT([b64], [1.0], [b 64-bugs@example.org])
AM_INIT_AUTOMAKE([subdir-objects])
AC_CONFIG_SRCDIR([src/b64.c])
AC_CONFIG_HEADERS([config.h])
AC_CONFIG_MACRO_DIRS([m4])
--snip--
AC_CONFIG_FILES([Makefile])

AC_OUTPUT
```

Listing 13-2: configure.ac: *Required changes to autoscan-generated* configure.ac

I've added the subdir-objects option to the AM_INIT_AUTOMAKE macro as part of creating a nonrecursive Automake build system. I've also added the AC_CONFIG_MACRO_DIRS macro to keep things clean.[2]

At this point, we should be able to run autoreconf -i, followed by configure and make, to build the project:

```
$ autoreconf -i
aclocal: warning: couldn't open directory 'm4': No such file or directory
configure.ac:12: installing './compile'
configure.ac:6: installing './install-sh'
configure.ac:6: installing './missing'
Makefile.am: installing './INSTALL'
Makefile.am: installing './COPYING' using GNU General Public License v3 file
Makefile.am:    Consider adding the COPYING file to the version control
system
Makefile.am:    for your code, to avoid questions about which license your
project uses
Makefile.am: installing './depcomp'
$
$ ./configure && make
checking for a BSD-compatible install... /usr/bin/install -c
```

---

2. I'm setting you up a bit here because Gnulib works better if your project uses a dedicated macro directory.

```
checking whether build environment is sane... yes
checking for a thread-safe mkdir -p... /bin/mkdir -p
--snip--
config.status: creating Makefile
config.status: creating config.h
config.status: executing depfiles commands
make  all-am
make[1]: Entering directory '/.../b64'
depbase=`echo src/b64.o | sed 's|[^/]*$|.deps/&|;s|\.o$||'`;\
gcc -DHAVE_CONFIG_H -I.     -g -O2 -MT src/b64.o -MD -MP -MF $depbase.Tpo -c
-o src/b64.o src/b64.c &&\
mv -f $depbase.Tpo $depbase.Po
gcc  -g -O2   -o src/b64 src/b64.o
make[1]: Leaving directory '/.../b64'
$
$ src/b64
$ b64 - convert data to and from base64 strings.
$
```

We're now ready to start adding Gnulib functionality to this project.
The first thing we need to do is use gnulib-tool to import the base64
module into our project. Assuming you've correctly cloned the Gnulib
git project and added a soft link to gnulib-tool to a directory in your
PATH (*$HOME/bin*, perhaps, if that directory is in your PATH), execute the
following command from the root of the *b64* project directory structure:

Git tag 13.1
```
$ gnulib-tool --import base64
Module list with included dependencies (indented):
    absolute-header
  base64
    extensions
    extern-inline
    include_next
    memchr
    snippet/arg-nonnull
    snippet/c++defs
    snippet/warn-on-use
    stdbool
    stddef
    string
File list:
  lib/arg-nonnull.h
  lib/base64.c
  lib/base64.h
  --snip--
  m4/string_h.m4
  m4/warn-on-use.m4
  m4/wchar_t.m4
Creating directory ./lib
Creating directory ./m4
Copying file lib/arg-nonnull.h
Copying file lib/base64.c
Copying file lib/base64.h
--snip--
```

```
Copying file m4/string_h.m4
Copying file m4/warn-on-use.m4
Copying file m4/wchar_t.m4
Creating lib/Makefile.am
Creating m4/gnulib-cache.m4
Creating m4/gnulib-comp.m4
Creating ./lib/.gitignore
Creating ./m4/.gitignore
Finished.

You may need to add #include directives for the following .h files.
  #include "base64.h"

Don't forget to
  - add "lib/Makefile" to AC_CONFIG_FILES in ./configure.ac,
  - mention "lib" in SUBDIRS in Makefile.am,
  - mention "-I m4" in ACLOCAL_AMFLAGS in Makefile.am,
  - mention "m4/gnulib-cache.m4" in EXTRA_DIST in Makefile.am,
  - invoke gl_EARLY in ./configure.ac, right after AC_PROG_CC,
  - invoke gl_INIT in ./configure.ac.
$
```

The lists elided in this console example can get quite long when using a module that has many dependencies on other Gnulib modules. The *base64* module only directly depends on the *stdbool* and *memchr* modules; however, the dependency list shows additional transitive dependencies. You can see the direct dependencies of a module before committing yourself to it by examining its dependency list on the *MODULES* page at *gnu.org*[3] or by reading the *modules/base64* file in your clone of the Gnulib repository.

Some of the transitive dependencies required by the base64 module include modules designed to make base64 much more portable to a wide variety of platforms. The *string* module, for example, provides a wrapper around your system's *string.h* header file that provides additional commonly available string functionality or fixes bugs on some platforms.

You can see from the output that a couple of directories were created—*m4* and *lib*—and then some supporting M4 macro files were added to the *m4* directory and some source and build files were added to the *lib* directory.

NOTE   *If you're working in a git repository, gnulib-tool adds .gitignore files to the m4 and lib directories so files that can be regenerated or recopied don't get checked in automatically when you run a command like git add -A. Instead, you'll see that the only files added are lib/.gitignore, m4/.gitignore, and m4/gnulib-cache.m4. All other files can be regenerated (or recopied) after you've finished configuring your project with the desired Gnulib modules.*

Finally, near the end of the output, gnulib-tool provides you with some concise instructions on how to use the base64 module you added. First, as

---

3. The base64 module page is found at *https://www.gnu.org/software/gnulib/MODULES .html#module=base64*.

per these instructions, we need to add *lib/Makefile* to our `AC_CONFIG_FILES` list in *configure.ac*. Later in the same list, we find additional instructions for more general modifications to *configure.ac*. Listing 13-3 shows all of the changes we should make to *configure.ac*, according to these instructions.

```
--snip--
# Checks for programs.
AC_PROG_CC
gl_EARLY

# Checks for libraries.

# Checks for header files.

# Initialize Gnulib.
gl_INIT

# Checks for typedefs, structures, and compiler characteristics.

# Checks for library functions.

AC_CONFIG_FILES([Makefile lib/Makefile])

AC_OUTPUT
```

Listing 13-3: configure.ac: Changes required by Gnulib

Some of the instructions also indicate changes required to the top-level *Makefile.am* file in our project. Listing 13-4 highlights these changes.

```
ACLOCAL_AMFLAGS = -I m4
EXTRA_DIST = m4/gnulib-cache.m4
SUBDIRS = lib

bin_PROGRAMS = src/b64
src_b64_SOURCES = src/b64.c
```

Listing 13-4: Makefile.am: Changes required by Gnulib

Your project should continue to build after making these changes. We'll have to run `autoreconf -i` to include additional files that are now required by the Gnulib macros we added to *configure.ac*.

When we imported the base64 module, the output from `gnulib-tool` indicated that we may need to add an include directive for *base64.h*. At the moment, we don't need such a directive because our code doesn't actually use any of base64's functionality. We're about to change that, but each module has its own set of include directives, so the steps I'm about to show you are related specifically to the base64 module. Other modules will have similar steps, but they'll be specific to the modules you choose to use. The documentation for each module tells you how to access the public interface for the module—that is, which header files to include.

While the documentation is not particularly clear on this point, you don't actually have to link any module-specific libraries into your project because the *lib/Makefile.am* file builds all imported modules' source files and adds the resulting objects to a static library called *libgnu.a*. This is a customized version of the Gnulib library, containing only the modules you pulled into your project. Since Gnulib is a source code library, there are no binary files (outside of the one built in the *lib* directory) required by projects consuming Gnulib functionality. Therefore, the procedure for linking to Gnulib functionality is the same for all Gnulib modules.

Let's add some of base64's functionality to our project to see what's involved in actually using this module. Make the changes highlighted in Listing 13-5 to your *src/b64.c* file.

Git tag 13.2

```
#include "config.h"
#include <stdio.h>
#include <stdlib.h>
#include <string.h>
#include <stdbool.h>
#include <errno.h>
#include <unistd.h>

#include "base64.h"

#define BUF_GROW 1024

static void exit_with_help(int status)
{
    printf("b64 - convert data TO and FROM base64 (default: TO).\n");
    printf("Usage: b64 [options]\n");
    printf("Options:\n");
    printf(" -d    base64-decode stdin to stdout.\n");
    printf(" -h    print this help screen and exit.\n");
    exit(status);
}

static char *read_input(FILE *f, size_t *psize)
{
    int c;
    size_t insize, sz = 0;
    char *bp = NULL, *cp = NULL, *ep = NULL;

    while ((c = fgetc(f)) != EOF)
    {
        if (cp >= ep)
        {
            size_t nsz = sz == 0 ? BUF_GROW : sz * 2;
            char *np = realloc(bp, nsz);
            if (np == NULL)
            {
                perror("readin realloc");
                exit(1);
            }
        }
```

```
                cp = np + (cp - bp);
                bp = np;
                ep = np + nsz;
                sz = nsz;
            }
            *cp++ = (char) c;
        }
        *psize = cp - bp;
        return bp;
    }

    static int encode(FILE *f)
    {
        size_t insize;
        char *outbuf, *inbuf = read_input(f, &insize);
        size_t outsize = base64_encode_alloc(inbuf, insize, &outbuf);
        if (outbuf == NULL)
        {
            if (outsize == 0 && insize != 0)
            {
                fprintf(stderr, "encode: input too long\n");
                return 1;
            }
            fprintf(stderr, "encode: allocation failure\n");
        }
        fwrite(outbuf, outsize, 1, stdout);
        free(inbuf);
        free(outbuf);
        return 0;
    }

    static int decode(FILE *f)
    {
        size_t outsize, insize;
        char *outbuf, *inbuf = read_input(f, &insize);
        bool ok = base64_decode_alloc(inbuf, insize, &outbuf, &outsize);
        if (!ok)
        {
            fprintf(stderr, "decode: input not base64\n");
            return 1;
        }
        if (outbuf == NULL)
        {
            fprintf(stderr, "decode: allocation failure\n");
            return 1;
        }
        fwrite(outbuf, outsize, 1, stdout);
        free(inbuf);
        free(outbuf);
        return 0;
    }

    int main(int argc, char *argv[])
    {
```

```
    int c;
    bool tob64 = true;

    while ((c = getopt(argc, argv, "dh")) != -1)
    {
        switch (c)
        {
            case 'd':
                tob64 = false;
                break;
            case 'h':
            default:                        exit_with_help(c == 'h' ? 0 : 1);
        }
    }
    return tob64 ? encode(stdin) : decode(stdin);
}
```

Listing 13-5: src/b64.c: Changes required to incorporate base64 functionality

I've provided the entire file in Listing 13-5 because there are only a few lines of the original code remaining. This program was designed to act as a Unix filter, reading input data from stdin and writing output data to stdout. To read from and write to files, just use command line redirection.

I should mention a few noteworthy points about this program. First, it uses a buffer growth algorithm in the read_input function. Much of this code can be replaced with a call to another Gnulib module function, x2nrealloc. The online documentation is sparse about the use of this method, or even the fact that it exists—perhaps because the xalloc interface has been around in various forms for many years. You can find the *xalloc.h* header file in the Gnulib source under the *lib* directory. There are long comments in there containing example usages of many of the functions, including the x2nrealloc function.

Another advantage of using xalloc functionality for all your allocation needs is that its allocation functions automatically check for NULL return values and abort your program with an appropriate error message on memory allocation failures. If you desire more control over the abort process, you can add a function to your code called xalloc_die (no arguments, no return value) that will be called by xalloc functions if it exists. You can use this hook to perform any cleanup needed before your program exits. Why not let you decide whether or not to exit? You're out of memory—what are you really going to do? Such out-of-memory conditions don't happen often in today's world of multi-terabyte-sized address spaces, but they still have to be checked for. The xalloc functions make doing so a little less painful.

Finally, unlike many filters, this program will likely crash if you feed it a file containing a gigabyte of data because it buffers the entire input in an allocated memory block, which it resizes as it reads data from stdin. The reason for this is that the default use of the base64 module is not designed to handle streaming data. It requires the entire buffer up front. There is, however, a base64_encode_alloc_ctx method that allows you to encode small chunks of your input text in an iterative fashion. I'll leave it as an exercise for you, the reader, to change this program to make use of this form of the base64 module.

To make this code build correctly, you'll need to change *Makefile.am* as shown in Listing 13-6.

```
ACLOCAL_AMFLAGS = -I m4
EXTRA_DIST = m4/gnulib-cache.m4
SUBDIRS = lib

bin_PROGRAMS = src/b64
src_b64_SOURCES = src/b64.c
src_b64_CPPFLAGS = -I$(top_builddir)/lib -I$(top_srcdir)/lib
src_b64_LDADD = lib/libgnu.a
```

Listing 13-6: Makefile.am: Changes required to use the base64 module in source

The src_b64_CPPFLAGS directive adds directories to the compiler's include search path so it can find any header files added with selected Gnulib modules. The src_b64_LDADD directive appends *lib/libgnu.a* to the linker command line. Both of these directives should be familiar at this point.

Let's build and run the b64 program. As I mentioned previously, you'll want to run autoreconf -i first, to pick up any changes required by Gnulib additions to the project.

```
$ autoreconf -i
--snip--
$ ./configure && make
--snip--
$ echo hi | src/b64
aGkK$ echo -n aGkK | src/b64 -d
hi
$
```

I used echo to pipe some text into the b64 filter, which outputs the base64 equivalent of that text: "aGkK". Note there's no line-feed character at the end of the output. The b64 filter outputs only the base64 text version of the input data. I then used echo -n to pipe the base64 text back into the filter, using the -d flag to decode to the original input data. The output is the original text, including a terminating line-feed character. By default, echo appends a line-feed character to the end of any text you hand it; therefore, the original encoded text includes a terminating line-feed character. The -n option tells echo to suppress the line-feed character. If you don't use -n, the decode will fail with an error indicating the input data is not valid base64 text because echo added a line-feed character to it, which is not part of the base64 text.[4]

One thing that's not clear from the Gnulib documentation is that, in keeping with the general philosophy of never committing files or data that

---

4. You can also use printf from the bash command line instead of echo for more direct control over terminating line-feed characters.

can be easily regenerated, Gnulib's *.gitignore* files keep imported module source code from being committed to your repository. There are a couple of reasons for this. First, Gnulib source code already lives in a repository—that of Gnulib itself. There's no point in proliferating copies of the Gnulib source code around the internet by storing it in every repository that consumes it.

Another reason for not storing it in your project repository is that bug fixes are always being supplied by users and maintainers. Each time you update your Gnulib work area and build your project, you could be getting a better version of the modules you're using.

Let's say you're finished for the day and you want to leave your work area in a nice clean state. You type git `clean -xfd` and wipe out everything not staged or already committed. The next day you come back and type `autoreconf -i`, followed by `configure && make`, but you find that your project won't build; there are files missing from the *m4* and *lib* directories that seemed pretty important the day before. In fact, you discover, only the *m4/gnulib-cache.m4* file remains as a subtle reminder to you that your project ever had anything to do with Gnulib.

As it happens, that *gnulib-cache.m4* file is all you really need. It tells gnulib-tool which modules you've imported. To get it all back again, execute gnulib-tool with the `--update` option. This causes gnulib-tool to recopy current versions of all the relevant Gnulib files back into your project.

NOTE *The use of the --update option with gnulib-tool will not update your Gnulib work area from its remote repository. Rather, it only updates your project's use of Gnulib modules with the files that currently exist in your Gnulib work area. If you really want to use a particular past version of a set of Gnulib modules, you can check out a revision of the Gnulib repository from the past and then run gnulib-tool --update to pull in the current set of files from your Gnulib work area.*

The `--update` option can also be used to copy updated versions of files after you've updated your Gnulib work area with git.

To help you remember to use gnulib-tool `--update` in projects that use Gnulib, the Gnulib manual suggests that you create a `bootstrap.sh` script (and flag it executable) containing at least the lines shown in Listing 13-7.

Git tag 13.3
```
#!/bin/sh
gnulib-tool --update
autoreconf -i
```

Listing 13-7: bootstrap.sh: A project bootstrap script for b64

It would be really nice if autoreconf was smart enough to notice that you've used Gnulib modules and just call gnulib-tool `--update` for you. I suspect that's on the feature list for a future release of Autoconf. For the present, however, you'll need to remember to run this command to pull in Gnulib files when you clone your project repository into a new work area or after you've asked git to make your current work area pristine.

Summary

In this chapter, I discussed how to add Gnulib modules to your Autotools-based projects. I believe I've given you enough of a taste of Gnulib to pique your interest in this resource. The Gnulib manual is well written and easy to grasp (though a bit shy of full documentation) once you have a handle on the basics.

The next step is for you to go to the Gnulib modules page and browse the functionality available to you. The header files and source code for the modules are also available for viewing from that page and in the *modules* and *lib* directories of the repository. Feel free to check them out.

The maintainers can always use help with documentation. Once you've used a module and become comfortable with it, see if its documentation could use some updating and consider becoming a contributor. You can use the Gnulib mailing list[5] as a resource, both for questions you may have about the use of Gnulib and for patches for the documentation and source code.[6]

---

5. The mailing lists for Gnulib can be found at *https://savannah.gnu.org/mail/?group=gnulib*.

6. As is commonly done with most open source projects, you should post a comment regarding enhancements you'd like to make in order to get community feedback and sign-off before you submit a code patch.

14

FLAIM: AN AUTOTOOLS EXAMPLE

Uncle Abner said . . . a person that started in to carry
a cat home by the tail was gitting knowledge
that was always going to be useful to him.
—*Mark Twain*, Tom Sawyer Abroad

So far in this book, I've taken you on a
whirlwind tour of the main features of
Autoconf, Automake, and Libtool, as well
as other tools that work well with the Autotools.
I've done my best to explain them in a manner that
is not only simple to digest but also easy to retain—
especially if you've had the time and inclination to follow along with my
examples on your own. I've always believed that no form of learning comes
anywhere close to the learning that happens while doing.

In this chapter and the next, we'll continue learning about the Autotools by studying the process I used to convert an existing, real-world, open
source project from a complex handcoded makefile to a complete GNU
Autotools build system. The examples I provide in these chapters illustrate
the decisions I had to make during the conversion process as well as some
concrete uses of Autotools features, including a few that I haven't yet presented in previous chapters. These two chapters will round out our study
of the Autotools by presenting real solutions to real problems.

The project I chose to convert is called *FLAIM*, which stands for *FLexible Adaptable Information Management.*

What Is FLAIM?

FLAIM is a highly scalable database-management library written in C++ and built on its own thin portability layer called the FLAIM toolkit. Some readers may recognize FLAIM as the database used by both Novell[1] eDirectory and the Novell GroupWise server. FLAIM originated at WordPerfect in the late 1980s, and it became part of Novell's software portfolio during the Novell/WordPerfect merger in 1994. Novell eDirectory used a spin-off of a then-late version of FLAIM to manage directory information bases that contain over a billion objects, and GroupWise used a much earlier spin-off to manage various server-side databases.

Novell made the FLAIM source code available as an open source project licensed under the GNU Lesser General Public License (LGPL) version 2[2] in 2006. The FLAIM project is currently hosted by SourceForge .net, and it is the result of 25 years of development and hardening in various WordPerfect and Novell products and projects.[3]

Why FLAIM?

While FLAIM is far from a mainstream OSS project, it has several qualities that make it a perfect example for showing how to convert a project to use the Autotools. For one, FLAIM is currently built using a handcoded GNU makefile that contains over 2,000 lines of complex make script. The FLAIM makefile contains a number of GNU Make–specific constructs, and thus you can only process this makefile using GNU Make. Individual (but nearly identical) makefiles are used to build the *flaim, xflaim,* and *flaimsql* database libraries, and the FLAIM toolkit (*ftk*), as well as several utility and sample programs on Linux, various flavors of Unix, Windows, and NetWare.

The existing FLAIM build system targets several different flavors of Unix, including AIX, Solaris, and HP-UX, as well as Apple's macOS. It also targets multiple compilers on these systems. These features make FLAIM ideal for this sample conversion project because I can show you how to handle differences in operating systems and toolsets in the new *configure.ac* files.

The existing build system also contains rules for many of the standard Automake targets, such as distribution tarballs. Additionally, it provides rules for building binary installation packages, as well as RPMs for systems that can build and install RPM packages. It even provides targets for

---

1. Novell was acquired by MicroFocus in 2010.

2. See the website for the GNU Lesser General Public License, version 2.1, at *https://www.gnu.org/licenses/lgpl-2.1.html.*

3. You can read more about the history and development of FLAIM at *http://sourceforge.net/projects/flaim/.*

building Doxygen[4] description files, which it then uses to build source documentation. I'll spend a few paragraphs showing you how you can add these types of targets to the infrastructure provided by Automake.

The FLAIM toolkit is a portability library that third-party projects can incorporate and consume independently. We can use the toolkit to demonstrate Autoconf's ability to manage separate subprojects as optional subdirectories within a project. If the user already has the FLAIM toolkit installed on their build machine, they can use the installed version or, optionally, override it with a local copy. On the other hand, if the toolkit is not installed, then the local, subdirectory-based copy will be used by default.

The FLAIM project also provides code to build both Java and C# language bindings, so I'll delve into those esoteric realms a bit. I won't go into great detail on building either Java or C# applications, but I will cover how to write *Makefile.am* files that generate both Java and C# programs and language-binding libraries.

The FLAIM project makes good use of unit tests. These are built as individual programs that run without command line options, so I can easily show you how to add real-world unit tests to the new FLAIM Autotools build system using Automake's trivial test framework.

The FLAIM project and its original build system employ a reasonably modular directory layout, making it rather simple to convert to an Autotools modular build system. A simple pass of the `diff` utility over the directory tree should suffice.

Logistics

When the first edition of this book was published in 2010, FLAIM had just been released as an open source project on SourceForge.net using Subversion to manage its source code repository. Since that time, the FLAIM project has become, more or less, inactive. No one I'm aware of is actively using the code base. As I am the only remaining maintainer of the source code, I've made a GitHub repository for FLAIM specifically for Chapters 14 and 15 of this second edition of this book. You can find this repository at the NSP-Autotools area on GitHub under the FLAIM project.[5] I've updated the information in this chapter to be relevant to FLAIM's storage in a git repository.

The source code repository for this chapter follows a somewhat different style than that for preceding chapters. The original Autotools build system changes I made to the FLAIM SourceForge.net project are buried beneath, and intermixed with, several dozen unrelated changes. Rather than spend hours separating out these changes in an effort to provide you with proper before and after snapshots of the FLAIM code base, I simply

---

4. See *http://www.doxygen.nl/*.

5. See *https://github.com/NSP-Autotools/FLAIM/*.

chose to commit the final FLAIM code, with its Autotools build system, to the GitHub project.[6]

Do not be discouraged about FLAIM's current activity status—it continues to provide a wide variety of opportunities to learn about Autotools build system techniques in real-world projects.

An Initial Look

Let me start by saying that converting FLAIM from GNU makefiles to an Autotools build system is not a trivial project. It took me a couple of weeks, and much of that time was spent determining exactly what to build and how to do it—in other words, analyzing the legacy build system. Another significant portion of my time was spent converting aspects that lay on the outer fringes of Autotools functionality. For example, I spent *much* more time converting build system rules for building C# language bindings than I did converting rules for building the core C++ libraries.

The first step in this conversion project is to analyze FLAIM's existing directory structure and build system. What components are actually built, and which components depend on which others? Can individual components be built, distributed, and consumed independently? These types of component-level relationships are important because they'll often determine how you'll lay out your project directory structure.

The FLAIM project is actually several small projects under one umbrella project within its repository. There are three separate and distinct database products: *flaim*, *xflaim*, and *flaimsql*. The flaim subproject is the original FLAIM database library used by eDirectory and GroupWise. The xflaim project is a hierarchical XML database developed for internal projects at Novell; it is optimized for path-oriented, node-based access. The flaimsql project is an SQL layer on top of the FLAIM database. It was written as a separate library in order to optimize the lower-level FLAIM API for SQL access. This project was an experiment that, frankly, isn't quite finished (but it does compile).

The point is that all three of these database libraries are separate and unrelated to each other, with no interlibrary dependencies. Since they may easily be used independently of one another, they can actually be shipped as individual distributions. You could consider each an open source project in its own right. This, then, will become one of my primary goals: to allow the FLAIM open source project to be easily broken up into smaller open source projects that can be managed independently of one another.

---

6. If you're interested in digging into the conversion process details, refer to the FLAIM Subversion repository commit history from r1056 through r1112. The most significant Autotools-related changes start at the earliest commits in this range, but minor Autotools-related build system modifications are made throughout the entire range. See *https://sourceforge.net/p/flaim/code/1112/log/?path=*. The actual code committed to the GitHub repository is version r1112, with some modifications for this edition of this book.

The FLAIM toolkit is also an independent project. While it's tailored specifically for the FLAIM database libraries, providing just the system service abstractions required for a DBMS, it depends on nothing but itself, and thus it may easily be used as the basis for portability within other projects without dragging along any unnecessary database baggage.[7]

The original FLAIM project was laid out in its repository as follows:

```
$ tree -d --charset=ascii FLAIM
FLAIM
|-- flaim
|   |-- debian
|   |-- docs
|   |-- sample
|   |-- src
|   `-- util
|-- ftk
|   |-- debian
|   |-- src
|   `-- util
|-- sql
|   `-- src
--snip--
`-- xflaim
    |-- csharp
    --snip--
    |-- java
    --snip--
    |-- sample
    --snip--
    |-- src
    `-- util
--snip--
```

The complete tree is fairly broad and somewhat deep in places, including significant utilities, tests, and other such binaries that are built by the legacy build system. At some point during the trek down into this hierarchy, I simply had to stop and consider whether it was worth converting that additional utility or layer. (If I hadn't done that, this chapter would be twice as long and half as useful.) To this end, I've decided to convert the following elements:

- The database libraries
- The unit and library interface tests
- The utilities and other such high-level programs found in various *util* directories
- The Java and C# language bindings found in the *xflaim* library

I'll also convert the C# unit tests, but I won't go into the Java unit tests, because I'm already converting the Java language bindings using

---

7. As you might guess, the FLAIM toolkit's file I/O abstraction is highly optimized.

Automake's JAVA primary. Since Automake provides no help for C#, I have to provide everything myself anyway, so I'll convert the entire C# code base. This will provide an example of writing the code for an entirely unsupported Automake product class.

Getting Started

As stated earlier, my first true design decision was how to organize the original FLAIM project into subprojects. As it turns out, the existing directory layout is almost perfect. I've created a master *configure.ac* file in the top-level *flaim* directory, which is just under the repository root directory. This topmost *configure.ac* file acts as a sort of Autoconf control file for each of the four lower-level projects: ftk, flaim, flaimsql, and xflaim.

I've managed the database library dependencies on the FLAIM toolkit by treating the toolkit as a pure external dependency defined by the make variables FTKINC and FTKLIB. I've conditionally defined these variables to point to one of a few different sources, including installed libraries and even locations given in user-specified configuration script options.

Adding the configure.ac Files

In the following directory layout, I've used an annotation column to indicate the placement of individual *configure.ac* files. Each of these files represents a project that may be packaged and distributed independently.

```
$ tree -d --charset=ascii FLAIM
FLAIM                           configure.ac (flaim-projects)
|-- flaim                       configure.ac (flaim)
|   |-- debian
|   |-- docs
|   |-- sample
|   |-- src
|   `-- util
|-- ftk                         configure.ac (ftk)
|   |-- debian
|   |-- src
|   `-- util
|-- sql                         configure.ac (flaimsql)
|   `-- src
--snip--
`-- xflaim                      configure.ac (xflaim)
    |-- csharp
    --snip--
    |-- java
    --snip--
    |-- sample
    --snip--
    |-- src
    `-- util
--snip--
```

My next task was to create these *configure.ac* files. The top-level file was trivial, so I created it by hand. The project-specific files were more complex, so I allowed the autoscan utility to do the bulk of the work for me. Listing 14-1 shows the top-level *configure.ac* file.

```
#                                              -*- Autoconf -*-
# Process this file with autoconf to produce a configure script.

AC_PREREQ([2.69])
❶ AC_INIT([flaim-projects], [1.0])
❷ AM_INIT_AUTOMAKE([-Wall -Werror foreign])
❸ AM_PROG_AR
❹ LT_PREREQ([2.4])
  LT_INIT([dlopen])

❺ AC_CONFIG_MACRO_DIRS([m4])
❻ AC_CONFIG_SUBDIRS([ftk flaim sql xflaim])
  AC_CONFIG_FILES([Makefile])
  AC_OUTPUT
```

Listing 14-1: configure.ac: The umbrella project's Autoconf input file

This *configure.ac* file is short and simple because it doesn't do much; nevertheless, there are some new and important concepts here. I invented the name flaim-projects and the version number 1.0 at ❶. These are not likely to change unless really dramatic changes take place in the project directory structure or the maintainers decide to ship a complete bundle of the subprojects.

NOTE *For your own projects, consider using the optional third argument to the AC_INIT macro. You can add an email or web address here to indicate to users where they can submit a bug report. The contents of this argument show up in* configure *output.*

The most important aspect of an umbrella project like this is the AC_CONFIG_SUBDIRS macro at ❻, which I have yet to cover in this book. The argument is a whitespace-separated list of the subprojects to be built, where each is a complete *GCS*-compliant project in its own right. Here's the prototype for this macro:

```
AC_CONFIG_SUBDIRS(dir1[ dir2 ... dirN])
```

It allows the maintainer to set up a hierarchy of projects in much the same way that Automake SUBDIRS configures the directory hierarchy for Automake within a single project.

Because the four subprojects contain all the actual build functionality, this *configure.ac* file acts merely as a control file, passing all specified configuration options to each of the subprojects in the order they're given in the macro's argument. The FLAIM toolkit project must be built first since the other projects depend on it.

Automake in the Umbrella Project

Automake usually requires the existence of several text files in the top-level project directory, including the *AUTHORS, COPYING, INSTALL, NEWS, README,* and *ChangeLog* files. It would be nice not to have to deal with these files in the umbrella project. One way to accomplish this is to simply not use Automake in the umbrella project. I'd either have to write my own *Makefile.in* template for this directory or use Automake just once to generate a *Makefile.in* template that I could then check into the repository as part of the project, along with the *install-sh* and *missing* scripts added by automake `--add-missing` (or `autoreconf -i`). Once these files were in place, I could remove AM_INIT_AUTOMAKE from the master *configure.ac* file.

Another option would be to keep Automake and simply use the foreign option in AM_INIT_AUTOMAKE (which I did at ❷) in the macro's optional parameter. This parameter contains a string of whitespace-separated options that tell Automake how to act in lieu of specific Automake command line options. When automake parses the *configure.ac* file, it notes these options and enables them as if they'd been passed on the command line. The foreign option tells Automake that the project will not entirely follow GNU standards, and thus Automake will not require the usual GNU project text files.

I chose the latter of the two methods because I might want to alter the list of subordinate projects at some point and I don't want to have to tweak a generated *Makefile.in* template by hand. I've also passed the `-Wall` and `-Werror` options in this list, which indicate that Automake should enable all Automake-specific warnings and report them as errors. These options have nothing to do with the user's compilation environment—only Automake processing.

Why Add the Libtool Macros?

Why include those expensive Libtool macros at ❹? Well, even though I don't do anything with Libtool in the umbrella project, the lower-level projects expect a containing project to provide all the necessary scripts, and the LT_INIT macro provides the *ltmain.sh* script. If you don't initialize Libtool in the umbrella project, tools like autoreconf, which actually looks in the *parent* directory to determine if the current project is itself a subproject, will fail when they can't find scripts that the current project's *configure.ac* file requires.

For instance, autoreconf expects to find a file called *../ltmain.sh* within the ftk project's top-level directory. Note the reference to the parent directory here: autoreconf noticed, by examining the parent directory, that ftk was actually a subproject of a larger project. Rather than install all the auxiliary scripts multiple times, the Autotools generate code that looks for scripts in the project's parent directory. This is done in an effort to reduce the number of copies of these scripts that are installed into multiproject packages.[8] If I don't use `LT_INIT` in the umbrella project, I can't successfully run `autoreconf` in the subprojects, because the *ltmain.sh* script won't be in the project's parent directory.

Adding a Macro Subdirectory

The `AC_CONFIG_MACRO_DIRS` macro at ❺ indicates the name of a subdirectory in which the `aclocal` utility can find all project-specific M4 macro files. Here's the prototype:

`AC_CONFIG_MACRO_DIRS(`*macro-dir*`)`

The *.m4* macro files in this directory are ultimately referenced with an `m4_include` statement in the aclocal-generated *aclocal.m4* file, which `autoconf` reads. This macro replaces the original *acinclude.m4* file with a directory containing individual macros or smaller sets of macros, each defined in its own *.m4* file.[9]

I've indicated by the parameter to `AC_CONFIG_MACRO_DIRS` that all of the local macro files to be added to *aclocal.m4* are in a subdirectory called *m4*. As a bonus, when `autoreconf -i` is executed, and then when it executes the required Autotools with their respective *add-missing* options, these tools will note the use of this macro in *configure.ac* and add any required system macro files that are missing to the *m4* directory.

The reason I chose to use `AC_CONFIG_MACRO_DIRS` here is that Libtool will not add its additional macro files to the project if you haven't enabled the macro directory option in this manner. Instead, it will complain that you should add these files to *acinclude.m4* yourself.[10]

Since this is a fairly complex project and I wanted the Autotools to do this job for me, I decided to use this macro-directory feature. Future releases of the Autotools will likely require this form because it's considered the more modern way of adding macro files to *aclocal.m4*, as opposed to using a single user-generated *acinclude.m4* file.

One final thought on this macro: if you look for it in the Autoconf manual, you won't find it—at least not yet, because it's not an Autoconf macro but

---

8. I don't think it's worth breaking hierarchical modularity in this manner, and to this degree, just to manage this strange child-to-parent relationship; `libtoolize` could have easily created and consumed these files within each project, and the space the files consume is hardly worth the effort that the Autotools go through to ensure there is only one copy of them in a distribution archive.

9. This entire system of combining M4 macro files into a single *aclocal.m4* file is a Band-Aid for a system that was not originally designed for more than one macro file. In my opinion, it could use a major overhaul by doing away with `aclocal` entirely and having Autoconf simply read the macro files in the specified (or defaulted) macro directory, along with other macro files found in system locations.

10. I found that my project didn't require any of the macros in the Libtool system macro files, but Libtool complained anyway.

an Automake macro. It's prefixed with AC_ because it was always intended that a future release of Autoconf would take on this macro. It's more functional than its singular predecessor, which *is* documented in the Autoconf manual, but the functionality was not needed until Automake came along. In fact, I have it on pretty good authority (the pre-release Autoconf *ChangeLog*) that ownership will change hands when Autoconf 2.70 is published.

The one item that we haven't yet covered here is the AM_PROG_AR macro at ❸. This is a newer Automake macro. The first edition of this book didn't use it. When I updated the Autotools, suddenly autoreconf complained that I needed it, so I added it and the complaint went away. The Autoconf manual says simply that you need it if you want to use an archiver (ar) that has an unusual interface (such as Microsoft lib). The fact is, the real complainer here was Libtool, which seems to have a habit of complaining about not including features of the other Autotools that it thinks you should be using. I added it to silence the warning.

The Top-Level Makefile.am File

The only other point to be covered regarding the umbrella project is the top-level *Makefile.am* file, shown in Listing 14-2.

```
❶ ACLOCAL_AMFLAGS = -I m4

❷ EXTRA_DIST = README.W32 tools win32

❸ SUBDIRS = ftk flaim sql xflaim

❹ rpms srcrpm:
        for dir in $(SUBDIRS); do \
          (cd $$dir && $(MAKE) $(AM_MAKEFLAGS) $@) || exit 1; \
        done

.PHONY: rpms srcrpm
```

Listing 14-2: Makefile.am: The umbrella project Automake input file

According to the Automake documentation, the ACLOCAL_AMFLAGS variable at ❶ should be defined in the top-level *Makefile.am* file of any project that uses AC_CONFIG_MACRO_DIR (singular) in its *configure.ac* file. The flags specified on this line tell aclocal where it should look for macro files when it's executed by rules defined in *Makefile.am*. The format of this option is similar to that of a C-compiler command line include (-I) directive; you can specify other aclocal command line options as well.

This variable used to be required when using a macro subdirectory with the older AC_CONFIG_MACRO_DIR, but with the advent of the newer AC_CONFIG_MACRO_DIRS, you no longer need this variable, as it generates code that allows Automake to understand which options it should pass to aclocal. Unfortunately, Libtool just can't help but pipe up during autoreconf when it sees you using a macro directory without this variable in your *Makefile.am* files. I'm hoping this noise will go away when Autoconf takes ownership of the newer macro (with a subsequent release of Libtool, of course).

The Autotools use this variable in two unrelated places. The first is in a make rule generated to update the *aclocal.m4* file from all of its various input sources. This rule and its supporting variable definitions are shown in Listing 14-3, which is a code snippet copied from an Autotools-generated makefile.

```
ACLOCAL_M4 = $(top_srcdir)/aclocal.m4
ACLOCAL=${SHELL} .../flaim-ch8-10/missing --run aclocal-1.10
ACLOCAL_AMFLAGS = -I m4
$(ACLOCAL_M4): $(am__aclocal_m4_deps)
        cd $(srcdir) && $(ACLOCAL) $(ACLOCAL_AMFLAGS)
```

Listing 14-3: The make rule and the variables used to update aclocal.m4 *from its various dependencies*

The ACLOCAL_AMFLAGS definition is also used during execution of autoreconf, which scans the top-level *Makefile.am* file for this definition and passes the value text directly to aclocal on the command line. Be aware that autoreconf does no variable expansion on this string, so if you add shell or make variable references to the text, they won't be expanded when autoreconf executes aclocal.

Returning to Listing 14-2, I've used the EXTRA_DIST variable at ❷ to ensure that a few additional top-level files get distributed—these files and directories are specific to the Windows build system. This isn't critical to the umbrella project, since I don't intend to create distributions at this level, but I like to be complete.

The SUBDIRS variable at ❸ duplicates the information in the *configure.ac* file's AC_CONFIG_SUBDIRS macro. I tried creating a shell substitution variable and exporting it with AC_SUBST, but it didn't work—when I ran autoreconf, I got an error indicating that I should use literals in the AC_CONFIG_SUBDIRS macro argument.

The rpms and srcrpm targets at ❹ allow the end user to build RPM packages for RPM-based Linux systems. The shell commands in this rule simply pass the user-specified targets and variables down to each of the lower-level projects in succession, just as we did with our handcoded makefiles and *Makefile.in* templates in Chapters 3, 4, and 5.

When passing control to lower-level makefiles in the manner shown in the commands for these RPM targets, you should strive to follow this pattern. Passing the expansion of AM_MAKEFLAGS allows lower-level makefiles access to the same make flags defined in the current or parent makefile. However, you can add more functionality to such recursive make code. To see how Automake passes control down to lower-level makefiles for its own targets, open an Automake-generated *Makefile.in* template and search for the text "$(am__recursive_targets):". The code beneath this target shows exactly how Automake does it. While it looks complex at first glance, the code performs only two additional tasks. First, it ensures that continue-after-error functionality (make -k) works properly. Second, it ensures that the current directory (.) is handled properly if found in the SUBDIRS variable.

This brings me to my final point about this code: if you choose to write your own recursive targets in this manner (and we'll see other examples of this later when we discuss conversion of the flaim build system), you should either avoid using a dot in the SUBDIRS variable or enhance the shell code to handle this special case. If you don't, your users will likely find themselves in an endless recursion loop when they attempt to make one of these targets. For a more extensive treatise on this topic, see "Item 2: Implementing Recursive Extension Targets" on page 505.

The FLAIM Subprojects

I used autoscan to generate a starting point for the ftk project. The autoscan utility is a bit finicky about where it will look for information. If your project doesn't contain a makefile named exactly *Makefile*, or if your project already contains an Autoconf *Makefile.in* template, autoscan will not add any information about required libraries to the *configure.scan* output file. It has no way of determining this information except to look into your old build system, and it won't do this unless conditions are just right.

Given the complexity of the ftk project's legacy makefile, I was quite impressed with autoscan's ability to parse it for library information. Listing 14-4 shows a portion of the resulting *configure.scan* file.

```
--snip--
AC_PREREQ([2.69])
AC_INIT(FULL-PACKAGE-NAME, VERSION, BUG-REPORT-ADDRESS)
AC_CONFIG_SRCDIR([src/ftktext.cpp])
AC_CONFIG_HEADERS([config.h])

# Checks for programs.
AC_PROG_CXX
AC_PROG_CC
AC_PROG_INSTALL

# Checks for libraries.
# FIXME: Replace `main' with a function in `-lc':
AC_CHECK_LIB([c], [main])
# FIXME: Replace `main' with a function in...
AC_CHECK_LIB([crypto], [main])
--snip--
AC_CONFIG_FILES([Makefile])
AC_OUTPUT
```

Listing 14-4: A portion of the output from autoscan when run over the ftk project directory structure

The FLAIM Toolkit configure.ac File

After this *configure.scan* file was modified and renamed, the resulting *configure.ac* file contained many new constructs, which I'll discuss in the next few sections. In order to facilitate the discussion, I split this file into two parts, the first half of which is shown in Listing 14-5.

```
#                                          -*- Autoconf -*-
# Process this file with autoconf to produce a configure script.
AC_PREREQ([2.69])
❶ AC_INIT([FLAIMTK],[1.2],[flaim-users@lists.sourceforge.net])
❷ AM_INIT_AUTOMAKE([-Wall -Werror])
AM_PROG_AR
LT_PREREQ([2.4])
LT_INIT([dlopen])

❸ AC_LANG([C++])

❹ AC_CONFIG_MACRO_DIRS([m4])
❺ AC_CONFIG_SRCDIR([src/flaimtk.h])
AC_CONFIG_HEADERS([config.h])

# Checks for programs.
AC_PROG_CXX
AC_PROG_INSTALL

# Checks for optional programs.
❻ FLM_PROG_TRY_DOXYGEN

# Configure options: --enable-debug[=no].
❼ AC_ARG_ENABLE([debug],
    [AS_HELP_STRING([--enable-debug],
      [enable debug code (default is no)])],
    [debug="$withval"], [debug=no])

# Configure option: --enable-openssl[=no].
AC_ARG_ENABLE([openssl],
    [AS_HELP_STRING([--enable-openssl],
      [enable the use of openssl (default is no)])],
    [openssl="$withval"], [openssl=no])

# Create Automake conditional based on the DOXYGEN variable
❽ AM_CONDITIONAL([HAVE_DOXYGEN], [test -n "$DOXYGEN"])
#AM_COND_IF([HAVE_DOXYGEN], [AC_CONFIG_FILES([docs/doxyfile])])
❾ AS_IF([test -n "$DOXYGEN"], [AC_CONFIG_FILES([docs/doxyfile])])
--snip--
```

Listing 14-5: ftk/configure.ac: The first half of the ftk project's configure.ac *file*

At ❶, you will see that I substituted real values for the placeholders autoscan left in the AC_INIT macro. I added calls to AM_INIT_AUTOMAKE, LT_PREREQ, and LT_INIT at ❷, and I added a call to AC_CONFIG_MACRO_DIRS at ❹. (For now, just ignore the AM_PROG_AR macro—I'll explain it later in this chapter.)

NOTE *I didn't use the foreign keyword in AM_INIT_AUTOMAKE this time. Since it's a real open source project, the FLAIM developers will (or at least, should) want these files. I used the touch command to create empty versions of the GNU project text files,[11] except for COPYING and INSTALL, which autoreconf adds.*

A new construct at ❸ is the AC_LANG macro, which indicates the programming language (and thus, the compiler) that Autoconf should use when generating compilation tests in configure. I've passed C++ as the parameter so Autoconf will compile these tests using the C++ compiler via the CXX variable, rather than the default C compiler via the CC variable. I then deleted the AC_PROG_CC macro call, since the source code for this project is written entirely in C++.

I changed the AC_CONFIG_SRCDIR file argument at ❺ to one that made more sense to me than the one randomly chosen by autoscan.

The FLM_PROG_TRY_DOXYGEN macro at ❻ is a custom macro that I wrote. Here's the prototype:

```
FLM_PROG_TRY_DOXYGEN([quiet])
```

I'll cover the details of how this macro works in Chapter 16. For now, just know that it manages a precious variable called DOXYGEN. If the variable is already set, this macro does nothing; if the variable is not set, it scans the system search path for a doxygen program, setting the variable to the program name if it finds one. I'll explain Autoconf precious variables when we get to the xflaim project.

At ❼, I added a couple of configuration options to configure's command line parser with AC_ARG_ENABLE. I'll discuss the details of these calls more completely as we come to other new constructs that use the variables these macros define.

Automake Configuration Features

Automake provides the AM_CONDITIONAL macro I used at ❽; it has the following prototype:

```
AM_CONDITIONAL(variable, condition)
```

---

11. Of course, it's silly to distribute empty GNU text files. The thought here is that the project maintainer will fill these files with appropriate information about building, installing, and using the project. If you never intend to populate these files with quality instructions, then you're better off simply using the foreign option to disable them entirely.

The *variable* argument is an Automake conditional name that you can use in your *Makefile.am* files to test the associated condition. The *condition* argument is a *shell condition*—a bit of shell script that could be used as the condition in a shell if-then statement. In fact, this is exactly how the macro uses the *condition* argument internally, so it must be formatted as a proper if-then statement *condition* expression:

```
if condition; then...
```

The `AM_CONDITIONAL` macro always defines two Autoconf substitution variables named *variable*_TRUE and *variable*_FALSE. If *condition* is true, *variable*_TRUE is empty and *variable*_FALSE is defined as a hash mark (#), which indicates the beginning of a comment in a makefile. If *condition* is false, the definitions of these two substitution variables are reversed; that is, *variable*_FALSE is empty, and *variable*_TRUE becomes the hash mark. Automake uses these variables to conditionally comment out portions of your makefile script that are defined within Automake conditional statements.

This instance of `AM_CONDITIONAL` defines the conditional name `HAVE_DOXYGEN`, which you can use in the project's *Makefile.am* files to do something conditionally, based on whether or not doxygen can be executed successfully (via the `DOXYGEN` variable). Any lines of make script found within a test for truth in *Makefile.am* are prefixed with @*variable*_TRUE@ in the Automake-generated *Makefile.in* template. Conversely, any lines found within an Automake conditional test for falseness are prefixed with @*variable*_FALSE@. When config. status generates *Makefile* from *Makefile.in*, these lines are either commented out (prefixed with hash marks) or not, depending on the truth or falseness of the condition.

There's just one caveat with using `AM_CONDITIONAL`: you cannot call it conditionally (for instance, within a shell if-then-else statement) in the *configure.ac* file. You can't define substitution variables conditionally—you can define their contents differently based on the specified condition, but the variables themselves are either defined or not at the time Autoconf creates the `configure` script. Since Automake-generated template files are created long before the user executes `configure`, Automake must be able to rely on the existence of these variables, regardless of how they're defined.

Within the `configure` script, you may want to perform other Autoconf operations based on the value of Automake conditionals. This is where the (commented) Automake-provided `AM_COND_IF` macro at ❾ comes into play.[12] Its prototype is as follows:

```
AM_COND_IF(conditional-variable, [if-true], [if-false])
```

---

12. The `AM_COND_IF` macro was introduced in Automake 1.11, but there was a merge error in the 1.10.2 branch of Automake that caused information about `AM_COND_IF` to be inadvertently added to the documentation for version 1.10.2. If you have a version of Automake older than 1.11, you will not be able to use this macro, even though the 1.10.2 documentation shows that it is available. The code shown in the ftk project's *configure.ac* file is a reasonable workaround.

If *conditional-variable* is defined as true by a previous call to `AM_CONDITIONAL`, the *if-true* shell script (including any Autoconf macro calls) is executed. Otherwise, the *if-false* shell script is executed.

Now let's suppose, for example, that you want to conditionally build a portion of your project directory structure—say, the *xflaim/docs/doxygen* directory—based on the Automake conditional `HAVE_DOXYGEN`. Perhaps you are appending the subdirectory in question onto the `SUBDIRS` variable within an Automake conditional statement in your *Makefile.am* file (I'm actually doing this, as you'll see in "The FLAIM Toolkit Makefile.am File" on page 388). Since `make` won't be building this portion of the project directory structure if the condition is false, there's certainly little reason to have `config.status` process the *doxyfile.in* template within that directory during configuration. Therefore, you might use the code shown in Listing 14-6 in your *configure.ac* file.

```
--snip--
AM_CONDITIONAL([HAVE_DOXYGEN], [test -n "$DOXYGEN"])
AM_COND_IF([HAVE_DOXYGEN], [AC_CONFIG_FILES([docs/doxyfile])])
#AS_IF([test -n "$DOXYGEN"], [AC_CONFIG_FILES([docs/doxyfile])])
--snip--
```

Listing 14-6: ftk/configure.ac: Using `AM_COND_IF` to conditionally configure a template

With this code in place, configure simply will not process the *doxyfile.in* template at all within the *docs* directory if doxygen isn't installed on the user's system.

NOTE   *The* docs/Makefile.in *template should not be included here because the* dist *target must be able to process all directories in the project—whether or not they're conditionally built—during execution of build targets such as* all *and* clean. *Thus, you should never conditionally process* Makefile.in *templates within* configure.ac. *However, you can certainly process other types of templates conditionally.*

The line following the line at ❾ is an alternative method of accomplishing the same thing using *M4sh*—a macro library built into Autoconf that's designed to make it easier to write portable Bourne shell script. Here is the prototype:

```
AS_IF(test1, [run-if-true], ..., [run-if-false])
```

The optional, elided parameters between the second and last ones shown are pairs of *testN* and *run-if-true* arguments. Ultimately, this macro works much like an `if-then-elif...` shell statement with a user-specified number of `elif` conditions.

Listing 14-7 shows the second half of ftk's *configure.ac* file.

```
--snip--
# Configure for large files, even in 32-bit environments
❶ AC_SYS_LARGEFILE
```

```
      # Check for pthreads
❷ AX_PTHREAD(
      [AC_DEFINE([HAVE_PTHREAD], [1],
        [Define if you have POSIX threads libraries and header files.])
      LIBS="$PTHREAD_LIBS $LIBS"
      CFLAGS="$CFLAGS $PTHREAD_CFLAGS"
      CXXFLAGS="$CXXFLAGS $PTHREAD_CXXFLAGS"])


❸ # Checks for libraries.
   AC_SEARCH_LIBS([initscr], [ncurses])
   AC_CHECK_HEADER([curses.h],,[echo "*** Error: curses.h not found - install
   curses devel package."; exit 1])
   AC_CHECK_LIB([rt], [aio_suspend])
   AS_IF([test "x$openssl" = xyes],
❹    [AC_DEFINE([FLM_OPENSSL], [1], [Define to use openssl])
      AC_CHECK_LIB([ssl], [SSL_new])
      AC_CHECK_LIB([crypto], [CRYPTO_add])
      AC_CHECK_LIB([dl], [dlopen])
      AC_CHECK_LIB([z], [gzopen])])

❺ # Checks for header files.
   AC_HEADER_RESOLV
   AC_CHECK_HEADERS([arpa/inet.h fcntl.h limits.h malloc.h netdb.h netinet/in.h
   stddef.h stdlib.h string.h strings.h sys/mount.h sys/param.h sys/socket.h sys/
   statfs.h sys/statvfs.h sys/time.h sys/vfs.h unistd.h utime.h])

   # Checks for typedefs, structures, and compiler characteristics.
   AC_CHECK_HEADER_STDBOOL
   AC_C_INLINE
   AC_TYPE_INT32_T
   AC_TYPE_MODE_T
   AC_TYPE_PID_T
   AC_TYPE_SIZE_T
   AC_CHECK_MEMBERS([struct stat.st_blksize])
   AC_TYPE_UINT16_T
   AC_TYPE_UINT32_T
   AC_TYPE_UINT8_T

   # Checks for library functions.
   AC_FUNC_LSTAT_FOLLOWS_SLASHED_SYMLINK
   AC_FUNC_MALLOC
   AC_FUNC_MKTIME
   AC_CHECK_FUNCS([atexit fdatasync ftruncate getcwd gethostbyaddr gethostbyname
   gethostname gethrtime gettimeofday inet_ntoa localtime_r memmove memset mkdir
   pstat_getdynamic realpath rmdir select socket strchr strrchr strstr])

   # Configure DEBUG source code, if requested.
❻ AS_IF([test "x$debug" = xyes],
      [AC_DEFINE([FLM_DEBUG], [1], [Define to enable FLAIM debug features])])

❼ --snip--

❽ AC_CONFIG_FILES([Makefile
                    docs/Makefile
                    obs/Makefile
```

```
                        obs/flaimtk.spec
                        src/Makefile
                        util/Makefile
                        src/libflaimtk.pc])

   AC_OUTPUT

   # Fix broken libtool
   sed 's/link_all_deplibs=no/link_all_deplibs=yes/' libtool >libtool.tmp && \
     mv libtool.tmp libtool

❾ cat <<EOF

       FLAIM toolkit ($PACKAGE_NAME) version $PACKAGE_VERSION
       Prefix.........: $prefix
       Debug Build....: $debug
       Using OpenSSL..: $openssl
       C++ Compiler...: $CXX $CXXFLAGS $CPPFLAGS
       Linker.........: $LD $LDFLAGS $LIBS
       Doxygen........: ${DOXYGEN:-NONE}

   EOF
```

Listing 14-7: ftk/configure.ac: *The second half of the ftk project's* configure.ac *file*

At ❶, I've called the AC_SYS_LARGEFILE macro. If the user has a 32-bit system, this macro ensures that appropriate C-preprocessor definitions (and possibly compiler options) that force the use of 64-bit file addressing (also called *large files*) are added to the *config.h.in* template. With these variables in place, C-library large-address-aware file I/O functions become available to the project source code. FLAIM, as a database system, cares very much about this feature.

In the last few years, 32-bit general-purpose computer systems have become less popular as companies like Intel and Microsoft have made media statements concerning future versions of their products that will no longer support 32-bit address spaces. However, market pressures caused by the millions of existing 32-bit systems have cause them to back off a bit on the rhetoric and return to a more pragmatic perspective. Nevertheless, 32-bit PCs are on their way out the door in the not-too-distant future. Even so, Linux will continue to run on 32-bit systems because many embedded systems still get significant benefits from using smaller, less-power-hungry 32-bit microprocessors.

Doing Threads the Right Way

There is another new construct, AX_PTHREAD, at ❷. In the Jupiter project, I simply linked the jupiter program with the *pthreads* library via the -lpthread linker flag. But frankly, this is the wrong way to use *pthreads*.

In the presence of multiple threads of execution, you must configure many of the standard C-library functions to act in a thread-safe manner. You can do this by ensuring that one or more preprocessor definitions are visible to all of the standard library header files as they're being compiled into the program. These C-preprocessor definitions must be defined on the compiler command line, and they're not standardized between compiler vendors.

Some vendors provide entirely different standard libraries for building single-threaded versus multithreaded programs, because adding thread safety to a library reduces performance to a degree. Compiler vendors believe (correctly) that they're doing you a favor by giving you different versions of the standard library for these purposes. In this scenario, it's necessary to tell the linker to use the correct runtime libraries.

Unfortunately, every vendor does multithreading in its own way, from compiler options to library names to preprocessor definitions. But there is a reasonable solution to the problem: the GNU Autoconf Archive[13] provides a macro called `AX_PTHREAD` that checks out a user's compiler and provides the correct flags and options for a wide variety of platforms.

This macro is very simple to use:

```
AX_PTHREAD(action-if-found[, action-if-not-found])
```

It sets several environment variables, including `PTHREAD_CFLAGS`, `PTHREAD_CXXFLAGS`, and `PTHREAD_LIBS`. It's up to the caller to use these variables properly by adding shell code to the *action-if-found* argument. If all of your project's code is multithreaded, things are simpler: you need only append these variables to, or consume them from within, the standard `CFLAGS`, `CXXFLAGS`, and `LIBS` variables. The FLAIM project code base is completely multi-threaded, so I chose to do this.

If you examine the contents of the *ax_pthread.m4* file in the *ftk/m4* directory, you might expect to find a large case statement that sets options for every compiler and platform combination known to man—but that's not the Autoconf way.

Instead, the macro incorporates a long list of known *pthreads* compiler options, and the generated configure script uses the host compiler to compile a small *pthreads* program with each one of these options in turn. The flags that are recognized by the compiler, and that therefore properly build the test program, are added to the `PTHREAD_CFLAGS` and `PTHREAD_CXXFLAGS` variables. This way, `AX_PTHREAD` stands a good chance of continuing to work properly, even in the face of significant changes to compiler options in the future—and this *is* the Autoconf way.

Getting Just the Right Libraries

I deleted the *FIXME* comments (see *configure.scan* in Listing 14-4 on page 378) above each of the `AC_CHECK_LIB` macro calls at ❸ in Listing 14-7. I started to replace the main placeholders in these macros with actual library function names, but then I began to wonder if all of those libraries were really necessary. I wasn't as concerned about autoscan's abilities as I was about the veracity of the original makefile. In handcoded build systems, I've

---

13. See *https://www.gnu.org/software/autoconf-archive/*.

occasionally noticed that the author will cut and paste sets of library names from one makefile to another until the program builds without missing symbols.[14]

Instead of blindly continuing this trend, I chose to simply comment out all of the calls to `AC_CHECK_LIB` to see how far I could get in the build, adding them back in one at a time as required to resolve missing symbols. Unless your project consumes literally hundreds of libraries, this will only take a few extra minutes. I like to link only the libraries that are necessary for my project; it speeds up the link process and, when done religiously, provides a good form of project-level documentation.

The *configure.scan* file contained 14 such calls to `AC_CHECK_LIB`. As it turned out, the FLAIM toolkit on my 64-bit Linux system only required three of them—*pthread, ncurses,* and *rt*—so I deleted the remaining entries and swapped out the placeholder parameters for real functions in the *ncurses* and *rt* libraries. In retrospect, it appears that my gambit paid off rather handsomely, because I dropped from 14 libraries to 2. The third library was the POSIX Thread (*pthreads*) library, which is added via the `AX_PTHREAD` macro I discussed in the previous section.

I also converted the *ncurses* `AC_CHECK_LIB` call to `AC_SEARCH_LIBS` because I suspect that future FLAIM platforms may use different library names for *curses* functionality. I'd like to prepare the build system to have additional libraries searched on these platforms. The *ncurses* library is an optional library on most platforms, so I added the `AC_CHECK_HEADER` macro to check for *curses.h*, display a message in the *action-if-not-found* (third) argument that the user should install the *curses-development* package, and exit the configuration process with an error. The rule is to find problems early, during configuration, rather than during compilation.

Maintainer-Defined Command Line Options

The next four libraries are checked within an Autoconf conditional statement at ❹. This statement is based on the end user's use of the `--enable -openssl` command line argument, which `AC_ARG_ENABLE` provides (see ❼ in Listing 14-5 on page 379).

I use `AS_IF` here instead of a shell `if-then` statement because, if any of the macros called within the conditional statement require additional macros to be expanded in order to operate correctly, `AS_IF` will ensure that these dependencies are expanded first, outside of the conditional statement. As well as being part of the *M4sh* library, the `AS_IF` macro is part of the Autoconf auto-dependency framework (also discussed in detail in "Autoconf and M4" on page 439).

In this case, the `openssl` variable is defined to either yes or no based on the default value given to `AC_ARG_ENABLE` and on the end user's command line choices.

---

14. For some reason, this activity is especially prevalent when libraries are being built, although programs are not immune to it.

The AC_DEFINE macro, called in the first argument of AS_IF, ensures that the C-preprocessor variable FLM_OPENSSL is defined in the *config.h* header file. The AC_CHECK_LIB macros then ensure that -lssl, -lcrypto, -ldl, and -lz strings are added to the LIBS variable, but only if the openssl variable is set to yes. We don't want to insist that the user have those libraries installed unless they have asked for features that need them.

You can get as sophisticated as you want when dealing with maintainer-defined command line options such as --enable-openssl. But be careful: some levels of automation can surprise your users. For instance, automatically enabling the option because your checks found that the OpenSSL libraries were installed and accessible can be a bit disconcerting.

I left all the header file and library function checks at ❺, as specified by autoscan, because a simple text scan through the source code for header files and function names is probably pretty accurate.

Notice, however, that autoscan did not put *all* of the header files used by ftk source code into the AC_CHECK_HEADERS argument. The autoscan utility's algorithm is simple but effective: it adds all header files included conditionally by your source code. This approach assumes that any header file you include conditionally might be included differently on different platforms due to portability issues. While this approach is usually correct, it's not always correct, so you should look at each of the headers added, find the conditional inclusion in your source code, and make a more intelligent assessment of whether or not it should be added to AC_CHECK _HEADERS in *configure.ac*.

A good example in this project is the conditional inclusion of *stdlib.h*. As it happens, *stdlib.h* is included for Windows builds, and it's also included for Unix builds. It is not, however, included for NetWare builds. Regardless, it doesn't really need to be checked for in AC_CHECK_HEADERS for two reasons. First, it's widely standardized across platforms, and second, this build system is specifically designed for Unix systems.[15] The point is, you should carefully examine what autoscan does for you to determine if it should be done in your project.

At ❻, we see the conditional (AS_IF) use of AC_DEFINE based on the contents of the debug variable. This is another environment variable that's conditionally defined based on the results of a command line parameter given to configure. The --enable-debug option sets the debug variable to yes, which ultimately enables the FLM_DEBUG C-preprocessor definition within *config.h*. Both FLM_OPENSSL and FLM_DEBUG were already used within the FLAIM project source code. Using AC_DEFINE in this manner allows the end user to determine which features are compiled into the libraries.

I left a fairly large chunk of code out of the listing at ❼ that deals with compiler and tool optimizations, which I'll present in the next chapter. This code is identical in all of the projects' *configure.ac* files.

Finally, I added references to the makefiles in the *docs*, *obs*, *src*, and *util* directories, as well as the *obs/flaimtk.spec* and *src/libflaimtk.pc* files at ❽ to the AC_CONFIG_FILES macro call, and then I added my usual cat statement at ❾

---

15. I left it in my call to AC_CHECK_HEADERS so I could discuss it here.

near the bottom for some visual verification of my configuration status. For now, just ignore the sed command right above the cat statement. I'll cover that in "Transitive Dependencies" on page 401.

The FLAIM Toolkit Makefile.am File

If we ignore the commands for Doxygen- and RPM-specific targets (for now), the *ftk/Makefile.am* file is fairly trivial. Listing 14-8 shows the entire file.

```
ACLOCAL_AMFLAGS = -I m4

EXTRA_DIST = GNUMakefile README.W32 debian netware win32

❶ if HAVE_DOXYGEN
    DOXYDIR = docs
  endif

  SUBDIRS = src util obs $(DOXYDIR)

❷ doc_DATA = AUTHORS ChangeLog COPYING INSTALL NEWS README

  RPM = rpm

❸ rpms srcrpm: dist
        (cd obs && $(MAKE) $(AM_MAKEFLAGS) $@) || exit 1
        rpmarch=`$(RPM) --showrc | \
          grep "^build arch" | sed 's/\(.*: \)\(.*\)/\2/'`; \
        test -z "obs/$$rpmarch" || \
        ( mv obs/$$rpmarch/* . && rm -rf obs/$$rpmarch )
        rm -rf obs/$(distdir)

❹ #dist-hook:
  #        rm -rf `find $(distdir) -name .svn`

  .PHONY: srcrpm rpms
```

Listing 14-8: ftk/Makefile.am: The entire contents of the FLAIM toolkit's top-level makefile

In this file you'll find the usual ACLOCAL_AMFLAGS, EXTRA_DIST, and SUBDIRS variable definitions, but you can also see the use of an Automake conditional at ❶. The if statement allows me to append another directory (*docs*) to the SUBDIRS list, but only if the doxygen program is available (according to configure). I used a separate variable here (DOXYDIR), but the Automake conditional could just as well have surrounded a statement that directly appends the directory name (doc) to the SUBDIRS variable using the Automake += operator.

NOTE *Don't confuse Automake conditionals with GNU Make conditionals, which use the keywords ifeq, ifneq, ifdef, and ifndef. If you try to use an Automake conditional in* Makefile.am *without a corresponding AM_CONDITIONAL statement in* configure.ac, *Automake will complain about it. When this construct is used properly, Automake converts it to something that make understands before make sees it.*

Another new construct (at least in a top-level *Makefile.am* file) is the use of the doc_DATA variable at ❷. The FLAIM toolkit provides some extra documentation files in its top-level directory that I'd like to have installed. By using the doc prefix on the DATA primary, I'm telling Automake that I'd like these files to be installed as data files in the $(docdir) directory, which ultimately resolves to the $(prefix)/*share/doc* directory, by default.

Files mentioned in DATA variables that don't already have special meaning to Automake are not automatically distributed (that is, they're not added to distribution tarballs), so you have to manually distribute them by adding them to the files listed in the EXTRA_DIST variable.

NOTE *I did not have to list the standard GNU project text files in EXTRA_DIST because they're always distributed automatically. However, I did have to mention theses files in the doc_DATA variable, because Automake makes no assumptions about which files you want to install.*

I'll defer a discussion of the RPM targets at ❸ to the next chapter.

Automake -hook and -local Rules

Automake recognizes two types of integrated extensions, which I call -local targets and -hook targets. Automake recognizes and honors -local extensions for the following standard targets:

all	install-data	installcheck
check	install-dvi	installdirs
clean	install-exec	maintainer-clean
distclean	install-html	mostlyclean
dvi	install-info	pdf
html	install-pdf	ps
info	install-ps	uninstall

Appending -local to any of these in your *Makefile.am* files will cause the associated commands to be executed *before* the standard target. Automake does this by generating the rule for the standard target so that the -local version is one of its dependencies (if it exists).[16] In "Cleaning Your Room" on page 404, I'll show an example of this concept using a clean-local target.

The -hook targets are a bit different in that they are executed *after* the corresponding standard target is executed.[17] Automake does this by adding another command to the end of the standard target command list. This command merely executes $(MAKE) on the containing makefile, with the

---

16. Automake -local targets can be somewhat problematic when using parallel make (make -j), because parallel make cannot guarantee that dependencies are processed in the order in which they're listed: they may be executed in parallel. This is arguably a design flaw in Automake, but it's far too late to fix it at this point.

17. There are exceptions to this rule. In fact, the dist-hook target is actually executed after the distdir target, rather than after the dist target. Basically, the hook rules are executed where they make the most sense.

-hook target as the command line target. Thus, the -hook target is executed at the end of the standard target commands in a recursive fashion.

The following standard Automake targets support -hook versions:

dist   install-data   uninstall
distcheck   install-exec

Automake automatically adds all existing -local and -hook targets to the .PHONY rule within the generated makefile.

In the first edition of this book, I used the dist-hook target at ❹ in *Makefile.am* (now commented out) to adjust the distribution directory after it's built but before make builds a distribution archive from its contents. The rm command removed extraneous files and directories that became part of the distribution directory as a result of my adding entire directories to the EXTRA_DIST variable. When you add directory names to EXTRA_DIST (*debian*, *netware*, and *win32*, in this case), everything in those directories is added to the distribution—even hidden repository control files and directories.[18]

Listing 14-9 is a portion of the generated *Makefile* that shows how Automake incorporates dist-hook into the final makefile. The relevant portions are highlighted.

```
--snip--
distdir: $(DISTFILES)
        ... # copy files into distdir
        $(MAKE) $(AM_MAKEFLAGS) top_distdir="$(top_distdir)" \
            distdir="$(distdir)" dist-hook
        ... # change attributes of files in distdir
--snip--
dist dist-all: distdir
        tardir=$(distdir) && $(am__tar) | GZIP=$(GZIP_ENV) gzip -c \
            >$(distdir).tar.gz
        $(am__remove_distdir)
--snip--
.PHONY: ... dist-hook ...
--snip--
dist-hook:
        rm -rf `find $(distdir) -name .svn`
--snip--
```

Listing 14-9: The results of defining the dist-hook target in ftk/Makefile.am

NOTE   *Don't be afraid to dig into the generated makefiles to see exactly what Automake is doing with your code. While there is a fair amount of ugly shell code in the make commands, most of it is safe to ignore. You're usually more interested in the make rules that Automake is generating, and it's easy to separate these out.*

18. The source code accompanying the first edition of this book was stored in a Subversion repository. This edition's source code is hosted by GitHub, which, of course, uses git. Since git does not have repository control files and directories scattered throughout the user source directory structure, this hook was no longer necessary for this project. However, I felt the information was important enough to leave this section in the chapter.

Designing the ftk/src/Makefile.am File

I now need to create *Makefile.am* files in the *src* and *utils* directories for the FLAIM toolkit project. I want to ensure that all of the original functionality is preserved from the old build system as I'm creating these files. Basically, this includes:

- Properly building the ftk shared and static libraries
- Properly specifying installation locations for all installed files
- Setting the ftk shared-library version information correctly
- Ensuring that all remaining unused files are distributed
- Ensuring that platform-specific compiler options are used

The template shown in Listing 14-10 should cover most of these points, so I'll be using it for all of the FLAIM library projects, with appropriate additions and subtractions, based on the needs of each individual library.

```
EXTRA_DIST = ...

lib_LTLIBRARIES = ...
include_HEADERS = ...

xxxxx_la_SOURCES = ...
xxxxx_la_LDFLAGS = -version-info x:y:z
```

Listing 14-10: A framework for the src and utils directory Makefile.am files

The original *GNUMakefile* told me that the library was named *libftk.so*. This is a bad name for a library on Linux, because most of the three-letter library names are already taken. Thus, I made an executive decision and renamed the *ftk* library to *flaimtk*.

Listing 14-11 shows most of the final *ftk/src/Makefile.am* file.

```
❶ EXTRA_DIST = ftknlm.h

❷ lib_LTLIBRARIES = libflaimtk.la
❸ include_HEADERS = flaimtk.h

❹ pkgconfigdir = $(libdir)/pkgconfig
   pkgconfig_DATA = libflaimtk.pc

❺ libflaimtk_la_SOURCES = \
   ftkarg.cpp \
   ftkbtree.cpp \
   ftkcmem.cpp \
   ftkcoll.cpp \
   --snip--
   ftksys.h \
   ftkunix.cpp \
   ftkwin.cpp \
```

```
ftkxml.cpp
```

❻ `libflaimtk_la_LDFLAGS = -version-info 0:0:0`

Listing 14-11: ftk/src/Makefile.am: *The entire file contents, minus a few dozen source files*

I added the Libtool library name, *libflaimtk.la,* to the `lib_LTLIBRARIES` list at **❷** and changed the *xxxxx* portions of the remaining macros in Listing 14-10 to `libflaimtk`. I could have entered all the source files by hand, but I noticed while reading the original makefile that it used the GNU `make` function macro `$(wildcard src/*.cpp)` to build the file list from the contents of the *src* directory. This tells me that all of the *.cpp* files within the *src* directory are required (or at least consumed) by the library. To get the file list into *Makefile.am,* I used a simple shell command to concatenate it to the end of the *Makefile.am* file (assuming I'm in the *ftk/src* directory):

```
$ printf '%s \\\n' *.cpp >> Makefile.am
```

This left me with a single-column, backslash-terminated, alphabetized list of all of the *.cpp* files in the *ftk/src* directory at the bottom of *ftk/src/Makefile.am.*

NOTE *Do not forget the single quotes around the* printf *argument, which are necessary to keep the first pair of backslashes from being interpreted by the shell as escape characters during generation of the list. Regardless of quoting,* printf *understands and interprets the \n character properly.*

I moved the list up to just below the `libflaimtk_la_SOURCES` line at **❺**, added a backslash character after the equal sign, and removed the one after the last file. Another formatting technique is to simply wrap the line with a backslash and a carriage return approximately every 70 characters, but I prefer to put each file on a separate line, especially early in the conversion process, so I can easily extract files from or add files to the lists as needed. Leaving the files on separate lines also gets you the benefit of having source lists be easier to compare when reviewing differences in pull-request reviews and other `diff`-style output.

I had to manually examine each header file in the *src* directory in order to determine its use in the project. There were only four header files, and, as it turns out, the only one the FLAIM toolkit does *not* use on Unix and Linux platforms is *ftknlm.h,* which is specific to the NetWare build. I added this file to the `EXTRA_DIST` list at **❶** so it would be distributed; just because the build doesn't use it doesn't mean that users won't want or need it.[19]

The (newly renamed) *flaimtk.h* file is the only public header file, so I moved it into the `include_HEADERS` list at **❸**. The other two files are used

---

19. I could have simply added this header file to the `libflaimtk_la_SOURCES` variable, because header files added to `SOURCES` variables are merely added to the distribution. But doing so would have hidden from observers the fact that this header file is not used in the Unix build in any way.

internally in the library build, so I left them in the `libflaimtk_la_SOURCES` list. Had this been my own project, I would have moved *flaimtk.h* into an *include* directory off the project root directory, but remember that one of my goals here was to limit changes to the directory structure and the source code. Moving this header file is a philosophical decision that I decided to leave to the maintainers.[20]

Finally, I noticed in the original makefile that the last release of the *ftk* library published an interface version of 4.0. However, since I changed the name of the library from *libftk* to *libflaimtk*, I reset this value to 0.0 because it's a different library. I replaced *x:y:z* with 0:0:0 in the -version-info option at ❻ within the `libflaimtk_la_LDFLAGS` variable.

NOTE   *A version string of 0:0:0 is the default, so I could have removed the argument entirely and achieved the same result. However, including it gives new developers some insight into how to change the interface version in the future.*

I added the `pkgconfigdir` and `pkgconfig_DATA` variables at ❹ in order to provide support for installing pkg-config metadata files for this project. For more on the pkg-config system, see Chapter 10.

Moving On to the ftk/util Directory

Properly designing *Makefile.am* for the *util* directory requires examining the original makefile again for more products. A quick glance at the *ftk/util* directory showed that there was only one source file: *ftktest.cpp*. This appeared to be some sort of testing program for the *ftk* library, but I know that the FLAIM developers use it all the time in various ways besides simply for testing a build. So I had a design decision to make here: should I build this as a normal program or as a check program?

Check programs are only built when `make check` is executed, and they're never installed. If I want `ftktest` built as a regular program, but not installed, I have to use the `noinst` prefix rather than the usual `bin` prefix in the program list variable.

In either case, I probably want to add `ftktest` to the list of tests that are executed during `make check`, so the two questions here are (1) whether I want to automatically run `ftktest` during `make check` and (2) whether I want to install the `ftktest` program. Given that the FLAIM toolkit is a mature product, I opted to build `ftktest` during `make check` and leave it uninstalled.

Listing 14-12 shows my final *ftk/util/Makefile.am* file.

```
FTK_INCLUDE = -I$(top_srcdir)/src
FTK_LTLIB = ../src/libflaimtk.la

check_PROGRAMS = ftktest
```

---

20. As mentioned earlier, I'm the only effective maintainer at this point, but this was not always the case, and it's still a valid rule of thumb to follow when suggesting changes to other peoples' code in the form of git pull requests or mailing list patches.

```
ftktest_SOURCES = ftktest.cpp
ftktest_CPPFLAGS = $(FTK_INCLUDE)
ftktest_LDADD = $(FTK_LTLIB)

TESTS = ftktest
```

Listing 14-12: ftk/util/Makefile.am: The final contents of this file

I hope that by now you can see the relationship between TESTS and
check_PROGRAMS. To be blunt, there really is *no* relationship between the files
listed in check_PROGRAMS and those listed in TESTS. The check target simply
ensures that check_PROGRAMS are built before the TESTS programs and scripts
are executed. TESTS can refer to anything that can be executed without command line parameters. This separation of duties makes for a very clean and
flexible system.

And that's it for the FLAIM toolkit library and utilities. I don't know
about you, but I'd much rather maintain this small set of short files than
a single 2,200-line makefile!

Designing the XFLAIM Build System

Now that I've finished with the FLAIM toolkit, I'll move on to the xflaim
project. I'm choosing to start with xflaim, rather than flaim, because it
supplies the most build features that can be converted to the Autotools,
including the Java and C# language bindings (which I won't actually discuss in detail until the next chapter). After xflaim, covering the remaining
database projects would be redundant, because the processes are identical,
if not a little simpler. However, you can find the other build system files in
this book's GitHub repositories.

I generated the *configure.ac* file using autoscan once again. It's important
to use autoscan in each of the individual projects, because the source code
for each project is different and will thus cause different macros to be written into each *configure.scan* file.[21] I then used the same techniques I used on
the FLAIM toolkit to create xflaim's *configure.ac* file.

The XFLAIM configure.ac File

After hand-modifying the generated *configure.scan* file and renaming it
configure.ac, I found it to be similar in many ways to the toolkit's *configure.ac*
file. It's fairly long, so I'll show you only the most significant differences in
Listing 14-13.

```
--snip--
❶ # Checks for optional programs.
FLM_PROG_TRY_CSC
FLM_PROG_TRY_CSVM
```

---

21. During review of this chapter for the second edition of this book, I also used autoupdate to
update my older *configure.ac* files to the latest Autotools best practices.

```
    FLM_PROG_TRY_JNI
    FLM_PROG_TRY_JAVADOC
    FLM_PROG_TRY_DOXYGEN

❷ # Configure variables: FTKLIB and FTKINC.
    AC_ARG_VAR([FTKLIB], [The PATH wherein libflaimtk.la can be found.])
    AC_ARG_VAR([FTKINC], [The PATH wherein flaimtk.h can be found.])
    --snip--
❸ # Ensure that both or neither is specified.
    if (test -n "$FTKLIB" && test -z "$FTKINC") || \
       (test -n "$FTKINC" && test -z "$FTKLIB"); then
       AC_MSG_ERROR([Specify both FTK library and include paths, or neither.])
    fi

    # Not specified? Check for FTK in standard places.
    if test -z "$FTKLIB"; then
❹    # Check for FLAIM toolkit as a sub-project.
      if test -d "$srcdir/ftk"; then
        AC_CONFIG_SUBDIRS([ftk])
        FTKINC='$(top_srcdir)/ftk/src'
        FTKLIB='$(top_builddir)/ftk/src'
      else
❺    # Check for FLAIM toolkit as a superproject.
        if test -d "$srcdir/../ftk"; then
          FTKINC='$(top_srcdir)/../ftk/src'
          FTKLIB='$(top_builddir)/../ftk/src'
        fi
      fi
    fi

❻ # Still empty? Check for *installed* FLAIM toolkit.
    if test -z "$FTKLIB"; then
      AC_CHECK_LIB([flaimtk], [ftkFastChecksum],
        [AC_CHECK_HEADERS([flaimtk.h])
          LIBS="-lflaimtk $LIBS"],
        [AC_MSG_ERROR([No FLAIM toolkit found. Terminating.])])
    fi

❼ # AC_SUBST command line variables from FTKLIB and FTKINC.
    if test -n "$FTKLIB"; then
      AC_SUBST([FTK_LTLIB], ["$FTKLIB/libflaimtk.la"])
      AC_SUBST([FTK_INCLUDE], ["-I$FTKINC"])
    fi

❽ # Automake conditionals
    AM_CONDITIONAL([HAVE_JAVA], [test "x$flm_prog_have_jni" = xyes])
    AM_CONDITIONAL([HAVE_CSHARP], [test -n "$CSC"])
    AM_CONDITIONAL([HAVE_DOXYGEN], [test -n "$DOXYGEN"])
    #AM_COND_IF([HAVE_DOXYGEN], [AC_CONFIG_FILES([docs/doxygen/doxyfile])])
    AS_IF([test -n "$DOXYGEN"], [AC_CONFIG_FILES([docs/doxygen/doxyfile])])
    --snip--
    AC_OUTPUT
```

```
# Fix broken libtool
sed 's/link_all_deplibs=no/link_all_deplibs=yes/' libtool >libtool.tmp && \
  mv libtool.tmp libtool

cat <<EOF

  ($PACKAGE_NAME) version $PACKAGE_VERSION
  Prefix.........: $prefix
  Debug Build....: $debug
  C++ Compiler...: $CXX $CXXFLAGS $CPPFLAGS
  Linker.........: $LD $LDFLAGS $LIBS
  FTK Library....: ${FTKLIB:-INSTALLED}
  FTK Include....: ${FTKINC:-INSTALLED}
  CSharp Compiler: ${CSC:-NONE} $CSCFLAGS
  CSharp VM......: ${CSVM:-NONE}
  Java Compiler..: ${JAVAC:-NONE} $JAVACFLAGS
  JavaH Utility..: ${JAVAH:-NONE} $JAVAHFLAGS
  Jar Utility....: ${JAR:-NONE} $JARFLAGS
  Javadoc Utility: ${JAVADOC:-NONE}
  Doxygen........: ${DOXYGEN:-NONE}

EOF
```

Listing 14-13: xflaim/configure.ac: The most significant portions of this Autoconf input file

First, notice that I've invented a few more FLM_PROG_TRY_* macros at ❶. Here I'm checking for the existence of the following programs: a C# compiler, a C# virtual machine, a Java compiler, a JNI header and stub generator, a Javadoc generation tool, a Java archive tool, and Doxygen. I've written separate macro files for each of these checks and added them to my *xflaim/m4* directory.

As with the FLM_PROG_TRY_DOXYGEN macro used in the toolkit, each of these macros attempts to locate the associated program, but these macros don't fail the configuration process if they can't find the program. I want to be able to use these programs if they're available, but I don't want to require the user to have them in order to build the base libraries.

You'll find a new macro, AC_ARG_VAR, at ❷. Like the AC_ARG_ENABLE and AC_ARG_WITH macros, AC_ARG_VAR allows the project maintainer to extend the command line interface of the configure script. This macro is different, however, in that it adds a public variable, rather than a command line option, to the list of public variables that configure cares about. In this case, I'm adding two public variables, FTKINC and FTKLIB. These will show up in configure's help text under the section "Some influential environment variables." The *GNU Autoconf Manual* calls these variables *precious*. All of my FLM_PROG_TRY_* macros use the AC_ARG_VAR macro internally to make the associated variables both public and precious.[22]

---

22. These variables are also automatically substituted into the *Makefile.in* templates that Automake generates. However, I don't really need this substitution functionality, because I'm going to build other variables out of these ones and I'll want the derived variables, instead of the public variables, to be substituted.

NOTE *The lines of code from ❷ through ❼ are found in the GitHub repository under xflaim/m4/flm_ftk_search.m4. By the end of Chapter 15, all discrepancies are resolved between the listings in this chapter and the files in the GitHub repository.*

The large chunk of code beginning at ❸ actually uses these variables to set other variables used in the build system. The user can set the public variables in the environment, or they can specify them on `configure`'s command line in this manner:

```
$ ./configure FTKINC="$HOME/dev/ftk/include" ...
```

First, I'll check to see that either both or neither of the `FTKINC` and `FTKLIB` variables is specified. If only one of them is given, I have to fail with an error. The user isn't allowed to tell me where to find only *half* the toolkit; I need both the header file and the library.[23] If neither of these variables is specified, I search for them at ❹ by looking for a subdirectory of the xflaim project directory called *ftk*. If I find one, I'll configure that directory as a subproject to be processed by Autoconf, using the `AC_CONFIG_SUBDIRS` macro.[24] I'll also set both of these variables to point to the appropriate relative locations within the ftk subproject.

If I don't find *ftk* as a subdirectory, I'll look for it in the parent directory at ❺. If I find it there, I'll set the variables appropriately. This time, I don't need to configure the located *ftk* directory as a subproject, because I'm assuming that the xflaim project is itself a subproject of the umbrella project. If I don't find *ftk* as either a subproject or a sibling project, I'll use the standard `AC_CHECK_LIB` and `AC_CHECK_HEADERS` macros at ❻ to see if the user's host has the toolkit library installed. In that case, I need only add `-lflaimtk` to the `LIBS` variable. Also in that case, the header file will be in the standard location: usually */usr(/local)/include*. The default functionality of the optional third argument to `AC_CHECK_LIB` would automatically add the library reference to the `LIBS` variable, but since I've overridden this default functionality, I have to manually add the toolkit library reference to `LIBS`.

If I don't find the library, I give up with an error message indicating that xflaim can't be built without the FLAIM toolkit. However, after making it through all these checks, if the `FTKLIB` variable is no longer empty, I use `AC_SUBST` at ❼ to publish the `FTK_INCLUDE` and `FTK_LTLIB` variables, which contain derivations of `FTKINC` and `FTKLIB` appropriate for use as command line options to the preprocessor and the linker.

NOTE *Chapter 16 converts the large chunk of code between ❸ and ❽ into a custom M4 macro called FLM_FTK_SEARCH.*

---

23. It is, of course, possible to allow the user to specify only `FTKINC` or `FTKLIB` if you're willing to check for a relative installation path of the other component. For instance, if the user specified `FTKINC=`*/path/to/include/ftk*, you could write shell code that tries to find *libflaimtk.so* in `$FTKINC/../../lib`. It's a bit more shell code, but it may be worth your time to make it easier on the end user.

24. You can use this macro conditionally and multiple times within the same *configure.ac* file.

The remaining code at ❽ calls AM_CONDITIONAL for Java, C#, and Doxygen in a manner similar to the way I handled Doxygen in the ftk project. These macros are configured to generate warning messages indicating that the Java or C# portions of the xflaim project will not be built if those tools can't be found, but I allow the build to continue in any case.

Creating the xflaim/src/Makefile.am File

I'm skipping the *xflaim/Makefile.am* file, because it's nearly identical to *ftk/Makefile.am*. Instead, we'll move on to *xflaim/src/Makefile.am*, which I wrote by following the same design principles used with the *ftk/src* version. It looks very similar to its ftk counterpart, with one exception: according to the original build system makefile, the Java native interface (JNI) and C# native language–binding sources are compiled and linked right into the *xflaim* shared library.

This is not an uncommon practice, and it's quite useful because it alleviates the need for extra library objects built specifically for these languages. Essentially, the *xflaim* shared library exports native interfaces for these languages, which are then consumed by their corresponding native wrappers. [25]

I'm going to ignore these language bindings for now, but later, when I'm finished with the entire xflaim project, I'll turn my attention back to properly hooking them into the library. With this exception then, the *Makefile.am* file shown in Listing 14-14 looks almost identical to its ftk counterpart.

```
if HAVE_JAVA
  JAVADIR = java
  JNI_LIBADD = java/libxfjni.la
endif

if HAVE_CSHARP
  CSDIR = cs
  CSI_LIBADD = cs/libxfcsi.la
endif

SUBDIRS = $(JAVADIR) $(CSDIR)

pkgconfigdir = $(libdir)/pkgconfig
pkgconfig_DATA = libxflaim.pc

lib_LTLIBRARIES = libxflaim.la
include_HEADERS = xflaim.h

libxflaim_la_SOURCES = \
```

---

25. There are a few platform-specific problems to be aware of when you're building JNI libraries into native libraries in this manner. Apple's macOS version 10.4 and older seem to require that JNI libraries be named with a *.jnilib* extension; if they aren't, the JVM won't load these files, so the xflaim Java bindings won't work correctly on these systems. These are very old releases of macOS however.

```
btreeinfo.cpp \
f_btpool.cpp \
f_btpool.h \
--snip--
rfl.h \
scache.cpp \
translog.cpp

libxflaim_la_CPPFLAGS = $(FTK_INCLUDE)
libxflaim_la_LIBADD = $(JNI_LIBADD) $(CSI_LIBADD) $(FTK_LTLIB)
libxflaim_la_LDFLAGS = -version-info 3:2:0
```

Listing 14-14: xflaim/src/Makefile.am: The xflaim project src directory Automake input file

I've conditionally defined the contents of the SUBDIRS variable here based on variables defined by corresponding Automake conditional statements in *configure.ac*. When make all is executed, the SUBDIRS variable conditionally recurses into the *java* and *cs* subdirectories. But when make dist is executed, a hidden DIST_SUBDIRS variable (which is created by Automake from *all of the possible contents* of the SUBDIRS variable) references all directories appended, either conditionally or unconditionally, to SUBDIRS.[26]

NOTE *The library interface version information was extracted from the original makefile.*

Turning to the xflaim/util Directory

The *util* directory for xflaim is a bit more complex. According to the original makefile, it generates several utility programs as well as a convenience library that is consumed by these utilities.

It was somewhat more difficult to find out which source files belong to which utilities and which were not used at all. Several of the files in the *xflaim/util* directory are not used by any of the utilities. Do we distribute these extra source files? I chose to do so, because they were already being distributed by the original build system and adding them to the EXTRA_DIST list makes it obvious to later observers that they aren't used.

Listing 14-15 shows a portion of the *xflaim/util/Makefile.am* file; the parts that are missing are redundant.

```
EXTRA_DIST = dbdiff.cpp dbdiff.h domedit.cpp diffbackups.cpp xmlfiles

XFLAIM_INCLUDE = -I$(top_srcdir)/src
XFLAIM_LDADD = ../src/libxflaim.la

❶ AM_CPPFLAGS = $(XFLAIM_INCLUDE) $(FTK_INCLUDE)
LDADD = libutil.la $(XFLAIM_LDADD)
```

---

26. When you think about it, I believe you'll agree that this is some pretty tricky code. Automake has to unravel the values of the make variables used in SUBDIRS, which are defined within Automake conditional statements.

```
## Utility Convenience Library

noinst_LTLIBRARIES = libutil.la

libutil_la_SOURCES = \
 flm_dlst.cpp \
 flm_dlst.h \
 flm_lutl.cpp \
 flm_lutl.h \
 sharutil.cpp \
 sharutil.h

## Utility Programs

bin_PROGRAMS = xflmcheckdb xflmrebuild xflmview xflmdbshell

xflmcheckdb_SOURCES = checkdb.cpp
xflmrebuild_SOURCES = rebuild.cpp

xflmview_SOURCES = \
 viewblk.cpp \
 view.cpp \
 --snip--
 viewmenu.cpp \
 viewsrch.cpp

xflmdbshell_SOURCES = \
 domedit.h \
 fdomedt.cpp \
 fshell.cpp \
 fshell.h \
 xshell.cpp

## Check Programs

check_PROGRAMS = \
 ut_basictest \
 ut_binarytest \
 --snip--
 ut_xpathtest \
 ut_xpathtest2
```

❷ ```
check_DATA = copy-xml-files.stamp
check_HEADERS = flmunittest.h

ut_basictest_SOURCES = flmunittest.cpp basictestsrv.cpp
```
❸ ```
--snip--
ut_xpathtest2_SOURCES = flmunittest.cpp xpathtest2srv.cpp
```

```
## Unit Tests

TESTS = \
 ut_basictest \
```

```
--snip--
  ut_xpathtest2

  ## Miscellaneous rules required by Check Programs

❹ copy-xml-files.stamp:
          cp $(srcdir)/xmlfiles/*.xml .
          echo Timestamp > $@

❺ clean-local:
          rm -rf ix2.*
          rm -rf bld.*
          rm -rf tst.bak
          rm -f *.xml
          rm -f copy-xml-files.stamp
```

Listing 14-15: xflaim/util/Makefile.am: The xflaim project's util directory Automake input file

In this example, you can see by the elided sections that I left out several long lists of files and products. This makefile builds 22 unit tests, but because they're all identical, except for naming differences and the source files from which they're built, I only left the descriptions for two of them (at ❸).

I've defined the file-global AM_CPPFLAGS and LDADD variables at ❶ in order to associate the XFLAIM and FTK include and library files with each of the projects listed in this *Makefile.am* file. This way, I don't have to explicitly append this information to each product.

Transitive Dependencies

Notice, however, that the AM_CPPFLAGS variable uses both the XFLAIM_INCLUDE and FTK_INCLUDE variables—the xflaim utilities clearly require information from both sets of header files. So why doesn't the LDADD variable reference the *ftk* library? This is because Libtool manages transitive dependencies for you and does so in a very portable manner, because some systems don't have a native mechanism for managing transitive dependencies. Because I reference *libxflaim.la* through XFLAIM_LDADD, and because *libxflaim.la* lists *libflaimtk.la* as a dependency, Libtool is able to provide the transitive reference for me on the utility programs' linker command lines.

For a clearer picture of this, examine the contents of *libxflaim.la* (in your build directory under *xflaim/src*—you will have to build the project first; run autoreconf -i; ./configure && make). You'll find a few lines near the middle of the file that look very much like the contents of Listing 14-16.

```
--snip--
# Libraries that this one depends upon.
dependency_libs=' .../flaim/build/ftk/src/libflaimtk.la -lrt -lncurses'
--snip--
```

Listing 14-16: The portion of xflaim/src/libxflaim.la that shows dependency libraries

The path information for *libflaimtk.la* is listed here so we don't have to specify it in the LDADD statement for the xflaim utilities.[27]

Like Libtool, the GNU linker and the Linux loader can manage transitive dependencies (TDs). This is done by having ld incorporate these indirect dependencies into the ELF binaries it generates when appropriate linker command line options are used. Libtool's mechanism relies on a recursive search of a hierarchy of *.la* files, whereas Linux's native mechanism simply recursively searches the library hierarchy at build time and embeds all required library references directly into the built program or library. The loader then sees and uses these references at load time. A nice aspect of using such native TD management is that, if a library is updated in a newer version of a package, the loader will immediately begin to reference the updated secondary symbols from the new library's updated reference list, and projects built against that library will immediately begin using the new version's transitive dependencies.

Recently, some distro vendors have decided it's worth taking advantage of this feature on their platforms. The problem is, Libtool's TD management reduces the perceived advantages of using ld's (and the system loader's) internal TD management—it gets in the way, so to speak. To solve this issue, these vendors have decided to release a modified version of Libtool on their platforms, wherein its TD management feature is effectively disabled. The result is that you must now specify all direct and indirect libraries on the linker (libtool) command line or modify your build system to use the non-portable native TD management linker options.

Since native TD management is not supported on all platforms, and Libtool's text file–based approach is completely portable, we often rely heavily on Libtool to do the right thing when indirect dependencies are required while linking our programs and libraries on systems that don't have a native TD management system. When you use a "distro-crippled" Libtool package to build projects designed to take advantage of Libtool's TD management features, your build simply fails at the link stage with "missing DSO" (dynamic shared object) messages.[28]

The sed command in *configure.ac* searches for the text link_all_deplibs=no in the *libtool* script and replaces it with link_all_deplibs=yes. It's in there twice, and the sed command will replace both occurrences. AC_OUTPUT executes config.status, which generates the *libtool* script in the project directory, so the sed command must follow AC_OUTPUT to be effective.

---

27. When *libxflaim.la* is installed into */usr/<local/>lib* next to *libxflaim.so.**, Libtool modifies the installed version of this file so it references the installed versions of the libraries rather than the libraries in the build directory structure. This allows builds against installed versions of your libraries to also take advantage of this feature.

28. When I wrote the first edition of this book, I relied on Libtool's ability to automatically link transitive dependencies in the makefiles for the flaim and xflaim *utils* directory. These utilities link against libflaim and, therefore, transitively against libflaimtk. When I tried to build the FLAIM source code with late versions of the Autotools on a different platform, I found suddenly that I could no longer link some of these utilities. There's quite a bit of traffic on the libtool mailing list about this controversial issue.

NOTE *It doesn't hurt to use this sed command, even on systems that do not exhibit the problem—sed simply won't find anything to replace in your libtool script. Be aware, however, that if your package is picked up for distribution by a Linux vendor that uses internal TD management, they'll probably ship your package with these sorts of commands "patched out."*

Of course, another option is to forgo the use of automatic transitive dependency management entirely by simply specifying all of the link dependencies you know you'll need on the linker's command line for every program or library you build. Pkg-config actually does this for you anyway, so if you can rely on pkg-config for all your library management needs, then your projects are simply not affected by this issue. This can be done manually in the flaim, xflaim, and flaimsql projects by adding $(FTK_LTLIB) to the LDADD variables as, for example, at ❶ in Listing 14-15.

Feel free to try this by commenting out the sed commands in *configure.ac* and then rebuilding the project. Assuming you're building on a platform where Libtool has been modified, your build will fail at the point where the flaim and xflaim projects try to link their utilities only against the *libflaim.la* and *libxflaim.la*. To make it work again, update your LDADD variables as I mentioned earlier.

Stamp Targets

In creating this makefile, I ran across another minor problem that I hadn't anticipated. At least one of the unit tests seemed to require that some XML data files be present in the directory from which the test is executed. The test failed, and when I dug into it, I noticed that it failed while trying to open these files. Looking around a bit lead me to the *xflaim/util/xmldata* directory, which contained several dozen XML files.

I needed to copy those files into the build hierarchy's *xflaim/util* directory before I could run the unit tests. I know that products prefixed with check are built before TESTS are executed, so it occurred to me that I might list these files at ❷ in a check_DATA PLV. The check_DATA variable refers to a file called *copy-xml-files.stamp*, which is a special type of file target called a *stamp* target. Its purpose is to replace a group of unspecified files, or a non-file-based operation, with one single, representative file. This stamp file is used to indicate to the build system that all the XML data files have been copied into the *util* directory. Automake often uses stamp files in its own generated rules.

The rule for generating the stamp file at ❹ also copies the XML data files into the test execution directory. The echo statement simply creates a file named *copy-xml-files.stamp* that contains a single word: *Timestamp*. The file may contain anything (or nothing at all). The important point here is that the file exists and has a time and date associated with it. The make utility uses this information to determine whether the copy operation needs to be executed. In this case, since *copy-xml-files.stamp* has no dependencies, its mere existence indicates to make that the operation has already been done. Delete the stamp file to get make to perform the copy operation on the next build.

This is a sort of hybrid between a true file-based rule and a phony target. Phony targets are always executed—they aren't real files, so make has no way of determining whether the associated operation should be performed based on file attributes. The timestamps of file-based rules can be checked against their dependency lists to determine whether they should be re-executed. Stamp rules like this are executed only if the stamp file is missing, because there are no dependencies against which the target's time and date should be compared.[29]

Cleaning Your Room

All files placed in the build directory should be cleaned up when the user enters make clean at the command prompt. Since I placed XML data files into the build directory, I also need to clean them up. Files listed in DATA variables are not cleaned up automatically, because DATA files are not necessarily generated. Sometimes the DATA primary is used to list static project files that need to be installed. I "created" a bunch of XML files and a stamp file, so I needed to remove these during make clean. To this end, I added the clean-local target at ❺, along with its associated rm commands.

NOTE *Be careful when deleting files copied from the source tree into the corresponding location in the build tree—you may inadvertently delete source files when building from within the source tree. You can compare $(srcdir) to ". " within make commands to see if the user is building in the source tree.*

There is another way to ensure that files created using your own make rules get cleaned up during execution of the clean target. You can define the CLEANFILES variable to contain a whitespace-separated list of files (or wildcard specifications) to be removed. I used a clean-local target in this case, because the CLEANFILES variable has one caveat: it won't remove directories, only files. Each of the rm commands that removes a wildcard file specification refers to at least one directory. I'll show you a proper use of CLEANFILES in Chapter 15.

Regardless of how well your unit tests clean up after themselves, you still might want to write clean rules that attempt to clean up intermediary test files. That way, your makefiles will clean up droppings from interrupted

---

29. Stamp files have the inherent problem of not properly specifying the true relationship between targets and their dependencies—a critical requirement of a proper update. Regardless, a stamp file is sometimes the only reasonable way to accomplish a task within a makefile. One special case is to properly handle rules that generate multiple output or product files. GNU make has special pattern rule syntax for dealing with situations where multiple output files are generated by a single rule, but Automake tries hard not to depend on GNU make extensions. The use of stamp files in this case represents a workaround for a missing feature of POSIX make. Automake also uses stamp files when not doing so would cause a very large file set to become part of a target's dependency list. Since there are inherent negative side effects associated with stamp files, Automake reserves their use for these sorts of special cases.

tests and debug runs.[30] Remember that the user may be building in the source directory. Try to make your wildcards as specific as possible so you don't inadvertently remove source files.

I use the Automake-supported `clean-local` target here as a way to extend the `clean` target. The `clean-local` target is executed as a dependency of (and thus executed before) the `clean` target, if it exists. Listing 14-17 shows the corresponding code from the Automake-generated *Makefile.in* template, so you can see how this infrastructure is wired up. The interesting bits are highlighted.

```
--snip--
clean: clean-am
❶ clean-am: clean-binPROGRAMS clean-checkPROGRAMS \
    clean-generic clean-libtool clean-local \
    clean-noinstLTLIBRARIES mostlyclean-am
--snip--
❷ .PHONY: ... clean-local...
--snip--
clean-local:
        rm -rf ix2.*
        rm -rf bld.*
        rm -rf tst.bak
        rm -f *.xml
        rm -f copy-xml-files.stamp
--snip--
```

Listing 14-17: xflaim/util/Makefile.in: The clean rules generated by Automake from xflaim/util/Makefile.am

Automake noted that I had a target named `clean-local` in *Makefile.am*, so it added `clean-local` to the dependency list for `clean-am` at ❶ and then added it to the `.PHONY` variable at ❷. Had I not written a `clean-local` target, these references would have been missing from the generated *Makefile*.

Summary

Well, those are the basics. If you've followed along and understood what we did in this chapter, then you should be able to convert nearly any project to use an Autotools-based build system. For more details on the topics covered here, I refer you to the Autotools manuals. Often just knowing the name of a concept so you can easily find it in the manual or in an online search is worth a great deal.

In Chapter 15, I'll cover the stranger aspects of converting this project, including the details of building Java and C# code, adding compiler-specific optimization flags and command line options, and even building RPM packages using user-defined `make` targets in your *Makefile.am* files.

---

30. You might also provide a debug option or an environment variable that causes your tests to leave these droppings behind so they can be examined during debugging.

15

FLAIM PART II: PUSHING THE ENVELOPE

What we do in college is to get over our little-mindedness.
Education—to get it you have to hang around till you catch on.
—*Robert Lee Frost*[1]

It's a well-understood principle that no matter how many books you read, or how many lectures you attend, or how many queries you present on mailing lists, you'll still be left with unanswered questions. It's estimated that half of the world's population has access to the internet today.[2] There are thousands of terabytes of information available from your desktop. Nevertheless, it seems every project has one or two issues that are just different enough from all others that even internet searches are fraught with futility.

To reduce the potential frustration of learning the Autotools, this chapter continues with the FLAIM build system conversion project by tackling some of the less common features of FLAIM's build system requirements.

---

1. Jay Parini, *Robert Frost: A Life*, p 185 (noted in his journals), citation from endnote 12.

2. See world internet usage statistics news and world population stats at *https://www.internetworldstats.com/stats.htm*.

My hope is that by presenting solutions to some of these less common problems, you'll become familiar with the underlying framework provided by the Autotools. Such familiarity provides the insight needed to bend the Autotools to your own unique requirements.

The *xflaim* library provides Java and C# language bindings. Automake provides rudimentary support for building Java sources but currently provides no built-in support for building C# sources. In this chapter, I'll show you how to use Automake's built-in Java support to build the Java language bindings in xflaim, and then I'll show you how to write your own make rules for the C# language bindings.

We'll round out this chapter, and finish up the FLAIM conversion project, with discussions of using native compiler options, building generated documentation, and adding your own top-level recursive make targets.

Building Java Sources Using the Autotools

The *GNU Automake Manual* presents information on building Java sources in two different ways. The first is the traditional and widely understood method of compiling Java source code into Java byte code, which can then be executed within the Java virtual machine (JVM). The second way is the lesser-known method of compiling Java source code directly into native machine code using the GNU Compiler for Java (gcj) frontend to the GNU compiler tool suite. The object files containing this machine code can then be linked together into native executable programs using the standard GNU linker. Probably due to lack of interest, and to the JVM's having been vastly improved over the years, the GCJ project is no longer being maintained. It's therefore likely that all support for this mechanism will soon be entirely dropped from the Autotools.

In this chapter, I'll focus on the former—building Java class files from Java source files using the Automake built-in JAVA primary. We'll also explore the necessary extensions required to build and install *.jar* files.

Autotools Java Support

Autoconf has little, if any, built-in support for Java. For example, it provides no macros that locate Java tools in the end user's environment.[3] Automake's built-in support for building Java classes is minimal, but getting it to work is not really that difficult if you're willing to dig in a bit. The biggest stumbling block is conceptual more than functional. You have to work a little to align your understanding of the Java build process with that of the Automake designers.

Automake provides a built-in primary (JAVA) for building Java sources, but it does not provide any preconfigured installation location prefixes for

---

3. The GNU Autoconf Archive (*https://www.gnu.org/software/autoconf-archive/*) has plenty of user-contributed macros that can help your configuration process set you up to build Java applications from Automake scripts.

installing Java classes. However, the usual place to install Java classes and *.jar* files is in the $(datadir)/*java* directory, so creating a proper prefix is as simple as using the Automake prefix extension mechanism of defining a variable suffixed with *dir*, as shown in Listing 15-1.

```
--snip--
javadir = $(datadir)/java
java_JAVA = file_a.java file_b.java ...
--snip--
```

Listing 15-1: Defining a Java installation directory in a Makefile.am file

Now, you don't often want to install Java sources, which is what you will accomplish when you define your JAVA primary in this manner. Rather, you want the *.class* files to be installed, or more likely a *.jar* file containing all of your *.class* files. It's generally more useful, therefore, to define the JAVA primary with the noinst prefix. Additionally, files in a JAVA primary list are not distributed by default, so you may even want to use the dist super-prefix, as shown in Listing 15-2.

```
dist_noinst_JAVA = file_a.java file_b.java...
```

Listing 15-2: Defining a list of non-installed Java files that are distributed

When you define a list of Java source files in a variable containing the JAVA primary, Automake generates a make rule that builds that list of files all in one command, using the syntax shown in Listing 15-3.[4]

```
--snip--
JAVAROOT = .
JAVAC = javac
CLASSPATH_ENV = CLASSPATH=$(JAVAROOT):$(srcdir)/$(JAVAROOT):\
  $${CLASSPATH:+":$$CLASSPATH"}
--snip--
all: all-am
--snip--
all-am: Makefile classnoinst.stamp $(DATA) all-local
--snip--
classnoinst.stamp: $(am__java_sources)
        @list1='$?'; list2=; if test -n "$$list1"; then \
        for p in $$list1; do \
          if test -f $$p; then d=; else d="$(srcdir)/"; fi; \
          list2="$$list2 $$d$$p"; \
        done; \
```

---

4. It's difficult to design a set of make rules to build individual *.class* files from corresponding *.java* files. The reasons for this include the fact that the name of a particular *.class* file can't be determined without parsing the corresponding source file. Additionally, due to inner and anonymous class definitions, multiple *.class* files, whose names are based on class names, can be generated from a single Java source file. Fortunately, it's orders of magnitude faster to compile an entire set of Java source files on one command line than to compile Java sources individually, based on individual source file timestamps.

```
❶ echo '$(CLASSPATH_ENV) $(JAVAC) -d $(JAVAROOT) \
        $(AM_JAVACFLAGS) $(JAVACFLAGS) '"$$list2"; \
    $(CLASSPATH_ENV) $(JAVAC) -d $(JAVAROOT) \
        $(AM_JAVACFLAGS) $(JAVACFLAGS) $$list2; \
    else :; fi
❷ echo timestamp > $@
--snip--
```

Listing 15-3: This long shell command was taken from a Makefile generated by Automake.

Most of the code you see in these commands exists solely to prepend the $(srcdir) prefix onto each file in the user-specified list of Java sources in order to properly support VPATH builds. This code uses a shell for statement to split the list into individual files, prepend the $(srcdir), and then reassemble the list.[5]

The part that actually does the work of building the Java sources is found in two lines (four wrapped lines, actually)[6] near the bottom at ❶.

Automake generates a stamp file at ❷ because the single $(JAVAC) command generates several *.class* files from the *.java* files. Rather than choosing one of these files at random, Automake generates and uses a stamp file as the target of the rule, which causes make to ignore the relationships between individual *.class* files and their corresponding *.java* files. That is, if you delete a *.class* file, the rules in the makefile will not cause it to be rebuilt. The only way to cause the re-execution of the $(JAVAC) command is to either modify one or more of the *.java* files, thereby causing their timestamps to become newer than that of the stamp file, or delete the stamp file entirely.

The variables used in the build environment and on the command line include JAVAROOT, JAVAC, JAVACFLAGS, AM_JAVACFLAGS, and CLASSPATH_ENV. Each variable may be specified in the *Makefile.am* file.[7] If a variable is not specified, the defaults shown in Listing 15-3 are used instead.

All *.java* files specified in a JAVA primary variable are compiled using a single command line, which may pose a problem on systems with limited command line lengths. If you encounter such a problem, you can either break up your Java project into multiple Java source directories or develop your own make rules for building Java classes. (When I discuss building C# code in "Building the C# Sources" on page 418, I demonstrate how to write such custom rules.)

---

5. It's interesting to note that this file list–munging process could have been done in a half line of GNU Make–specific code, but Automake is designed to generate makefiles that can be executed by many older make programs.

6. I added additional wrapping to this example from the text I originally obtained from the generated Makefile to better format it for the listing. The original Makefile text had only one long wrapped line.

7. Technically, JAVAC, JAVACFLAGS, and CLASSPATH_ENV are reserved for the user to specify on the configure command line, but these variables often don't have reasonable defaults. The way to play nicely in such situations is to specify them in such a way as to allow the user to override any defaults you define within your build system files. You can do this, for instance, by checking for a value in the variables before setting them to a default or by using GNU Make–specific syntax to set them if not already in the environment.

The CLASSPATH_ENV variable sets the Java CLASSPATH environment variable so that it contains $(JAVAROOT), $(srcdir)/$(JAVAROOT), and then any class path that may have been configured in the environment by the end user.

The JAVAROOT variable is used to specify the location of the project's Java root directory within the project's build tree, where the Java compiler will expect to find the start of generated package directory hierarchies belonging to your project.

The JAVAC variable contains javac by default, with the assumption that javac can be found in the system path. The AM_JAVACFLAGS variable may be set in *Makefile.am*, though the non-Automake version of this variable (JAVACFLAGS) is considered a user variable and thus shouldn't be set in makefiles.

This is all fine as far as it goes, but it doesn't go nearly far enough. In this relatively simple Java project, we still need to generate Java native interface (JNI) header files using the javah utility as well as a *.jar* file from the *.class* files built from the Java sources. Unfortunately, Automake-provided Java support doesn't even begin to handle these tasks, so we'll have to do the rest with handcoded make rules. We'll begin with Autoconf macros to ensure that we have a good Java build environment.

Using ac-archive Macros

The GNU Autoconf Archive supplies community-contributed Autoconf macros that come close to what we need in order to ensure that we have a good Java development environment. In this particular case, I downloaded the latest source package and just hand-installed the *.m4* files that I needed into the *xflaim/m4* directory.[8]

Then I modified the files (including their names) to work the way my FLM_PROG_TRY_DOXYGEN macro works. I wanted to locate any existing Java tools, but I also wanted be able to continue without them if necessary. Though it has gotten much better in the last 10 years, given the politics surrounding the existence of Java tools in Linux distributions, this is probably a wise approach.

I created the following macros within corresponding Java-related *.m4* files:

- FLM_PROG_TRY_JAVAC is defined in *flm_prog_try_javac.m4*.
- FLM_PROG_TRY_JAVAH is defined in *flm_prog_try_javah.m4*.
- FLM_PROG_TRY_JAVADOC is defined in *flm_prog_try_javadoc.m4*.
- FLM_PROG_TRY_JAR is defined in *flm_prog_try_jar.m4*.
- FLM_PROG_TRY_JNI is defined in *flm_prog_try_jni.m4*.

With a bit more effort, I was also able to create the C# macros I needed to accomplish the same tasks for the C# language bindings:

- FLM_PROG_TRY_CSC is defined in *flm_prog_try_csc.m4*.
- FLM_PROG_TRY_CSVM is defined in *flm_prog_try_csvm.m4*.

---

8. The original Autoconf Archive macros I used were ax_jni_include_dir, ax_prog_jar, ax_prog_javac, ax_prog_javah, and ax_prog_javadoc. What I ultimately ended up with was a mix of code from these original macros.

Listing 15-4 shows the portion of the xflaim *configure.ac* file that invokes these Java and C# macros.

```
--snip--
# Checks for optional programs.
FLM_PROG_TRY_CSC
FLM_PROG_TRY_CSVM
FLM_PROG_TRY_JNI
FLM_PROG_TRY_JAVADOC
--snip--
# Automake conditionals.
AM_CONDITIONAL([HAVE_JAVA], [test "x$flm_prog_have_jni" = xyes])
AM_CONDITIONAL([HAVE_CSHARP], [test -n "$CSC"])
--snip--
```

Listing 15-4: xflaim/configure.ac: The portion of this file that searches for Java and C# tools

These macros set the CSC, CSVM, JAVAC, JAVAH, JAVADOC, and JAR variables to the location of their respective C# and Java tools and then substitute them into the xflaim project's *Makefile.in* templates using AC_SUBST. If any of these variables are already set in the user's environment when the configure script is executed, their values are left untouched, thus allowing the user to override the values that would have been set by the macros.

I discuss the internal operation of these macros in Chapter 16.

Canonical System Information

The only non-obvious bit of information you need to know about using macros from the GNU Autoconf Archive is that many of them rely on the built-in Autoconf macro, AC_CANONICAL_HOST. Autoconf provides a way to automatically expand any macros used internally by a macro definition right before the definition so that required macros are made available immediately. However, if AC_CANONICAL_HOST is not used before certain macros (including LT_INIT), autoreconf will generate about a dozen warning messages.

To eliminate these warnings, I added AC_CANONICAL_TARGET to my xflaim-level *configure.ac* file, immediately after the call to AC_INIT. The AC_CANONICAL _SYSTEM macro, and the macros that it calls (AC_CANONICAL_BUILD, AC_CANONICAL_HOST, and AC_CANONICAL_TARGET), are designed to ensure that the $build, $host, and $target environment variables are defined by configure to contain appropriate values describing the user's build, host, and target systems, respectively. Because I'm not doing any cross-compiling in this build system, I only needed to invoke AC_CANONICAL_TARGET.

These variables contain canonical values for the build, host, and target CPU, vendor, and operating system. Values like these are very useful to extension macros. If a macro can assume these variables are set properly, then it saves quite a bit of code duplication in the macro definition.

The values of these variables are calculated using the helper scripts `config .guess` and `config.sub`, which are distributed with Autoconf.[9] The `config.guess` script uses a combination of `uname` commands to ferret out information about the build system, then uses that information to derive a set of canonical values for CPU, vendor, and operating system. The `config.sub` script is used to reformat build, host, and target information specified by the user on the `configure` command line into a canonical value. The host and target values default to that of the build, unless you override them with command line options to `configure`. Such an override might be used when cross-compiling. (See "Item 6: Cross-Compiling" on page 517 for a more detailed explanation of cross-compiling within the Autotools framework.)

The xflaim/java Directory Structure

The original xflaim source layout had the Java JNI and C# native sources located in directory structures outside of *xflaim/src*. The JNI sources were in *xflaim/java/jni*, and the C# native sources were in *xflaim/csharp/xflaim*. While Automake can generate rules for accessing files outside the current directory hierarchy, it seems silly to put these files so far away from the only library they can really belong to. Therefore, in this case, I broke my own rule about not rearranging files and moved the contents of these two directories beneath *xflaim/src*. I named the JNI directory *xflaim/src/java* and the C# native sources directory *xflaim/src/cs*. The following diagram illustrates this new directory hierarchy:

```
flaim
  xflaim
    src
      cs
        wrapper
      java
        wrapper
          xflaim
```

As you can see, I also added a *wrapper* directory beneath the *java* directory, in which I rooted the xflaim wrapper package hierarchy. Since the Java xflaim wrapper classes are part of the Java xflaim package, they must be located in a directory called *xflaim*. Nevertheless, the build happens in the wrapper directory. There are no build files found in the *wrapper/xflaim* directory or any directories below that point.

---

9. Although these files are distributed with Autoconf, they are constantly being updated— far more often than Autoconf releases occur. It's worth adding the latest of these files to your projects directly by downloading them from the *GNU config* project at *http://savannah .gnu.org/projects/config/*. You can access them directly from the Savannah git repository at *http://git.savannah.gnu.org/cgit/config.git/tree/*.

NOTE *No matter how deep your package hierarchy is, you will still build the Java classes in the* wrapper *directory, which is the* JAVAROOT *directory for this project. Autotools Java projects consider the* JAVAROOT *directory the build directory for the* java *package.*

The xflaim/src/Makefile.am File

At this point, the *configure.ac* file is doing about all it can to ensure that I have a good Java build environment, in which case my build system will be able to generate my JNI wrapper classes and header files and build my C++ JNI sources. If my end user's system doesn't provide these tools, they simply won't be able to build or link the JNI language bindings to the *xflaim* library on that host.

Have a look at the *xflaim/src/Makefile.am* file shown in Listing 15-5 and examine the portions that are relevant to building the Java and C# language bindings.

```
if HAVE_JAVA
  JAVADIR = java
  JNI_LIBADD = java/libxfjni.la
endif

if HAVE_CSHARP
  CSDIR = cs
  CSI_LIBADD = cs/libxfcsi.la
endif

SUBDIRS = $(JAVADIR) $(CSDIR)
--snip--
libxflaim_la_LIBADD = $(JNI_LIBADD) $(CSI_LIBADD) $(FTK_LTLIB)
--snip--
```

Listing 15-5: xflaim/src/Makefile.am: The portion of this makefile that builds Java and C# sources

I've already explained the use of the conditionals to ensure that the *java* and *cs* directories are only built if the proper conditions are met. You can now see how this fits into the build system I've created so far.

Notice that I'm conditionally defining two new library variables. If I can build the Java language bindings, the *java* subdirectory will be built, and the JNI_LIBADD variable will refer to the library that is built in the *java* directory. If I can build the C# language bindings, the *cs* subdirectory will be built, and the CSI_LIBADD variable will refer to the library that is built in the *cs* directory. In either case, if the required tools are not found by configure, the corresponding variable will remain undefined. When an undefined make variable is referenced, it expands to nothing, so there's no harm in using it in libxflaim_la_LIBADD.

Building the JNI C++ Sources

Now turn your attention to the *xflaim/src/java/Makefile.am* file shown in Listing 15-6.

```
SUBDIRS = wrapper

XFLAIM_INCLUDE = -I$(srcdir)/..

noinst_LTLIBRARIES = libxfjni.la

libxfjni_la_SOURCES = \
  jbackup.cpp \
  jdatavector.cpp \
  jdb.cpp \
  jdbsystem.cpp \
  jdomnode.cpp \
  jistream.cpp \
  jniftk.cpp \
  jniftk.h \
  jnirestore.cpp \
  jnirestore.h \
  jnistatus.cpp \
  jnistatus.h \
  jostream.cpp \
  jquery.cpp

libxfjni_la_CPPFLAGS = $(XFLAIM_INCLUDE) $(FTK_INCLUDE)
```

Listing 15-6: xflaim/src/java/Makefile.am: This makefile builds the JNI sources.

Again, I want the *wrapper* directory to be built first (the dot at the end of the SUBDIRS list is implied), before the *xflaim* library, because the *wrapper* directory will build the class files and JNI header files required by the JNI convenience library sources. Building this directory is not conditional. If I've made it this far into the build hierarchy, I know I have all the Java tools I need. This *Makefile.am* file simply builds a convenience library containing my JNI C++ interface functions.

Because of the way Libtool builds both shared and static libraries from the same sources, this convenience library will become part of both the *xflaim* shared and static libraries. The original build system makefile accounted for this by linking the JNI and C# native interface objects only into the shared library (where they make sense).

NOTE   *The fact that these libraries are added to both the shared and static* xflaim *libraries is not really a problem. Objects in a static library remain unused in applications or libraries linking to the static library, as long as functions and data in those objects remain unreferenced, though this is a bit of a wart on my new build system.*

The Java Wrapper Classes and JNI Headers

Finally, *xflaim/src/java/wrapper/Makefile.am* takes us to the heart of the matter. I've tried many different configurations for building Java JNI wrappers, and this one always comes out on top. Listing 15-7 shows the *wrapper* directory's Automake input file.

```
JAVAROOT = .

❶ jarfile = $(PACKAGE)jni-$(VERSION).jar
❷ jardir = $(datadir)/java
pkgpath = xflaim
jhdrout = ..

$(jarfile): $(dist_noinst_JAVA)
        $(JAR) cvf $(JARFLAGS) $@ $(pkgpath)/*.class

❸ jar_DATA = $(jarfile)

java-headers.stamp: $(classdist_noinst.stamp)
        @list=`echo $(dist_noinst_JAVA) | sed -e 's|\.java||g' -e 's|/|.|g'`;\
          echo "$(JAVAH) -cp . -jni -d $(jhdrout) $(JAVAHFLAGS) $$list"; \
          $(JAVAH) -cp . -jni -d $(jhdrout) $(JAVAHFLAGS) $$list
    ❹ @echo "JNI headers generated" > java-headers.stamp

❺ all-local: java-headers.stamp

❻ CLEANFILES = $(jarfile) $(pkgpath)/*.class java-headers.stamp\
    $(jhdrout)/xflaim_*.h

dist_noinst_JAVA = \
 $(pkgpath)/BackupClient.java \
 $(pkgpath)/Backup.java \
 --snip--
 $(pkgpath)/XFlaimException.java \
 $(pkgpath)/XPathAxis.java
```

Listing 15-7: xflaim/src/java/wrapper/Makefile.am: The wrapper directory's Makefile.am file

At the top of the file, I've set the JAVAROOT variable to dot (.), because I want Automake to be able to tell the Java compiler that this is where the package hierarchy begins. The default value for JAVAROOT is $(top_builddir), which would incorrectly have the wrapper class belong to the *xflaim.src.java .wrapper.xflaim* package.

I create a variable at ❶ called jarfile, which derives its value from $(PACKAGE _TARNAME) and $(PACKAGE_VERSION). (Recall from Chapter 3 that this is also how the distdir variable is derived, from which the name of the tarball comes.) A make rule indicates how the *.jar* file should be built. Here, I'm using the JAR variable, whose value was calculated by the FLM_PROG_TRY_JNI macro in the configure script.

I define a new installation variable at ❷ called jardir where *.jar* files are to be installed, and I use that variable as the prefix for a DATA primary at ❸. Automake considers files that fit the Automake *where_HOW* scheme (with a defined *where*dir) as either architecture-independent data files or platform-specific executables. Installation location variables (those ending in dir) that begin with `bin`, `sbin`, `libexec`, `sysconf`, `localstate`, `lib`, or `pkglib` or that contain the string "exec" are considered platform-specific executables and are installed during execution of the `install-exec` target. Automake considers files installed in any other locations data files. These are installed during execution of the `install-data` target. The well-known installation locations such as *bindir*, *sbindir*, and so on are already taken, but if you want to install custom architecture-dependent executable files, just ensure that your custom installation location variable contains the string "exec," as in `myspecialexecdir`.

I use another stamp file at ❹ in the rule that builds the JNI header files from the *.class* files for the same reasons that Automake uses a stamp file in the rule that it uses to build *.class* files from *.java* source files.

This is the most complex part of this makefile, so I'll break it into smaller pieces.

The rule states that the stamp file depends on the source files listed in the `dist_noinst_JAVA` variable. The command is a bit of complex shell script that strips the *.java* extensions from the file list and converts all the slash characters into periods. The reason for this is that the `javah` utility wants a list of class names, not a list of filenames. The `$(JAVAH)` command then accepts this entire list as input in order to generate a corresponding list of JNI header files. The last line, of course, generates the stamp file.

Finally at ❺, I hook my `java-headers.stamp` target into the `all` target by adding it as a dependency to the `all-local` target. When the `all` target (the default for all Automake-generated makefiles) is executed in this makefile, *java-headers.stamp* will be built, along with the JNI headers.

NOTE *It's a good idea to add custom rule targets as dependencies to the Automake-provided hook and local targets, rather than directly associating commands with these hook and local targets. By doing this, the commands for individual tasks on those targets remain isolated and thus easier to maintain.*

I add the *.jar* file, all of the *.class* files, the *java-headers.stamp* file, and all of the generated JNI header files to the `CLEANFILES` variable at ❻ so that Automake will clean them up when `make clean` is executed. Again, I can use the `CLEANFILES` variable here because I'm not trying to delete any directories.

The final step in writing any such custom code is to ensure that the `distcheck` target still works, because when we generate our own products, we have to ensure that the `clean` target properly removes them all.

Finally, I should mention that the rule to build the *.jar* file, near the top of Listing 15-7, relies on a wildcard to pick up all the *.class* files in the *xflaim* directory. The Autotools purposely avoid such wildcards for many reasons, including the very valid reason that you may inadvertently pick up files that were built by a previous build that are no longer relevant to your project after changes eliminate those sources from the project. For Java, the only way to specify the exact *.class* files that should go into the *.jar* file is to parse all the *.java* files and derive a list of *.class* files that would be built from those sources. I made a judgment call here and decided that using a wildcard was worth the possible problems doing so may cause. I also used wildcards in the `CLEANFILES` variable near the bottom of Listing 15-7. Of course, the same potential problems exist here—you could remove a file that is present but no longer associated with the build.

A Caveat About Using the JAVA Primary

The one important caveat to using the `JAVA` primary is that you may define only one `JAVA` primary variable per *Makefile.am* file. The reason for this is that multiple classes may be generated from a single *.java* file, and the only way to know which classes came from which *.java* file would be for Automake to parse the *.java* files (which is ridiculous, and arguably the primary reason why build tools like *Apache Ant* and *Maven* were developed). Rather than do this, Automake allows only one `JAVA` primary per file, so all *.class* files generated within a given build directory are installed in the location specified by the single `JAVA` primary variable prefix.[10]

NOTE *The system I've designed will work fine for this case, but it's a good thing I don't need to install my JNI header files, because I have no way of knowing what they're called from within my* Makefile.am *file!*

You should by now be able to see the problems that the Autotools have with Java. In fact, these problems are not so much related to the design issues in the Autotools as they are to design issues within the Java language itself, as you'll see in the next section.

Building the C# Sources

Returning to the *xflaim/src/cs* directory brings us to a discussion of building sources for a language for which Automake has no support: C#. Listing 15-8 shows the *Makefile.am* file that I wrote for the *cs* directory.

```
SUBDIRS = wrapper

XFLAIM_INCLUDE = -I$(srcdir)/..

noinst_LTLIBRARIES = libxfcsi.la

libxfcsi_la_SOURCES = \
  Backup.cpp \
```

---

10. It seems that I've broken this rule by assuming in my `java-headers.stamp` rule that the source for class information is the list of files specified in the `dist_noinst_JAVA` variable. In reality, I should probably be looking in the current build directory for all *.class* files found after the rules for the JAVA primary are executed. However, this goes against the general Autotools philosophy of only building or using prespecified sources for a build step. Since I've already broken this rule a couple of times, we'll live with what we have for the present.

```
DataVector.cpp \
Db.cpp \
DbInfo.cpp \
DbSystem.cpp \
DbSystemStats.cpp \
DOMNode.cpp \
IStream.cpp \
OStream.cpp \
Query.cpp

libxfcsi_la_CPPFLAGS = $(XFLAIM_INCLUDE) $(FTK_INCLUDE)
```

Listing 15-8: xflaim/src/cs/Makefile.am: *The contents of the* cs *directory's Automake input file*

Not surprisingly, this looks almost identical to the *Makefile.am* file found in the *xflaim/src/java* directory because I'm building a simple convenience library from C++ source files found in this directory, just as I did in the *java* directory. As in the Java version, this makefile first builds a subdirectory called *wrapper*.

Listing 15-9 shows the full contents of the *wrapper/Makefile.am* file.

```
EXTRA_DIST = xflaim cstest sample xflaim.ndoc

xfcs_sources = \
 xflaim/BackupClient.cs \
 xflaim/Backup.cs \
 --snip--
 xflaim/RestoreClient.cs \
 xflaim/RestoreStatus.cs

cstest_sources = \
 cstest/BackupDbTest.cs \
 cstest/CacheTests.cs \
 --snip--
 cstest/StreamTests.cs \
 cstest/VectorTests.cs

TESTS = cstest_script

AM_CSCFLAGS = -d:mono -nologo -warn:4 -warnaserror+ -optimize+
#AM_CSCFLAGS += -debug+ -debug:full -define:FLM_DEBUG
```

❶ ```
all-local: xflaim_csharp.dll

clean-local:
        rm -f xflaim_csharp.dll xflaim_csharp.xml cstest_script\
          cstest.exe libxflaim.so
        rm -f Output_Stream
        rm -rf abc backup test.*

install-exec-local:
        test -z "$(libdir)" || $(MKDIR_P) "$(DESTDIR)$(libdir)"
        $(INSTALL_PROGRAM) xflaim_csharp.dll "$(DESTDIR)$(libdir)"
```
```

```
        install-data-local:
                test -z "$(docdir)" || $(MKDIR_P) "$(DESTDIR)$(docdir)"
                $(INSTALL_DATA) xflaim_csharp.xml "$(DESTDIR)$(docdir)"

        uninstall-local:
                rm -f "$(DESTDIR)$(libdir)/xflaim_csharp.dll"
                rm -f "$(DESTDIR)$(docdir)/xflaim_csharp.xml"

❷ xflaim_csharp.dll: $(xfcs_sources)
                @list1='$(xfcs_sources)'; list2=; if test -n "$$list1"; then \
                  for p in $$list1; do \
                    if test -f $$p; then d=; else d="$(srcdir)/"; fi; \
                    list2="$$list2 $$d$$p"; \
                  done; \
                  echo '$(CSC) -target:library $(AM_CSCFLAGS) $(CSCFLAGS) -out:$@\
                    -doc:$(@:.dll=.xml) '"$$list2";\
                  $(CSC) -target:library $(AM_CSCFLAGS) $(CSCFLAGS) \
                    -out:$@ -doc:$(@:.dll=.xml) $$list2; \
                else :; fi

        check_SCRIPTS = cstest.exe cstest_script

❸ cstest.exe: xflaim_csharp.dll $(cstest_sources)
                @list1='$(cstest_sources)'; list2=; if test -n "$$list1"; then \
                  for p in $$list1; do \
                    if test -f $$p; then d=; else d="$(srcdir)/"; fi; \
                    list2="$$list2 $$d$$p"; \
                  done; \
                  echo '$(CSC) $(AM_CSCFLAGS) $(CSCFLAGS) -out:$@ '"$$list2"'\
                    -reference:xflaim_csharp.dll'; \
                  $(CSC) $(AM_CSCFLAGS) $(CSCFLAGS) -out:$@ $$list2 \
                    -reference:xflaim_csharp.dll; \
                else :; fi

❹ cstest_script: cstest.exe
                echo "#!/bin/sh" > cstest_script
                echo "$(top_builddir)/libtool --mode=execute \
       ❺ -dlopen=../../libxflaim.la $(CSVM) cstest.exe" >> cstest_script
                chmod 0755 cstest_script
```

Listing 15-9: xflaim/src/cs/wrapper/Makefile.am: The full contents of the C# makefile

The default target for *Makefile.am* is all, the same as that of a normal
non-Automake makefile. Again, I've hooked my code into the all target
by implementing the all-local target, which depends on a file named
xflaim_csharp.dll.[11]

---

11. This executable filename may be a bit confusing to those who are new to C#. In essence,
Microsoft, the creator of C#, designed the C# virtual machine to execute Microsoft native
(or almost native) binaries. In porting the C# virtual machine to Unix, the Mono team (the
Linux C# compiler project) decided against breaking Microsoft's naming conventions so
that Microsoft-generated portable C# programs could be executed by the Mono C# virtual
machine implementation. Nevertheless, C# still suffers from problems that need to be managed occasionally by name-mapping configuration files.

The C# sources are built by the commands under the *xflaim_csharp.dll* target at ❷, and the *xflaim_csharp.dll* binary depends on the list of C# source files specified in the xfcs_sources variable. The commands in this rule are copied from the Automake-generated *java/wrapper/Makefile* and are slightly modified to build C# binaries from C# source files (as highlighted in the listing). This isn't intended to be a lesson in building C# sources; the point here is that the default target is automatically built by creating a dependency between the all-local target and your own targets at ❶.

This *Makefile.am* file also builds a set of unit tests in C# that assess the C# language bindings. The target of this rule is *cstest.exe* (❸), which ultimately becomes a C# executable. The rule states that *cstest.exe* depends on *xflaim_csharp.dll* and the source files. I've again copied the commands from the rule for building *xflaim_csharp.dll* (as highlighted) and modified them for building the C# programs.

Ultimately, upon building the check target, the Automake-generated makefile will attempt to execute the scripts or executables listed in the TESTS variable. The idea here is to ensure that all necessary components are built before these files are executed. I've tied into the check target by defining check-local and making it depend on my test code targets.

The cstest_script at ❹ is a shell script built solely to execute the *cstest.exe* binary within the C# virtual machine. The C# virtual machine is found in the CSVM variable, which was defined in configure by the code generated by the FLM_PROG_TRY_CSVM macro.

The cstest_script depends only on the cstest.exe program. However, the *xflaim* library either must be present in the current directory or must be in the system library search path. We gain maximum portability here by using Libtool's *execute* mode to add the *xflaim* library to the system library search path before executing the C# virtual machine at ❺.

Manual Installation

Since in this example I'm doing everything myself, I have to write my own installation rules. Listing 15-10 reproduces only the installation rules in the *Makefile.am* file from Listing 15-9.

```
--snip--
install-exec-local:
        test -z "$(libdir)" || $(MKDIR_P) "$(DESTDIR)$(libdir)"
        $(INSTALL_PROGRAM) xflaim_csharp.dll "$(DESTDIR)$(libdir)"

install-data-local:
        test -z "$(docdir)" || $(MKDIR_P) "$(DESTDIR)$(docdir)"
        $(INSTALL_DATA) xflaim_csharp.xml "$(DESTDIR)$(docdir)"

uninstall-local:
        rm -f "$(DESTDIR)$(libdir)/xflaim_csharp.dll"
        rm -f "$(DESTDIR)$(docdir)/xflaim_csharp.xml"
--snip--
```

Listing 15-10: xflaim/src/cs/wrapper/Makefile.am: The installation rules of this makefile

According to the rules defined in the *GNU Coding Standards*, the installation targets do not depend on the binaries they install, so if the binaries haven't been built yet, I may have to exit from *root* to my user account to build the binaries with `make all` first.

Automake distinguishes between installing programs and installing data. However, there's only one `uninstall` target. The rationale seems to be that you might want to do an `install-exec` operation per system in your network, but only one shared `install-data` operation. Uninstalling a product requires no such separation, because uninstalling data multiple times is typically harmless.

Cleaning Up Again

As usual, things must be cleaned up properly in order to make distribution checks happy. The `clean-local` target handles this nicely, as shown in Listing 15-11.

```
--snip--
clean-local:
        rm -f xflaim_csharp.dll xflaim_csharp.xml cstest_script \
          cstest.exe libxflaim.so
        rm -f Output_Stream
        rm -rf abc backup test.*
--snip--
```

Listing 15-11: xflaim/src/cs/wrapper/Makefile.am: The clean rules defined in this makefile

Configuring Compiler Options

The original GNU Make build system provided a number of command line build options. By specifying a list of auxiliary targets on the `make` command line, the user could indicate that they wanted a debug or release build, force a 32-bit build on a 64-bit system, generate generic SPARC code on a Solaris system, and so on. This was a turnkey approach to build systems that is quite common in commercial code.

In open source projects, and particularly in Autotools-based build systems, the more common practice is to omit much of this rigid framework, allowing the user to set their own options in the standard user variables: `CC`, `CPP`, `CXX`, `CFLAGS`, `CXXFLAGS`, `CPPFLAGS`, and so on.[12]

---

12. The strange thing is that commercial software is developed by industry experts, whereas open source software is often built and consumed by hobbyists. And yet the *experts* are the ones using the menu-driven rigid-options framework, while the hobbyists have the flexibility to manually configure their compiler options the way they want. I suppose the most reasonable explanation for this is that commercial software relies on carefully crafted builds that must be able to be duplicated—usually by people who didn't write the original build system. Open source hobbyists would rather not give up the flexibility afforded by a more policy-driven approach.

Probably the most compelling argument for the Autotools approach to option management is that it's policy driven and the rigid frameworks used by commercial software vendors can easily be implemented in terms of the much more flexible policy-driven Autotools framework. For example, a *config.site* file might be used to provide site-wide options for all Autotools-based builds done at a particular site. A simple script can be used to configure various environment-based options before calling configure, or these options may even be passed to configure or make directly within such a script. The Autotools policy-driven approach offers the flexibility to be as configurable as a developer might want or as tight as required by management.

Ultimately, we'd like to have FLAIM project options conform to the Autotools policy-driven approach; however, I didn't want to lose the research effort involved in determining the hardcoded native compiler options specified in the original makefile. To this end, I've added back in *some* of the options to the *configure.ac* file that were supported by the original build system, but I've left others out. Listing 15-12 shows the end result of these efforts. This code enables various native compiler options, optimizations, and debugging features on demand, based on the contents of some of the user variables.

```
--snip--
# Configure supported platforms' compiler and linker flags
❶ case $host in
   sparc-*-solaris*)
     LDFLAGS="$LDFLAGS -R /usr/lib/lwp"
     case $CXX in
       *g++*) ;;
       *)
         if "x$debug" = xno; then
           CXXFLAGS="$CXXFLAGS -xO3"
         fi
         SUN_STUDIO=`$CXX -V | grep "Sun C++"`
         if "x$SUN_STUDIO" = "xSun C++"; then
           CXXFLAGS="$CXXFLAGS -errwarn=%all -errtags\
             -erroff=hidef,inllargeuse,doubunder"
         fi ;;
     esac ;;

   *-apple-darwin*)
     AC_DEFINE([OSX], [1], [Define if building on Apple OSX.]) ;;

   *-*-aix*)
     case $CXX in
       *g++*) ;;
       *) CXXFLAGS="$CXXFLAGS -qstrict" ;;
     esac ;;

   *-*-hpux*)
     case $CXX in
       *g++*) ;;
       *)
         # Disable "Placement operator delete
```

```
        # invocation is not yet implemented" warning
        CXXFLAGS="$CXXFLAGS +W930" ;;
    esac ;;
esac
--snip--
```

Listing 15-12: xflaim/configure.ac: The portion of this file that enables compiler-specific options

Remember that this code depends on the earlier use of the AC_CANONICAL _SYSTEM (or AC_CANONICAL_TARGET) macro, which sets build, host, and target environment variables to canonical string values that indicate CPU, vendor, and operating system.

In Listing 15-12, I used the host variable in the case statement at ❶ to determine the type of system for which I was building. This case statement determines if the user is building on Solaris, Apple Darwin, AIX, or HP-UX by looking for substrings in host that are common to all variations of these platforms. The config.guess and config.sub files are your friends here. If you need to write code like this for your project, examine these files to find common traits for the processes and systems for which you'd like to set various compiler and linker options.

NOTE   *In each of these cases (except for the definition of the OSX preprocessor variable on Apple Darwin systems), I'm really only setting flags for native compilers. The GNU compiler tools seem to be able to handle any code without the need for additional compiler options. It's worth reiterating here that the Autotools feature-present approach to setting options once again wins. Maintenance is reduced dramatically when you don't have to support large case statements for an ever-growing list of supported hosts and tool sets.*

Hooking Doxygen into the Build Process

I want to generate documentation as part of my build process, if possible. That is, if the user has doxygen installed, the build system will use it to build Doxygen documentation as part of the make all process.

The original build system has both static and generated documentation. The static documentation should always be installed, but the Doxygen documentation can only be built if the doxygen program is available on the host. Thus, I always build the *docs* directory, but I use the AM_CONDITIONAL macro to conditionally build the *docs/doxygen* directory.

Doxygen uses a configuration file (often called *doxyfile*) to configure literally hundreds of Doxygen options. This configuration file contains some information that is known to the configuration script. This sounds like the perfect opportunity to use an Autoconf-generated file. To this end, I've written an Autoconf template file called *doxyfile.in* that contains most of what a normal Doxygen input file would contain, as well as a few Autoconf substitution variable references. The relevant lines in this file are shown in Listing 15-13.

```
--snip--
PROJECT_NAME                = @PACKAGE_NAME@
--snip--
PROJECT_NUMBER              = @PACKAGE_VERSION@
--snip--
STRIP_FROM_PATH             = @top_srcdir@
--snip--
INPUT                       = @top_srcdir@/src/xflaim.h
--snip--
```

Listing 15-13: xflaim/docs/doxygen/doxyfile.in: The lines in this file that contain Autoconf variables

There are many other lines in this file, but they are all identical to the output file, so I've omitted them for the sake of space and clarity. The key here is that config.status will replace these substitution variables with their values as defined in *configure.ac* and by Autoconf itself. If these values change in *configure.ac*, the generated file will be rewritten with the new values. I've added a conditional reference for *xflaim/docs/doxygen/doxyfile* to the AC_CONFIG_FILES list in xflaim's *configure.ac* file. That's all it takes.

Listing 15-14 shows the *xflaim/docs/doxygen/Makefile.am* file.

```
❶ docpkg = $(PACKAGE_TARNAME)-doxy-$(PACKAGE_VERSION).tar.gz

❷ doc_DATA = $(docpkg)

❸ $(docpkg): doxygen.stamp
          tar chof - html | gzip -9 -c >$@

   doxygen.stamp: doxyfile
          $(DOXYGEN) $(DOXYFLAGS) $<
          echo Timestamp > $@

❹ install-data-hook:
          cd $(DESTDIR)$(docdir) && tar xf $(docpkg)

   uninstall-data-hook:
          cd $(DESTDIR)$(docdir) && rm -rf html

❺ CLEANFILES = doxywarn.txt doxygen.stamp $(docpkg)

   clean-local:
          rm -rf html
```

Listing 15-14: xflaim/docs/doxygen/Makefile.am: The full contents of this makefile

Here, I create a package name at ❶ for the tarball that will contain the Doxygen documentation files. This is basically the same as the distribution tarball for the xflaim project, except that it contains the text -doxy after the package name.

I define a doc_DATA variable at ❷ that contains the name of the Doxygen tarball. This file will be installed in the $(docdir) directory, which by default is $(datarootdir)/*doc*/$(PACKAGE_TARNAME), and $(datarootdir) is configured by Automake as $(prefix)/*share*, by default.

NOTE   *The* DATA *primary brings with it significant Automake functionality—installation is managed automatically. While I must build the Doxygen documentation package, the* DATA *primary automatically hooks the* all *target for me so that my package is built when the user executes* make *or* make all.

I use another stamp file at ❸ because Doxygen generates literally hundreds of *.html* files from the source files in my project. Rather than attempt to figure out a rational way to assign dependencies, I've chosen to generate one stamp file and then use that to determine whether the documentation is out-of-date.[13]

I also decided that it would be nice to unpack the documentation archive into the package *doc* directory. Left up to Automake, the tarball would make it into the proper directory at installation time, but that's as far as it would go. I needed to be able to hook the installation process to do this, and this is the perfect use for an Automake -hook target. I use the install-data-hook target at ❹ because the -hook targets allow you to perform extra user-defined shell commands after the operation that's being hooked has completed. Likewise, I use uninstall-hook to remove the *html* directory created when the *.tar* file was extracted during installation. (There is no distinction between uninstalling platform-specific and platform-independent files, so there is only one hook for uninstalling files.)

To clean my generated files, I use a combination of the CLEANFILES variable at ❺ and a clean-local rule just to demonstrate that it can be done.

Adding Nonstandard Targets

Adding a new nonstandard target is a little different than hooking an existing target. In the first place, you don't need to use AM_CONDITIONAL and other Autoconf tests to see if you have the tools you need. Instead, you can do all conditional testing from the *Makefile.am* file because you control the entire command set associated with the target, although this isn't recommended practice. (It's always preferable to ensure that the build environment is configured correctly from the configure script.) In cases where make targets can only be expected to work under certain conditions, or on certain platforms, it's a good idea to provide checks within the target to ensure that the operation requested can actually be performed.

---

13. In fact, the only source file in this project that currently contains Doxygen markup is the *xflaim.h* header file, but that could easily change, and it certainly won't hold true for all projects. Additionally, Doxygen generates hundreds of *.html* files, and this entire set of files represents the target of a rule to build the documentation. The stamp file stands in for these files as the target of the rule.

To start with, I create a directory within each project root directory called *obs* to contain the *Makefile.am* file for building RPM package files. (*OBS* is an acronym for *openSUSE Build Service*, an online package-building service.)[14]

Building RPM package files is done using a configuration file, called a *spec* file, which is very much like the *doxyfile* used to configure Doxygen for a specific project. As with the *doxyfile*, the RPM spec file references information that configure knows about the package. So, I wrote an *xflaim.spec.in* file, adding substitution variables where appropriate, and then added another file reference to the AC_CONFIG_FILES macro. This allows configure to substitute information about the project into the spec file. Listing 15-15 shows the relevant portion of the *xflaim.spec.in* file in bold.

```
Name: @PACKAGE_TARNAME@
BuildRequires: gcc-c++ libstdc++-devel flaimtk-devel gcc-java gjdoc fastjar
mono-core doxygen
Requires: libstdc++ flaimtk mono-core java >= 1.4.2
Summary: XFLAIM is an XML database library.
URL: http://sourceforge.net/projects/flaim/
Version: @PACKAGE_VERSION@
Release: 1
License: GPL
Vendor: Novell, Inc.
Group: Development/Libraries/C and C++
Source: %{name}-%{version}.tar.gz
BuildRoot: %{_tmppath}/%{name}-%{version}-build
--snip--
```

Listing 15-15: x flaim/obs/xflaim.spec.in: *The portion of this file that illustrates using Autoconf variables*

Notice the use of the variables @PACKAGE_TARNAME@ and @PACKAGE_VERSION@ in this listing. Although the tar name is not likely to change much over the life of this project, the version will change often. Without the Autoconf substitution mechanism, I'd have to remember to update this version number whenever I updated the version in the *configure.ac* file. Listing 15-16 shows the *xflaim/obs/Makefile.am* file, which actually does the work of building the RPMs.

```
rpmspec = $(PACKAGE_TARNAME).spec

rpmmacros =\
 --define="_rpmdir $${PWD}"\
 --define="_srcrpmdir $${PWD}"\
 --define="_sourcedir $${PWD}/.."\
 --define="_specdir $${PWD}"\
 --define="_builddir $${PWD}"
```

14. See *http://build.opensuse.org/*. This is a service that I fell in love with almost as soon as it came out. I've had some experience building distro packages, and I can tell you, it's far less painful with the OBS than it is using more traditional techniques. Furthermore, packages built with the OBS can be published automatically on the OBS website (*http://software .opensuse.org/search/*) for public consumption immediately after they're built.

```
RPMBUILD = rpmbuild
RPMFLAGS = --nodeps --buildroot="$${PWD}/_rpm"
```

❶ rpmcheck:
```
        if ! ($(RPMBUILD) --version) >/dev/null 2>&1; then \
          echo "*** This make target requires an rpm-based Linux
distribution."; \
          (exit 1); exit 1; \
        fi

srcrpm: rpmcheck $(rpmspec)
        $(RPMBUILD) $(RPMFLAGS) -bs $(rpmmacros) $(rpmspec)

rpms: rpmcheck $(rpmspec)
        $(RPMBUILD) $(RPMFLAGS) -ba $(rpmmacros) $(rpmspec)

.PHONY: rpmcheck srcrpm rpms
```

Listing 15-16: xflaim/obs/Makefile.am: *The complete contents of this makefile*

Building RPM packages is rather simple, as you can see. The targets provided by this makefile include srcrpm and rpms. The rpmcheck target at ❶ is used internally to verify that RPMs can be built in the end user's environment.

In order to find out which targets in a lower-level *Makefile.am* file are supported by a top-level build, look at the top-level *Makefile.am* file. As Listing 15-17 shows, if the target is not passed down, that target must be intended for internal use only, within the lower-level directory.

```
--snip--
RPM = rpm

rpms srcrpm: dist
❶       (cd obs && $(MAKE) $(AM_MAKEFLAGS) $@) || exit 1
        rpmarch=`$(RPM) --showrc | grep "^build arch" | \
          sed 's/\(.*: \)\(.*\)/\2/'`; \
        test -z "obs/$$rpmarch" || \
          ( mv obs/$$rpmarch/* . && rm -rf /obs/$$rpmarch )
        rm -rf obs/$(distdir)
--snip--
.PHONY: srcrpm rpms
```

Listing 15-17: xflaim/Makefile.am: *If the target is not passed down, it's an internal target.*

As you can see from the command at ❶ in Listing 15-17, when a user targets rpms or srcrpm from the top-level build directory, the commands are recursively passed down to *obs/Makefile.* The remaining commands simply remove droppings left behind by the RPM build process that are simpler to remove at this level. (Try building an RPM package sometime, and you'll see what I mean!)

Notice, too, that both of these top-level makefile targets depend on the dist target because the RPM build process requires the distribution tarball. Adding the tarball as a dependency of the rpms target simply ensures that the distribution tarball is there when the rpmbuild utility needs it.

Summary

While using the Autotools, you have many details to manage—most of which, as they say in the open source software world, *can wait for the next release*! Even as I committed this code to the FLAIM project repository, I noticed details that could be improved. The takeaway lesson here is that a build system is never really finished. It should be incrementally improved over time, as you find time in your schedule to work on it. And it can be rewarding to do so.

I've shown you a number of new features that have not been covered in earlier chapters, and there are many more features that I cannot begin to cover in this book. Study the Autotools manuals to become truly proficient. At this point, it should be pretty simple for you to pick up that additional information yourself.

16

USING THE M4 MACRO PROCESSOR WITH AUTOCONF

*By the time you've sorted out a complicated idea
into little steps that even a stupid machine can deal with,
you've learned something about it yourself.*
—*Douglas Adams,* Dirk Gently's Holistic Detective Agency

The M4 macro processor is simple to use, yet hard to comprehend. The simplicity comes from the fact that it does just one thing very well. I'll wager that you or I could write the base functionality of M4 in a C program in just a few hours. At the same time, two aspects of M4 make it rather difficult to understand immediately.

First, the exceptions introduced by *special cases* that M4 deals with when it processes input text make it hard to grasp all of its rules immediately, though this complexity is easily mastered with time, patience, and practice. Second, the stack-based, pre-order recursive nature of M4's text-processing model is difficult for the human mind to comprehend. Humans tend to process information breadth first, comprehending complete levels of a problem or data set, one level at a time, whereas M4 processes text in a depth-first fashion.

This chapter covers what I consider the bare minimum you need to know to write Autoconf input files. I can't do justice to M4 in a single chapter of this book, so I'll cover just the highlights. For more detail, read the

GNU M4 Manual.[1] If you've already had some experience with M4, try the examples in that manual and then try solving a few text problems of your own using M4. A small amount of such experimentation will vastly improve your understanding of M4.

M4 Text Processing

Like many other classic Unix tools, M4 is written as a standard input/output (stdio) filter. That is, it accepts input from standard input (stdin), processes it, and then sends it to standard output (stdout). Input text is read in as a stream of bytes and converted to *tokens* before processing. Tokens consist of comments, names, quoted strings, and single characters that are not part of a comment, name, or quoted string.

The *default* quote characters are the backtick (`` ` ``) and the single quote (`'`).[2] Use the backtick to start a quoted string and the single quote character to terminate one:

```
`A quoted string'
```

M4 comments are similar to quoted strings in that each one is processed as a single token. Each comment is delimited by a hash mark (#) and a newline (\n) character. Thus, all text following an *unquoted* hash mark, up to and including the next newline character, is considered part of a comment.

Comments are not stripped from the output as they are in other computer language preprocessors, such as the C-language preprocessor. Rather, they are simply passed through without further processing.

The following example contains five tokens: a name token, a space character token, another name token, a second space character token, and finally, a single comment token:

```
Two names # followed by a comment
```

A *name* is any sequence of letters, digits, and underscore characters that does not begin with a digit. Thus, the first line of the following example contains two digit character tokens, followed by a name token, whereas the second line contains only a single name token:

```
88North20th_street
_88North20th_street
```

---

1. See the Free Software Foundation's *GNU M4 - GNU Macro Processor* at *https://www.gnu.org/software/m4/manual/*.

2. Why? Seems rather strange to use such different characters for quotes, but the fact is M4 requires two separate and distinct characters for the open and close quotes in order to perform correctly. When M4 was first written, these two characters looked rather symmetrical with respect to each other; the modern fonts we use today tend to make the choice look silly.

Note that whitespace characters (horizontal and vertical tabs, form feeds, carriage returns, spaces, and newlines) are specifically not part of a name, so whitespace characters may (and often do) act as name or other token delimiters. However, such whitespace delimiters are not discarded by M4, as they often are by a computer language compiler's parser. They're simply passed through from the input stream directly to the output stream without further modification.

Defining Macros

M4 provides a variety of built-in macros, many of which are critical to the proper use of this tool. For instance, it would be very difficult to get any useful functionality out of M4 if it didn't provide a way of defining macros. M4's macro-definition macro is called define.

The define macro is simple to describe:

```
define(macro[, expansion])
```

The define macro expects at least one parameter, even if it's empty. If you supply only one parameter, then instances of the macro name that are found in the input text are simply deleted from the output text:

```
$ m4
define(`macro')

Hello macro world!
❶ Hello  world!
<ctrl-d>$
```

Note in the output text at ❶ that there are two spaces between Hello and world! All tokens except names that map to defined macros are passed from the input stream to the output stream without modification, with one exception: whenever any quoted text outside of comments is read from the input stream, one level of quotes is removed.

Another subtle aspect of the define macro is that its expansion is the empty string. Thus, the output of the preceding definition is simply the trailing carriage return after the definition in the input string.

Names, of course, are candidates for macro expansion. If a name token is found in the symbol table, it is replaced with the macro expansion, as shown in the following example:

```
$ m4
❶ define(`macro', `expansion')
❷
macro ``quoted' macro text'
❸ expansion `quoted' macro text
<ctrl-d>$
```

The second output line at ❸ shows us that the first token (the name macro) is expanded and the outer level of quotes around ``quoted' macro text' are removed by M4. The blank line at ❷ following the macro definition comes from the newline character I entered into the input stream when I pressed the ENTER key after the macro definition at ❶. Since this newline character is not part of the macro definition, M4 simply passes it through to the output stream. This can be a problem when defining macros in input text because you could end up with a slew of blank lines in the output text, one for each macro defined in the input text. Fortunately, there are ways around this problem. For example, I could simply not enter that newline character, as shown here:

```
$ m4
define(`macro', `expansion')macro
expansion
<ctrl-d>$
```

That solves the problem, but it doesn't take a genius to see that this can lead to some readability issues. If you have to define your macros in this manner so that they don't affect your output text, you'll have a few run-on sentences in your input text!

As a solution for this problem, M4 provides another built-in macro called dnl,[3] which causes all input text up to and including the next newline character to be discarded. It's common to find dnl used in *configure.ac*, but it's even more common to find it used in *.m4* macro definition files consumed by Autoconf while processing *configure.ac* files.

Here's an example of the proper use of dnl:

```
$ m4
define(`macro', `expansion')dnl
macro
expansion
<ctrl-d>$
```

There are a few dozen built-in M4 macros, all of which provide functionality that can't be obtained in any other way within M4. Some redefine fundamental behavior in M4.

For example, the changequote macro is used to change the default quote characters from backtick and single quote to whatever you want. Autoconf uses a line like this near the top of the input stream to change the M4 quotes to the left and right square bracket characters, like so:

```
changequote(`[',`]')dnl
```

Why would the Autoconf designers do this? Well, it's quite common in shell code to find unbalanced pairs of single quote characters. In shell code, both backtick and single quote are common in expressions that use the same character to both start and end an expression. This is confusing to M4, which requires open and close quote characters to be distinct from

---

3. The dnl macro name is actually an acronym that stands for "discard to next line."

each other in order to properly process its input stream. You'll recall from Chapter 4 that the input text to Autoconf is shell script, which means that there's a good chance Autoconf will run into an unbalanced pair of M4 quotes in every input file it reads. This can lead to errors that are very difficult to track down, because they have more to do with M4 than they do with Autoconf. It's far less likely that the input shell script will contain an unbalanced pair of square bracket characters.

Macros with Arguments

Macros may also be defined to accept arguments, which may be referenced in the expansion text with $1, $2, $3, and so on. The number of arguments passed can be found in the variable $#, and $@ can be used to pass all arguments of one macro call onto another. When using arguments in a macro call, there can be no intervening whitespace between the macro name and the opening parenthesis. Here's an example of a macro that's defined and then called in various ways:

```
$ m4
define(`with2args', `The $# arguments are $1 and $2.')dnl
❶ with2args
The 0 arguments are  and .
with2args()
The 1 arguments are  and .
❷ with2args(`arg1')
The 1 arguments are arg1 and .
with2args(`arg1', `arg2')
The 2 arguments are arg1 and arg2.
with2args(`arg1', `arg2', `arg3')
The 3 arguments are arg1 and arg2.
❸ with2args (`arg1', `arg2')
The 0 arguments are  and . (arg1, arg2)
<ctrl-d>$
```

In this example, the first call at ❶ is a macro call without arguments. The second call is a macro call with one empty argument. Such calls treat the parameters as if empty arguments were actually passed.[4] In both cases, the macro expands to "The N arguments are and ." (note the double space between the last two words and the space between the last word and the period), but the "N" is 0 in the first call and 1 in the second. Therefore, the empty set of parentheses in the second call carries an empty single argument. The next three calls, beginning at ❷, pass one, two, and three arguments, respectively. As you can see by the resulting outputs of these three calls, parameters in the expansion text that reference missing arguments are treated as empty, while arguments passed without corresponding references are simply ignored.

---

4. Actually, in the call without parentheses, as you can see, $# is zero, while in the call with empty parentheses, $# is one. However, in both cases, both referenced parameters ($1 and $2) contain the empty string.

The last call, at ❸, is a bit different. Notice that it contains a space between the macro name and the opening parenthesis. The initial output of this call is similar to that of the first call, but following that initial output, we find what appears to be a minor variation on the originally intended argument list (the quotes are missing). This is a macro call *without arguments*. Since it's not actually part of the macro call, M4 treats the argument list simply as text on the input stream. Thus, it's copied directly to the output stream, minus one level of quotes.

When passing arguments in macro calls, be aware of whitespace around arguments. The rules are simple: unquoted *leading* whitespace is removed from arguments, and *trailing* whitespace is always preserved, whether quoted or not. Of course, *whitespace* here refers to carriage returns and newline characters, as well as spaces and tabs. Here's an example of calling a macro with variations in leading and trailing whitespace:

```
$ m4
define(`with3args', `The three arguments are $1, $2, and $3.')dnl
❶ with3args(arg1,
          arg2,
          arg3)
The three arguments are arg1, arg2, and arg3.
❷ with3args(arg1
          ,arg2
          ,arg3
          )
The three arguments are arg1
          , arg2
          , and arg3.
<ctrl-d>$
```

In this example, I purposely omitted the quotes around the macro arguments in the calls at ❶ and ❷ in order to reduce confusion. The call at ❶ has only leading whitespace in the form of newlines and tab characters, while the call at ❷ has only trailing whitespace. I'll cover quoting rules shortly, at which point you'll see clearly how quoting affects whitespace in macro arguments.

The Recursive Nature of M4

Now we consider the recursive nature of the M4 input stream. Whenever a name token is expanded by a macro definition, the expansion text is pushed back onto the input stream for complete reprocessing. This recursive reprocessing continues to occur as long as there are macro calls found in the input stream that generate text.

Here's an example:

```
$ m4
define(`macro', `expansion')dnl
macro ``quoted' text'
expansion `quoted' text
<ctrl-d>$
```

Here, I define a macro called *macro* and then present this macro name on the input stream, followed by additional text, some of which is quoted and some of which is double quoted.

The process used by M4 to parse this example is shown in Figure 16-1.

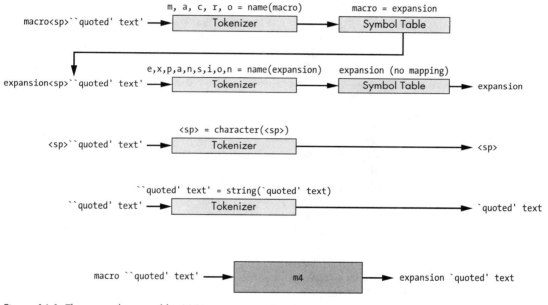

Figure 16-1: The procedure used by M4 to process an input text stream

In the bottom line of the figure, M4 is generating a stream of output text (expansion `quoted' text) from a stream of input text (macro ``quoted' text'). The diagram above this line shows how M4 actually generates the output text from the input text. When the first token (macro) is read in the top line, M4 finds a matching symbol in the symbol table, pushes it onto the input stream on the second line, and then restarts the input stream. Thus, the very next token read is another name token (expansion). Since this name is not found in the symbol table, the text is sent directly to the output stream. The third line sends the next token from the input stream (a space character) directly to the output stream. Finally, in the fourth line, one level of quotes is removed from the quoted text (``quoted' text'), and the result (`quoted' text) is sent to the output stream.

Infinite Recursion

As you might guess, there are some potentially nasty side effects of this process. For example, you can accidentally define a macro that is infinitely recursive. The expansion of such a macro would lead to a massive amount of unwanted output, followed by a stack overflow. This is easy to do:

```
$ m4
define(`macro', `This is a macro')dnl
macro
This is a This is a This is a This is a This is a This is a... <ctrl-c>
$
```

This happens because the macro name expands into text containing the macro's own name, which is then pushed back onto the input stream for reprocessing. Consider the following scenario: *What would have been the result if I'd left the quotes off of the expansion text in the macro definition? What would have happened if I'd added another set of quotes around the expansion text?* To help you discover the answers to these questions, let's turn next to M4 quoting rules.

Quoting Rules

Proper quoting is critical. You have probably encountered situations where your invocations of Autoconf macros didn't work as you expected. The problem is often a case of under-quoting, which means you omitted a required level of quotes around some text.

You see, each time text passes through M4, a layer of quotes is stripped off. Quoted strings are not names and are therefore not subject to macro expansion, but if a quoted string passes through M4 twice, the second time through, it's no longer quoted. As a result, individual words within that string are no longer part of a string but instead are parsed as name tokens, which are subject to macro expansion. To illustrate this, enter the following text at a shell prompt:

```
$ m4
  define(`def', `DEF')dnl
❶ define(`abc', `def')dnl
  abc
  DEF
❷ define(`abc', ``def'')dnl
  abc
  def
❸ define(`abc', ```def''')dnl
  abc
  `def'
  <ctrl-d>$
```

In this example, the first time abc is defined (at ❶), it's quoted once. As M4 processes the macro definition, it removes a layer of quotes. Thus, the expansion text is stored in the symbol table without quotes, but it's pushed back onto the input stream and therefore is transformed into DEF due to the first macro definition.

As you can see, the second definition of abc (at ❷) is double quoted, so when the definition is processed and the outer layer of quotes is stripped off, we would expect the expansion text in the symbol table to contain at least one set of quotes, and it does. Then why don't we see quotes around the output text? Remember that when macros are expanded, the expansion text is pushed onto the front of the input stream and reparsed using the usual rules. Thus, while the text of the second definition is stored quoted in the symbol table, as it's reprocessed upon use, the second layer of quotes is removed between the input and output streams.

The difference between ❶ and ❷ in this example is that the expansion text of ❷ is treated as quoted text by M4, rather than as a potential macro name. The quotes are removed during definition, but the enclosed text is not considered for further expansion because it's still quoted.

In the third definition of abc (at ❸), we finally see the result we were perhaps hoping to obtain: a quoted version of the output text. The expansion text is entered into the symbol table double quoted, because the outermost set of quotes is stripped off during processing of the definition. Then, when the macro is used, the expansion text is reprocessed and the second set of quotes is stripped off, leaving one set in the final output text.

If you keep these rules in mind as you work with macros within Autoconf (including both definitions and calls), you'll find it easier to understand why things may not work the way you think they should. The *GNU M4 Manual* provides a simple rule of thumb for using quotes in macro calls: for each layer of nested parentheses in a macro call, use one layer of quotes.

Autoconf and M4

The autoconf program is a rather simple shell script. About 80 percent of the shell code in the script exists simply to ensure that the shell is functional enough to perform the required tasks. The remaining 20 percent parses command line options. The last line of the script executes the autom4te program, a Perl script that acts as a wrapper around the m4 utility. Ultimately, autom4te calls m4 like this:

```
/usr/bin/m4 --nesting-limit=1024 --gnu --include=/usr/share/autoconf \
--debug=aflq --fatal-warning --debugfile=autom4te.cache/traces.0t \
--trace=AC_CANONICAL_BUILD ... --trace=sinclude \
--reload-state=/usr/share/autoconf/autoconf.m4f aclocal.m4 configure.ac
```

As you can see, the three files that M4 is processing are */usr/share /autoconf/autoconf.m4f*, *aclocal.m4*, and *configure.ac*, in that order.

NOTE   *The .m4f extension on the master Autoconf macro file signifies a frozen M4 input file—a sort of precompiled version of the original .m4 file. When a frozen macro file is processed, it must be specified after a --reload-state option, in order to make M4* *aware that it's not a normal input file. State is built cumulatively within M4 over all input files, so any macros defined by* aclocal.m4, *for instance, are available during the processing of* configure.ac.

The ellipsis between the two --trace options in the command line above is a placeholder for more than 100 such --trace options. It's a good thing the shell can handle long command lines!

The master Autoconf macro file, *autoconf.m4*, merely includes (using the m4_include macro) the other dozen or so Autoconf macro files, in the correct order, and then does a small amount of housekeeping before leaving M4 ready to process user input (via *configure.ac*). The *aclocal.m4* file is our project's macro file, built originally by the aclocal utility or handwritten for projects that don't use Automake. By the time *configure.ac* is processed, the M4 environment has been configured with hundreds of Autoconf macro definitions, which may be called as needed by *configure.ac*. This environment includes not only the recognized AC_* macros but also a few lower layers of Autoconf-provided macros that you may use to write your own macros.

One such lower layer is *m4sugar*,[5] which provides a nice clean namespace in which to define all of the Autoconf macros, as well as several improvements and additions to the existing M4 macros.

Autoconf modifies the M4 environment in a few ways. First, as mentioned earlier, it changes the default quote characters from the backtick and single quote characters to the open and close square bracket characters. In addition, it configures M4 built-in macros such that most are prefixed with m4_, thereby creating a unique namespace for M4 macros. Thus, the M4 define macro becomes m4_define, and so on.[6]

Autoconf provides its own version of m4_define called AC_DEFUN. You should use AC_DEFUN instead of m4_define because it ensures that certain environmental constraints important to Autoconf are in place when your macro is called. The AC_DEFUN macro supports a prerequisite framework, so you can specify which macros are required to have been called before your macro may be called. This framework is accessed by using the AC_REQUIRE macro to indicate your macro's requirements at the beginning of your macro definition, like so:

```
# Test for option A
# ----------------
AC_DEFUN([TEST_A],
[AC_REQUIRE([TEST_B])dnl
test "$A" = "yes" && options="$options A"])
```

The rules for writing Autoconf macros using AC_DEFUN and the prerequisite framework are outlined in Chapter 9 of the *GNU Autoconf Manual*. Before you write your own macros, read Chapters 8 and 9 of that manual.

---

5. This is a hybrid palindromic acronym: *Readability And Greater Understanding Stands 4 M4Sugar.*

6. A notable exception is dnl. This macro is thankfully not renamed to m4_dnl.

Writing Autoconf Macros

Why would we want to write Autoconf macros in the first place? One reason is that a project's *configure.ac* file might contain several instances of similar sets of code and we need the configure script to perform the same set of high-level operations on multiple directories or file sets. By converting the process into a macro, we reduce the number of lines of code in the *configure.ac* file, thereby reducing the number of possible points of failure. Another reason might be that an easily encapsulated bit of *configure.ac* code could be useful in other projects, or even to other people.

NOTE *The GNU Autoconf Archive provides many sets of related macros to solve common Autoconf problems. Anyone may contribute to the archive by emailing their macros to the project maintainer. There are frequent tarball releases available for free from the project website.[7]*

Simple Text Replacement

The simplest type of macro is one that replaces text verbatim, with no substitutions. An excellent example of this is found in the FLAIM project, where the flaim, xflaim, and sql projects' configure scripts attempt to locate the ftk (FLAIM toolkit) project library and header file. Since I already discussed the operation of this code in Chapter 14, I'll only cover it briefly here as it relates to writing Autoconf macros, but I provide the relevant bit of *configure.ac* code in Listing 16-1 for convenience.[8]

```
--snip--
# Configure FTKLIB, FTKINC, FTK_LTLIB and FTK_INCLUDE
AC_ARG_VAR([FTKLIB], [The PATH wherein libflaimtk.la can be found.])
AC_ARG_VAR([FTKINC], [The PATH wherein flaimtk.h can be found.])

# Ensure that both or neither FTK paths were specified.
if { test -n "$FTKLIB" && test -z "$FTKINC"; } || \
   { test -z "$FTKLIB" && test -n "$FTKINC"; }; then
  AC_MSG_ERROR([Specify both FTKINC and FTKLIB, or neither.])
fi

# Not specified? Check for FTK in standard places.
if test -z "$FTKLIB"; then
  # Check for FLAIM toolkit as a sub-project.
  if test -d "$srcdir/ftk"; then
    AC_CONFIG_SUBDIRS([ftk])
    FTKINC='$(top_srcdir)/ftk/src'
    FTKLIB='$(top_builddir)/ftk/src'
  else
    # Check for FLAIM toolkit as a superproject.
```

---

7. See the GNU Autoconf Archive at *https://www.gnu.org/software/autoconf-archive/*.

8. Note that the FLAIM git repository for Chapters 14 and 15 includes changes made by this chapter as well. So in this chapter, when I refer to a snippet of *configure.ac* from those chapters, I mean it literally—from a code listing in the book, rather than from the repository.

```
       if test -d "$srcdir/../ftk"; then
         FTKINC='$(top_srcdir)/../ftk/src'
         FTKLIB='$(top_builddir)/../ftk/src'
       fi
     fi
fi

# Still empty? Check for *installed* FLAIM toolkit.
if test -z "$FTKLIB"; then
  AC_CHECK_LIB([flaimtk], [ftkFastChecksum],
    [AC_CHECK_HEADERS([flaimtk.h])
     LIBS="-lflaimtk $LIBS"],
    [AC_MSG_ERROR([No FLAIM toolkit found. Terminating.])])
fi

# AC_SUBST command line variables from FTKLIB and FTKINC.
if test -n "$FTKLIB"; then
  AC_SUBST([FTK_LTLIB], ["$FTKLIB/libflaimtk.la"])
  AC_SUBST([FTK_INCLUDE], ["-I$FTKINC"])
fi
--snip--
```

Listing 16-1: xflaim/configure.ac: The ftk search code from the xflaim project

This code is identical in flaim, xflaim, and sql, though it may be modified in the future for one reason or another, so keeping it embedded in all three *configure.ac* files is redundant and error prone.

Even if we were to convert this code to a macro, we'd still have to put a copy of the macro file into each of the projects' *m4* directories. However, we could later edit only one of these macro files and copy it from the authoritative location into the other projects' *m4* directories, or even use symlinks in git rather than copies to ensure there is truly only a single copy of the *.m4* file. This would be a better solution than having all of the code embedded in all three *configure.ac* files.

By converting this code to a macro, we can keep it in one place where portions of it cannot be confused for code that is not related to the process of locating the FLAIM toolkit library and header file. This happens quite often during later maintenance of a project's *configure.ac* file, as additional code designed for other purposes is dropped between chunks of code belonging to sequences like this.

Let's try converting this code into a macro. Our first attempt might look like Listing 16-2. (I've omitted a large chunk in the middle that is identical to the original code, for the sake of brevity.)

```
AC_DEFUN([FLM_FTK_SEARCH],
[AC_ARG_VAR([FTKLIB], [The PATH wherein libflaimtk.la can be found.])
AC_ARG_VAR([FTKINC], [The PATH wherein flaimtk.h can be found.])
--snip--
# AC_SUBST command line variables from FTKLIB and FTKINC.
if test -n "$FTKLIB"; then
  AC_SUBST([FTK_LTLIB], ["$FTKLIB/libflaimtk.la"])
```

```
    AC_SUBST([FTK_INCLUDE], ["-I$FTKINC"])
fi])
```

Listing 16-2: xflaim/m4/flm_ftk_search.m4: A first attempt at encapsulating ftk search code

In this pass, I've simply cut and pasted the entire *configure.ac* code sequence verbatim into the macro-body argument of a call to AC_DEFUN. The AC_DEFUN macro is defined by Autoconf and provides some additional functionality over the m4_define macro provided by M4. This additional functionality is strictly related to the prerequisite framework provided by Autoconf.

NOTE *Be aware that AC_DEFUN must be used (rather than m4_define) in order for the macro definition to be found by aclocal in your external macro definition files. You must use AC_DEFUN if your macro definitions are in external files, but for simple macros defined within* configure.ac *itself, you can use* m4_define.

Notice the use of M4 quoting around both the macro name (FLM_FTK _SEARCH) and the entire macro body. To illustrate the problems with not using these quotes in this example, consider how M4 would process the macro definition without the quotes. If the macro name were left unquoted, not much damage would be done, unless the macro happened to already be defined. If the macro were already defined, M4 would treat the macro name as a call with no parameters, and the existing definition would replace the macro name as M4 was reading the macro definition. (In this case, because of the unique name of the macro, there's not much chance that it's already defined, so I could have left the macro name unquoted with little effect, but it's good to be consistent.)

On the other hand, the macro body contains a fair amount of text and even Autoconf macro calls. Had we left the body unquoted, these macro calls would be expanded during the reading of the definition rather than during the later use of the macro, as we had intended.

Because the quotes are present, M4 stores the macro body as provided, with no additional processing during the reading of the definition other than to remove the outermost layer of quotes. Later, when the macro is called, the body text is inserted into the input stream in place of the macro call, with one layer of quotes removed, and only then are the embedded macros expanded.

This macro requires no arguments because the same text is used identically in all three *configure.ac* files. The effect on *configure.ac* is to replace the entire chunk of code with the name of the macro, as shown in Listing 16-3.

```
--snip--
# Add jni.h include directories to include search path
AX_JNI_INCLUDE_DIR
for JNI_INCLUDE_DIR in $JNI_INCLUDE_DIRS; do
  CPPFLAGS="$CPPFLAGS -I$JNI_INCLUDE_DIR"
done
```

❶ # Configure FTKLIB, FTKINC, FTK_LTLIB, and FTK_INCLUDE
FLM_FTK_SEARCH

```
# Check for Java compiler.
--snip--
```

Listing 16-3: xflaim/configure.ac: Replacing the ftk search code with the new macro call

When writing a macro from existing code, consider the inputs to the existing chunk of code and the outputs provided by the code. Inputs will become possible macro arguments, and outputs will become documented effects. In Listing 16-3, we have no inputs and thus no arguments, but what are the documentable effects of this code?

The comment at ❶ over the macro call in Listing 16-3 alludes to these effects. The FTKLIB and FTKINC variables are defined, and the FTK_LTLIB and FTK_INCLUDE variables are defined and substituted using AC_SUBST.

Documenting Your Macros

A proper macro definition provides a header comment that documents possible arguments, results, and potential side effects of the macro, as shown in Listing 16-4.

```
# FLM_FTK_SEARCH
# --------------
# Define AC_ARG_VAR (user variables), FTKLIB, and FTKINC,
# allowing the user to specify the location of the flaim toolkit
# library and header file. If not specified, check for these files:
#
#   1. As a sub-project.
#   2. As a super-project (sibling to the current project).
#   3. As installed components on the system.
#
# If found, AC_SUBST FTK_LTLIB and FTK_INCLUDE variables with
# values derived from FTKLIB and FTKINC user variables.
# FTKLIB and FTKINC are file locations, whereas FTK_LTLIB and
# FTK_INCLUDE are linker and preprocessor command line options.
#
# Author:   John Calcote <john.calcote@gmail.com>
# Modified: 2009-08-30
# License:  AllPermissive
#
AC_DEFUN([FLM_FTK_SEARCH],
--snip--
```

Listing 16-4: xflaim/m4/flm_ftk_search.m4: Adding a documentation header to the macro definition

This header comment documents both the effects of this macro and the way it operates, giving the user a clear picture of the sort of functionality they'll get when they call it. The *GNU Autoconf Manual* indicates that such macro definition header comments are stripped from the final output; if you search the configure script for some text in the comment header, you'll see that it's missing.

Regarding coding style, the *GNU Autoconf Manual* suggests that it is good macro definition style to place the macro body's closing square-bracket quote and the closing parenthesis alone on the last line of the macro definition, along with a comment containing only the name of the macro being defined, as shown in Listing 16-5.

```
--snip--
AC_SUBST([FTK_INCLUDE], ["-I$FTKINC"])
❶ fi[]dnl
])# FLM_FTK_SEARCH
```

Listing 16-5: xflaim/m4/flm_ftk_search.m4: Suggested macro body closing style

The *GNU Autoconf Manual* also suggests that, if you don't like the extra carriage return that the use of this format adds to the generated `configure` script, you can append the text `[]dnl` to the last line of the macro body, as shown at ❶ in Listing 16-5. The use of `dnl` causes the trailing carriage return to be ignored, and the open and close square brackets are simply empty Autoconf quotes that are stripped out during processing of later macro calls. The quotes (square brackets) are used to separate `fi` and `dnl` so they're recognized by M4 as two separate words.

NOTE   *The* GNU Autoconf Manual *defines a very complete naming convention for macros and their containing files. I've chosen simply to prefix all macro names and their containing files that are strictly related to the project with a project-specific prefix—in this case,* FLM_ (flm_)*.*

M4 Conditionals

Now that you know how to write basic M4 macros, we'll consider what it means to allow M4 to decide which text should be used to replace your macro call, based on arguments passed in the call.

Calling a Macro with and Without Arguments

Take a look at Listing 16-6, which is my first attempt at writing the `FLM_PROG _TRY_DOXYGEN` macro that was first used in Chapter 14. This macro was designed with an optional argument, which isn't apparent from its use in Chapter 14 because the FLAIM code called the macro without arguments. Let's examine the definition of this macro. In the process, we'll discover what it means to call it with and without arguments.

```
# FLM_PROG_TRY_DOXYGEN([quiet])
# -----------------------------
# FLM_PROG_TRY_DOXYGEN tests for an existing doxygen source
# documentation program. It sets or uses the environment
# variable DOXYGEN.
#
# If no arguments are given to this macro, and no doxygen
# program can be found, it prints a warning message to STDOUT
# and to the config.log file. If the quiet argument is passed,
# then only the normal "check" line is displayed. Any other-token
# argument is considered by autoconf to be an error at expansion
```

```
# time.
#
# Makes the DOXYGEN variable precious to Autoconf. You can
# use the DOXYGEN variable in your Makefile.in files with
# @DOXYGEN@.
#
# Author: John Calcote <john.calcote@gmail.com>
# Modified: 2009-08-30
# License: AllPermissive
#
AC_DEFUN([FLM_PROG_TRY_DOXYGEN],
❶ [AC_ARG_VAR([DOXYGEN], [Doxygen source doc generation program])dnl
❷ AC_CHECK_PROGS([DOXYGEN], [doxygen])
❸ m4_ifval([$1],,
❹ [if test -z "$DOXYGEN"; then
    AC_MSG_WARN([doxygen not found - continuing without Doxygen support])
  fi])
])# FLM_PROG_TRY_DOXYGEN
```

Listing 16-6: ftk/m4/flm_prog_try_doxygen.m4: A first attempt at FLM_PROG_TRY_DOXYGEN

First, we see a call to the AC_ARG_VAR macro at ❶, which is used to make the DOXYGEN variable precious to Autoconf. Making a variable precious causes Autoconf to display it within the configure script's help text as an influential environment variable. The AC_ARG_VAR macro also makes the specified variable an Autoconf substitution variable. At ❷, we come to the heart of this macro—the call to AC_CHECK_PROGS. This macro checks for a doxygen program in the system search path, but it only looks for the program (passed in the second argument) if the variable (passed in the first argument) is empty. If this variable is not empty, AC_CHECK_PROGS assumes that the end user has already specified the proper program in the variable in the user's environment, and it does nothing. In this case, the DOXYGEN variable is populated with *doxygen* if the doxygen program is found in the system search path. In either case, a reference to the DOXYGEN variable is substituted into template files by Autoconf. (Since we just called AC_ARG_VAR on DOXYGEN, this step is redundant but harmless.)

The call to m4_ifval at ❸ brings us to the point of this section. This is a conditional macro defined in Autoconf's *m4sugar* layer—a layer of simple macros designed to make writing higher-level Autoconf macros easier. M4 conditional macros are designed to generate one block of text if a condition is true and another if the condition is false. The purpose of m4_ifval is to generate text based on whether its first argument is empty. If its first argument is not empty, the macro generates the text in its second argument. If its first argument is empty, the macro generates the text in its third argument.

The FLM_PROG_TRY_DOXYGEN macro works with or without an argument. If no arguments are passed, FLM_PROG_TRY_DOXYGEN will print a warning message that the build is continuing without Doxygen support if the doxygen program is not in the system search path. On the other hand, if the quiet option is passed to FLM_PROG_TRY_DOXYGEN, no message will be printed if the doxygen program is not found.

In Listing 16-6, m4_ifval generates no text (the second argument is empty) if the first argument contains text. The first argument is $1, which refers to the contents of the first argument passed to FLM_PROG_TRY_DOXGEN. If no arguments are given to our macro, $1 will be empty, and m4_ifval will generate the text in its third argument shown at ❹. On the other hand, if we pass quiet (or any text, for that matter) to FLM_PROG_TRY_DOXYGEN, $1 will contain quiet, and m4_ifval will generate nothing.

The shell code in the third argument (at ❹) checks to see if the DOXYGEN variable is still empty after the call to AC_CHECK_PROGS. If it is, it calls AC_MSG_WARN to display a configuration warning.

Adding Precision

Autoconf provides a macro called m4_if, a renamed version of the M4 built-in ifelse macro. The m4_if macro is similar in nature to *m4sugar's* m4_ifval. Listing 16-7 shows how we might use ifelse in place of m4_ifval, if we didn't have *m4sugar* macros to work with.

```
--snip--
ifelse ([$1],,
[if test -z "$DOXYGEN"; then
AC_MSG_WARN([Doxygen program not found - continuing without Doxygen])
fi])
--snip--
```

Listing 16-7: Using ifelse instead of m4_ifval

The macros appear to be identical in function, but this appearance is only circumstantial; the parameters are used differently. In this case, if the first argument ($1) is the same as the second argument (the empty string), the contents of the third argument ([if test -z ...]) are generated. Otherwise, the contents of the fourth (nonexistent) argument are generated because omitted arguments are treated as if the empty string had been passed. Therefore, the following two macro invocations are identical:

```
m4_ifval([$1],[a],[b]])
ifelse([$1],[],[b],[a])
```

FLM_PROG_TRY_DOXYGEN treats any text in its argument as if quiet was passed. In order to facilitate future enhancements to this macro, we should limit the allowed text in this argument to something that makes sense; otherwise, users could abuse this parameter and we'd be stuck supporting whatever they pass for the sake of backward compatibility. The m4_if macro can help us out here. This macro is quite powerful because it accepts an unlimited number of arguments. Here are its basic prototypes:

```
m4_if(comment)
m4_if(string-1, string-2, equal[, not-equal])
m4_if(string-1, string-2, equal-1, string-3, string-4, equal-2,
    ...[, not-equal])
```

If only one parameter is passed to m4_if, that parameter is treated as a comment because there's not much that m4_if can do with one argument. If three or four arguments are passed, the description I gave for ifelse in Listing 16-7 is also accurate for m4_if. However, if five or more arguments are passed, the fourth and fifth become the comparison strings for a second else-if clause. The last argument in an arbitrarily long set of triples is generated if the last two comparison strings are different.

We can use m4_if to ensure that quiet is the only acceptable option in the list of options accepted by FLM_PROG_TRY_DOXYGEN. Listing 16-8 shows one possible implementation.

```
--snip--
m4_if([$1],,
[if test -z "$DOXYGEN"; then
  AC_MSG_WARN([doxygen not found - continuing without Doxygen support])
fi], [$1], [quiet],, [m4_fatal([Invalid option in FLM_PROG_TRY_DOXYGEN])])
--snip--
```

Listing 16-8: Restricting the argument options allowed by FLM_PROG_TRY_DOXYGEN

In this case, we want a message to be printed if doxygen is missing in all cases except when the quiet option is given as the first argument passed into our macro. In Listing 16-8, I've given FLM_PROG_TRY_DOXYGEN the ability to detect cases when something other than quiet or the empty string is passed in this parameter and to do something specific in response. Listing 16-9 shows the resulting pseudocode generated by the expansion of FLM_PROG_TRY_DOXYGEN.

```
if $1 == '' then
    Generate WARNING if no doxygen program is found
else if $1 == 'quiet' then
    Don't generate any messages
else
    Generate a fatal "bad parameter" error at autoconf (autoreconf) time
end
```

Listing 16-9: Pseudocode for Listing 16-8's use of the m4_if macro

Let's examine exactly what's going on in Listing 16-8. If arguments one ([$1]) and two ([]) are the same, a warning message is generated when doxygen is not found. If arguments four ([$1]) and five ([quiet]) are the same, nothing is generated; otherwise, arguments four and five are different, and a fatal error (via m4_fatal) is generated by Autoconf when it's executed against the calling *configure.ac* file. It's very simple, once you see how it works *and* once you get the bugs worked out—which brings us nicely to our next topic.

Diagnosing Problems

One of the most significant stumbling blocks that people run into at this point is not so much a lack of understanding of how these macros work but a lack of attention to detail. There are several places where things can go wrong when writing even a simple macro like this. For example, you might have any of the following problems:

- Space between a macro name and the opening parenthesis
- Unbalanced brackets or parentheses
- The wrong number of parameters
- A misspelled macro name
- Incorrectly quoted arguments to a macro
- A missing comma in a macro's parameter list

M4 is rather unforgiving of such mistakes. Worse, its error messages can be even more cryptic than those of make.[9] If you get strange errors and you think your macro should be working, your best diagnostic method is to scan the definition very carefully looking for the preceding conditions. These mistakes are easy to make, and in the end most problems come down to some combination of them.

Another very useful debugging tool is the m4_traceon and m4_traceoff macro pair. The macro signatures are as follows:

```
m4_traceon([name, ...])
m4_traceoff([name, ...])
```

All arguments are optional. When given, the arguments should be a comma-separated list of macro names you'd like M4 to print to the output stream as these names are encountered in the input stream. If you omit the arguments, M4 will print the name of every macro it expands.

A typical trace session in M4 looks something like this:

```
$ m4
define(`abc', `def')dnl
define(`def', `ghi')dnl
traceon(`abc', `def')dnl
abc
❶ m4trace: -1- abc
m4trace: -1- def
```

---

9. The reason for such cryptic messages in both make and M4 is that it's very difficult for these programs to determine the proper context for an error, if the parsing context is drastically different with and without the error. In make, for example, a missing tab character on a command is problematic simply because commands are only commands by virtue of the tab character. Without it, the line looks to make like some other type of construct—perhaps a rule or a macro definition. The same is true of M4. When a comma is missing, for instance, M4 has so little context to go on that it appears as if two intended parameters are simply one parameter. M4 doesn't even complain—it simply processes the errant call as if there were one less parameter than intended (but it has no way of knowing the caller's intention).

```
ghi
traceoff(`abc', `def')dnl
```
❷ `m4trace: -1- traceoff`
`<ctrl-d>$`

The number between dashes in the output lines at ❶ and ❷ indicates the nesting level, which is usually 1. The value of the trace facility is that you can easily see when the traced macros are expanded within the context of the output text generated. The M4 tracing facility can also be enabled from the command line with the -t or --trace option:

```
$ m4 --trace=abc
```

Or more appropriately for this discussion:

```
$ autoconf --trace=FLM_PROG_TRY_DOXYGEN
```

The latter has the added benefit of allowing you to specify a format for the trace output. For more insight into the use of the format portion of the option, try entering autom4te --help at the command prompt. For more information on the use of the M4 trace options, refer to Chapter 7 (specifically, Section 7.2) of the *GNU M4 Manual*.

NOTE *The Autotools rely heavily on tracing for more than just debugging. Various of the Autotools and their supporting utilities use traces on* configure.ac *to gather information used in other stages of the configuration process. (Recall the 100+ trace options on the* m4 *command line.) For more information on tracing within Autoconf, refer to Section 3.4 of the* GNU Autoconf Manual, *titled "Using* autoconf *to Create* configure.*"*

Summary

Using M4 is deceptively complex. On the surface it appears simple, but as you get deeper into it, you find ways of using it that almost defy comprehension. Nonetheless, the complexities are not insurmountable. As you become proficient with M4, you'll find that your way of thinking about certain problems changes. It's worth gaining some M4 proficiency for that reason alone. It's like adding a new tool to your software-engineering toolbox.

A powerful M4 concept I did not cover, but that you should be aware of, is *iteration*. Normally, we think of iteration in terms of loops, but M4 has no actual looping constructs. Rather, iteration is managed through recursion. For details, refer to the manual's discussion of the forloop and foreach macros.

Because the very foundation of Autoconf is M4, becoming proficient with M4 will give you more insight into Autoconf than you might think. The more about M4 you know, the more about Autoconf you'll understand at a glance.

17

USING THE AUTOTOOLS
WITH WINDOWS

*"Well, Steve, I think there's more than one way of looking at it. I think
it's more like we both had this rich neighbor named Xerox and I broke
into his house to steal the TV set and found out that you had already
stolen it."* —Bill Gates, quoted in Steve Jobs *by Walter Isaacson*

Autoconf generates configure scripts containing hundreds of lines of Bourne shell
code. If that statement doesn't make you
wonder how we could ever use the Autotools
with Windows, you should probably re-read it until it
does. In fact, the only way Autoconf *can* be used is
with an actual Bourne shell and a subset of Unix tools like grep, awk, and sed.
So before we can even get started, we need to ensure that we have a proper
execution environment.

When I started working on the first edition of this book, there were few
options that provided the required environment for building Windows software with the Autotools. During the last 10 years, that story has changed.
Today, an entire gamut of options is available to developers, depending on
whether your goal is to build Windows applications on Linux or Windows.

In the last decade, Windows has been viewed by the GNU community
as a more important target than it has in the past. Significant efforts have
been made recently to ensure that GNU source code at least considers

Windows as a target environment. This attitude shift has provided important source-level support for making Cygwin and its sibling environments manage clean ports of GNU packages to Windows.

Environment Options

Since our goal is to build native Windows software using GNU tools, including specifically the Autotools, we're naturally going to have to consider systems that provide various levels of POSIX environment functionality.

At one end of the spectrum, we have actual Linux installations, which may take any one of several forms, including bare-metal dedicated machine installations and virtual machines running on KVM, Xen, or VMware ESX servers or on a Windows machine running Microsoft HyperV, VMware Workstation, or Oracle's VirtualBox. There are also Mac options for running virtual machines, and macOS itself provides a reasonably POSIX-compliant environment. We could also use Windows Subsystem for Linux (WSL).

A full Linux installation obviously provides the most POSIX-compliant environment for building software using GNU tools. To actually generate Windows software on a Linux system, we have to configure a cross-compile. That is, we have to build software that's not designed to run on the build system.

At the other end of that spectrum, we have various POSIX environment emulators running within Windows applications. The "application" in these cases is almost always a Bash shell running in some sort of shell host process or terminal, but these environments are more or less compatible with a true Linux build environment. The flavors we have to pick from today include Cygwin, MinGW, and MSys2.

A final option—and one we won't spend much time on—is that of cross-compiling Windows software on other types of systems, including mainframes and supercomputers. If you want to see a Windows program compile fast, you should watch it happen on a Cray XC50 with an SSD or RAM disk. Since GNU software can run on pretty much any Unix system that has a Bourne shell, we can cross-compile software on it for any platform, including Windows. After you've cross-compiled on Linux, moving the process to a different POSIX-compliant platform is relatively simple.

Tool Chain Options

Once we've chosen an environment, we'll then need to select a tool chain in which to build our native software for Windows. Generally speaking, the environment you choose limits your tool chain options. For example, if you select a full Linux installation, your only tool chain option is to install a cross-compiler for Windows—probably *mingw-w64*. Don't knock it until you've tried it—this is a really good option because it does a pretty reasonable job of building Windows software.

The biggest problem you'll find here is the inconvenience of having to copy your software over to a Windows system in order to test it. In fact, running tests as part of your build is pretty much a nonstarter, as you can't execute your products on your build machine.[1] I've seen such cross-compilation testing done by having a remote copy and execution stage as part of the build system's test phase, but doing this tends to make your build brittle because it requires additional environment configuration that's not part of a normal package build process.

Getting Started

I'll present a full cross-section of options for building Windows software using GNU tools. We'll start by using a Windows cross-compiler tool chain on native Linux and then check out Windows Subsystem for Linux, and finally move on to the remaining Windows-based options, presented in the order they were created. We'll first check out Cygwin on a Windows 10 system. Next, we'll try MinGW and finally finish up with MSys2. By the time you reach the end of this chapter, you should be very comfortable with these processes.

For Windows-based systems, I'll presume you're running a reasonably recent copy of Windows 10. I installed *Windows 10 Build 1803* (released April 30, 2018) in a virtual machine under Oracle's VirtualBox on my Linux Mint system. You can take this path, or you can use a "bare metal" (nonvirtual) installation of Windows 10. The manner in which you choose to run Windows and, to a lesser extent, the exact version you choose to run are really not significant issues here.

NOTE   *The majority of this book centers on the use of free and open source software (FOSS). Microsoft Windows is, of course, not free software. You should pay for any copy of Windows—or any other non-free software—you choose to use.[2]*

I've also installed Git for Windows[3] on my Windows system and cloned the b64 project from Chapter 13 and Gnulib from the Savannah Git server. We won't be making any significant changes to the b64 project source code, except to make it work in a given environment where necessary.

When you install Git for Windows, you'll have the option of downloading a 32- or 64-bit version in one of two varieties—as an installer or as a portable package. The installer style installs Git on your Windows system in the usual fashion and may be uninstalled from the Windows installed-programs panel.

---

1. You might also consider using Wine, a Windows execution environment on Linux. Wine is, in many respects, a counterpart to the Linux emulation environments I spend the better part of this chapter discussing. Since Wine is purely a reverse-engineering effort, it has its own set of problems in the form of Win32 API emulation bugs, so attempting to test Windows software under Wine can be more problematic than simply copying the program over to a real Windows system. Still, it's worth considering. For details, see *https://www.winehq.org/*.

2. Of course, there's nothing wrong with taking full advantage of free trial periods offered by software vendors.

3. See *https://git-scm.com/download/win*.

The portable style requires no installation and can be executed directly from its expanded archive. Select an installer or a portable package option for your Windows system.

If you chose to use an installer, during the installation process you'll be asked how you want Git to treat your source file line endings. I generally avoid the first option, which is to "check out" using Windows-style line endings but "commit" using Unix-style line endings. You might want to use this option if you're planning to use Notepad as your editor (not advisable). I generally select the option to check out and commit as is. Git has no business modifying your source files as they pass through it. Just configure your editor to recognize and manage line endings the way you like.

Cross-Compiling for Windows on Linux

Since we're already running Linux, let's start our investigation of the options right here at home.

Installing a Windows Cross Tool Chain

The first thing we'll need to do is install a Windows cross-compiler tool chain (often referred to simply as a "cross tool chain" or as "cross tools") on our Linux system. The most widely available one is mingw-w64 for Linux, which can build native Windows programs and libraries that look very much like they were generated by Microsoft tools.

On my Linux Mint system, I searched the internet for *Linux Mint mingw-w64*; the top result was my goal. You can generally use your system's package manager to find and install this package because mingw-w64 is pretty popular. On CentOS and other Red Hat–based systems, try yum search mingw-w64. For Debian-based systems like Ubuntu and Mint, try apt-cache search mingw-w64.

Be aware when you run these package searches that you may get back a long result list composed of a few dozen real packages and one or two meta packages. It's better to select one of the meta packages so you get all of the required real packages in one shot. I highly recommend you search the internet for your distro name and *mingw-w64* in order to get some background on which package you should install using your package manager. A little research up front can save you a lot of headache later.

For example, on my Debian-based system, I got these results from an apt-cache search:

```
$ apt-cache search mingw-w64
--snip--
g++-mingw-w64 - GNU C++ compiler for MinGW-w64
g++-mingw-w64-i686 - GNU C++ compiler for MinGW-w64 targeting Win32
g++-mingw-w64-x86-64 - GNU C++ compiler for MinGW-w64 targeting Win64
gcc-mingw-w64 - GNU C compiler for MinGW-w64
gcc-mingw-w64-base - GNU Compiler Collection for MinGW-w64 (base package)
gcc-mingw-w64-i686 - GNU C compiler for MinGW-w64 targeting Win32
gcc-mingw-w64-x86-64 - GNU C compiler for MinGW-w64 targeting Win64
```

```
--snip--
mingw-w64 - Development environment targeting 32- and 64-bit Windows
mingw-w64-common - Common files for Mingw-w64
mingw-w64-i686-dev - Development files for MinGW-w64 targeting Win32
mingw-w64-tools - Development tools for 32- and 64-bit Windows
mingw-w64-x86-64-dev - Development files for MinGW-w64 targeting Win64
--snip--
$
```

The actual results list contained dozens of entries, but according to a quick internet search, I found the only package I really needed to install was *mingw-w64* (highlighted); a meta-package referencing actual packages that install the GCC C and C++ compilers for generating 32- and 64-bit Windows software; and a *binutils* package containing the librarian, linker, and other common development tools. Some package management systems divide this set of packages differently, allowing you the option of installing gcc and g++ separately or of installing 32-bit and 64-bit code generators separately. Installing this package on my system displays the following output:

```
$ sudo apt-get install mingw-w64
[sudo] password for jcalcote:
Reading package lists... Done
Building dependency tree
Reading state information... Done
The following additional packages will be installed:
  binutils-mingw-w64-i686 binutils-mingw-w64-x86-64 g++-mingw-w64 g++-
mingw-w64-i686 g++-mingw-w64-x86-64 gcc-mingw-w64 gcc-mingw-w64-base
gcc-mingw-w64-i686 gcc-mingw-w64-x86-64
  mingw-w64-common mingw-w64-i686-dev mingw-w64-x86-64-dev
Suggested packages:
  gcc-7-locales wine wine64
The following NEW packages will be installed:
  binutils-mingw-w64-i686 binutils-mingw-w64-x86-64 g++-mingw-w64 g++-
mingw-w64-i686 g++-mingw-w64-x86-64 gcc-mingw-w64 gcc-mingw-w64-base
gcc-mingw-w64-i686 gcc-mingw-w64-x86-64
  mingw-w64 mingw-w64-common mingw-w64-i686-dev mingw-w64-x86-64-dev
0 upgraded, 13 newly installed, 0 to remove and 31 not upgraded.
Need to get 127 MB of archives.
After this operation, 744 MB of additional disk space will be used.
Do you want to continue? [Y/n] Y
--snip--
$
```

Testing the Build

Once you've found and installed the proper cross tool chain, you're ready to start building Windows software. I've chosen something simple, but not trivial—the b64 project from Chapter 13. It uses Gnulib, so it has a convenience library. Gnulib aims for portability, so we can assess how good it is with Windows portability, at least for the few modules b64 uses.

To build for another platform, you need to configure the project for cross-compilation. For a full explanation of cross-compiling using the

Autotools, see Item 6 in Chapter 18. For now, just be aware that the configuration options you'll need are --build and --host. The first option describes the system on which you'll be building the software, and the second option describes the system on which the generated software will be executed.

In order to discover our build platform, we can run the config.guess script installed into the root of our project by automake (via autoreconf). To do this, we'll need to bootstrap the project for a regular build so that config.guess gets installed.[4] Let's do that within the *b64* directory itself:

```
$ cd b64
$ ./bootstrap.sh
Module list with included dependencies (indented):
    absolute-header
  base64
    extensions
    extern-inline
--snip--
configure.ac:12: installing './compile'
configure.ac:20: installing './config.guess'
configure.ac:20: installing './config.sub'
configure.ac:6: installing './install-sh'
configure.ac:6: installing './missing'
Makefile.am: installing './depcomp'
$
$ ./config.guess
x86_64-pc-linux-gnu
$
```

Running config.guess is how configure determines the default value to use for the --build option, so it will always be correct. Determining the value we should use for the --host option is just a bit more difficult. We need to find the prefix on our cross tool chain, because the --host option value is what configure uses to find the correct tool chain and to set up our CC and LD variables.

This can be done in a few different ways. You can use your system's package manager to determine what files were installed when you installed the mingw-w64 meta package, or you can look in your */usr/bin* directory to see what the compiler is named—this usually works, and actually *does* work for this tool chain, but sometimes cross tool chains are installed into a completely different directory, so I'll use my package manager. Your package manager has similar options, but you can follow along directly with my usage if you happen to be on a Debian-based system:

```
$ dpkg -l | grep mingw
ii  binutils-mingw-w64-i686 ...
ii  binutils-mingw-w64-x86-64 ...
--snip--
ii  gcc-mingw-w64-i686 ...
ii  gcc-mingw-w64-x86-64 ... GNU C compiler for MinGW-w64 targeting Win64
```

---

4. You can also run config.guess from its installed location at */usr/share/automake-1.15*.

```
ii  mingw-w64 ... Development environment targeting 32- and 64-bit Windows
--snip--
$ dpkg -L gcc-mingw-w64-x86-64
--snip--
/usr/lib/gcc/x86_64-w64-mingw32
/usr/lib/gcc/x86_64-w64-mingw32/5.3-win32
/usr/lib/gcc/x86_64-w64-mingw32/5.3-win32/libgcc_s_seh-1.dll
/usr/lib/gcc/x86_64-w64-mingw32/5.3-win32/libgcov.a
--snip--
$
```

The first command lists all of the installed packages on my system and filters the list through grep, searching for anything associated with *mingw*. The equivalent rpm command on Red Hat–based systems would be rpm -qa | grep mingw. The package I'm looking for will be related to the GCC C compiler and x86_64 development. It will likely look very similar, if not exactly the same, on your system.

The second command lists the files installed by that package. Here, I'm looking for the compiler, standard C library, headers, and other target-specific files. The equivalent rpm command would be rpm -ql mingw64-gcc.[5] The tag I'm searching for is x86_64-w64-mingw32. It should look similar in structure (but not content) to the value printed previously by our execution of ./config.guess. This is the value that should be used with the --host option on the configure command line. It's used by configure as a prefix for gcc, and a careful examination of your package manager output will show that there was indeed a program called x86_64-w64-mingw32-gcc installed into your */usr/bin* directory.

Now let's use the information we've gathered to build b64 for Windows. From within the *b64* directory, create a subdirectory called *w64* (or whatever you like) and change into it; this will be the build directory we'll use to build a 64-bit Windows version of b64. Run ../configure with options to target Windows, as follows (assuming we're still in the *b64* directory):[6]

```
$ mkdir w64
$ cd w64
$ ../configure --build=x86_64-pc-linux-gnu --host=x86_64-w64-mingw32
checking for a BSD-compatible install... /usr/bin/install -c
checking whether build environment is sane... yes
checking for x86_64-w64-mingw32-strip... x86_64-w64-mingw32-strip
checking for a thread-safe mkdir -p... /bin/mkdir -p
checking for gawk... gawk
checking whether make sets $(MAKE)... yes
```

---

5. If you're not using Fedora, you may have a different package name. Use the one you find in your package search.

6. You don't actually need to be in the *b64* directory. Your remote build directory can be anywhere on your filesystem, as long as you can provide a proper relative path back to b64's *configure* script. For instance, you can run from a sibling directory to *b64*, in which case you'd execute configure as ../b64/configure rather than ../configure, as we're doing here. The only exception is if your relative path back to *b64* contains spaces. Autoconf-generated configure scripts don't really like spaces in this context.

```
checking whether make supports nested variables... yes
checking for x86_64-w64-mingw32-gcc... x86_64-w64-mingw32-gcc
checking whether the C compiler works... yes
checking for C compiler default output file name... a.exe
checking for suffix of executables... .exe
checking whether we are cross compiling... yes
checking for suffix of object files... o
--snip--
configure: creating ./config.status
config.status: creating Makefile
config.status: creating lib/Makefile
config.status: creating config.h
config.status: executing depfiles commands
$
```

I've highlighted some of the important output lines from configure when cross-compiling. If you make a mistake entering the --host value on the command line, you'll see output similar to the following:

```
$ ../configure --build=x86_64-pc-linux-gnu --host=x86_64-w64-oops
checking for a BSD-compatible install... /usr/bin/install -c
checking whether build environment is sane... yes
checking for x86_64-w64-oops-strip... no
checking for strip... strip
configure: WARNING: using cross tools not prefixed with host triplet
--snip--
$
```

The warning is telling you that it could not find a strip program called x86_64-w64-oops-strip (not surprising). Most cross tool chains come with a properly prefixed version of strip because this is one of the tools in the *binutils* package, so this is a reasonable test. If configure can't find a prefixed version of the tools, it falls back to using the base names of the tools, which may be perfectly fine if your cross tools are named by their base names but simply stored in a different directory (which you've presumably added to your PATH).

Now that we've configured the build for cross-compilation, everything else works exactly the same as a regular Linux build. Try running make:

```
$ make
make  all-recursive
make[1]: Entering directory '/.../b64/w64'
Making all in lib
make[2]: Entering directory '/.../b64/w64/lib'
--snip--
make  all-recursive
make[3]: Entering directory '/.../b64/w64/lib'
make[4]: Entering directory '/.../b64/w64/lib'
depbase=`echo base64.o | sed 's|[^/]*$|.deps/&|;s|\.o$||'`;\
x86_64-w64-mingw32-gcc -DHAVE_CONFIG_H -I. -I../../lib -I..     -g -O2 -MT
base64.o -MD -MP -MF $depbase.Tpo -c -o base64.o ../../lib/base64.c &&\
mv -f $depbase.Tpo $depbase.Po
```

```
rm -f libgnu.a
x86_64-w64-mingw32-ar cr libgnu.a base64.o
x86_64-w64-mingw32-ranlib libgnu.a
make[4]: Leaving directory '/.../b64/w64/lib'
make[3]: Leaving directory '/.../b64/w64/lib'
make[2]: Leaving directory '/.../b64/w64/lib'
make[2]: Entering directory '/.../b64/w64'
x86_64-w64-mingw32-gcc -DHAVE_CONFIG_H -I. -I.. -I./lib -I../lib   -g -O2 -MT
src/src_b64-b64.o -MD -MP -MF src/.deps/src_b64-b64.Tpo -c -o src/src_b64-
b64.o `test -f 'src/b64.c' || echo '../'`src/b64.c
mv -f src/.deps/src_b64-b64.Tpo src/.deps/src_b64-b64.Po
x86_64-w64-mingw32-gcc  -g -O2   -o src/b64.exe src/src_b64-b64.o lib/libgnu.a
make[2]: Leaving directory '/.../b64/w64'
make[1]: Leaving directory '/.../b64/w64'
$
$ ls -1p src
b64.exe
src_b64-b64.o
$
```

I've highlighted the lines that indicate we're using the *mingw-w64* cross
tool chain to build b64. A listing of the *src* directory shows our Windows
executable, b64.exe.

To be complete, let's copy this program over to a Windows system and
give it a try. As mentioned previously, I have Windows 10 installed in a
virtual machine on my Linux system so I can simply run it in place from
a Windows-mapped drive (Z:, in my case):

```
Z:\...\b64\w64\src>dir /B
src_b64-b64.o
b64.exe
Z:\...\b64\w64\src>type ..\..\bootstrap.sh | b64.exe
IyEvYmluL3NoCmdudWxpYi10b29sIC0tdXBkYXRlCmF1dG9yZWNvbmYgLWkK
Z:\...\b64\w64\src>set /p="IyEvYmluL3NoCmdudWxpYi10b29sIC0tdXBkYXRlCmF1dG9yZWN
vbmYgLWkK" <nul | b64.exe -d
#!/bin/sh
gnulib-tool --update
autoreconf -i

Z:\...\b64\w64\src>
```

NOTE   *Don't be concerned about the* set /p *command—it's just a tricky way of echoing text
to the Windows console without a trailing newline, since* cmd.exe*'s echo statement has
no option to suppress the trailing newline.*

I'm not going to try to tell you that you'll never experience problems
building Windows software this way. You will, but they'll be porting issues
related primarily to a few POSIX system calls made directly by your project's
source code. I will, however, say that whatever problems you do run into will
be a proper subset of those you'd experience if you tried to use Microsoft
tools to build this package. In addition to any source-code-porting issues

you might find (they'll still be there, even with Microsoft tools), you'd also have to work the kinks out of hand-configured Visual Studio solution and project files or Microsoft nmake files. For some projects, it's worth the extra effort to be able to access the additional fine-grained tuning available when using Microsoft tools. For others, such tuning is not that important; building these projects for Windows on a Linux system works very well.

Windows Subsystem for Linux

Before we leave the Linux world behind, let's examine the Windows Subsystem for Linux (WSL) as an option for building Windows software using GNU tools.

You can obtain a flavor of Linux for WSL by downloading the version you want to use from the Windows Store. Before doing this, however, you must enable the optional Windows Subsystem for Linux feature. You can do this either from the **Windows Features** panel (type **windows features** into the Cortana search bar and select the top result) or from a PowerShell command prompt opened as *Administrator*.

From the **Windows Features** panel, scroll down until you find the entry for **Windows Subsystem for Linux**, check the associated checkbox, and click **OK**. Alternatively, from a PowerShell prompt (as *Administrator*), enter the following command and follow the prompts:

```
PS C:\Windows\system32> Enable-WindowsOptionalFeature -Online -FeatureName
Microsoft-Windows-Subsystem-Linux
```

Installing the Windows Subsystem for Linux will require a system restart.

Now open the Windows Store and search for "Windows Subsystem for Linux," select the "Run Linux on Windows" search result, and select the Linux flavor you want to install. On my system, installing the Ubuntu 18.04 flavor downloaded about 215MB and installed an "Ubuntu 18.04" icon in my Start menu.

Upon first execution, the terminal window displayed text indicating that the system was being installed:

```
Installing, this may take a few minutes...
Please create a default UNIX user account. The username does not need to match
your Windows username.
For more information visit: https://aka.ms/wslusers
Enter new UNIX username: jcalcote
Enter new UNIX password:
Retype new UNIX password:
passwd: password updated successfully
Installation successful!
To run a command as administrator (user "root"), use "sudo <command>".
See "man sudo_root" for details.

$
```

Use the `mount` command to see how Microsoft integrates the Windows and Linux filesystems:

```
$ mount
rootfs on / type lxfs (rw,noatime)
sysfs on /sys type sysfs (rw,nosuid,nodev,noexec,noatime)
proc on /proc type proc (rw,nosuid,nodev,noexec,noatime)
none on /dev type tmpfs (rw,noatime,mode=755)
devpts on /dev/pts type devpts (rw,nosuid,noexec,noatime,gid=5,mode=620)
none on /run type tmpfs (rw,nosuid,noexec,noatime,mode=755)
none on /run/lock type tmpfs (rw,nosuid,nodev,noexec,noatime)
none on /run/shm type tmpfs (rw,nosuid,nodev,noatime)
none on /run/user type tmpfs (rw,nosuid,nodev,noexec,noatime,mode=755)
binfmt_misc on /proc/sys/fs/binfmt_misc type binfmt_misc (rw,noatime)
C: on /mnt/c type drvfs (rw,noatime,uid=1000,gid=1000)
$
```

The key item of interest (highlighted) here is the fact that your Windows *C:* drive is mounted under */mnt/c* on this Linux installation.

On the other side of the coin, the Linux root filesystem is installed into your Windows filesystem in the hidden user-specific *AppData* directory. For example, I found my Ubuntu 18.04 installation's root filesystem at *C:\Users\your-username\AppData\Local\Packages\CanonicalGroupLimited .UbuntuonWindows_79rhkp1fndgsc\LocalState\rootfs.*[7]

If you selected a Debian-based distribution, start by updating your system software repository cache with `sudo apt-get update`. Then you can install development tools like GCC and the Autotools in the usual manner for the distro you selected:

```
$ sudo apt-get install gcc make autoconf automake libtool libtool-bin
```

If you're at all familiar with Ubuntu and apt, you'll see there's no significant difference between the output of the preceding command on WSL Ubuntu 18.04 and a native installation of Ubuntu 18.04. That's because you're really running Ubuntu 18.04 on Windows here.

As with a regular installation of Ubuntu 18.04, if you create a *bin* directory in your home directory and then open a new Ubuntu 18.04 terminal window, you'll find your personal *bin* directory at the beginning of your PATH. Do this now so you can create a symlink, *~/bin/gnulib-tools*, that refers to the `gnulib-tool` in your Windows clone of Gnulib, as we did when we built b64 on Linux:

```
$ ln -s /mnt/c/Users/.../gnulib/gnulib-tool ~/bin/gnulib-tool
```

---

7. Clearly, Microsoft did not intend for users to play around with this content from a Windows command or PowerShell prompt, or from a Explorer window.

Change into the */mnt/c/Users/.../b64* directory and run ./bootstrap.sh, followed by ./configure && make to build b64:

```
$ cd /mnt/c/Users/.../b64
$ ./bootstrap.sh
--snip--
$ ./configure && make
--snip--
$ cd src
$ ./b64 <../../../b64/bootstrap.sh
IyEvYmluL3NoCmdudWxpYi10b29sICOtdXBkYXRlCmF1dG9yZWNvbmYgLWkK$
$ printf "IyEvYmluL3NoCmdudWxpYi10b29sICOtdXBkYXRlCmF1dG9yZWNvbmYgLWkK" | ./
b64 -d
#!/bin/sh
gnulib-tool --update
autoreconf -i
$
```

Wonderful! Except that this is a Linux program, not a Windows program:

```
$ objdump -i b64
BFD header file version (GNU Binutils for Ubuntu) 2.30
elf64-x86-64
 (header little endian, data little endian)
  i386
elf32-i386
 (header little endian, data little endian)
  i386
--snip--
$
```

Attempting to run this program from a Windows command prompt will result in the Windows equivalent of a blank stare. You see, what you really have here with WSL is an inexpensive form of virtual machine guest with some built-in filesystem integration. That's not to say it's not useful. It's very handy to have Linux closely integrated with Windows for many purposes.

So what can we do? Our only option is to do the same thing we did on our native Linux installation earlier—install mingw-w64 and cross-compile. The process is identical, so I won't reiterate the details. Refer to that discussion in "Cross-Compiling for Windows on Linux" on page 454 for instructions.

Cygwin

The Cygwin project was established in 1995 by Cygnus Solutions as an effort to create tool chains using GNU software for the various embedded environments for which the company was hired to provide development tools.

Cygwin's general philosophy is that GNU packages should be able to be compiled for Windows without any modifications to the source code at all. Time is money to a support company, and any time not spent modifying source code is money in the bank. They wanted their engineers to *use* GNU tools in these environments, not spend their time porting them.

So how do they do this? Well, most GNU packages are written in C and use the C standard library for accessing most of the system functionality they require. The C standard library—being standardized—is portable by definition. Additionally, GNU projects strive for portability—at least among Unix flavors. Nevertheless, there is a subset of POSIX system functionality of which many GNU packages avail themselves, including POSIX threads (pthreads) and system calls like fork, exec, and mmap. While recent C and C++ standards now include a threading API, those other system calls are very specific to Unix and Linux. In fact, they have no direct counterparts on Windows that align well enough to use without some adapter code between the caller and the Windows API.

For a simple example, when you get right down to the bare metal, the two kernels work fundamentally differently with respect to how processes are created. Windows uses the Win32 CreateProcess function to create a new process and load a program into it in a single step. Unix, on the other hand, uses the fork and exec system calls to respectively clone an existing process and replace the contents of the clone with another program.

It's actually fairly easy to replace the fork-exec pair with a call to CreateProcess. The true difficulties arise when fork is used independently of exec, and this does happen occasionally.[8] There is simply no way to make CreateProcess do only half its job.[9] Many GNU programs don't use fork without exec, but some important ones do. Mapping these calls to the Windows API is difficult at best, and it's often impossible without significant structural changes to the source code.

Cygnus therefore elected to create a shim library of POSIX system call functionality. This library is called *cygwin1.dll*, and programs built using Cygwin are linked to this library and therefore depend on it at runtime. More to the point, every standard library call and most system calls pass though *cygwin1.dll* so that porting to a new platform without existing tools is an easy process.

You can detect if a Windows program was built for the Cygwin platform by simply looking for *cygwin1.dll* in its dependency list.[10] But the Cygwin platform is not the only target that Cygwin supports. The mingw-w64 tool

---

8. It has been a common paradigm for many years in Unix server software to fork a parent server process to handle a client request, without using exec to load a different program into the child's address space. The fork process is very fast, creating a copy-on-write clone of the parent in a new address space. This gives the parent and child desired address-space isolation and equal footing with the kernel scheduler, without the overhead of loading and initializing a new program from disk.

9. Even if "half its job" involved cloning an existing process—which it does not.

10. The official Microsoft tool for examining such details is the dumpbin.exe utility program that ships with Visual Studio. Use dumpbin /dependents some.dll.

chains have been ported to Cygwin and may be used as cross-compilers in Cygwin to build native Windows software, just as we did on Linux.

In 1999, Red Hat purchased Cygnus Solutions, and Cygwin has been maintained by various Red Hat employees and outside volunteers since then. Because of this maturity, Cygwin's package repository is very large, and its Windows POSIX environment is one of the most complete implementations available. Cygwin is one of the most-used systems for porting GNU and other software to Windows.

Installing Cygwin

To install Cygwin, download the installer from Cygwin's website at *https://www.cygwin.com*. The installer is called setup-x86_64.exe. Cygwin's installer does not use the Windows installation database; you can remove Cygwin merely by deleting its installation directory.

A unique and useful aspect of the Cygwin installer is that it caches its downloaded packages at a location of your choice on your filesystem. This cache can then be used as a standalone installation source for later installations.

Running the installer presents a setup wizard, the opening page of which is shown in Figure 17-1.

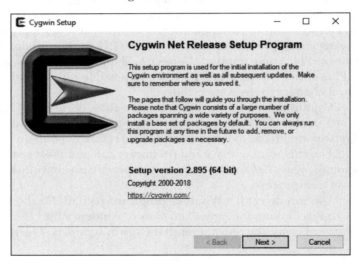

Figure 17-1: The initial copyright screen of the Cygwin64 setup program

Click **Next** to move to the second page of the setup wizard, shown in Figure 17-2.

You're asked here to select how you want to obtain packages. Your options include the internet or a local installation directory. You may also elect to download files from the internet but not install them, which is useful for building a local installation source for installing multiple systems from the same cache of downloaded files. Select **Install from Internet** and click **Next** to continue to the next page, shown in Figure 17-3.

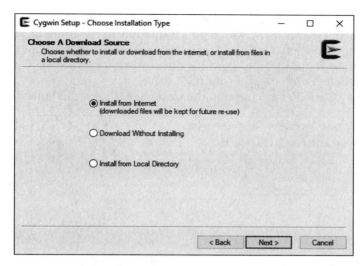

Figure 17-2: The installation type screen of the Cygwin64 setup program

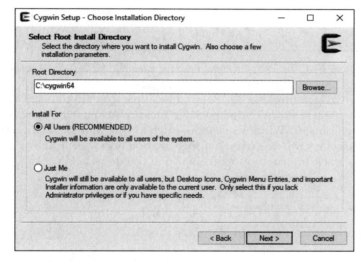

Figure 17-3: The installation location screen of the Cygwin64 setup program

Here, you're asked where you want to install Cygwin. The default location is *C:\cygwin64*, and it's recommended that you just stick with this default location, though Cygwin does a better job than some of the other installers of managing the required system changes if you do choose to install in a nondefault location.

If you're a Windows power user, you're very likely feeling that gut-wrenching desire right now to change the default location to something more reasonable on Windows. I admonish you not to do this. The problem is that you're trying to view Cygwin as an application and, while it technically is one, it can also be viewed as being somewhat akin to a full Linux virtual machine installation. It provides a foreign (to Windows) development environment, which puts it squarely in the camp of a sibling operating system to Windows. From this perspective, it's perhaps a bit easier to understand why the *Cygwin64* directory deserves a special place next to the *Windows* directory on your hard drive.

Click **Next** to move to the next page, shown in Figure 17-4.

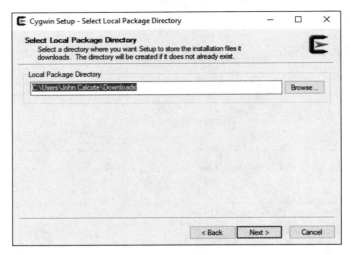

Figure 17-4: The local package directory screen of the Cygwin64 setup program

You're now asked to select a local package directory. This is the directory where downloaded package files are stored. Choose a reasonable location on your Windows system—such as your *Downloads* directory. A *cygwin* directory will be created at this location and will contain a subdirectory for each internet source from which you download packages. Click **Next** to move to the next screen, shown in Figure 17-5.

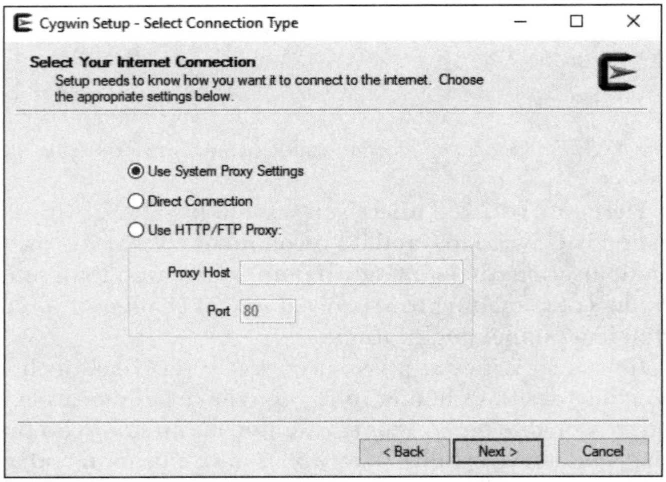

Figure 17-5: The proxy settings screen of the Cygwin64 setup program

You may now select or modify your proxy settings. Usually, you can just use the default system proxy settings. Those who use a proxy in their work or home environments will be used to configuring this for internet applications and will know what to do with the options here. Click **Next** to move to the next screen, shown in Figure 17-6.

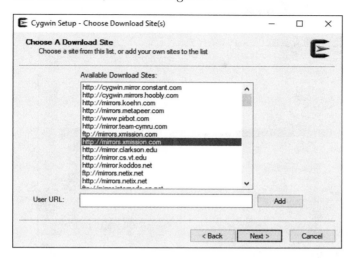

Figure 17-6: The package download source screen of the Cygwin64 setup program

Select a package download source. As with a Linux distribution, there are multiple sites you can use as a package source for Cygwin. Select one that's geographically close to you for the fastest installation and then click **Next** to begin downloading the package catalog, shown in Figure 17-7.

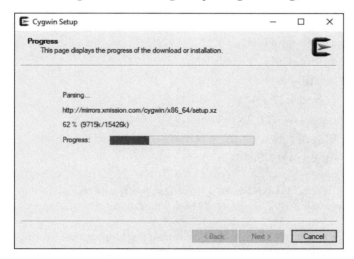

Figure 17-7: The package catalog download screen of the Cygwin64 setup program

Downloading and parsing the package catalog from the selected source site takes only a few seconds, and then the package manager main screen is displayed, as shown in Figure 17-8.

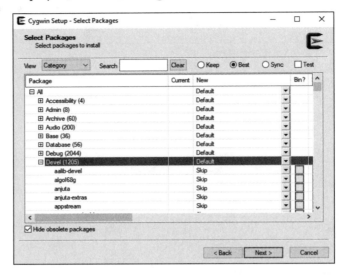

Figure 17-8: The package manager screen of the Cygwin64 setup program

I've expanded the All root element and the Devel category to show you the first few packages in this category, sorted alphabetically. Select the following additional packages in the Devel category by clicking the down arrow on the right end of the New column and choosing the highest version number available in the list for each package (with a few exceptions):

- autoconf2.5 (2.69-3)
- automake (10-1)
- automake1.15 (1.15.1-1)
- binutils (2.29-1)
- gcc-core (7.4.0-1)
- gcc-g++ (7.4.0-1 - optional)
- libtool (2.4.6-6)
- make (4.2.1-2)

NOTE   *As you scroll though the list of packages, you'll note that some have been preselected for you. Do not deselect any of the default packages.*

Each entry in this list has a base package name followed by a package version in parentheses. The versioning system is similar to that of a standard Linux distribution. The upstream source package version is suffixed with a dash, followed by a packager's version. For example, the *autoconf2.5* package has a source package version of 2.69 and a packager's version of 3. The packager's version is specific to the distribution—in this case, Cygwin.

Cygwin uses a rolling release mechanism, meaning that Cygwin packages are updated somewhat independently as newer source package versions become available and as the Cygwin maintainers consume them. The versions I've listed here were current at the time of this writing. Your most recent version numbers may be newer. Select the most recent rather than the ones I've listed. Feel free to use the search box at the top of the dialog to quickly find the packages in the list. Once you've selected these additional packages, click **Next** to continue to the next page, shown in Figure 17-9.

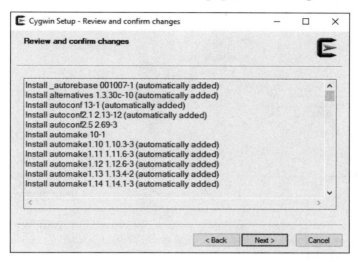

Figure 17-9: The download confirmation screen of the Cygwin64 setup program

After reviewing the list here to ensure you've selected the desired set, click **Next** to start the download process, shown in Figure 17-10.

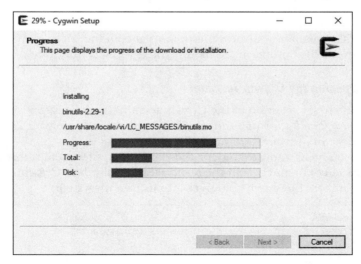

Figure 17-10: The package download progress screen of the Cygwin64 setup program

Since you elected to install only a few packages, this should not take long. Once the process completes, click **Next** to continue to the next screen, shown in Figure 17-11.

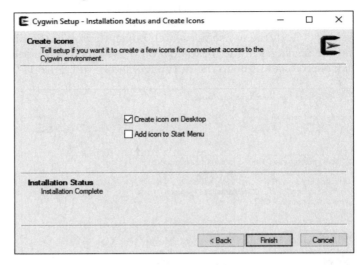

Figure 17-11: The icon selection screen of the Cygwin64 setup program

Select where you'd like icons to be created on your Windows system. You have two checkbox options here: Desktop and Start Menu. If you elect not to add any icons, you can still run the Cygwin terminal program by executing *C:\cygwin64\cygwin.bat* from an Explorer window or from a command or PowerShell prompt.

Click **Finish** to close the package manager. When you want to modify your Cygwin environment by adding or removing packages, or updating your existing packages to newer versions, just run setup-x86_64.exe again.[11] You'll need to go through all the same initial screens, but the package manager will remember your previous options and all the packages you currently have installed, allowing you to modify the existing configuration as you desire.

Opening the Cygwin Terminal

The first execution of the Cygwin terminal indicates that skeleton *.bashrc, .bash_profile, .inputrc,* and *.profile* files are copied into your */home/*username directory within the Cygwin filesystem.

The best way to understand the Cygwin filesystem is to execute the mount command within the terminal to display how Cygwin maps your Windows filesystem resources into its own filesystem:

```
$ mount
C:/cygwin64/bin on /usr/bin type ntfs (binary,auto)
```

---

11. You can subscribe to the relatively low-volume *cygwin-announce* mailing list to receive emails announcing when new packages come out. See *https://www.cygwin.com/lists.html*.

```
C:/cygwin64/lib on /usr/lib type ntfs (binary,auto)
C:/cygwin64 on / type ntfs (binary,auto)
C: on /cygdrive/c type ntfs (binary,posix=0,user,noumount,auto)
Z: on /cygdrive/z type vboxsharedfolderfs (binary,posix=0,user,noumount,auto)
$
```

Cygwin auto-mounts *C:\cygwin64*, *C:\cygwin64\bin*, and *C:\cygwin64\ lib* to */*, */usr/bin*, and */usr/lib*, respectively. It also auto-mounts all of your Windows drive roots to directories named by the drive letter under the */cygdrive* directory. I have my Windows operating system installed on the *C:* drive, and I have the *Z:* drive mapped to my Linux host though VirtualBox's shared folder system. Therefore, I have full access to both my Windows filesystem and my Linux host filesystem from within Cygwin's POSIX environment.[12] I also have access to Cygwin's entire filesystem from Windows, via the *C:\Cygwin64* directory.

Testing the Build

Because Cygwin provides access to your Windows environment within its own POSIX environment, you can simply run a previously installed stand-alone copy of Git for Windows directly from the Cygwin shell prompt. An even better option, but one that only works inside of the Cygwin terminal, is to install Cygwin's version of git from its package manager. Why is this option better? Because Cygwin's git package understands Cygwin's filesystem conventions better than the Windows version does. For instance, the Windows version will sometimes create files with the wrong permissions when viewed from a POSIX environment.

Unlike MinGW and Msys2, Cygwin can manage symlinks correctly within the Cygwin filesystem. Recall from Chapter 13, and earlier in this chapter, that we need to create a symlink to the `gnulib-tool` utility somewhere in our PATH so that b64's `bootstrap.sh` script is able to find Gnulib. Let's do that now in the Cygwin terminal. Fill in the elided section of the following command with the proper path to your clone of Gnulib:

```
$ ln -s /cygdrive/c/.../gnulib/gnulib-tool /usr/bin/gnulib-tool
```

This command creates a symbolic link in Cygwin's */usr/bin* directory, referring to the gnulib-tool program in the root of the Gnulib work area you cloned.

By default, Cygwin creates symlinks as text files flagged with the Windows *System* (S) attribute, making them invisible to normal Windows directory listing commands and within Windows File Explorer. If you examine the contents of a Cygwin symlink file, you'll find it contains a magic

---

12. A warning for those who are thinking of building on that VirtualBox shared folder: I've found it's fine for copying single files back and forth, but when you start doing a lot of I/O across these virtual machine shared folder channels, you get anomalous behavior; for example, files get half written, scripts quit in the middle, and so on. Since Windows' NFS client is terrible, it's better to set up Samba on your Linux system and just use the standard Windows network client to access your Linux filesystem.

cookie, !<symlink>, followed by the path to the target filesystem entry in UTF-16 format (beginning with the little-endian byte order mark, 0xFFFE).

You can configure Cygwin to create true Windows symbolic links by exporting a CYGWIN environment variable containing the text winsymlinks:nativestrict. However, if you do this, you must then run your Cygwin terminal as Administrator, because creating Windows native symbolic links requires administrative rights by default. Recent versions of Windows 10 allow native symlinks to be created without elevated privileges if you're willing to switch your system into so-called "developer mode."

All that said, Cygwin's own system of managing symlinks works really well, as long as the tools interpreting the links are built for the Cygwin platform. In fact, to see the contents of a Cygwin symlink file, you have to use a non-Cygwin tool because Cygwin tools will simply follow the symlink file, rather than open the file, even from a Windows command prompt!

Now, let's build b64 for Windows. We'll start by changing directories within the Cygwin terminal to the b64 work area you cloned on your Windows system and running the bootstrap.sh script to pull in our Gnulib dependencies and to run autoreconf -i:

```
$ cd /cygdrive/c/.../b64
$ ./bootstrap.sh
Module list with included dependencies (indented):
    absolute-header
  base64
--snip--
configure.ac:12: installing './compile'
configure.ac:20: installing './config.guess'
configure.ac:20: installing './config.sub'
configure.ac:6: installing './install-sh'
configure.ac:6: installing './missing'
Makefile.am: installing './depcomp'
$
```

And now we can simply run configure and make. We'll do this from within a subdirectory structure so we can reuse this work area for other build types later. Note there's no need to specify --build or --host options here to set up a cross-compile. We're running "native" Cygwin tools, which automatically build Cygwin programs designed to run on the host platform:

```
$ mkdir -p cw-builds/cygwin
$ cd cw-builds/cygwin
$ ../../configure
--snip--
checking for C compiler default output file name... a.exe
checking for suffix of executables... .exe
--snip--
checking build system type... x86_64-unknown-cygwin
```

```
checking host system type... x86_64-unknown-cygwin
--snip--
configure: creating ./config.status
config.status: creating Makefile
config.status: creating lib/Makefile
config.status: creating config.h
config.status: executing depfiles commands
$
$ make
make  all-recursive
make[1]: Entering directory '/cygdrive/c/.../cw-builds/cygwin'
--snip--
make[2]: Entering directory '/cygdrive/c/.../cw-builds/cygwin'
gcc -DHAVE_CONFIG_H -I. -I../../b64  -I./lib -I../../b64/lib  -g -O2 -MT src/
src_b64-b64.o -MD -MP -MF src/.deps/src_b64-b64.Tpo -c -o src/src_b64-b64.o
`test -f 'src/b64.c' || echo '../../b64/'`src/b64.c
mv -f src/.deps/src_b64-b64.Tpo src/.deps/src_b64-b64.Po
gcc  -g -O2   -o src/b64.exe src/src_b64-b64.o lib/libgnu.a
make[2]: Leaving directory '/cygdrive/c/.../cw-builds/cygwin'
make[1]: Leaving directory '/cygdrive/c/.../cw-builds/cygwin'
$
```

Finally, we'll test our new b64.exe program to see if it works on Windows. While the Cygwin terminal may look like Linux, it's really just a Linux-like way of accessing Windows, so you can execute Windows programs from the Cygwin terminal. This is nice because it allows us to use the Bash version of echo with its -n option to suppress the default linefeed during our testing:

```
$ cd src
$ ./b64.exe <../../bootstrap.sh
IyEvYmluL3NoCmdudWxpYi10b29sICOtdXBkYXRlCmF1dG9yZWNvbmYgLWkK
$
$ echo -n "IyEvYmluL3NoCmdudWxpYi10b29sICOtdXBkYXRlCmF1dG9yZWNvbmYgLWkK" |\
   ./b64 -d
#!/bin/sh
gnulib-tool --update
autoreconf -i
$
```

NOTE   *I did not use the .exe extension on the command to reverse the base64 encoding operation in this console listing. I wanted to show that, like Windows, Cygwin does not require the use of the extension on executable files.*

If you run a dependency checker like Visual Studio's dumpbin.exe or Cygwin's cygcheck utility, you'll find that this version of b64.exe depends heavily on *cygwin1.dll*, which must be shipped with your program. By default, Cygwin builds "Cygwin" software—software designed to run on the Cygwin platform, and an important part of the Cygwin platform is *cygwin1.dll* on Windows.

Building True Native Windows Software

You may also install the mingw-w64 tool chain and compile using the same techniques we used in "Cross-Compiling for Windows on Linux" on page 454. The mingw-w64 tool chain is available in the Cygwin package manager and is a Cygwin port of the same tool chain we installed earlier on Linux.

Let's do that now. Run the setup-x86_64.exe program again and skip through all the leading dialogs until you come to the package manager window. After initial installation, the default view shown by the package manager window is a list of pending updates of packages you've already installed. Depending on how long it has been since your initial installation, this list may even be empty. Select the **Full** option from the **View** drop-down box to return to the complete list of packages. Locate and select (under the **Devel** category) the following packages for installation. You may see newer version options than I've listed here; select the latest available to you. You can enter a prefix (**mingw64-**) in the **Search** box to narrow down the result list to a subset of packages containing those you want.

- mingw64-i686-gcc-core (7.4.0-1)
- mingw64-i686-gcc-g++ (7.4.0-1)
- mingw64-x86_64-gcc-core (7.4.0-1)
- mingw64-x86_64-gcc-g++ (7.4.0-1)

The first two of these packages are for generating 32-bit Windows software, and the last two are for generating 64-bit Windows software. Click **Next** to continue and install these additional packages.

NOTE   *You may notice on the summary screen that other packages you did not explicitly select are also getting installed. That's because these four are meta-packages, as described previously. If it has been a while since you initially installed Cygwin, you may also see updates for packages you previously installed.*

Create other subdirectories under *b64/cw-builds* for 32- and 64-bit mingw-w64 builds:

```
$ pwd
/cygdrive/c/.../cw-builds
$ mkdir mingw32 mingw64
$ cd mingw32
$
```

Let's start by building the 32-bit Windows program in the *mingw32* directory using the i686 variation of the mingw-w64 cross tool set:

```
$ cd mingw32
$ ../../configure --build=x86_64-unknown-cygwin --host=i686-w64-mingw32
--snip--
checking for C compiler default output file name... a.exe
checking for suffix of executables... .exe
```

```
--snip--
checking build system type... x86_64-unknown-cygwin
checking host system type... i686-w64-mingw32
--snip--
$
$ make
make all-recursive
--snip--
i686-w64-mingw32-gcc -DHAVE_CONFIG_H -I. -I../../b64 -I./lib -I../../lib -g
-O2 -MT src/src_b64-b64.o -MD -MP -MF src/.deps/src_b64-b64.Tpo -c -o src/
src_b64-b64.o `test -f 'src/b64.c' || echo '../../'`src/b64.c
mv -f src/.deps/src_b64-b64.Tpo src/.deps/src_b64-b64.Po
i686-w64-mingw32-gcc -g -O2 -o src/b64.exe src/src_b64-b64.o lib/libgnu.a
make[2]: Leaving directory '/cygdrive/c/.../cw-builds/mingw32'
make[1]: Leaving directory '/cygdrive/c/.../cw-builds/mingw32'
$
```

Though it may seem odd, you must use the --build and --host options on the configure command line here to cross-compile for Windows. The reason is the mingw-w64 tool chain is not the default tool chain on Cygwin. All you're really doing is telling configure where to find the nondefault tools you want to use. From a certain point of view, it actually is a cross-compile because you're building non-Cygwin software on the Cygwin platform.

Do the same for the 64-bit build:

```
$ cd ../mingw64
$ ../../configure --build=x86_64-unknown-cygwin --host=x86_64-w64-mingw32
--snip--
checking for C compiler default output file name... a.exe
checking for suffix of executables... .exe
--snip--
checking build system type... x86_64-unknown-cygwin
checking host system type... x86_64-w64-mingw32
--snip--
$
$ make
make all-recursive
--snip--
x86_64-w64-mingw32-gcc -DHAVE_CONFIG_H -I. -I../.. -I./lib -I../../lib -g
-O2 -MT src/src_b64-b64.o -MD -MP -MF src/.deps/src_b64-b64.Tpo -c -o src/
src_b64-b64.o `test -f 'src/b64.c' || echo '../../'`src/b64.c
mv -f src/.deps/src_b64-b64.Tpo src/.deps/src_b64-b64.Po
x86_64-w64-mingw32-gcc -g -O2 -o src/b64.exe src/src_b64-b64.o lib/libgnu.
amake[2]: Leaving directory '/cygdrive/c/.../cw-builds/mingw64'
make[1]: Leaving directory '/cygdrive/c/.../cw-builds/mingw64'
$
```

NOTE *It may seem strange that the 64-bit version of the gcc is called x86_64-w64-mingw32-gcc. What's with that 32 on the end of the cross-tool prefix? The reason is that mingw was originally a 32-bit Windows compiler, named specifically mingw32. The mingw32 project was eventually renamed to MinGW, but tools and package names are harder to change once they're in widespread use.*

Analyzing the Software

To really understand the differences between these builds, you'll need to obtain a tool for looking inside the b64.exe files we generated using each of these three tool sets. You can run the dumpbin.exe utility that comes with Visual Studio or Cygwin's cygcheck tool, if you like. I found a very nice tool called *Dependencies* on GitHub by user lucasg.[13]

First, let's look at the cygcheck output for all three versions of the program. We'll start in the *cw-builds* directory to give us easy access to all of them:

```
$ pwd
/cygdrive/c/Users/.../cw-builds
$
❶ $ cygcheck cygwin/src/b64.exe
C:\Users\...\cw-builds\cygwin\src\b64.exe
  C:\cygwin64\bin\cygwin1.dll
    C:\Windows\system32\KERNEL32.dll
      C:\Windows\system32\ntdll.dll
      C:\Windows\system32\KERNELBASE.dll
$
❷ $ cygcheck mingw32/src/b64.exe
C:\Users\...\cw-builds\mingw32\src\b64.exe
$
❸ $ cygcheck mingw64/src/b64.exe
C:\Users\...\cw-builds\mingw64\src\b64.exe
  C:\Windows\system32\KERNEL32.dll
    C:\Windows\system32\ntdll.dll
    C:\Windows\system32\KERNELBASE.dll
  C:\Windows\system32\msvcrt.dll
$
```

The concept being conveyed by the hierarchies here is that a library is a direct dependency of the library or program directly above it and an indirect dependency of ancestors farther up the chain. The Cygwin version at ❶ shows a dependency hierarchy with *cygwin1.dll* near the top, just under b64.exe, and with all other libraries as direct or indirect dependencies of that library. This implies that every system or library call made by b64.exe is being made directly to *cygwin1.dll*, which then calls the other libraries on its behalf.

The 64-bit mingw64 version at ❸ displays a similar hierarchy, except that the b64.exe program depends directly on *kernel32.dll* and *msvcrt.dll*. This is a native Windows program, by all accounts.

My version of the cygcheck utility has some problems with 32-bit native Windows software. You can see this at ❷, where the tool shows us only the

---

13. See *https://github.com/lucasg/Dependencies/releases/*. The *README.md* displayed on the project home page states that the Visual C++ Redistributable package must be installed on your Windows system to properly run the software. The author provides a link to that package at the top of the project description. Install the latest from the Microsoft site referenced by the link. The package provides a command line version and a GUI version of the tool, but I found the command line version was not quite as robust as the GUI version.

program, b64.exe, with no library dependencies. To see the true details of this version, let's switch to the Dependencies program I mentioned earlier. I've loaded all three versions of the program into one instance of Dependencies in Figure 17-12.

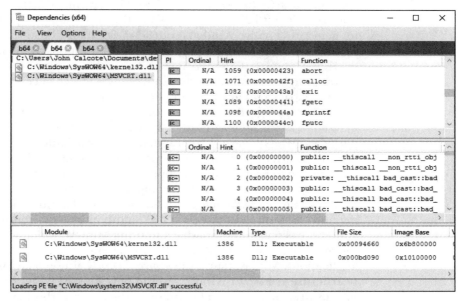

Figure 17-12: Modules and exports for b64.exe built as a 32-bit mingw-w64 program

Here, you can see that the 32-bit mingw-w64 version really does have library dependencies similar to those of the 64-bit mingw-w64 version. The 32-bit version uses *C:\Windows\SysWOW64\msvcrt.dll*, and the 64-bit version uses *C:\Windows\system32\msvcrt.dll*. The same is true of *kernel32.dll*.

There are additional subtle differences between the Cygwin version and the mingw-w64 versions. For a simple example, the Cygwin version imports getopt from *cygwin1.dll*. You'll perhaps recall that we used the POSIX getopt function to parse command line options in b64. You won't find getopt in *msvcrt.dll*, however, so where does it come from? The mingw-w64 tool chain provides a static archive of such POSIX functionality that ends up becoming a part of b64.exe.

MinGW: Minimalist GNU for Windows

In 1998, Colin Peters authored the initial release of what was then called *mingw32*. Later the numbers were dropped in order to avoid the implication that MinGW could only generate 32-bit software.[14]

MinGW initially offered only a Cygwin port of GCC. Sometime later, Jan-Jaap Van der Heijden created a native Windows port of GCC and added

---

14. It's somewhat ironic that 15 years later, MinGW is still only generating 32-bit software.

a *binutils* package and GNU make. MinGW has been a very popular alternative to Cygwin ever since, mainly because of its primary goal of creating software that closely resembles software generated by Microsoft tools. For reasonably portable C code, no libraries other than Windows system and Visual Studio runtime libraries (*msvcrt.dll*) are required. Remember that mingw-w64 was not available until 2013, so MinGW was the only available open source option for generating native Windows code for more than 10 years.

This concept is central to the philosophy espoused by the MinGW project. The goal of MinGW is to use only the standard C library as an abstraction layer and to modify source code where necessary to make other key packages available under MinGW.

There is a significant portion of GNU software, however, that makes use of the *pthreads* library. To accommodate this major set of GNU packages, MinGW gave in to pragmatism by providing a library called *pthreads-win32.dll*. This library shows up so often in the dependency list for software today that many people don't associate it with MinGW at all. Indeed, some portable software compiled using Microsoft's tools has even used *pthreads-win32.dll* independently, as a portable threading library, relying on POSIX threads in both POSIX and Windows environments.[15]

There is one major drawback to using MinGW, which is lately becoming more of an issue: MinGW still only generates 32-bit native Windows applications. Microsoft and Intel have recently announced jointly that some near-future version of Windows would only support 64-bit hardware. While 32-bit software will generally run on 64-bit systems, this may change with time. MinGW's package repository does not provide a port of its GCC compiler that generates 64-bit Windows object code, but there are third parties that make mingw-w64 available on the MinGW platform if you're willing to move away from MinGW's package manager. However, this sort of activity is discouraged because adding third-party packages to the environment can cause dependency problems the package manager can't resolve.

The MinGW community has survived on its name for many years, and the project has only recently started taking monetary donations to help with maintenance. Perhaps the additional financial support will spur the community into moving forward with these important upgrades.

Installing MinGW

In spite of MinGW generating only 32-bit Windows programs and libraries, it's worth looking at here because using it is so simple and effective for software that's already portable. Once you've begun to understand how MinGW and Msys work, moving to the other Windows-based POSIX platforms is a trivial task because all of them are based on some form of an Msys-like environment.

---

15. This is less of an issue today because the C11 and C++11 standards added direct support for threads. Prior to this, POSIX threads were used in C and C++ programs on POSIX systems like Linux, and Win32 native threads were used on Windows. With this improvement in these languages' standard libraries, even *pthreads-win32.dll* can be dispensed with for GNU software that's updated to use these newer standards directly.

We'll start by installing MinGW, which could not be simpler. Navigate in your favorite browser to *http://www.mingw.org*. Click the **Downloads** tab in the top menu bar. This link takes you to the *osdn.net*[16] download page for the MinGW project. Scroll down a bit (being careful to avoid the ridiculous large green-button advertisement links intended to look like legitimate download buttons). Under the gray bar labeled "Operating System: Windows," click the small blue button with the embedded Windows 10–like logo. Save the `mingw-get-setup.exe` program to a location on your hard drive.

Running this program presents you with a very simple dialog-based installer for the MinGW Installation Manager Setup Tool, as shown in Figure 17-13.

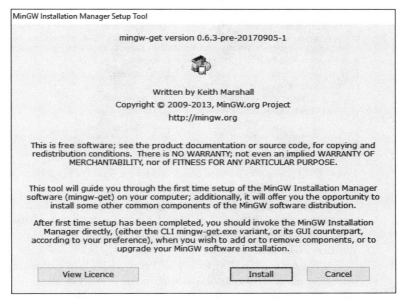

Figure 17-13: The initial dialog presented by the MinGW Installation Manager Setup Tool

This program actually installs the MinGW Installation Manager, a tool that, much like the Cygwin package manager, allows fine-grained control over the MinGW components that get installed or updated. Before the installation manager, the only option for updating MinGW was either to uninstall an existing full installation and then reinstall a new version from scratch, or try to upgrade, which was a hit-and-miss proposition at best.

In spite of the apparent out-of-date copyright range, the setup program and the installation manager itself were last refreshed (as of this writing) in September of 2017. The installation manager keeps a package catalog up-to-date, so you always have access to the very latest MinGW packages. For instance, a package containing GCC 8.2.0 was uploaded in August of 2018. By the time you read this, it will likely have been updated with an even newer version.

---

16. The original *SourceForge.JP* site.

Go ahead and click **Install**. You're presented with an options page, shown in Figure 17-14.

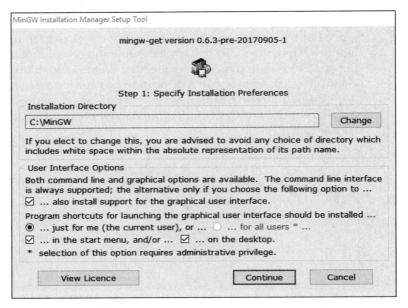

Figure 17-14: The options page presented by the MinGW Installation Manager Setup Tool

As with Cygwin, MinGW wants to be installed in a path off the root of the system drive, and also as with Cygwin, you need to consider MinGW a virtualized operating system. As such, it has need of a special place on your Windows filesystem.

Additionally, like Cygwin, you'll find MinGW is not installed using the Windows installation database and, hence, does not show up in the Windows installed programs panel. You can, in fact, completely remove MinGW from your Windows system by merely deleting the *C:\MinGW* directory.

NOTE *If you do decide to install into a different location, you'll need to carefully read the initial installation instructions on the MinGW website, because you'll need to make additional changes to files in the* C:\MinGW\msys\1.0\etc *directory after installation.*

Leave all options as they are and click **Continue**. The next screen you'll see is the download progress page, showing you that the latest installation manager program is being downloaded into the *C:\MinGW\libexec\mingw-get* directory. Figure 17-15 shows the state of this dialog once the catalog has been updated from the download source and the latest version of the installation manager has been downloaded and installed.

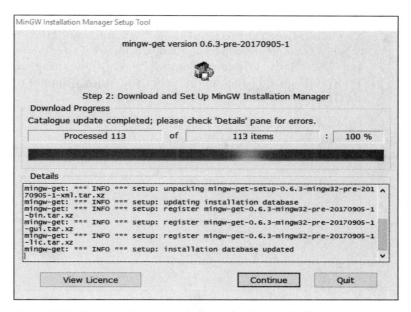

Figure 17-15: The download progress page for the MinGW Installation Manager
Setup Tool

Click **Continue** to open the installation manager, shown in Figure 17-16.

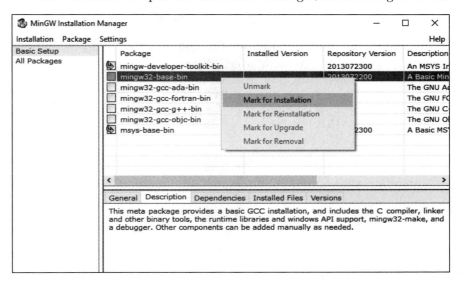

Figure 17-16: The installation manager main screen with package context menu

The packages you see in the Basic Setup panel (shown by default) are
actually meta-packages, or packages referring to large groups of actual
packages. To see real packages, you can select the **All Packages** option on
the left and then scroll through the list displayed on the right. When you're
ready to continue, return to the Basic Setup panel.

Selecting the *mingw-developer-toolkit-bin* meta-package will also automatically select the *msys-base-bin* meta-package. These two, plus the *mingw32-base-bin* meta-package, are all you need to compile C programs into 32-bit native Windows programs. Select these three packages, as shown in Figure 17-16, and then click the **Apply Changes** option from the **Installation** menu, as shown in Figure 17-17.

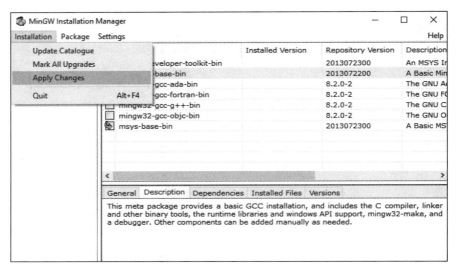

Figure 17-17: Applying selected changes in the installation manager

You're presented with a confirmation dialog titled "Schedule of Pending Actions," which allows you to apply the scheduled changes, defer these changes in order to return to the main window and modify the current list, or simply discard all changes. Select **Apply**, as shown in Figure 17-18.

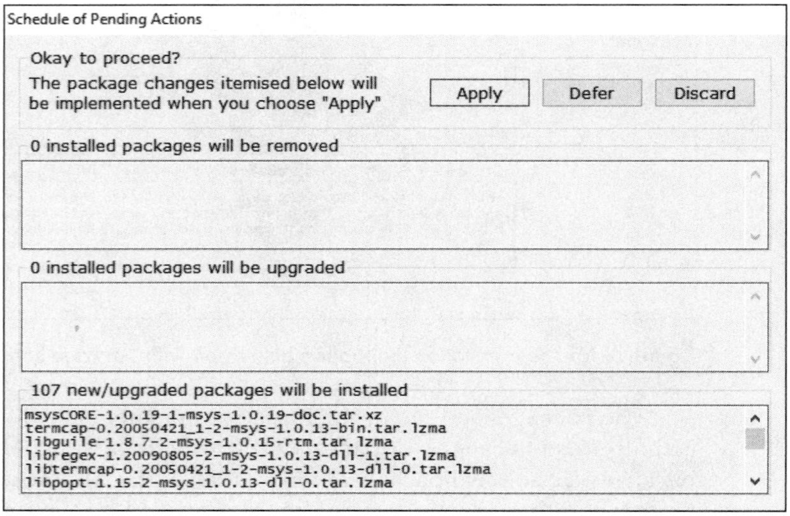

Figure 17-18: The installation manager's schedule of pending actions dialog

Finally, you're presented with the Download Package dialog, shown in Figure 17-19, in which each of the 112 packages you selected for download is displayed with a progress bar.

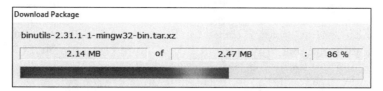

Download Package

binutils-2.31.1-1-mingw32-bin.tar.xz

| 2.14 MB | of | 2.47 MB | : | 86 % |

Figure 17-19: The installation manager's Download Package dialog

NOTE *Since the MinGW download site will have undoubtedly been updated after this writing, you may see a different number of packages to be installed in the bottom pane of the dialog shown in Figure 17-18.*

This may take a while, depending on your internet connection, so go grab a snack.

NOTE *If you get any package download errors, just click **OK** to dismiss the error dialog, wait for the successful download and installation of the remaining packages to complete, and then click **Apply Changes** from the **Installation** menu again to retry downloading and installing the failed packages. Only the failed packages will be redownloaded.*

Once all packages have been downloaded, they'll be installed into a standard Unix-like directory structure within the *C:\MinGW\msys\1.0* directory. Figure 17-20 shows the installation manager's Applying Scheduled Changes dialog, after it has installed each of the previously downloaded packages.

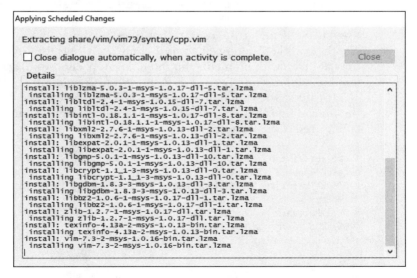

Applying Scheduled Changes

Extracting share/vim/vim73/syntax/cpp.vim

☐ Close dialogue automatically, when activity is complete.   Close

Details

```
install: liblzma-5.0.3-1-msys-1.0.17-dll-5.tar.lzma
  installing liblzma-5.0.3-1-msys-1.0.17-dll-5.tar.lzma
install: libltdl-2.4-1-msys-1.0.15-dll-7.tar.lzma
  installing libltdl-2.4-1-msys-1.0.15-dll-7.tar.lzma
install: libintl-0.18.1.1-1-msys-1.0.17-dll-8.tar.lzma
  installing libintl-0.18.1.1-1-msys-1.0.17-dll-8.tar.lzma
install: libxml2-2.7.6-1-msys-1.0.13-dll-2.tar.lzma
  installing libxml2-2.7.6-1-msys-1.0.13-dll-2.tar.lzma
install: libexpat-2.0.1-1-msys-1.0.13-dll-1.tar.lzma
  installing libexpat-2.0.1-1-msys-1.0.13-dll-1.tar.lzma
install: libgmp-5.0.1-1-msys-1.0.13-dll-10.tar.lzma
  installing libgmp-5.0.1-1-msys-1.0.13-dll-10.tar.lzma
install: libcrypt-1.1_1-3-msys-1.0.13-dll-0.tar.lzma
  installing libcrypt-1.1_1-3-msys-1.0.13-dll-0.tar.lzma
install: libgdbm-1.8.3-3-msys-1.0.13-dll-3.tar.lzma
  installing libgdbm-1.8.3-3-msys-1.0.13-dll-3.tar.lzma
install: libbz2-1.0.6-1-msys-1.0.17-dll-1.tar.lzma
  installing libbz2-1.0.6-1-msys-1.0.17-dll-1.tar.lzma
install: zlib-1.2.7-1-msys-1.0.17-dll.tar.lzma
  installing zlib-1.2.7-1-msys-1.0.17-dll.tar.lzma
install: texinfo-4.13a-2-msys-1.0.13-bin.tar.lzma
  installing texinfo-4.13a-2-msys-1.0.13-bin.tar.lzma
install: vim-7.3-2-msys-1.0.16-bin.tar.lzma
  installing vim-7.3-2-msys-1.0.16-bin.tar.lzma
```

Figure 17-20: The installation manager's Applying Scheduled Changes dialog

You may now close the installation manager program. There is one final step in preparing your installation of MinGW—creating a convenient desktop icon for the MinGW terminal, which is a somewhat outdated version of the Bash shell ported to Windows and running in a Windows Console Host (conhost.exe) process. MinGW installs a Windows batch file at *C:\MinGW\msys\1.0\msys.bat*. Execute this batch file to start the MinGW terminal that provides your POSIX build environment. I like to create a shortcut to this file on my desktop and change the icon for it to point to the *msys.ico* file found in the same directory.

Double-click the msys.bat file and start up the MinGW terminal. You'll find pwd shows that you're left in the */home/username* directory, where *username* is your Windows system user name.

As with Cygwin, the best way to understand the filesystem is to use mount to view the mount points in the MinGW filesystem:

```
$ mount
C:\Users\...\AppData\Local\Temp on /tmp type user (binmode,noumount)
C:\MinGW\msys\1.0 on /usr type user (binmode,noumount)
C:\MinGW\msys\1.0 on / type user (binmode,noumount)
C:\MinGW on /mingw type user (binmode)
c: on /c type user (binmode,noumount)
d: on /d type user (binmode,noumount)
z: on /z type user (binmode,noumount)
$
```

This output looks similar to its Cygwin counterpart, but there are a few differences. First, MinGW mounts your Windows user temporary directory as */tmp*. Second, both */usr* and / represent the same Windows directory, *C:\MinGW\msys\1.0*. Finally, *C:\MinGW* itself is mounted under */mingw*.

Windows drives are managed a bit differently also. Windows drive letters show up in the MinGW filesystem as they do in Cygwin, but they're listed directly under the root, rather than as a separate top-level directory. Another subtle difference here is that my *D:* drive is listed. It's a virtual optical drive, with no media mounted. MinGW chooses to show it even without media, while Cygwin only shows it with media.

If you cat the contents of the */etc/fstab* file, you can see that most of the preceding is hardcoded. The only mount point that's actually soft-configured is the */mingw* path:[17]

```
$ cat /etc/fstab
# /etc/fstab -- mount table configuration for MSYS.
# Please refer to /etc/fstab.sample for explanatory annotation.

# MSYS-Portable needs this "magic" comment:
# MSYSROOT=C:/MinGW/msys/1.0
```

_____

17. If you had decided to install MinGW into a non-default location, you would have needed to change the Win32_Path in this file to reflect the actual installation location; otherwise, the */mingw* directory would fail to mount.

```
# Win32_Path                             Mount_Point
#----------------------------------       -----------
C:/MinGW                                  /mingw
$
```

Testing the Build

We're now ready to try building the b64 project. First, we should clean up
the *b64* directory in order to demonstrate bootstrap.sh in this environment,
so change into the *b64* directory from your MinGW terminal and use git to
remove all artifacts. Then make a build directory structure for testing MinGW:

```
$ cd /c/Users/.../Documents/dev/b64
$ git clean -xfd
--snip--
$ mkdir -p mgw-builds/mingw
$
```

At this point, rather than lead you down a sure path to failure, I'll state
up front that you'll immediately run into problems with symbolic links. While
Cygwin has no trouble properly creating and using symlinks in its environment, the same is not true of MinGW. If you attempt to create a symlink from
/usr/bin/gnulib-tool to the .../gnulib/gnulib-tool program, you'll find the ln
-s command seems to work, but when you try to run bootstrap.sh, it fails to
find Gnulib. A closer examination shows that the symlink you thought you
created was actually just a copy. Well, a copy won't work, because gnulib-tool
uses its real location in the filesystem as the base of the Gnulib repository
and a copy of gnulib-tool in another location cannot do this.

To fix this problem, we'll have to adjust b64's bootstrap.sh program to
use a relative path to the actual gnulib-tool. I cloned gnulib right next to
b64, so I merely have to change bootstrap.sh so that it refers to ../gnulib/
gnulib-tool, rather than relying on its being accessible from the system PATH.
Use any editor to make changes similar to those highlighted in Listing 17-1
on your system.

```
#!/bin/sh
../gnulib/gnulib-tool --update
autoreconf -i
```

Listing 17-1: b64/bootstrap.sh: Changes required to allow MinGW to find Gnulib

NOTE   *After making these changes to bootstrap.sh, you should expect to see different output
when you run b64.exe against it.*

That will fix our Gnulib issues, but there's another problem lurking
here. While MinGW may have the very latest GCC tool chain, it doesn't stay
as current with the Autotools. We've been working with Autoconf 2.69 and
Automake 1.15.1, but as of this writing, MinGW only provides Autoconf 2.68

and Automake 1.11.1. Perhaps by the time you read this, these tools will have been updated, and you will not have to make the changes to *configure.ac* shown in Listing 17-2. Check your Autoconf and Automake versions before making these changes.

```
#                                                -*- Autoconf -*-
# Process this file with autoconf to produce a configure script.

AC_PREREQ([2.68])
AC_INIT([b64], [1.0], [b64-bugs@example.com])
AM_INIT_AUTOMAKE([subdir-objects])
AC_CONFIG_SRCDIR([src/b64.c])
AC_CONFIG_HEADERS([config.h])
AC_CONFIG_MACRO_DIR([m4])

# Checks for programs.
AC_PROG_CC
AM_PROG_CC_C_O
--snip--
AC_OUTPUT
```

Listing 17-2: b64/configure.ac: Changes required to work with Autoconf 2.68

The highlighted lines show the changes that need to be made. First, we need to reduce the lowest supported version in AC_PREREQ to allow Autoconf 2.68 to process this *configure.ac* file. Then we need to change the Automake macro, AC_CONFIG_MACRO_DIRS (plural), to its Autoconf counterpart, AC_CONFIG_MACRO_DIR. It works the same, except that the one that comes with Automake 1.15.1 makes it possible for us to forego the use of AC_LOCAL_AMFLAGS = -I m4 in our *Makefile.am* file. Luckily, I had already added that line into *Makefile.am* and just left it there, so this change is easy. Finally, Automake 1.15.1 consolidated the functionality of AM_PROG_CC_C_O into the AM_INIT_AUTOMAKE macro when the subdir-objects option is given. Moving back to Automake 1.11.1 requires us to change to the old format where we have to explicitly mention AM_PROG_CC_C_O and it must come after AC_PROG_CC.

After these changes, we can finally run bootstrap.sh to generate our configure script:

```
$ ./bootstrap.sh
Module list with included dependencies (indented):
    absolute-header
  base64
--snip--
configure.ac:13: installing `./compile'
configure.ac:21: installing `./config.guess'
configure.ac:21: installing `./config.sub'
configure.ac:6: installing `./install-sh'
configure.ac:6: installing `./missing'
lib/Makefile.am: installing `./depcomp'
$
```

Now change into the *mgw-builds/mingw* directory created earlier and run configure with a relative path back to *b64*:

```
$ cd mgw-builds/mingw
$ ../../configure
--snip--
checking for C compiler default output file name... a.exe
checking for suffix of executables... .exe
checking whether we are cross compiling... no
--snip--
checking build system type... i686-pc-mingw32
checking host system type... i686-pc-mingw32
--snip--
configure: creating ./config.status
config.status: creating Makefile
config.status: creating lib/Makefile
config.status: creating config.h
config.status: executing depfiles commands

$ make
--snip--
make[2]: Entering directory `/c/Users/.../mgw-builds/mingw'
gcc -DHAVE_CONFIG_H -I. -I../..  -I./lib -I../../lib  -g -O2 -MT src/src_b64-
b64.o -MD -MP -MF src/.deps/src_b64-b64.Tpo -c -o src/src_b64-b64.o `test -f
'src/b64.c' || echo '../../'`src/b64.c
mv -f src/.deps/src_b64-b64.Tpo src/.deps/src_b64-b64.Po
gcc  -g -O2   -o src/b64.exe src/src_b64-b64.o lib/libgnu.a
make[2]: Leaving directory `/c/Users/.../mgw-builds/mingw'
make[1]: Leaving directory `/c/Users/.../mgw-builds/mingw'
$
```

Opening b64.exe in DependenciesGUI.exe shows us it's a 32-bit Windows program that depends only on SysWOW64\kernel32.dll and SysWOW64\MSVCRT.dll.

Msys2

Msys2 was developed by a company called OneVision Software in 2013 using "clean room" techniques in order to loosen up the open source licensing requirements imposed by Cygwin and to provide a more modern alternative to the old out-of-date Msys environments used by Cygwin and MinGW.

MSys2 uses a OneVision 64-bit port of the MinGW tool chain called *mingw-w64*. However, like Cygwin, Msys2 provides its own library of POSIX system-level functionality called *msys-2.0.dll*. Msys2 provides a C standard library implemented in terms of this library. You can detect if a Windows program was built for the Msys2 platform in the same manner as previously described for the Cygwin platform.

Because Msys2 is more or less a feature-for-feature replacement of Cygwin, and since so many people are already used to the way Cygwin works, Msys2 has had a difficult time gaining traction, though it is used by some key players, including Git for Windows. Msys2 is advertised as an upgrade to Cygwin, but Msys2 merely provides different implementations of the same portability mechanisms used by Cygwin.

The one distinguishing characteristic that sets Msys2 apart from Cygwin is the fact that Msys2's open source license is much more lenient than that of Cygwin. While Cygwin uses GPL-based licensing, Msys2 uses only a standard 3-clause BSD license, making it a viable option for building proprietary Windows software using Linux tools.

The most significant offering in OneVision's system is the 64-bit port of the MinGW compiler—not that it runs on 64-bit platforms (which it does), but that it generates 64-bit Windows code. It's safe to say that the world of cross-compiled Windows code was expanded dramatically when this compiler was released. It has since been ported to many different platforms.

Msys2 does tie into the Windows installation database, so you may uninstall Msys2 from the Windows installed-programs panel.

What's Msys?

The term "Msys" has been misused for many purposes over the years. Some think it means "Unix on Windows" or, at the very least, the sense of such. All it really provides is potential. Msys, at its most basic, provides a Unix-compatible terminal program, a Bourne-like shell (usually Bash), and a base set of utilities. Some implementations have more and some have fewer utilities. Whatever implementation you're using, be it Cygwin, MinGW, or Msys2, the Msys component of these packages is there for you to build upon by installing additional packages to build up the environment the way you like.

When OneVision created Msys2, the company's vision was to start out small, allowing the user to build up the environment exactly the way they wanted. Msys2's version of Msys has very few packages preinstalled, while Cygwin has many. Msys2 and Cygwin base their terminal window on mintty.exe,[18] while MinGW bases its terminal on conhost.exe (Windows console host process), and you can tell this is so because the look and feel of the Cygwin and Msys2 terminals are very similar to each other, but significantly different from that of MinGW.

Installing Msys2

The installation procedure for Msys2 is pretty simple. Navigate in your web browser to the Msys2 home page at *http://www.msys2.org* and click the button at the top of the page for either the 32-bit (msys2-i686-*yyyymmdd*.exe) or the 64-bit (msys2-x86_64-*yyyymmdd*.exe) version of the Msys2 installer.[19] When the download completes, run the installer and you're presented with a dialog-based installation wizard. The welcome page is shown in Figure 17-21.

---

18. See *https://mintty.github.io*. You can actually swap out the terminal in MinGW for one based on mintty. See the mintty site for details.

19. I downloaded the 20180531 version of the 64-bit Msys2 installer for the examples in this chapter.

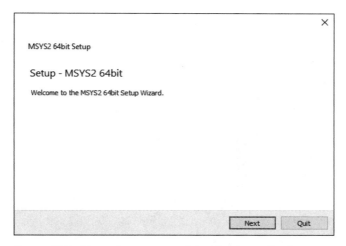

Figure 17-21: The welcome page of the Msys2 install utility

Click **Next** to move to the next page, shown in Figure 17-22.

Figure 17-22: The installation folder page of the Msys2 install utility

Select an installation location. Like Cygwin and MinGW, Msys2 wants to be installed at the root of the system drive. I recommend sticking with the default location for the same reasons I gave for the other two systems. Click **Next** to move to the next page, shown in Figure 17-23.

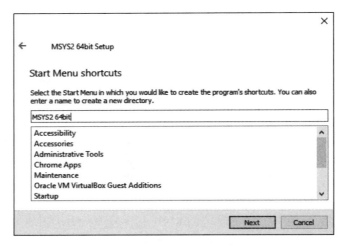

Figure 17-23: The shortcuts page of the Msys2 install utility

You can select the Windows Start Menu folder in which you'd like the installer to create its shortcuts. The default is sufficient. Click **Next** to move to the next page and begin installation, shown in Figure 17-24.

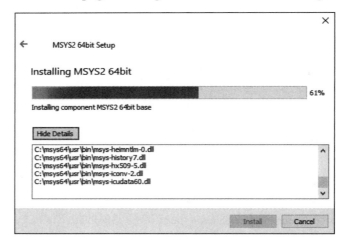

Figure 17-24: The installation progress page (showing details) of the Msys2 install utility

When installation has completed, click **Next** to move to the final page, shown in Figure 17-25.

Figure 17-25: The final page of the Msys2 install utility

You're given the option here of starting Msys2 upon completion. Click
Finish to exit the installer. Allow the installer to execute Msys2 as it exits
or go to the Windows 10 Start Menu, locate the **MSYS2 64bit** folder, and
click the **MSYS2 MSYS** entry. Both options start the Msys2 terminal window
in the same manner by executing the C:\msys64\msys2_shell.cmd script with a
command line option of -msys.

Unlike the other systems' installers, the Msys2 installer doesn't download packages from the internet. Rather, much like a Linux distribution
release, Msys2 installs a small base set of packages that get further out-of-
date as time goes by until, eventually, a new installer is made available by
the Msys2 maintainers. Therefore, the first thing you need to do is update
these installed base packages.

Msys2 uses a Windows port of the Arch Linux package manager,
Pacman, to provide access to repositories of packages ported to Msys2.
The basic installation provides relatively few packages, and it needs to be
updated with Pacman before additional packages can be installed.

With the terminal window open, we'll update the Msys2 system using
the command pacman -Syu. Pacman commands are uppercase, and options
to those commands are lowercase. The -S command is the remote repository "sync" command. The -u option of this command updates existing
packages from the remote repository. The -y option updates the catalog
from the repository before checking for updates. To get help on the commands, run pacman -h. To get help on the available options for a command,
add -h to the command line along with the command. For example, to get
help on available options for the -S command, run pacman -Sh.

Go ahead and update the system now, downloading the latest catalog first:

```
$ pacman -Syu
:: Synchronizing package databases...
 mingw32                                      530.8 KiB   495K/s 00:01
 mingw32.sig                                  119.0   B   116K/s 00:00
```

```
mingw64                                        532.0 KiB    489K/s 00:01
mingw64.sig                                    119.0   B    116K/s 00:00
msys                                           178.2 KiB    655K/s 00:00
msys.sig                                       119.0   B    0.00B/s 00:00
:: Starting core system upgrade...
warning: terminate other MSYS2 programs before proceeding
resolving dependencies...
looking for conflicting packages...

Packages (6) bash-4.4.023-1  filesystem-2018.12-1  mintty-1~2.9.5-1
             msys2-runtime-2.11.2-1  pacman-5.1.2-1  pacman-mirrors-20180604-2

Total Download Size:    19.04 MiB
Total Installed Size:   68.24 MiB
Net Upgrade Size:       11.96 MiB

:: Proceed with installation? [Y/n] Y
```

Press Y (or just ENTER to accept the default) at the prompt to continue updating the Msys2 system. You'll note the small number of packages being updated. The base Msys2 system provides almost nothing. The core system comes with Bash, a Unix-like filesystem emulator that sits on top of the Windows filesystem, mintty, the Msys2 Msys runtime, and the Pacman package manager, so it's not really a problem that the Msys2 installer installs out-of-date packages that need to be updated upon first use.

Pacman will download and install several packages, including Pacman:

```
:: Retrieving packages...
msys2-runtime-2.11.2-1-x86_64                    2.5 MiB   1012K/s 00:03
bash-4.4.023-1-x86_64                         1931.4 KiB   1003K/s 00:02
filesystem-2018.12-1-x86_64                     46.3 KiB    242K/s 00:00
mintty-1~2.9.5-1-x86_64                        296.7 KiB   1648K/s 00:00
pacman-mirrors-20180604-2-any                   17.1 KiB   2.09M/s 00:00
pacman-5.1.2-1-x86_64                           14.3 MiB   1010K/s 00:14
(6/6) checking keys in keyring
(6/6) checking package integrity
(6/6) loading package files
(6/6) checking for file conflicts
(6/6) checking available disk space
warning: could not get file information for opt/
:: Processing package changes...
(1/6) upgrading msys2-runtime
(2/6) upgrading bash
(3/6) upgrading filesystem
(4/6) upgrading mintty
(5/6) upgrading pacman-mirrors
(6/6) upgrading pacman
warning: terminate MSYS2 without returning to shell and check for updates again
warning: for example close your terminal window instead of calling exit
```

When you reach this point, Pacman itself needs to be updated, but it cannot update itself while it's running—an artifact of the way Windows manages running executable images. It displays a message indicating that you should close the terminal window by clicking the X in the upper-right corner and then restart Msys2 from the Windows Start Menu and continue the installation process.

NOTE *If you get a pop-up box warning you of running processes, just click **OK** to close the window anyway.*

Run `pacman -Su` again (no need to update the catalog again) to continue the update process and press ENTER when prompted again. This time, many more packages are updated. Press Y (or just ENTER) to continue and then wait for the update process to complete.

Before we continue installing additional packages, let's take a look at the filesystem by running the `mount` command:

```
$ mount
C:/msys64 on / type ntfs (binary,noacl,auto)
C:/msys64/usr/bin on /bin type ntfs (binary,noacl,auto)
C: on /c type ntfs (binary,noacl,posix=0,user,noumount,auto)
Z: on /z type vboxsharedfolderfs (binary,noacl,posix=0,user,noumount,auto)
$
```

You'll notice that the Msys2 base installation directory, *C:\msys64*, is mounted at */* and that *C:\msys64\usr\bin* is mounted at */bin*. As with the other systems, Windows drives are mounted as their drive letter. Msys2 mounts them off the root like MinGW. (Cygwin can be configured to do the same.) Like Cygwin, Msys2 doesn't mount optical drives without media, so you don't see my *D:* drive mounted here as */d*, but if I inserted a virtual disc and then executed `mount`, it would show up in the list.

Installing Tools

At this point, Msys2 is completely up-to-date and ready for you to add tools. But what tools should we add? I'd venture to guess this is the point where most uncommitted Msys2 explorers bail out. There is fairly complete documentation on the Msys2 wiki site, but you really need to read at least the introductory material in order to understand Msys2 the way you should.

While Cygwin and MinGW make it pretty obvious which path you should take at each turn, Msys2 simply offers options. You can build POSIX-only software for the Msys2 environment, or you can build native 32- or 64-bit Windows applications. Msys2 doesn't attempt to persuade you to go one way or the other by preinstalling software for a particular goal.

Msys2 provides three different terminal window shortcuts, each configured (using different command line options) to build software for one of the three targets Msys2 supports: Msys2 native applications, 32-bit Windows applications, or 64-bit Windows applications. This is not really any different from cross-compiling; it just targets a specific tool chain with a custom environment instead of using options on the `configure` command line.

Obviously, our goal here is not to build Msys2 software. Rather, we want to build Windows software, and Msys2 fully supports this goal. To find out what needs to be installed to meet our goals, read the Msys2 wiki page titled "Creating Packages."[20] According to this page, the following are the important Pacman groups:

base-devel   Required by all targets

msys2-devel   For building Msys2 native POSIX packages

mingw-w64-i686-toolchain   For building native 32-bit Windows software

mingw-w64-x86_64-toolchain   For building native 64-bit Windows software

For our purposes, the first and last of these packages will suffice. It's not necessarily obvious, but you don't get gcc and the *binutils* package unless you install the *msys2-devel* meta package. Like Cygwin, Msys2 sees its own platform, which generates applications that fully depend on *msys2.0.dll*, as owning the native Msys2 tool chain.

There is no GUI package installer in Msys2, so let's begin by exploring a few Pacman commands. Like most package management systems, the packages available to Pacman are grouped into logical sets. The -Sg option shows you a list of package groups:

```
$ pacman -Sg
kf5
mingw-w64-i686-toolchain
mingw-w64-i686
mingw-w64-i686-gimp-plugins
kde-applications
kdebase
mingw-w64-i686-qt4
mingw-w64-i686-qt
mingw-w64-i686-qt5
--snip--
$
```

To find out what packages are in a group, just add the group name to the end of the previous command line:

```
$ pacman -Sg mingw-w64-i686-toolchain
mingw-w64-i686-toolchain mingw-w64-i686-binutils
mingw-w64-i686-toolchain mingw-w64-i686-crt-git
mingw-w64-i686-toolchain mingw-w64-i686-gcc
mingw-w64-i686-toolchain mingw-w64-i686-gcc-ada
```

---

20. See *https://github.com/msys2/msys2/wiki/Creating-packages/*.

```
mingw-w64-i686-toolchain mingw-w64-i686-gcc-fortran
mingw-w64-i686-toolchain mingw-w64-i686-gcc-libgfortran
mingw-w64-i686-toolchain mingw-w64-i686-gcc-libs
mingw-w64-i686-toolchain mingw-w64-i686-gcc-objc
mingw-w64-i686-toolchain mingw-w64-i686-gdb
--snip--
$
```

When you use pacman -S to install a package and give it a group name, it shows you a list of group members and asks which of these members to install. If you simply press ENTER, it installs all of them. The --needed option ensures that only packages that are not already installed are downloaded. Without it, you'll download and install packages in your target groups that are already installed:

```
$ pacman -S --needed base-devel mingw-w64-i686-toolchain \
    mingw-w64-x86_64-toolchain
:: There are 56 members in group base-devel:
:: Repository msys
   1) asciidoc  2) autoconf  3) autoconf2.13  4) autogen  ...
--snip--
Enter a selection (default=all):
warning: file-5.35-1 is up to date -- skipping
warning: flex-2.6.4-1 is up to date -- skipping
--snip--
:: There are 17 members in group mingw-w64-x86_64-toolchain:
:: Repository mingw64
   1) mingw-w64-x86_64-binutils  2) mingw-w64-x86_64-crt-git  ...
--snip--
Enter a selection (default=all):
--snip--
Total Download Size:    185.24 MiB
Total Installed Size:  1071.62 MiB

:: Proceed with installation? [Y/n] Y
```

Press Y to download and then install the packages you requested. This could take a while, so I guess it's time for another snack.

Testing the Build

Once these packages have been installed, your Msys2 environment is ready to use. Let's change into the *b64* directory again, clean up, and create another build directory for Msys2 build testing.

If you've been working through this chapter and just came out of testing MinGW, your bootstrap.sh and *configure.ac* files have probably been modified. These changes will work fine in this environment. However, if you do decide to revert, only revert the *configure.ac* changes. The bootstrap.sh file should remain as is with the relative-path reference to gnulib-tool. If you skipped the section on MinGW, you'll need to modify bootstrap.sh by adding a relative path to the Gnulib repository work area to the execution of gnulib-tool. Msys2 has the same problem as MinGW in that it creates copies of the target, rather than working symlinks, so creating a symlink in */usr/bin* for gnulib-tool will not work here:[21]

```
$ cd /c/Users/.../Documents/dev/b64
$ git clean -xfd
$ mkdir -p ms2-builds/mw32 ms2-builds/mw64
$ ./bootstrap.sh
Module list with included dependencies (indented):
    absolute-header
  base64
--snip--
configure.ac:12: installing './compile'
configure.ac:21: installing './config.guess'
configure.ac:21: installing './config.sub'
configure.ac:6: installing './install-sh'
configure.ac:6: installing './missing'
Makefile.am: installing './depcomp'
$
```

Before we can build b64, we need to change terminals. You may have noticed that Msys2 configured three shortcuts in the **MSYS2 64bit** folder it created in the Windows Start Menu. Up to this point, we've been using the **MSYS2 MSYS** shortcut to start an Msys2 terminal. This was fine as long as we were just installing packages or if we were going to target the Msys2 platform.

We're building 64-bit native Windows software, so open a MinGW 64-bit terminal now. Then, from within that terminal window, change into the *b64/ms2-builds/mw64* directory and build b64:

```
$ cd /c/Users/.../Documents/dev/b64/ms2-builds/mw64
$ ../../configure
configure: loading site script /mingw64/etc/config.site
--snip--
checking for C compiler default output file name... a.exe
checking for suffix of executables... .exe
checking whether we are cross compiling... no
--snip--
checking build system type... x86_64-w64-mingw32
checking host system type... x86_64-w64-mingw32
--snip--
configure: creating ./config.status
config.status: creating Makefile
config.status: creating lib/Makefile
config.status: creating config.h
```

---

21. There are several conversations on the Msys2 mailing lists about this issue. There is a solution that allows Msys2 to create true Windows symlinks. Uncomment the line referring to winsymlinks in the *C:\msys2\msys2.ini* file (and also the *mingw32.ini* and *mingw64.ini* files in the same directory) and then execute the Msys2 terminal as a Windows Administrator to make it work properly.

```
config.status: executing depfiles commands
$
$ make
make  all-recursive
make[1]: Entering directory '/c/Users/.../ms2-builds/mw64'
--snip--
make[2]: Entering directory '/c/Users/.../ms2-builds/mw64'
gcc -DHAVE_CONFIG_H -I. -I../../b64  -I./lib -I../../lib   -g -O2 -MT src/
b64-b64.o -MD -MP -MF src/.deps/b64-b64.Tpo -c -o src/b64-b64.o `test -f 'src/
b64.c' || echo '../../'`src/b64.c
mv -f src/.deps/b64-b64.Tpo src/.deps/b64-b64.Po
gcc  -g -O2    -o src/b64.exe src/b64-b64.o lib/libgnu.a
make[2]: Leaving directory '/c/Users/.../ms2-builds/mw64'
make[1]: Leaving directory '/c/Users/.../ms2-builds/mw64'
$
```

The difference between these terminal windows is defined by the
values of various environment variables starting with MSYS. These variables
are used to configure the site configuration file referenced in the first
line of configure's output. See the Msys2 wiki for lots of details on how to
configure environments this way.

Running the b64.exe program through DependenciesGUI.exe, we can see
that it's a true native 64-bit Windows program, dependent only on Windows
system and Visual Studio runtime libraries.

Summary

After the whirlwind tour we just took, you should now have no trouble
building Windows software using GNU tools. Personally, I find the Cygwin
and Msys2 environments to be the most useful for many purposes.

Msys2 is a bit more modern and fresh, but they both serve as good
general-purpose platforms for the GNU tools. Msys2 also has the distinct
advantage of building pure Windows software wherever possible, while
Cygwin (unless using the mingw-w64 cross tools) builds apps that, although
they do run on Windows, rely heavily on the Cygwin system library. That's
not necessarily a showstopper for me. If there are other factors and Cygwin
comes out on top, then I don't mind the extra library, but the purist in me
leans toward a desire for no unnecessary third-party libraries.

Cygwin has the advantage of maturity. It's not so far behind Msys2
that it's unusable, but it's been around long enough that it has some nice
features, like integrating a working symbolic link mechanism into all of its
utilities so it can properly emulate POSIX symlinks. It also has many more
packages available for installation within the environment.

MinGW is a bit out-of-date, and it really needs to support 64-bit
Windows builds, but it's clean and small. With its new package manager,
it stacks up pretty well against the other two. All it really needs to move into
the big leagues is to embrace the newer 64-bit code generator in the *mingw-
w64* package. It might also be nice to upgrade to mintty from the conhost-
based console it uses.

18

A CATALOG OF TIPS AND REUSABLE SOLUTIONS FOR CREATING GREAT PROJECTS

Experience is a hard teacher because she gives the test first,
the lesson afterwards. —Vernon Sanders Law[1]

This chapter began as a catalog of reusable solutions—canned macros, if you will. But as I finished the chapters preceding this one, it became clear to me that I needed to broaden my definition of a *canned solution*. Instead of just cataloging interesting macros, this chapter lists several unrelated but important tips for creating great projects. Some of these are related to the GNU Autotools, but others are merely good programming practice with respect to open source and free software projects.

Item 1: Keeping Private Details out of Public Interfaces

At times, I've come across poorly designed library interfaces where a project's *config.h* file is required by the project's public header files. This presents a problem when more than one such library is required by a consumer. Which

---

1. Nathan, David H. (2000). *The McFarland Baseball Quotations Dictionary*. McFarland & Company.

config.h file should be included? Both have the same name, and chances are that both provide similar or identically named definitions.

When you carefully consider the purpose of *config.h*, you see that it makes little sense to expose it in a library's public interface (by including it in any of the library's public header files), because its purpose is to provide platform-specific definitions to a particular build of the library. On the other hand, the public interface of a portable library is, by definition, platform independent.

Interface design is a fairly general topic in computer science. This item focuses a bit more specifically on how to avoid including *config.h* in your public interfaces and, by extension, ensuring that you never install *config.h*.

When designing a library for consumption by other projects, you're responsible for not polluting your consumers' symbol spaces with useless garbage from your header files. I once worked on a project that consumed a library interface from another team. This team provided both a Windows and a Unix version of their library, with the header file being portable between the two platforms. Unfortunately, they didn't understand the definition of a clean interface. At some point in their public header files, they had a bit of code that looked like Listing 18-1.

```
#ifdef _WIN32
# include <windows.h>
#else
typedef void * HANDLE;
#endif
```

Listing 18-1: A poorly designed public header file that exposes platform-specific header files

Ouch! Did they really need to include *windows.h* just for the definition of HANDLE? No, and they probably should have used a different name for the handle object in their public interface because HANDLE is too generic and could easily conflict with a dozen other library interfaces. Something like XYZ_HANDLE or something more specific to the *XYZ* library would have been a better choice.

To properly design a library, first design the public interface to expose as little of the library's internals as is reasonable. Now, you'll have to determine the definition of *reasonable*, but it will probably involve a compromise between abstraction and performance.

When designing an API, start with the functionality you want to expose from your library; design functions that will maximize ease of use. If you find yourself trying to decide between a simpler implementation and a simpler user experience, always err on the side of ease of use for your consumers. They'll thank you by actually using your library. Of course, if the interface is already defined by a software standard, then much of your work is done for you. Often this is not the case, and you will have to make these decisions.

Next, try to abstract away internal details. Unfortunately, the C language doesn't make this easy to do because you often need to pass structure references in public APIs containing internal details of your implementation that consumers don't need to see. (C++ is actually worse in this area: C++ classes define public interfaces and private implementation details in the same class definition.)

Solutions in C

In C, a common solution for this problem is to define a public alias for a private structure in terms of a generic (void) pointer. Many developers don't care for this approach because it reduces type safety in the interface, but the loss of type safety is significantly offset by the increase in interface abstraction, as shown in Listings 18-2 and 18-3.

```
#include <abc_pub.h>

# include <config.h>

typedef struct
{
    /* private details */
} abc_impl;

int abc_func(abc * p)
{
    abc_impl * ip = (abc_impl *)p;
    /* use 'p' through 'ip' */
}
```

Listing 18-2: An example of a private C-language source file

```
typedef void abc;
int abc_func(abc * p);
```

Listing 18-3: abc_pub.h: A public header file describing a public interface (API)

Notice that the abstraction conveniently alleviates the need to include a bunch of really private definitions in the library's public interface.[2]

But there's a way that makes even better use of language syntax. In C, there's a little-known, and even less used, concept called a *forward declaration* that allows you to name the type in a public header file without actually defining it there. Listing 18-4 provides an example of a library's public header file that uses a forward declaration for the type used in the function declaration.

```
struct abc;
--snip--
int abc_func(struct abc * p);
```

Listing 18-4: Using a forward declaration in a public header file

---

2. The C language does not require a cast from a void pointer to any other pointer type, which is why you can assign the result of malloc to any pointer type without a cast. In Listing 18-2, I used an explicit cast to emphasize what I was doing, but it's actually redundant.

Of course, this use of struct abc assumes that some other function in your public interface returns pointers to objects of that type that you can then pass into abc_func. If your user is responsible for filling out the structure before passing its address, then this mechanism will obviously not work for you. Rather, its use here is for the sole purpose of hiding the internals of struct abc.

Solutions in C++

Forward declarations can also be used in C++, but not in the same manner. In C++, forward declarations are used more to minimize compile-time header file interdependencies than to hide implementation details in public interfaces. We can use other techniques, however, to hide implementation details from users.

In C++, hiding implementation details with interface abstraction can be done in a few different ways, which include using virtual interfaces and the *PIMPL (Private IMPLementation)* pattern.

The PIMPL Pattern

In the PIMPL pattern, implementation details are hidden behind a pointer to a private implementation class stored as private data within the public interface class, as shown in Listings 18-5 and 18-6.

```
#include <abc_pub.h>

# include <config.h>

class abc_impl
{
    /* private details */
};

int abc::func(void)
{
    /* use 'pimpl' pointer */
}
```

Listing 18-5: A private C++-language source file showing the proper use of the PIMPL pattern

```
❶ class abc_impl;

class abc {
❷ abc_impl * pimpl;
public:
  int func(void);
};
```

Listing 18-6: abc_pub.h: The public header file exposes few private details via the PIMPL pattern.

As mentioned previously, the C++ language also allows the use of a forward declaration (like the one at ❶) for any types used only through references or pointers (as at ❷) but never actually dereferenced in the public interface. Thus, the definition of the implementation class need not be exposed in the public interface because the compiler will happily compile the public interface header file without the definition of the private implementation class.

The performance tradeoff here generally involves the dynamic allocation of an instance of the private implementation class and then the access of class data indirectly through this pointer, rather than directly in the public structure. Notice that all internal details are now conveniently hidden and thus not required by the public interface.

C++ Virtual Interfaces

Another approach when using C++ is to define a public *interface* class, whose methods are declared *pure virtual*, with the interface implemented internally by the library. To access an object of this class, consumers call a public *factory* function, which returns a pointer to the implementation class in terms of the interface definition. Listings 18-7 and 18-8 illustrate the concept of C++ virtual interfaces.

```
#include <abc_pub.h>

# include <config.h>

class abc_impl : public abc {
    virtual int func(void) {
        int rv;
        // implementation goes here
        return rv;
    }
};
```

Listing 18-7: A private C++ language source file implementing a pure virtual interface

```
#define xyz_interface class

xyz_interface abc {
public:
  virtual int func(void) = 0;
};

❶ abc * abc_instantiate( /* abc_impl ctor params */ );
```

Listing 18-8: abc_pub.h: A public C++ language header file, providing only the interface definition

Here, I use the C++ preprocessor to define a new keyword, xyz_interface. By definition, xyz_interface is synonymous with class, so the terms may be used interchangeably. The idea here is that an interface doesn't expose any implementation details to the consumer. The public *factory* function abc_instantiate at ❶ returns a pointer to a new object of type abc_impl, except in terms of abc. Thus, nothing internal needs to be shown to the caller in the public header file.

It may seem like the virtual interface class method is more efficient than the PIMPL method, but the fact is that most compilers implement virtual function calls as tables of function pointers referred to by a hidden *vptr* address within the implementation class. As a result, you still end up calling all of your public methods indirectly through a pointer. The technique you choose to help hide your implementation details is more a matter of personal preference than performance.[3]

When I design a library, I first design a minimal, but complete, functional interface with as much of my internal implementation abstracted away as is reasonable. I try to use only standard library basic types, if possible, in my function prototypes and then include only the C or C++ standard header files required by the use of those types and definitions. This technique is the fastest way I've found to create a highly portable and maintainable interface.

If you still can't see the value in the advice offered by this item, then let me give you one more scenario to ponder. Consider what happens when a Linux distro packager decides to create a *devel* package for your library—that is, a package containing static libraries and header files, designed to be installed into the */usr/lib* and */usr/include* directories on a target system. Every header file required by your library must be installed into the */usr/include* directory. If your library's public interface requires the inclusion of your *config.h* file, then by extension your *config.h* file must be installed into the */usr/include* directory. Now consider what happens when multiple such libraries need to be installed. Which copy of *config.h* will win? Only one *config.h* file can exist in */usr/include*.

I've seen message threads on the Autotools mailing lists defending the need to publish *config.h* in a public interface and providing techniques for naming *config.h* in a package-specific manner. These techniques often involve some form of post-processing of this file to rename its macros so they don't conflict with *config.h* definitions installed by other packages. While this can be done, and while there are a few good reasons for doing so (usually involving a widely used legacy code base that can't be modified without breaking a lot of existing code), these situations should be considered the exception, not

---

3. The recently discovered Spectre microarchitecture flaw makes it possible that using PIMPL can be more efficient than using virtual interface classes. The reason is that indirection though function pointers (virtual methods) is harder for a processor to speculate correctly compared to indirection though data pointers. Thus, Spectre mitigations that prevent timing channels and data leaks though indirect function pointers may penalize virtual interfaces more severely.

the rule, because a well-designed project should not need to expose platform- and project-specific definitions in its public interface.

If your project simply can't live without *config.h* in its public interface, explore the nuances of the AC_CONFIG_HEADERS macro. Like all of the instantiating macros, this macro accepts a list of input files. The autoheader utility only writes the first input file in the list, so you can hand-create a second input file that contains definitions that you feel must be included in your public interface. Remember to name your public input file so as to reduce conflict with other packages' public interfaces.

NOTE   *Also, explore the AX_PREFIX_CONFIG_H macro, found in the Autoconf Macro Archive (see "Item 8: Using the Autoconf Archive Project" on page 528, which will add a custom prefix to all items found in config.h.*

Item 2: Implementing Recursive Extension Targets

An *extension target* is a make target that you write to accomplish some build goal that Automake doesn't automatically support. A *recursive extension target* is one that traverses your project directory structure, visiting every *Makefile.am* file in your Autotools build system and giving each one the opportunity to do some work when the extension target is made.

When you add a new top-level target to your build system, you have to either tie it into an existing Automake target or add your own make code to the desired target that traverses the subdirectory structure provided by Automake in your build system.

The SUBDIRS variable is used to recursively traverse all subdirectories of the current directory, passing requested build commands into the makefiles in these directories. This works great for targets that must be built based on configuration options, because after configuration, the SUBDIRS variable contains only those directories destined to be built.

However, if you need to execute your new recursive target in *all* subdirectories, regardless of any conditional configuration that might exclude a subdirectory specified in SUBDIRS, use the DIST_SUBDIRS variable instead.

There are various ways to traverse the build hierarchy, including some really simple one-liners provided by GNU make-specific syntax. But the most portable way is to use the technique that Automake itself uses, as shown in Listing 18-9.

```
my-recursive-target:
    ❶ $(preorder_commands)
        for dir in $(SUBDIRS); do \
          ($(am__cd) $$dir && $(MAKE) $(AM_MAKEFLAGS) $@) || exit 1; \
        done
    ❷ $(postorder_commands)

.PHONY: my-recursive-target
```

Listing 18-9: A makefile with a recursive target (WARNING: no support for "." in SUBDIRS)

At some point in the hierarchy, you'll need to do something useful besides calling down to lower levels. The preorder_commands macro at ❶ can be used to do things that must be done before recursing into lower-level directories. The postorder_commands macro at ❷ can likewise be used to do additional things once you return from the lower-level directories. Simply define either or both of these macros in any makefiles that need to do some pre-order or post-order processing for my-recursive-target.

For example, if you want to build some generated documentation, you might have a special target called doxygen. Even if you happen to be okay with building your documentation in the top-level directory, there may be times when you need to distribute the generation of your documentation to various directories within your project hierarchy. You might use code similar to that shown in Listing 18-10 in each *Makefile.am* file in your project.

```
      # uncomment if doxyfile exists in this directory
  ❶  # postorder_commands = $(DOXYGEN) $(DOXYFLAGS) doxyfile

    doxygen:
            $(preorder_commands)
        ❷  for dir in $(SUBDIRS); do \
            ❸  ($(am__cd) $$dir && $(MAKE) $(AM_MAKEFLAGS) $@) || exit 1; \
            done
            $(postorder_commands)

    .PHONY: doxygen
```

Listing 18-10: Implementing postorder_commands for a doxygen directory

For directories where *doxyfile* doesn't exist, you can comment out (or better yet, simply omit) the postorder_commands macro definition at ❶. In this case, the doxygen target will be harmlessly propagated to the next lower level in the build tree by the three lines of shell code at ❷.

The exit statement at the end of ❸ ensures that the build terminates when a lower-level makefile fails on the recursive target, propagating the shell error code (1) back up to each parent makefile until the top-level shell is reached. This is important; without it, the build may continue after a failure until a different error is encountered.

NOTE   *I chose not to use the somewhat less portable -C make command line option to change directories before running the sub-make operation. I also use an Automake macro called am__cd to change directories. This macro is defined to take the contents of the CDPATH environment variable into account to reduce extraneous output noise during a built. You can replace it with cd (or chdir). Examine an Automake-generated makefile to see how Automake defines this macro.*

If you choose to implement a completely recursive global target in this manner, you must include Listing 18-10 in every *Makefile.am* file in your project, even if that makefile has nothing to do with the generation of documentation. If you don't, make will fail on that makefile because no doxygen target exists within it. The commands may do nothing, but the target must exist.

If you want to do something simpler, such as pass a target down to a single subdirectory beneath the top-level directory (such as a *doc* directory just below the top), life becomes easier. Just implement the code shown in Listings 18-11 and 18-12.

```
doxygen:
    ❶ $(am__cd) doc && $(MAKE) $(AM_MAKEFLAGS) $@

.PHONY: doxygen
```

Listing 18-11: A top-level makefile that propagates a target to a single subdirectory

```
doxygen:
        $(DOXYGEN) $(DOXYFLAGS) doxyfile

.PHONY: doxygen
```

Listing 18-12: doc/Makefile.am: The code to handle the new target

The shell statement at ❶ in the top-level makefile in Listing 18-11 simply passes the target (doxygen) down to the desired directory (doc).

NOTE   *The variables DOXYGEN and DOXYFLAGS are assumed to exist by virtue of some macro or shell code executed within the* configure *script.*

Automake recursive targets are more sophisticated in that they also support make's -k command line option to continue building after errors. Additionally, Automake's recursive target implementation supports the use of the dot (.) in the SUBDIRS variable, which represents the current directory. You may also support these features, but if you do, your boilerplate recursive make shell code will be messier. For the sake of completeness, Listing 18-13 shows an implementation that supports these features. Compare this listing to Listing 18-9. The highlighted shell code shows the differences between these listings.

```
my-recursive-target:
        $(preorder_commands)
        @failcom='exit 1'; \
        for f in x $$MAKEFLAGS; do \
          case $$f in \
            *=* | --[!k]*);; \
      ❶ *k*) failcom='fail=yes';; \
          esac; \
        done; \
        for dir in $(SUBDIRS); do \
```

```
❷ if test "$$dir" != .; then \
     ($(am__cd) $$dir && $(MAKE) $(AM_MAKEFLAGS) $@) || eval \
        $$failcom; \
   fi; \
 done
 $(postorder_commands)

.PHONY: my-recursive-target
```

Listing 18-13: Adding make -k and a check for the current directory

At ❶, the case statement checks for a -k option in the MAKEFLAGS environment variable and, on finding it, sets the failcom shell variable to some innocuous shell code. If it's not found, then failcom is left at its default value, exit 1, which is then inserted where an exit should occur on error. The if statement within the for loop at ❷ simply skips the recursive call for the dot entry in SUBDIRS. As with the previous examples, for the current directory, the functionality of the recursive target is found entirely within the $(preorder_commands) and $(postorder_commands) macro expansions.

I've tried to show you in this item that you can do as much or as little as you like with your own recursive targets. Most of the implementation is simply shell code in the command.

Item 3: Using a Repository Revision Number in a Package Version

Version control is an important part of every project. Not only does it protect intellectual property, but it also allows the developer to back up and start again after a long series of mistakes. One advantage of version control systems like Git and Subversion is that the system assigns a unique revision number to every change to a project's repository. This means that any distribution of the project's source code can be logically tied to a particular repository revision number. This item presents a technique you can use to automatically insert a repository revision number into your package's Autoconf version string.

Arguments to Autoconf's AC_INIT macro must be static text. That is, they can't be shell variables, and Autoconf will flag attempts to use shell variables in these arguments as errors. This is all well and good until you want to calculate any portion of your package's version number during the configuration process.

I once tried to use a shell variable in AC_INIT's VERSION argument so that I could substitute my Subversion revision number into the VERSION argument when configure was executed. I spent a couple of days trying to figure out how to trick Autoconf into letting me use a shell variable as a *revision* field in my package's version number. Eventually, I discovered the trick shown in Listing 18-14, which I implemented in my *configure.ac* file and in my top-level *Makefile.am* file.

```
❶ SVNREV=`LC_ALL=C svnversion $srcdir 2>/dev/null`
❷ if ! svnversion || case $SVNREV in Unver*) true;; *) false;; esac;
  ❸ then SVNREV=`cat $srcdir/SVNREV`
  ❹ else echo $SVNREV>$srcdir/SVNREV
  fi
❺ AC_SUBST(SVNREV)
```

Listing 18-14: configure.ac: *Implementing a dynamic revision number as part of the package version*

Here, the shell variable SVNREV is set at ❶ to the output of the svnversion command, as executed on the project top-level directory. The output is a raw Subversion revision number—that is, *if* the code is executed in a true Subversion work area, which isn't always the case.

When a user executes this configure script from a distribution archive, Subversion may not even be installed on his workstation. Even if it is, the top-level project directory comes from the archive, not a Subversion repository. To handle these situations, the line at ❷ checks to see if svnversion can be executed or if the output from the first line starts with the first few letters of the phrase *Unversioned directory,* the result of executing the svnversion utility on a non-work-area directory.

If either of these cases is true, the SVNREV variable is populated at ❸ from the contents of a file called *SVNREV.* The project should be configured to ship the *SVNREV* file with a distribution archive containing the configuration code in Listing 18-14. This must be done because if svnversion generates a true Subversion repository revision number, that value is immediately written to the *SVNREV* file by the else clause of this if statement at ❹.

Finally, the call to AC_SUBST at ❺ substitutes the SVNREV variable into template files, including the project makefiles.

In the top-level *Makefile.am* file, I ensure that the *SVNREV* file becomes part of the distribution archive by adding it to the EXTRA_DIST list. Thus, when a distribution archive is created and published by the maintainer, it contains an *SVNREV* file with the source tree revision number used to generate the archive from this source code. The value in the *SVNREV* file is also used when an archive is generated from the source code in this tarball (via make dist). This is accurate because the original archive was actually generated from this particular revision of the Subversion repository.

Generally, it's not particularly important that a project's distribution archive be able to generate a proper distribution archive, but an Automake-generated archive can do so without this modification, so it should also be able to do so *with* it. Listing 18-15 highlights the relevant changes to the top-level *Makefile.am* file.

```
EXTRA_DIST = SVNREV
distdir = $(PACKAGE)-$(VERSION).$(SVNREV)
```

Listing 18-15: Makefile.am: *A top-level makefile configured for SVN revision numbers*

In Listing 18-15, the distdir variable controls the name of the distribution directory and the archive filename generated by Automake. Setting this variable in the top-level *Makefile.am* file affects the generation of the distribution archive, because that *Makefile.am* file is where this functionality is located in the final generated *Makefile*.

NOTE *Note the similarity of the SVNREV filename and the SVNREV make variable [$(SVNREV)] in Listing 18-15. Although they appear to be the same, the text added to the EXTRA_DIST line refers to the SVNREV file in the top-level project directory, while the text added to the distdir variable refers to a make variable.*

For most purposes, setting distdir in the top-level *Makefile.am* file should be sufficient. However, if you need distdir to be formatted correctly in another *Makefile.am* file in your project, just set it in that file as well.

The technique presented in this item does not automatically reconfigure the project to generate a new *SVNREV* file when you commit new changes (and so change the Subversion revision used in your build). I could have added this functionality with a few well-placed make rules, but that would have forced the build to check for commits with each new build.[4]

Listing 18-16 shows code similar to that in Listing 18-14, except this code works for Git, rather than Subversion.

```
GITREV=`git -C $srcdir rev-parse --short HEAD`
if [ -z "$GITREV" ];
  then GITREV=`cat $srcdir/GITREV`
  else echo $GITREV>$srcdir/GITREV
fi
AC_SUBST(GITREV)
```

Listing 18-16: configure.ac: Implementing a Git dynamic revision number

This version seems a little more intuitive to me because the git utility makes better use of the proper output channels for error conditions—the output of the command is sent to stderr if the current working directory is not a Git repository.

Of course, you should also modify the code from Listing 18-15 to reference the *GITREV* file instead of the *SVNREV* file.

Another great option, if you're already using Gnulib, is to use the *version-gen* module in that library. This module provides many nice features related to incorporating a version number into your build.

Item 4: Ensuring Your Distribution Packages Are Clean

Have you ever downloaded and unpacked an open source package and then tried to run ./configure && make, only to have it fail halfway through one of these steps? As you dug into the problem, perhaps you discovered missing

---

4. My work habits are such that I tend to regenerate a build tree from scratch before releasing a new distribution package, so this issue doesn't really affect me that much.

files in the archive. How sad to have this happen in an Autotools project, when the Autotools make it so easy to ensure that this simply doesn't happen.

To ensure that your distribution archives are always clean and complete, run the distcheck target on a newly created archive. Don't be satisfied with what you *believe* about your package. Allow Automake to run the distribution unit tests. I call these tests *unit tests* because they provide the same testing functionality for a distribution package that regular unit tests provide for your source code.

You'd never make a code change and ship a package without running your unit tests, would you? (If so, then you can safely skip this section.) Likewise, don't ship your archives without running the build system unit tests—run **make distcheck** on your project *before* posting your new archives. If the distcheck target fails, find out why and fix it. The payoff is worth the effort.

Item 5: Hacking Autoconf Macros

Occasionally you need a macro that Autoconf doesn't quite provide. That's when it pays to know how to copy and modify existing Autoconf macros.[5]

For example, here's a solution to a common Autoconf mailing list issue. A user wants to use AC_CHECK_LIB to capture a desired library in the LIBS variable. The catch is that this library exports functions with C++, rather than C linkage. AC_CHECK_LIB is not very accommodating when it comes to C++, primarily because AC_CHECK_LIB makes certain assumptions about symbols exported with C linkage that just don't apply to C++ symbols.

For example, the widely known (and standardized) rules of C linkage state that an exported C-linkage symbol (also known as the cdecl calling convention on Intel systems) is case sensitive and decorated with a leading underscore,[6] whereas a symbol exported with C++ linkage is *mangled* using nonstandard, vendor-defined rules. The decorations are based on the signature of the function—specifically, the number and types of parameters as well as the classes and/or namespaces to which the function belongs. But the exact scheme is not defined by the C++ standard.

Now, stop and consider under what circumstances you're likely to have symbols exported from a library using C++ linkage. There are two ways to export C++ symbols from a library. The first is to (either purposely or accidentally) export *global* functions without using the extern "C" linkage specification on your function prototypes. The second is to export entire classes—including public and protected methods and class data.

If you've accidentally forgotten to use extern "C" on your global functions, well, then, stop it. If you're doing it on purpose, then I wonder why? The only reason I can think of is that you want to export more than one

---

5. This technique is also an excellent way to learn your way around Autoconf-provided macros.

6. The cdecl keyword or attribute does not decorate the symbol with a leading underscore on some systems.

overload of a given function name. This seems a rather trivial reason to keep your C developers from being able to use your library.

If you're exporting classes, now that's another story. In this case, you're catering specifically to C++ users, which presents a real issue with AC_CHECK_LIB.

Autoconf provides a framework around the definition of AC_CHECK_LIB that allows for differences between C and C++. If you use the AC_LANG([C++]) macro before you call AC_CHECK_LIB, you'll generate a version of the test program that's specific to C++. But don't get your hopes up; the current implementation of the C++ version is simply a copy of the C version. I expect that a generic C++ implementation would be difficult at best to design.

But all is not lost. While a *generic* implementation would be difficult, as the project maintainer, you can easily write a project-specific version of the test code using AC_CHECK_LIB's test code.

First we need to find the definition of the AC_CHECK_LIB macro. A grep of the Autoconf macro directory (usually */usr/(local/)share/autoconf/autoconf*) should quickly locate the definition of AC_CHECK_LIB in the file called *libs.m4*. Because most macro definitions start with a comment header containing a hash mark and then the name of the macro and a single space, the following should work:

```
$ cd /usr/share/autoconf/autoconf
$ grep "^# AC_CHECK_LIB" *.m4
libs.m4:# AC_CHECK_LIB(LIBRARY, FUNCTION,
$
```

The definition of AC_CHECK_LIB is shown in Listing 18-17.[7]

```
# AC_CHECK_LIB(LIBRARY, FUNCTION,
#              [ACTION-IF-FOUND], [ACTION-IF-NOT-FOUND],
#              [OTHER-LIBRARIES])
--snip--
# freedom.
AC_DEFUN([AC_CHECK_LIB],
[m4_ifval([$3], , [AH_CHECK_LIB([$1])])dnl
AS_LITERAL_IF([$1], [AS_VAR_PUSHDEF([ac_Lib], [ac_cv_lib_$1_$2])],
    [AS_VAR_PUSHDEF([ac_Lib], [ac_cv_lib_$1''_$2])])dnl
AC_CACHE_CHECK([for $2 in -l$1], [ac_Lib],
    [ac_check_lib_save_LIBS=$LIBS
    LIBS="-l$1 $5 $LIBS"
❶ AC_LINK_IFELSE([AC_LANG_CALL([], [$2])],
        [AS_VAR_SET([ac_Lib], [yes])],
        [AS_VAR_SET([ac_Lib], [no])])
    LIBS=$ac_check_lib_save_LIBS])
    AS_VAR_IF([ac_Lib], [yes],
        [m4_default([$3], [AC_DEFINE_UNQUOTED(AS_TR_CPP(HAVE_LIB$1))
        LIBS="-l$1 $LIBS"
    ])],
```

---

7. This version of AC_CHECK_LIB is from Autoconf version 2.63. Portions of the macro were rewritten in version 2.64, but this version is a bit easier to understand and analyze.

```
    [$4])dnl
AS_VAR_POPDEF([ac_Lib])dnl
])# AC_CHECK_LIB
```

Listing 18-17: The definition of AC_CHECK_LIB, as found in libs.m4

This apparent quagmire is easily sorted out with a little analysis. The macro appears to accept up to five arguments (as shown in the comment header), the first two of which are required. The highlighted portion is the macro definition—the part we'll copy into our *configure.ac* file and modify to work with our C++ exports.

Recall from Chapter 16 that the placeholders for M4 macro definition parameters are similar to those of shell scripts: a dollar sign followed by a number. The first parameter is represented by $1, the second by $2, and so on. We need to determine which parameters are important to us and which ones to discard. We know that most calls to AC_CHECK_LIB pass only the first two arguments. The third and fourth parameters are optional and exist only so that you can change the macro's default behavior, depending on whether it locates the desired function in the specified library. The fifth parameter allows you to provide a list of additional linker command line arguments (usually additional library and library directory references) that are required to properly link the desired library so the test program will not fail for extraneous reasons.

Say we have a C++ library that exports a class's public data and methods. Our library is named *fancy*, our class is Fancy, and the method we're interested in is called execute—specifically the execute method that accepts two integer arguments. Thus, its signature would be

```
Fancy::execute(int, int)
```

When exported with C linkage, such a function would be presented to the linker merely as _execute (or simply as execute, without the leading underscore, on some platforms), but when it's exported with C++ linkage, all bets are off because of vendor-specific name mangling.

The only way to get the linker to find this symbol is to declare it in compiled source code with exactly this signature, but we don't supply enough information to AC_CHECK_LIB to properly declare the function signature in the test code. Here's the declaration required to tell the compiler how to properly mangle this method's name:

```
class Fancy { public: void execute(int,int); };
```

Assuming that we're looking for a function with C linkage called execute, the AC_CHECK_LIB macro generates a small test program like the one shown in Listing 18-18. I've highlighted our function name so you can easily see where the macro inserts it into the generated test code.

```
/* confdefs.h. */
#define PACKAGE_NAME ""
```

```
#define PACKAGE_TARNAME ""
#define PACKAGE_VERSION ""
#define PACKAGE_STRING ""
#define PACKAGE_BUGREPORT ""
/* end confdefs.h. */

/* Override any GCC internal prototype to avoid an error.
   Use char because int might match the return type of a GCC
   builtin and then its argument prototype would still apply. */
#ifdef __cplusplus
extern "C"
#endif

char execute ();
int
main ()
{
return execute ();
    ;
    return 0;
}
```

Listing 18-18: An Autoconf-generated check for the global C-language execute function

Except for these two uses of the specified function name, the entire test program is identical for every call to AC_CHECK_LIB. This macro creates a common prototype for all functions so that all functions are treated the same way. Clearly, however, not all functions accept no parameters and return a character, as defined in this code. AC_CHECK_LIB effectively lies to the compiler about the true nature of the function. The test only cares whether the test program can successfully be linked; it will never attempt to execute it (an operation that would fail spectacularly in most cases).

For C++ symbols, we need to generate a different test program—one that makes no assumptions about the signature of our exported symbol.

Looking back at ❶ in Listing 18-17, it appears as if the AC_LANG_CALL macro has something to do with the generation of the test code in Listing 18-18 because the output of AC_LANG_CALL is generated directly into the first argument of a call to AC_LINK_IFELSE; its first argument is source code to be tested with the linker. As it turns out, this macro, too, is a higher-level wrapper around another macro, AC_LANG_PROGRAM. Listing 18-19 shows the definitions of both macros.[8]

```
# AC_LANG_CALL(C)(PROLOGUE, FUNCTION)
# ----------------------------------
# Avoid conflicting decl of main.
m4_define([AC_LANG_CALL(C)],
❶ [AC_LANG_PROGRAM([$1
```

---

8. I'm showing you the AC_LANG_CALL(C) macro here—the C-specific version that the polymorphic wrapper, AC_LANG_CALL, actually calls. The (C) on the end is actually part of the macro name, and some special trickery must be used to actually call this macro, as it cannot be called directly. To see how this works, look at the definition of the AC_LANG_CALL wrapper macro.

```
      m4_if([$2], [main], , 
      [/* Override any GCC internal prototype to avoid an error. 
          Use char because int might match the return type of a GCC 
          builtin and then its argument prototype would still apply. */ 
      #ifdef __cplusplus 
      extern "C" 
      #endif 
❷ char $2 ();])], [return $2 ();])]) 

      # AC_LANG_PROGRAM(C)([PROLOGUE], [BODY]) 
      # ----------------------------------- 
      m4_define([AC_LANG_PROGRAM(C)], 
❸ [$1 
      m4_ifdef([_AC_LANG_PROGRAM_C_F77_HOOKS], [_AC_LANG_PROGRAM_C_F77_HOOKS])[]dnl 
      m4_ifdef([_AC_LANG_PROGRAM_C_FC_HOOKS], [_AC_LANG_PROGRAM_C_FC_HOOKS])[]dnl 
      int 
      main () 
      { 
      dnl Do *not* indent the following line: there may be CPP directives. 
      dnl Don't move the `;' right after for the same reason. 
❹ $2 
        ; 
        return 0; 
      }]) 
```

Listing 18-19: The definitions of AC_LANG_CALL and AC_LANG_PROGRAM

At ❶, AC_LANG_CALL(C) generates a call to AC_LANG_PROGRAM, passing the PROLOGUE argument in the first parameter. At ❸, this prologue (in the form of $1) is immediately sent to the output stream. If the second argument passed to AC_LANG_CALL(C) (FUNCTION) is not main, a C-style function prototype is generated for the function. At ❷, the text return $2 (); is passed as the BODY argument to AC_LANG_PROGRAM, which uses this text at ❹ to generate a call to the function. (Remember that this code will only be linked, never executed.)

For C++, we need to be able to define more of the test program so that it makes no assumptions about the prototype of our exported symbol, and AC_LANG_CALL is too specific to C, so we'll use the lower-level macro, AC_LANG_PROGRAM, instead. Listing 18-20 shows how we might rework AC_CHECK_LIB to handle the function Fancy::execute(int, int) from a library called *fancy*. I've highlighted the places where I've modified the original macro definition of Listing 18-17 on page 512.

```
AC_PREREQ([2.59]) 
AC_INIT([test], [1.0]) 

AC_LANG([C++]) 

# --- A modified version of AC_CHECK_LIB 
m4_ifval([], , [AH_CHECK_LIB([fancy])])dnl 
❶ AS_VAR_PUSHDEF([ac_Lib], [ac_cv_lib_fancy_execute])dnl 
❷ AC_CACHE_CHECK([whether -lfancy exports Fancy::execute(int,int)], [ac_Lib], 
[ac_check_lib_save_LIBS=$LIBS 
```

```
      LIBS="-lfancy $LIBS"
❸ AC_LINK_IFELSE([AC_LANG_PROGRAM(
      [[class Fancy {
          public: void execute(int i, int j);
      };]],
      [[ MyClass test;
        test.execute(1, 1);]])],
        [AS_VAR_SET([ac_Lib], [yes])],
        [AS_VAR_SET([ac_Lib], [no])])
      LIBS=$ac_check_lib_save_LIBS])
      AS_VAR_IF([ac_Lib], [yes],
        [AC_DEFINE_UNQUOTED(AS_TR_CPP(HAVE_LIBFANCY))
        LIBS="-lfancy $LIBS"
      ],
      [])dnl
      AS_VAR_POPDEF([ac_Lib])dnl
      # --- End of modified version of AC_CHECK_LIB

      AC_OUTPUT
```

Listing 18-20: Hacking a modified version of AC_CHECK_LIB *into* configure.ac

In Listing 18-20, I've replaced the parameter placeholders with library and function names at ❶ and ❷ and added the prologue and body of the program to be generated by AC_LANG_PROGRAM at ❸. I've also removed some extraneous text that specifically had to do with the optional parameters of AC_CHECK_LIB that I don't care about in my version.

This code is much longer and more difficult to understand than a simple call to AC_CHECK_LIB, so it just begs to be turned into a macro. I'll leave that to you as an exercise. Having read Chapter 16, you should be able to do this without too much difficulty. Note also that there is much room for optimization in this macro. As you become more proficient with M4, you'll undoubtedly find ways you can reduced the size and complexity of this reworked macro, while maintaining the desired functionality.

Providing Library-Specific Autoconf Macros

This item is about hacking Autoconf macros when you need special features not provided by the standard macros, but the example I used was specifically about looking for a particular function in a library. This is a special case of a more general issue: finding libraries that provide desired functionality.

If you're a library developer, consider providing downloadable Autoconf macros that test for the existence of your libraries, and perhaps version-specific functionality within them. By doing so, you make it easier for your users to ensure that their users have proper access to your libraries.

Such macros don't have to be general purpose in nature, because they're tailored to a specific library. Library-specific macros are much easier to write and can be more thorough in testing for the functionality of your library. As the author, you're more likely to understand all the nuances of various versions of your library, so your macros can be spot-on with respect to determining library characteristics that your users may need to differentiate.

Item 6: Cross-Compiling

Cross-compilation occurs when the *build system* (the system on which the binaries are built) and the *host system* (the system on which those binaries are meant to be executed) are not of the same types. For example, we're cross-compiling when we build Motorola 68000 binaries for an embedded system on a typical Intel x86 platform running GNU/Linux, or when we build Sparc binaries on a DEC Alpha system running Solaris. A far more common scenario is using your Linux system to build software designed to run on an embedded microcontroller.

The situation becomes even more complex if the software you're building, such as a compiler or linker, can generate software. In this case, the *target system* represents the system for which your compiler or linker will ultimately generate code. When such a build system involves three different architectures, it's often referred to as a *Canadian cross*.[9] In this case, a compiler or linker is built on architecture A to run on architecture B and generate code for architecture C. Another type of three-system build, called a *cross-to-native* build, involves building an architecture-A compiler *on* architecture A to run on architecture B. In this case, three architectures are involved, but the host and target architectures are the same. Once you master the concepts of dual-system cross-compilation, moving on to using a three-system cross-compile mode is fairly simple.

Autoconf generates configuration scripts that attempt to guess the build system type and then assume that the host system type is the same. Unless told otherwise with command line options, `configure` assumes that non-cross-compilation mode is in effect. When executed without command line options that specify the build or host system types, an Autoconf-generated configuration script can usually accurately determine system type and characteristics.

NOTE *Section 14, "Manual Configuration," of the* GNU Autoconf Manual *discusses how to put Autoconf into cross-compilation mode. Unfortunately, the information that you'll need in order to write proper* configure.ac *files for cross-compilation is spread throughout that manual in bits and pieces. Each macro with nuances specific to cross-compilation has a paragraph describing the effects of cross-compilation mode on that macro. Search the manual for "cross-comp" to find all the references.*

System types are defined in the *GNU Autoconf Manual* in terms of a three-part canonical naming scheme involving CPU, vendor, and operating system, in the form `cpu-vendor-os`. But the os portion can itself be a pair containing a kernel and system type (`kernel-system`). If you know a canonical name for a system, you can specify it in each of three parameters to `configure`, as follows:

- `--build=build-type`
- `--host=host-type`
- `--target=target-type`

---

9. The name comes from the fact that during early discussions of cross-compilation issues on the internet, Canada had three political parties.

These configure command line options, with correct canonical system type names, allow you to define the build, host, and target system types. (Defining the host system type to be the same as your build system type is redundant, because this is the default case for configure.)

One of the most challenging (and least documented) aspects of using these options is determining a proper canonical system name to use in these command line options. Nowhere in the *GNU Autoconf Manual* will you find a statement that tells you how to contrive a proper canonical name because canonical names are not unique for each system type. For instance, in most valid cross-compilation configurations, the vendor portion of the canonical name is simply ignored and can thus be set to anything.

When you use the AC_CANONICAL_SYSTEM macro early in your *configure.ac* file, you'll find two new Autoconf helper scripts added to your project directory (by automake --add-missing, which is also executed by autoreconf --install). Specifically, these helper scripts are config.guess and config.sub. The job of config.guess is to determine, through heuristics, the canonical system name for your user's system—the build system. You can execute this program yourself to determine an appropriate canonical name for your own build system. For instance, on my 64-bit Intel GNU/Linux system, I get the following output from config.guess:

```
$ /usr/share/automake-1.15/config.guess
x86_64-pc-linux-gnu
$
```

As you can see here, config.guess requires no command line options, although there are a few available. (Use the --help option to see them.) Its job is to guess your system type, mostly based on the output of the uname utility. This guess is used as a default system type that can be overridden by a user on the configure command line. When cross-compiling, you can use this value in your --build command line option.[10]

The task of the config.sub program is to accept an input string as a sort of alias for a system type that you're looking for and then to convert it to a proper Autoconf canonical name. But what is a valid alias? For a few clues, search for "Decode aliases" within config.sub. You'll likely find a comment above a bit of code whose job it is to decode aliases for certain CPU-COMPANY combinations. Here are a few examples executed from my system; you should find the same results on your system:

```
$ /usr/share/automake-1.15/config.sub i386
i386-pc-none
$ /usr/share/automake-1.15/config.sub i386-linux
i386-pc-linux-gnu
```

---

10. For normal two-system cross-compilation mode, you shouldn't have to specify the build system type, only the host system type. However, for historical and backward-compatibility reasons, always use the --build option when you use --host. Specify the build system type as your actual build system type (such as i686-pc-linux-gnu on an Intel x86 GNU/Linux system). This requirement will be relaxed in a future version of Autoconf.

```
$ /usr/share/automake-1.15/config.sub m68k
m68k-unknown-none
$ /usr/share/automake-1.15/config.sub m68k-sun
m68k-sun-sunos4.1.1
$ /usr/share/automake-1.15/config.sub alpha
alpha-unknown-none
$ /usr/share/automake-1.15/config.sub alpha-dec
alpha-dec-ultrix4.2
$ /usr/share/automake-1.15/config.sub sparc
sparc-sun-sunos4.1.1
$ /usr/share/automake-1.15/config.sub sparc-sun
sparc-sun-sunos4.1.1
$ /usr/share/automake-1.15/config.sub mips
mips-unknown-elf
$
```

As you can see, a lone CPU name is usually not quite enough information for config.sub to properly determine a useful canonical name for a desired host system.

Notice, too, that there are a few generic keywords that can sometimes provide enough information for cross-compilation, without actually providing true vendor or operating system names. For instance, unknown can be substituted for the vendor name in general, and none is occasionally appropriate for the operating system name. Clearly elf is a valid system name as well, and it can be enough in some circumstances for configure to determine which tool chain to use. However, by simply appending a proper vendor name to the CPU, you can allow config.sub can take a pretty good stab at coming up with the most likely operating system for that pair and then generate a useful canonical system type name.

Ultimately, the best way to determine a proper canonical system type name is to examine config.sub for something close to what you think you should be using for a CPU and a vendor name and then simply ask it. While this may seem like a shot in the dark, chances are good that if you've gotten to the point of writing a build system for a program that should be cross-compiled, you're probably already very familiar with the names of your host CPU, vendor, and operating system.

When cross-compiling, you'll most likely use tools other than the ones you normally use on your system or, at the very least, additional command line options on your normal tools. Such tools are usually installed in sets as packages. Another clue to a proper host system canonical name is the prefix of these tools' names. There's nothing magic in the way Autoconf handles cross-compilation. The host system canonical name is used directly to locate the proper tools by name in the system path. Thus, the host system canonical name you use will have to match the prefix on your tools.

Now let's examine a common scenario: building 32-bit code on a 64-bit machine of the same CPU architecture. Technically, this is a form of cross-compilation, and it's often a much simpler scenario than cross-compiling code for an entirely different machine architecture. Many GNU/Linux systems support both 32- and 64-bit execution. On these systems, you can often use your build system's tool chain to perform this task with special command line options. For example, to build C source code for a 32-bit Intel system on a 64-bit Intel system, you would simply use the following `configure` command line (I've highlighted the lines related to cross-compilation):[11]

```
$ ./configure CPPFLAGS=-m32 LDFLAGS=-m32
checking for a BSD-compatible install... /usr/bin/install -c
checking whether build environment is sane... yes
checking for a thread-safe mkdir -p... /bin/mkdir -p
checking for gawk... gawk
checking whether make sets $(MAKE)... yes
❶ checking build system type... x86_64-pc-linux-gnu
❷ checking host system type... x86_64-pc-linux-gnu
checking for style of include used by make... GNU
checking for gcc... gcc
checking for C compiler default output file name... a.out
checking whether the C compiler works... yes
❸ checking whether we are cross compiling... no
checking for suffix of executables...
checking for suffix of object files... o
--snip--
```

Notice at ❸ that, as far as `configure` is concerned, we are not cross-compiling because we haven't given `configure` any command line options instructing it to use a different tool chain than it would normally use. As you can see at ❶ and ❷, both the build and host system types are what you'd expect for a 64-bit GNU/Linux system. Additionally, because my system is a dual-mode system, it can execute test programs compiled with these flags. They'll run on the 64-bit CPU in 32-bit mode just fine.

NOTE   *Many systems require that you install the 32-bit tools before* gcc *will even recognize the* -m32 *flag. For example, Fedora systems require the installation of the* glibc-devel .i686 *package, and my Linux Mint (Ubuntu-based) system required me to install the* gcc-multilib *package.*

To be even more certain of a proper build on Linux systems, you can also use the `linux32` utility to change the personality of your 64-bit system to that of a 32-bit system, like this:

```
$ linux32 ./configure CPPFLAGS=-m32 LDFLAGS=-m32
--snip--
checking whether we are cross compiling... no
```

---

11. Why not use `CFLAGS`? Using `CPPFLAGS` (C-PreProcessor FLAGS) has two positive effects: it properly renders C-preprocessor tests that rely on bit size, and it allows C++ compilers (which would normally honor `CXXFLAGS` over `CFLAGS`) to correctly define the proper bit size as well. Another popular option is to specify `CC="gcc -m32"`, thereby changing the compiler type to that of a 32-bit compiler. I've added `-m32` to both `CPPFLAGS` and `LDFLAGS` so the linker will also be notified of the architecture change. If you add `-m32` to the `CC` variable, you don't need to do this because the linker is called via the compiler.

```
--snip--
checking build system type... i686-pc-linux-gnu
checking host system type... i686-pc-linux-gnu
--snip--
```

We use `linux32` here because some subscripts executed by `configure` may inspect `uname -m` to determine the build machine's architecture. The `linux32` utility ensures that these scripts properly see a 32-bit Linux system. You can test this yourself by running `uname` under `linux32`:

```
$ uname -m
x86_64
$ linux32 uname -m
i686
$
```

To get this sort of cross-compile to work on a Linux dual-mode system, you usually need to install one or more 32-bit development packages, as noted previously. If your project uses other system-level services, such as a graphical desktop, you will need the 32-bit versions of these libraries, as well.

Now let's do it the more conventional (dare I say, *canonical*?) way. Rather than add `-m32` to the `CPPFLAGS` and `LDFLAGS` variables, we'll set the build and host system types manually on the `configure` command line and see what happens. Again, I've highlighted the output lines related to cross-compilation:

```
$ ./configure --build=x86_64-pc-linux-gnu --host=i686-pc-linux-gnu
checking for a BSD-compatible install... /usr/bin/install -c
checking whether build environment is sane... yes
checking for a thread-safe mkdir -p... /bin/mkdir -p
checking for gawk... gawk
checking whether make sets $(MAKE)... yes
❶ checking for i686-pc-linux-gnu-strip... no
checking for strip... strip
❷ configure: WARNING: using cross tools not prefixed with host triplet
checking build system type... x86_64-pc-linux-gnu
checking host system type... i686-pc-linux-gnu
checking for style of include used by make... GNU
❸ checking for i686-pc-linux-gnu-gcc... no
checking for gcc... gcc
checking for C compiler default output file name... a.out
checking whether the C compiler works... yes
checking whether we are cross compiling... yes
checking for suffix of executables...
checking for suffix of object files... o
--snip--
```

Several key lines in this example indicate that, as far as `configure` is concerned, we're cross-compiling. The cross-compilation build environment is `x86_64-pc-linux-gnu`, while the host is `i686-pc-linux-gnu`.

But notice the highlighted WARNING text at ❷. My system doesn't have a tool chain that's dedicated to building 32-bit Intel binaries. Such a tool chain includes all of the same tools required to build the 64-bit versions of my products, but the 32-bit versions are prefixed with the canonical system name of the host system. If you don't have a properly prefixed tool chain installed and available in the system path, configure will default to using the build system tools—those without a prefix. This can work fine if your build system's tools can cross-compile to the host system with proper command line options and if you've also specified those options in CPPFLAGS and LDFLAGS.

Normally, you'd have to install a tool chain designed to build the correct type of binaries. In this example, a version of such tools could easily be provided by creating soft links and simple shell scripts that pass additional required flags. According to the configure script output at ❶ and ❸, I need to provide i686-pc-linux-gnu- prefixed versions of strip and gcc.

Generally, such foreign tool chains are installed into an auxiliary directory, which means you'd have to add that directory to your system PATH variable in order to allow configure to find them. For this example, I'll just create them in *~/bin*.[12] Once again I've highlighted the output text related to cross-compilation:

```
$ ln -s /usr/bin/strip ~/bin/i686-pc-linux-gnu-strip
$ echo '#!/bin/sh
> gcc -m32 "$@"' > ~/bin/i686-pc-linux-gnu-gcc
$ chmod +x ~/bin/i686-pc-linux-gnu-gcc
$ ./configure --build=x86_64-pc-linux-gnu --host=i686-pc-linux-gnu
checking for a BSD-compatible install... /usr/bin/install -c
checking whether build environment is sane... yes
checking for a thread-safe mkdir -p... /bin/mkdir -p
checking for gawk... gawk
checking whether make sets $(MAKE)... yes
checking for i686-pc-linux-gnu-strip... i686-pc-linux-gnu-strip
checking build system type... x86_64-pc-linux-gnu
checking host system type... i686-pc-linux-gnu
checking for style of include used by make... GNU
checking for i686-pc-linux-gnu-gcc... i686-pc-linux-gnu-gcc
checking for C compiler default output file name... a.out
checking whether the C compiler works... yes
checking whether we are cross compiling... yes
checking for suffix of executables...
checking for suffix of object files... o
checking whether we are using the GNU C compiler... yes
checking whether i686-pc-linux-gnu-gcc accepts -g... yes
--snip--
$ make
--snip--
```
❶ `libtool: compile: i686-pc-linux-gnu-gcc -DHAVE_CONFIG_H -I. -I.. -g -O2`
```
 -MT print.lo -MD -MP -MF .deps/print.Tpo -c print.c -fPIC -DPIC -o
--snip--
$
```

---

12. If you try this, be sure your *$HOME/bin* directory is in your search path.

This time, `configure` was able to find the proper tools. Notice that the compiler command at ❶ no longer contains the `-m32` flag. It's there, but it's hidden inside the `i686-pc-linux-gnu-gcc` script. As far as the Autotools are concerned, `i686-pc-linux-gnu-gcc` already knows how to build 32-bit binaries on a 64-bit system.

Cross-compilation is not for the average end user. As open source software developers, we use packages like the Autotools to ensure that our end users don't have to be experts in software development in order to build and install our packages. But cross-compilation requires a certain level of system configuration that is beyond the scope of what the Autotools generally expect of end users. Additionally, cross-compilation is used most often within specialized fields, such as tool chain or embedded systems development. End users in these areas usually *are* experts in software development.

There are a few places where cross-compilation can, and possibly should, be made available to the average end user. However, I strongly encourage you to be explicit and detailed in the instructions you provide your users in your *README* and *INSTALL* documents.

Item 7: Emulating Autoconf Text Replacement Techniques

Say your project builds a daemon that is configured at startup with values in a configuration text file. How does the daemon know where to find this file on startup? One way is to simply assume it's located in */etc*, but a well-written program will allow the user to configure this location when building the software. The system configuration directory has a variable location whose value can be specified on the `configure`, `make all`, or `make install` command line, as shown in the following examples:

```
$ ./configure --sysconfdir=/etc
--snip--
$ make all sysconfdir=/usr/mypkg/etc
--snip--
$ sudo make install sysconfdir=/usr/local/mypkg/etc
--snip--
```

All of these examples take advantage of command line functionality provided by Autotools build systems, so they must all be carefully taken into account when creating project and project build source files. Let's look at some examples that will explain how to do this.

Now, some conditions simply can't work. For instance, you can't pass a system configuration directory path into C source code from within the makefile when you build your program and then expect it to run correctly if you change where the configuration files are installed on the `make install` command line. Most end users won't pass anything on the command line, but you should still ensure that they can set prefix directories from the `configure` and `make` command lines.

This item is focused on placing command line prefix variable override information into the proper locations in your code and installed data files as late as possible in the build process.

Autoconf replaces text in AC_SUBST variables with the values of those variables as defined in configure at configuration time, but it doesn't replace the text with raw values. In an Autotools project, if you execute configure with a specific datadir, you get the following:

```
$ ./configure --datadir=/usr/share
--snip--
$ grep "datadir =" Makefile
pkgdatadir = $(datadir)/b64
❶ datadir = /usr/share
$
```

You can see at ❶ that the value of the shell variable datadir in configure is substituted exactly according to the command line instructions in the make variable datadir in *Makefile*. What's not obvious here is that the default value of datadir, both in the configure script and in the makefile after substitution, is relative to other variables within the build system. By not overriding datadir on the configure command line, we see that the default value in the makefile contains unexpanded shell variable references:

```
$ ./configure
--snip--
$ cat Makefile
--snip--
datadir = ${datarootdir}
datarootdir = ${prefix}/share
--snip--
prefix = /usr/local
--snip--
$
```

In Chapter 3 (see Listing 3-36), we saw that we could pass command line options to the preprocessor that would allow us to consume these sorts of path values within our source code. Listing 18-21 demonstrates this by passing a C-preprocessor definition in the CPPFLAGS variable for a hypothetical program called myprog.[13]

```
myprog_CPPFLAGS = -DSYSCONFDIR="\"@sysconfdir@\""
```

Listing 18-21: Pushing prefix variables into C source code in Makefile.am *or* Makefile.in

---

13. The escaped double quotes in this example are passed as part of the definition to the preprocessor and ultimately into the source code. The unescaped double quotes are stripped off by the shell as it passes the option on the compiler command line. The unescaped double quotes allow the value of the definition to contain spaces, which are not protected by the escaped double quotes because the shell doesn't recognize them as quotes.

A C source file might then contain the code shown in Listing 18-22.

```
--snip--
#ifndef SYSCONFDIR
# define SYSCONFDIR "/etc"
#endif
--snip--
const char * sysconfdir = SYSCONFDIR;
--snip--
```

Listing 18-22: Using the preprocessor-defined variables in C source code

Automake does nothing special with the line in Listing 18-21 between *Makefile.am* and *Makefile.in*, but the configure script converts the *Makefile.in* line into the *Makefile* line shown in Listing 18-23.

```
myprog_CPPFLAGS = -DSYSCONFDIR="\"${prefix}/etc\""
```

Listing 18-23: The resulting Makefile line after configure substitutes @sysconfdir@

When make passes this option on the compiler command line, it dereferences the variables to produce the following output command line (shown only in part here):

```
libtool: compile: gcc ... -DSYSCONFDIR=\"/usr/local/etc\" ...
```

There are a couple of problems with this approach. First, between configure and make, you lose the resolution of the sysconfdir variable because configure substitutes ${prefix}/*etc*, rather than ${sysconfdir}, for @sysconfdir@. The problem is that you can no longer set the value of sysconfdir on the make command line. To solve this problem, use the ${sysconfdir} make variable directly in your CPPFLAGS variable, as shown in Listing 18-24, rather than the Autoconf @sysconfdir@ substitution variable.

```
myprog_CPPFLAGS = -DSYSCONFDIR="\"${sysconfdir}\""
```

Listing 18-24: Using the make variable in CPPFLAGS instead of the Autoconf substitution variable

You can use this approach to specify a value for sysconfdir on both the configure and make command lines. Setting the variable on the configure command line defines a default value in *Makefile.in* (and subsequently in the generated *Makefile*), which can then be overridden on the make command line.

The problem with using different values on the make all and make install command lines is a bit more subtle. Consider what happens if you do the following:

```
$ make sysconfdir=/usr/local/myprog/etc
--snip--
$ sudo make install sysconfdir=/etc
--snip--
$
```

Here, you're basically lying to the compiler when you tell it that your configuration file will be installed in */usr/local/myprog/etc* during the build. The compiler will happily generate the code in Listing 18-22 so that it refers to this path; the second command line will then install your configuration file into */etc*, and your program will contain a hardcoded path to the wrong location. Unfortunately, there's little that you can do to correct this, because you've allowed your users to define these variables anywhere and because the *GNU Coding Standards* state the make install shouldn't recompile anything.

NOTE   *There are cases where different installation paths are given to the build and install processes on purpose. Recall the discussion of DESTDIR in "Getting Your Project into a Linux Distro" on page 67, wherein RPM packages are built and installed in a staging directory so that built products can be packaged in an RPM to be installed into the correct location later.*

Regardless of the potential pitfalls, being able to specify installation locations on the make command line is a powerful technique, but one that only works in makefiles because it relies heavily on make variable substitution within compiler command lines in your makefiles.

What if you want to replace a value in an installed data file that isn't processed by make on a shell command line? You could convert your data file into an Autoconf template and then simply reference the Autoconf substitution variable within that file.

In fact, we did just that in the *doxyfile.in* templates that we created for the FLAIM project in Chapter 15. However, this only worked in Doxygen input files because the class of variables used in those templates is always defined with complete absolute or relative paths by configure. That is, the values of @srcdir@ and @top_srcdir@ contain no additional shell variables. These variables are not installation directory (prefix) variables, which, with the exception of prefix itself, are always defined relative to other prefix variables.

You can, however, *emulate* the Autoconf substitution variable process within a makefile, allowing substitution variables to be used in installed data files. Listing 18-25 shows a template in which you might want to replace variables with path information normally found in the standard prefix variables during a build.

```
# Configuration file for myprog
logdir = @localstatedir@/log
--snip--
```

Listing 18-25: A sample config file template for myprog, *to be installed in* $(sysconfdir)

This template is for a program configuration file, which might normally be installed in the system configuration directory. We want the location of the program's log file, specified in this configuration file, to be determined at install time by the value of @localstatedir@. Unfortunately, configure would replace this variable with a string containing at least ${prefix}, which is not useful in a program configuration file. Listing 18-26 shows a *Makefile.am* file with additional make script to generate *myprog.cfg* by performing substitution on variables in *myprog.cfg.in*.

```
   EXTRA_DIST = myprog.cfg.in
❶ sysconf_DATA = myprog.cfg

❷ edit = sed -e 's|@localstatedir[@]|$(localstatedir)|g'
❸ myprog.cfg: myprog.cfg.in Makefile
           $(edit) $(srcdir)/$@.in > $@.tmp
           mv $@.tmp $@

   CLEANFILES = myprog.cfg
```

Listing 18-26: Substituting make variables into data files using sed in a makefile

In this *Makefile.am* file, I've defined a custom make target at ❸ to build the *myprog.cfg* data file. I've also defined a make variable called edit at ❷, which resolves to a partial sed command that replaces all instances of @localstatedir@ in the template file ($(srcdir)/*myprog.cfg.in*) with the value of the $(localstatedir) variable. Because make recursively processes variable replacements until all variable references are resolved, using make in this manner will ensure that you never leave any variable references in your final output. In the command where this variable is used, sed's output is redirected to the output file (*myprog.cfg*).[14]

The only nonobvious code in this example is the use of the square brackets around the trailing at sign (@) in the sed expression, which represent regular expression syntax indicating that any of the enclosed characters should be matched. Because there is only one enclosed character, this would seem to be a pointless complication, but the purpose of these brackets is to keep configure from replacing @localstatedir@ in the edit variable when it performs Autoconf variable substitution on this makefile. We want make to use this variable, not configure.

I assign *myprog.cfg* to the sysconf_DATA variable at ❶ to tie execution of this new rule into the framework provided by Automake. Automake will install this file into the system configuration directory after building it, if necessary.

The files in DATA primaries are added as dependencies to the all target via the internal all-am target. If *myprog.cfg* doesn't exist, make will look for a rule to build it. Since I have such a rule, make will simply execute that rule when I build the all target.

I've added the template file name *myprog.cfg.in* to the EXTRA_DIST variable at the top of Listing 18-26 because neither Autoconf nor Automake is aware of this file. In addition, I've added the generated file *myprog.cfg* to the CLEANFILES variable at the bottom of the listing because, as far as Automake is concerned, *myprog.cfg* is a distributed data file that should not be automatically deleted by make clean.

---

14. It's actually redirected to a temporary file called *myprog.cfg.tmp*, which is renamed atomically using mv in the next command to *myprog.cfg*. This is done so that parallel make (make -j) won't see the output file in a half-baked state.

NOTE *This example demonstrates a good reason for Automake to not automatically distribute files listed in* DATA *primaries. Sometimes such files are built in this manner. If built data files were automatically distributed, the* distcheck *target would fail because* myprog.cfg *was not available for distribution before building.*

In this example, I tied the building of *myprog.cfg* into the install process by adding it to the sysconf_DATA variable, and then I placed a dependency between *mydata.cfg.in* and *mydata.cfg*[15] to ensure that the installed file is built when make all is executed. You could also tie into a standard or custom build or installation target using appropriate -hook or custom targets.

No discussion of this topic would be complete without a mention of the Gnulib *configmake* module. If you're already using Gnulib and need to do something like what I've been talking about in this item, consider using *configmake*, which creates a *configmake.h* header file that can be included by your source files to provide access to all of the standard directory variables as C preprocessor macros. It's only useful for C code, so you'd still need the techniques I've shown you here for non-C-source code use cases (such as installed configuration files that need to reference prefix variable paths).

Item 8: Using the Autoconf Archive Project

In "Item 5: Hacking Autoconf Macros" on page 511, I demonstrated a technique for hacking Autoconf macros to provide functionality that's close to, but not exactly the same as, that of the original macro. When you need a macro that Autoconf doesn't provide, you can either write it yourself or look for one that someone else has written. This item is about the second option, and a perfect place to begin your search is the Autoconf Archive project.

As of this writing, the Autoconf Archive source project is hosted by GNU Savannah.[16] The original ac-archive project was the result of a merger between two older projects: one by Guido Draheim (at *http://ac-archive.sourceforge.net/*) and the other by Peter Simon (at *http://auto-archive.cryp.to*). The first of these sites is still online today, although it displays a huge red warning box indicating that you should submit updates to the GNU Autoconf Macro Archive at Savannah; the second has been taken down. There is some long history and not a few flame wars on email lists between these two projects. Ultimately, each project incorporated most of the contents of the other, but Peter Simon's is the one that was migrated into the Savannah repository, and the current home page is found at *https://www.gnu.org/software/autoconf-archive/*.[17]

---

15. Note the dependency on *Makefile* as well. If *Makefile* changes, the sed expression or command line may have changed, in which case *myprog.cfg* should be regenerated. As of this writing, make has no inherent functionality to tie particular commands within the makefile to a given target, so if the makefile changes in anyway, we must assume that it affects *myprog.cfg*.

16. See *http://git.savannah.nongnu.org/cgit/autoconf-archive.git*. GitHub supplies a mirror at *https://github.com/autoconf-archive/autoconf-archive*.

17. It appears that Guido has given up, because the last updates to his SourceForge project were made in August 2007.

The value in the archive is that private macros become public and public macros are incrementally improved by many users.

As of this writing, the macro archive contains over 500 macros not distributed with Autoconf, including the AX_PTHREAD macro discussed in "Doing Threads the Right Way" on page 384. The latest release of the archive can be checked out from the project's Savannah git site. The site indexes macros by category, author, and open source license, allowing you to choose macros based on specific criteria. You can also search for a macro by name or by entering any text that might be found in the macro's header comments.

If you find yourself in need of a macro that Autoconf doesn't appear to provide, check out the Autoconf Archive.

Item 9: Using Incremental Installation Techniques

Some people have requested that make install be made smart enough to install only files that are not already installed or that are newer than installed versions of the same files.

This feature is available by default to users by passing the -C command line option to install-sh. It can be enabled directly by end users by using the following syntax on the make command line during execution of make install:

```
$ make install "INSTALL=/path/to/install-sh -C"
```

If you think your users will benefit from this option, consider adding some information about its proper use to the *INSTALL* file that ships with your project. Don't you just love features you don't have to implement?

Item 10: Using Generated Source Code

Automake requires that all source files used within a project be statically defined within the project's *Makefile.am* files, but sometimes the contents of source files need to be generated at build time.

There are two ways to deal with generated sources (more specifically, generated header files) in your projects. The first involves the use of an Automake-provided crutch for developers not interested in the finer points of make. The second involves writing proper dependency rules to allow make to understand the relationships between your source files and your products. I'll cover the crutch first, and then we'll get into the details of proper dependency management in *Makefile.am* files.

Using the BUILT_SOURCES Variable

When you have a header file that's generated as part of your build process, you can tell Automake to generate rules that will always create this file first, before attempting to build your products. To do this, add the header file to the Automake BUILT_SOURCES variable, as shown in Listing 18-27.

```
bin_PROGRAMS = program
program_SOURCES = program.c program.h
nodist_program_SOURCES = generated.h
BUILT_SOURCES = generated.h
CLEANFILES = generated.h
generated.h: Makefile
        echo "#define generated 1" > $@
```

Listing 18-27: Using BUILT_SOURCES to deal with generated source files

The `nodist_program_SOURCES` variable ensures that Automake will not generate rules that try to distribute this file; we want it to be built when the end user runs `make`, not shipped in the distribution package.

Without a user-provided clue, Automake-generated makefiles have no way of knowing that the rule for *generated.h* should be executed before *program.c* is compiled. I call `BUILT_SOURCES` a "crutch" because it simply forces the rules used to generate the listed files to execute first, and only when the user makes the `all` or `check` target. The rules created using `BUILT_SOURCES` aren't even executed if you attempt to make the `program` target directly. With that said, let's look at what's going on under the covers.

Dependency Management

There are two distinct classes of source files in a C or C++ project: those explicitly defined as dependencies within your makefile and those referenced only indirectly through, for instance, preprocessor inclusion.

You can hardcode all of these dependencies directly into your makefiles. For instance, if *program.c* includes *program.h*, and if *program.h* includes *console.h* and *print.h*, then *program.o* actually depends on all of these files, not just *program.c*. And yet, a normal handcoded makefile explicitly defines only the relationships between the *.c* files and the program. For a truly accurate build, `make` needs to be told about all of these relationships using a rule like the one shown in Listing 18-28.

```
program: program.o
        $(CC) $(CFLAGS) $(LDFLAGS) -o $@ program.o

❶ program.o: program.c program.h console.h print.h
        $(CC) -c $(CPPFLAGS) $(CFLAGS) -o $@ program.c
```

Listing 18-28: Rules describing the complete relationship between files

The relationship between *program.o* and *program.c* is often defined by an *implicit* rule, so the rule at ❶ in Listing 18-28 is often broken into two separate rules, as shown in Listing 18-29.

```
program: program.o
        $(CC) $(CFLAGS) $(LDFLAGS) -o $@ program.o

❶ %.o: %.c
```

```
        $(CC) -c $(CPPFLAGS) $(CFLAGS) -o $@ $<
```

❷ program.o: program.h console.h print.h

Listing 18-29: An implicit rule for C source files, defined as a GNU make pattern rule

In Listing 18-29, the GNU make-specific _pattern rule_ at ❶ tells make that the associated command can generate a file ending in _.o_ from a file of the same base name ending in _.c_.[18] Thus, whenever make needs to find a rule to generate a file ending in _.o_ that's listed as a dependency in one of your rules, it searches for a _.c_ file with the same base name. If it finds one, it applies this rule to rebuild the _.o_ file from the corresponding _.c_ file if the timestamp on the _.c_ file is newer than that of the existing _.o_ file or if the _.o_ file is missing.

There is a documented set of implicit pattern rules built into make, so you don't generally have to write such rules. Still, you must somehow tell make about the indirect[19] dependencies between the _.o_ file and any included _.h_ files. These dependencies cannot simply be implied with a built-in rule because there are no implicit relationships between these files that are based on file naming conventions, such as the relationship between _.c_ and _.o_ files. The relationships are manually coded into the source and header files as inclusions.

As I mentioned in Chapter 3, writing such rules is tedious and error prone, because during development (and even maintenance, to a lesser degree), the myriad relationships between source and header files can change all the time and the rules must be updated carefully with each change to keep the build accurate. The C preprocessor is much better suited to automatically writing and maintaining these rules for you.

A Two-Pass System

There are two ways to use the preprocessor to manage dependencies. The first is to create a two-pass system, wherein the first pass just builds the dependencies and the second pass compiles the source code based on those dependencies. This is done by defining rules that use certain preprocessor commands to generate make dependency rules, as shown in Listing 18-30.[20]

---

18. Simple GNU make pattern rules like this can also be implemented in standard Unix make using double suffix rules. For instance, the line %.o: %.c could be replaced with .c.o:.

19. I use the term _indirect_ here to mean that the _.o_ file depends on the _.h_ file _through_ the _.c_ file. That is, the _.o_ file is built from the _.h_ file by virtue of the fact that it's included by the _.c_ file. Technically, the _.o_ file's dependency on the _.h_ file is just as direct as that of the _.c_ file, because when the compiler picks up where the preprocessor leaves off, there are no _.h_ files—only a single file composed of the _.c_ file and all included header files—a _translation unit_, in the vernacular.

20. Microsoft has apparently never felt the need to support the make utility to the same degree that Unix compiler vendors have, instead relying heavily on its IDEs to create properly defined dependency graphs for project builds. Thus, while the preprocessor option used here is generally portable among Unix compilers, Microsoft compilers simply have no support for this sort of feature.

```
program: program.o
        $(CC) $(CFLAGS) $(LDFLAGS) -o $@ program.o

%.o: %.c
        $(CC) $(CPPFLAGS) -c $(CFLAGS) -o $@ $<

❶ %.d: %.c
        $(CC) -M $(CPPFLAGS) $< >$@

❷ sinclude program.d
```

Listing 18-30: Building automatic dependencies directly

In Listing 18-30, the pattern rule at ❶ specifies the same sort of relationship between *.d* and *.c* files as the one shown at ❶ in Listing 18-29 does for *.o* and *.c* files. The sinclude statement here at ❷ tells make to include another makefile, and GNU make is smart enough not only to ensure that all makefiles are included before the primary dependency graph is analyzed but also to look for rules to build them.[21] Running make on this makefile produces the following output:

```
$ make
cc -M program.c >program.d
cc -c -o program.o program.c
cc -o program program.o
$
$ cat program.d
program.o: program.c /usr/include/stdio.h /usr/include/features.h \
/usr/include/sys/cdefs.h /usr/include/bits/wordsize.h \
❶ --snip--
/usr/include/bits/pthreadtypes.h /usr/include/alloca.h program.h \
console.h print.h
$
$ touch console.h && make
cc -c -o program.o program.c
cc -o program program.o
$
```

As you can see here, the rule to generate *program.d* is executed first, as make attempts to include that file. The elided section at ❶ refers to the many system header files traversed while recursively scanning the included set of headers. The file contains a dependency rule similar[22] to the one we wrote

---

21. Only GNU make is smart enough to silently include dependency files with sinclude. Other brands of make provide only include, which will fail if any of the included makefiles are missing. GNU make is also the only version smart enough to re-execute itself when it notices the build system has been updated.

22. The GNU toolset supports several non-portable extensions to the classic -M option. For example, the -MM option has the wonderful effect of not bothering to add system header files to generated dependency lists. So, the long list of system headers omitted in the example need not be present at all if portability is not a concern. The -MD and -MMD options used in the examples are not portable either.

at ❷ in Listing 18-29. (The reference to *program.c* is missing in our hand-coded rule's dependency list because it's redundant, though harmless.) You can also see from the console example that touching one of these included files now properly causes the *program.c* source file to be rebuilt.

The problems with the mechanism outlined in Listing 18-30 include the fact that the entire source tree must be traversed twice: once to check for and possibly generate the dependency files and then again to compile any modified source files.

Another problem is that if one header includes another, and the second header is modified, the object file will be updated but not the dependency file included by make. The next time the second-level header is modified, neither the object nor the dependency file will be updated. Deleted header files also cause problems: the build system doesn't recognize that the deleted file was purposely removed, so it complains that files referenced in the existing dependencies are missing.

Doing It in One Pass

A more efficient way to handle automatic dependencies is to generate the dependency files as a side effect of compilation. Listing 18-31 shows how this can be done by using the non-portable -MMD GNU extension compiler option (highlighted in the listing).

```
program: program.o
        $(CC) $(CFLAGS) $(LDFLAGS) -o $@ program.o

%.o: %.c
❶       $(CC) -MMD $(CPPFLAGS) -c $(CFLAGS) -o $@ $<

❷ sinclude program.d
```

Listing 18-31: Generating dependencies as a side effect of compilation

Here, I've removed the second pattern rule (originally shown at ❶ in Listing 18-30) and added a -MMD option to the compiler command line at ❶ in Listing 18-31. This option tells the preprocessor to generate a *.d* file of the same base name as the *.c* file that it's currently compiling. When make is executed on a clean work area, the sinclude statement at ❷ silently fails to include the missing *program.d* file, but it doesn't matter because all of the object files will be built the first time anyway. During subsequent incremental builds, the previously built *program.d* is included, and its dependency rules take effect during those builds.

Built Sources Done Right

The one-pass method just described is roughly the one that Automake uses to manage automatic dependencies, when possible. The problems with this approach are most often manifested when working with generated sources, including both *.c* files and *.h* files. For instance, let's expand the example shown in Listing 18-31 a bit to contain a generated header file called *generated.h*, included by *program.h*. Listing 18-32 shows a first attempt at this modification. Additions to Listing 18-31 are highlighted in this listing.

```
program: program.o
        $(CC) $(CFLAGS) $(LDFLAGS) -o $@ program.o

%.o: %.c
        $(CC) -MMD $(CPPFLAGS) -c $(CFLAGS) -o $@ $<

generated.h: Makefile
        echo "#define generated" >$@

sinclude program.d
```

Listing 18-32: A makefile that works with a generated header file dependency

In this case, when we execute make, we find that the lack of an initial dependency file works against us:

```
$ make
cc -MMD -c -o program.o program.c
In file included from program.c:4:
program.h:3:23: error: generated.h: No such file or directory
make: *** [program.o] Error 1
$
```

Because there is no initial secondary dependency information, make doesn't know it needs to run the commands for the *generated.h* rule yet, because *generated.h* only depends on *Makefile*, which hasn't changed. To fix this problem in a *Makefile.am* file, we could just list *generated.h* in the BUILT_SOURCES variable, as we did in Listing 18-27 on page 530. This would add *generated.h* as the first dependency of the all and check targets, thereby forcing them to be built first in the likely event the user happens to enter make, make all, or make check.[23]

The proper way to handle this problem is very simple, and it works every time in both makefiles and *Makefile.am* files: write a dependency rule between *program.o* and *generated.h*, as shown in the updated makefile in Listing 18-33. The highlighted line contains the additional rule.

```
program: program.o
        $(CC) $(CFLAGS) $(LDFLAGS) -o $@ program.o

%.o: %.c
        $(CC) -MMD $(CPPFLAGS) -c $(CFLAGS) -o $@ $<

program.o: generated.h
```

---

23. Note that you can't rely on dependency order for build order with parallel make (make -j).

```
generated.h: Makefile
        echo "#define generated" >$@

sinclude program.d
```

Listing 18-33: Adding a hardcoded dependency rule for a generated header file

The new rule tells make about the relationship between *program.o* and *generated.h*:

```
$ make
echo "#define generated" >generated.h
cc -MMD -c -o program.o program.c
cc -o program program.o
$
$ make
make: 'program' is up-to-date.
$
❶ $ touch generated.h && make
cc -MMD -c -o program.o program.c
cc -o program program.o
$
❷ $ touch Makefile && make
echo "#define generated" >generated.h
cc -MMD -c -o program.o program.c
cc -o program program.o
$
```

Here, touching *generated.h* (at ❶) causes program to be updated. Touching *Makefile* (at ❷) causes *generated.h* to be re-created first.

To implement the dependency rule shown in Listing 18-33 in an Automake *Makefile.am* file, you'd use the highlighted rule shown in Listing 18-34.

```
bin_PROGRAMS = program
program_SOURCES = program.c program.h
nodist_program_SOURCES = generated.h
program.$(OBJEXT): generated.h
CLEANFILES = generated.h
generated.h: Makefile
        echo "#define generated 1" > $@
```

Listing 18-34: Replacing BUILT_SOURCES with a proper dependency rule

This is exactly the same code shown previously in Listing 18-27 on page 530, except that we've replaced the BUILT_SOURCES variable with a proper dependency rule. The advantage of this method is that it always works as it should; *generated.h* will always be built exactly when it needs to be, regardless of the target specified by the user.[24]

If you had tried to generate a C source file rather than a header file, you'd find that you didn't even need the additional dependency rule because *.o* files implicitly depend on their *.c* files. However, you must still list your generated *.c* file in the `nodist_program_SOURCES` variable to keep Automake from trying to distribute it.

NOTE *When you define your own rule, you suppress any rules that Automake may generate for that product. In the case of a specific object file, this is not likely to be a problem, but keep this Automake idiosyncrasy in mind when defining rules.*

As you can see, all you really need to properly manage generated sources is a correctly written set of dependency rules as well as appropriate `nodist_*_SOURCES` variables. The `make` utility and the Autotools provide the required framework in the form of built-in `make` functionality, macros, and variables. You just have to put them together correctly. For example, in the *GNU Automake Manual*, see Section 8.1.2, which discusses program linking.[25] This section refers to the `EXTRA_prog_DEPENDENCIES` variable as a mechanism for extending Automake's generated dependency graph for a specific target.

Item 11: Disabling Undesirable Targets

Sometimes the Autotools do too much for you. Here's an example from the Automake mailing list:

> I use automake in one of my projects along with texinfo. That project has documentation full of images. As you probably know, `make pdf` makes a PDF document from JPGs and PNGs, whereas `make dvi` requires EPSs. However, EPS images are insanely large (in this case like 15 times larger than JPGs).
>
> The problem is that running `make distcheck` results in error since the EPS images that should be there aren't there and `make distcheck` tries to run `make dvi` everywhere. I would like to run `make pdf` instead, or at least to disable building DVI. Is there any way to accomplish that?

---

24. This technique fails when you try to use program-specific Automake flags. For example, if you use `program_CFLAGS`, Automake generates a different set of rules for building the objects associated with the `program` and it munges the object name to contain the program name. This way, these special objects won't be confused with ones generated for other products from the same sources, but your handcoded dependency rules won't line up with the object filenames generated by the compiler. For more information, see the documentation for the `AC_PROG_CC_C_O` macro in the *GNU Autoconf Manual*.

25. See *https://www.gnu.org/software/automake/manual/html_node/Linking.html* in the March 11, 2018, version of the online GNU Automake manual.

First a little background information: The Automake `TEXINFOS` primary makes several documentation targets available to the end user, including `info`, `dvi`, `ps`, `pdf`, and `html`. It also provides several installation targets, including `install-info`, `install-dvi`, `install-ps`, `install-pdf`, and `install-html`. Of these targets, only `info` is automatically built with `make` or `make all`, and only `install-info` is executed with `make install`.[26]

However, it appears that the `distcheck` target also builds at least the `dvi` target, as well. The problem just outlined is that the poster doesn't provide the Encapsulated PostScript (EPS) graphics files required to build the DVI documentation, so the `distcheck` target fails because it can't build documentation that the poster doesn't want to support anyway.

To fix this issue, you would simply provide your own version of the target that does nothing, as shown in Listing 18-35.

```
--snip--
info_TEXINFOS = zardoz.texi
❶ dvi: # do nothing for make dvi
```

Listing 18-35: Disabling the dvi target in a Makefile.am that specifies TEXINFOS primaries

With the one-line addition at ❶, `make distcheck` is back in business. Now, when it builds the `dvi` target, it succeeds because it does nothing.

Other Automake primaries provide multiple additional targets as well. If you only want to support a subset of these targets, you can effectively disable the undesired targets by providing one of your own. If you'd like to be a bit more vocal about the disabling override, simply include an `echo` statement as a command that tells the user that your package doesn't provide DVI documentation, but be careful not to execute anything that might fail in this override, or your user will be right back in the same boat.

Item 12: Watch Those Tab Characters!

Having made the transition to Automake, you're not using raw makefiles anymore, so why should you still care about TAB characters? Remember that *Makefile.am* files are simply stylized makefiles. Ultimately, every line in a *Makefile.am* file will be either consumed directly by Automake, and then transformed into true `make` syntax, or copied directly into the final makefile. This means that TAB characters matter within *Makefile.am* files.

---

26. The *.info* files generated by the `info` target are automatically distributed, so your users don't have to have *texinfo* installed.

Consider this example from the Automake mailing list:

```
lib_LTLIBRARIES = libfoo.la
libfoo_la_SOURCES = foo.cpp
if WANT_BAR
❶    libfoo_la_SOURCES += a.cpp
else
❷    libfoo_la_SOURCES += b.cpp
endif

AM_CPPFLAGS = -I${top_srcdir}/include
libfoo_la_LDFLAGS = -version-info 0:0:0
```

I have been reading both autoconf and automake manuals and as far as I can see, the above should work. However the files (a.cpp or b.cpp) [are] always added at the bottom of the generated Makefile and are therefore not used in the compilation. No matter what I try, I cannot get even the above code to generate a correct makefile, but obviously I am doing something wrong.

The answer, provided by another poster, was simple and accurate, if not terse to a fault:

Remove the indentation.

The trouble here is that the two lines within the Automake conditional at ❶ and ❷ are indented with TAB characters.

You may recall from "Automake Configuration Features" on page 380, where I discussed the implementation of Automake conditionals, that text within conditionals is prefixed with an Autoconf substitution variable that is ultimately transformed into either an empty string or a hash mark. The implication here is that these lines are essentially either left as is or commented out within the final makefile. The commented lines really don't concern us, but you can clearly see that if the uncommented lines in the makefile begin with the TAB character, Automake will treat them as commands, rather than as definitions, and sort them accordingly in the final makefile. When make processes the generated makefile, it will attempt to interpret these lines as orphan commands.

NOTE   *Had the original poster used spaces to indent the conditional statements, they'd have had no problem.*

The moral of the story: watch those TAB characters!

Item 13: Packaging Choices

The ultimate goal of a package maintainer is to make it easy for the end user. System-level packages never have this problem because they don't rely on anything that's not part of the core operating system. But higher-level packages often rely on multiple subpackages, some of which are more pervasive than others.

For example, consider the Subversion project. If you download the latest source archive from the Subversion project website, you'll find that it comes in two flavors. The first contains only the Subversion source code, but if you unpack and build this project, you'll find that you'll need to download and install the Apache runtime and runtime utility (*apr* and *apr-utils*) packages, the *zlib-devel* package, and the *sqlite-devel* package. At this point, you can build Subversion, but to enable secure access to repositories via HTTPS, you'll also need *neon* or *serf* and *openssl*.

The Subversion project maintainers felt that community adoption of Subversion was important enough to go the extra mile, so to speak. To help you out in your quest to build a functional Subversion package, they've provided a second package called *subversion-deps*, which contains a source-level distribution of some of Subversion's more important requirements.[27] Simply unpack the *subversion-deps* source package in the same directory where you unpacked your *subversion* source package. The root directory in the *subversion-deps* package contains only subdirectories—one for each of these source-level dependencies.

You can choose to add source packages to your projects' build systems in the same manner. Of course, the process is much simpler if you're using Automake. You need only call AC_CONFIG_SUBDIRS for subdirectories containing add-on projects in your build tree. AC_CONFIG_SUBDIRS quietly ignores missing subproject directories. I showed you an example of this process in Chapter 14, where I built the FLAIM toolkit as a subproject if it existed as a subdirectory within any of the higher-level FLAIM project directories.

Which packages should you ship with your package? The key lies in determining which packages your consumers are least likely to be able to find on their own.

---

27. You'll still have to download and install the *openssl-devel* package for your GNU/Linux distribution, or else download, build, and install a source-level distribution of OpenSSL in order to build HTTPS support into your Subversion client. The reason for this is that the tricky nature of various countries' import and export laws surrounding OpenSSL make it rather difficult for anyone but the project maintainers to distribute OpenSSL.

Wrapping Up

I hope you find these solutions—indeed, this book—useful on your quest to create a really great user experience with your open source projects. I began this book with the statement that people often start out hating the Autotools because they don't understand the purpose of the Autotools. By now, you should have a fairly well-developed sense of this purpose. If you were disinclined to use the Autotools before, then I hope I've given you reason to reconsider.

Recall the famously misquoted line from Albert Einstein: "Everything should be made as simple as possible, but no simpler."[28] Not all things can be made so simple that anyone can master them with little training. This is especially true when it comes to processes that are designed to make life simpler for others. The Autotools offer the ability for experts—programmers and software engineers—to make open source software more accessible to end users. Let's face it—this process is less than trivial, but the Autotools attempt to make it as simple as possible.

---

28. See *http://en.wikiquote.org/wiki/Talk:Albert_Einstein*. What Einstein actually said was "The supreme goal of all theory is to make the irreducible basic elements as simple and as few as possible without having to surrender the adequate representation of a single datum of experience."

INDEX

interfaces
    designing, 499–505
    plug-in, 181, 221–223
    versioning, 209
internal versioning, 210
International Components for Unicode (ICU), 315
internationalization
    dynamic messages, 296–325
    overview, 295–296
    static messages, 325–329
intN_t definitions, 139–140
-Ipath directives, 112
iteration, 450

J

Java
    Autotools support for, 17, 408–411
    Eclipse and, 181
Java native interface (JNI), 398, 411, 415–417
JAVA primary, 159, 418
Java virtual machine (JVM), 17, 408
Jupiter project
    adding functionality, 87–90
    adding shared libraries, 188–207
    directory structure, 37–38
    installation, 56–62
    location variables, 66–67
    nonrecursive build systems, 175–177
    optional features, 132–138
    source distribution archive, 50–54
    VPATH build functionality, 91–94
JVM (Java virtual machine), 17, 408

K

Katz, Phil, 3
Kernighan, Brian, 80
key-value tags, 276–277
--keyword option, 335

L

language packs, 294
language selection, 332–334
LANGUAGE variable, 332–333
Lattarini, Stefano, 146
lazy binding, 183–184
LC_ADDRESS, 324
LC_ALL, 297

LC_COLLATE, 297, 309–314
LC_CTYPE, 297, 314–315
LC_IDENTIFICATION, 325
LC_MEASUREMENT, 325
LC_MESSAGES, 324, 333
LC_MONETARY, 297
LC_NAME, 324
LC_NUMERIC, 297
LC_PAPER, 324
LC_TELEPHONE, 324
LC_TIME, 297
LD_PRELOAD variable, 182–183
leading control characters, 53–54
Lerdorf, Rasmus, 293
LGPLv2+ and LGPLv3+, 354
*_LIBADD variables, 190
libcrypto.pc files, 289–290
libraries
    building, 169–171
    checking for, 123–125
    convenience libraries, 164–169
LIBRARIES primary, 158
library versioning, 216–220
library_LIBADD variable, 168
library-specific macros, 516
@LIBS@ substitution variable, 125–126
libssl.pc files, 288–290
libthreads library, 123
Libtool
    macros, 374–376
    overview, 14, 179–180
    purpose of, 24–26
libtoolize shell script, 25
libxyz.so, 81
LINGUAS files, 349–350
linker names, 212
links, 181
Linux distros, 67–69
Linux installations, 452, 454–462
Linux versioning, 210–212
Lirzin, Mathieu, 146
LISP primary, 158
localeconv function, 296, 298
LOCALE_DIR variable, 340
locales
    generating and installing, 303–307
    LC_COLLATE, 309–314
    LC_CTYPE, 314–315
    POSIX standard, 315–324
    setting and using, 298–303
    time and date, 307–308
    X/Open standard, 315–324

P

package argument, 100
package maintainers, 539
@PACKAGE_BUGREPORT@, 100
paper sizes, 324
parallel make, 74
parallel-tests option, 149
pattern rules, 531
.pc.in templates, 282–283
per-makefile option variables, 169
.PHONY rule, 51
phony targets, 49, 54–55
PIC (position-independent code),
    200–204
pic-only option, 193–194
PIMPL (Private IMPLementation)
    pattern, 502–503
pkg prefix, 156
pkgconf project, 273
pkg-config
    Autoconf macros, 290–292
    clones, 273
    configure.ac, 287–290
    functional fields, 279–282
    --help option, 274
    informational fields, 278–279
    key-value tags, 277–278
    M4 utility, 274
    metadata files, 276–282
    --modversion option, 279
    overview, 272–276
    .pc files, 282–286
    Requires and Requires.private
        fields, 281
    --variable option, 276
platforms, targeting, 36
plug-in interfaces, 181, 221–223
PLV modifiers, 161–162
po directory, 334–337
portable object template (*.pot*) files,
    334–337
POSIX standard, 315–324, 452
POSIX threads (*pthread*) library,
    123–131
POSIX-compliant platforms, 15
postorder_commands macro, 506
--prefix option, 10
prefix overrides, 69–71
prefix variables, 65–66, 156

preorder_commands macro, 506
preprocessor directives, 82
preprocessor-defined variables,
    524–525
primaries, 158–160
print.h, 46–47
product list variables (PLVs), 155
product option variables
    (POVs), 167–169
*product*_CFLAGS, 168
*product*_CPPFLAGS, 168
*product*_CXXFLAGS, 168
*product*_LDFLAGS, 168
program, 44–45
*program*_LDADD, 168
programming languages, choice of,
    16–17
PROGRAMS primary, 158
prog-to-check-for program, 119–120, 122
project names, 37
PSV modifiers, 161–162
pthread libraries, 123–131, 384
public interfaces, 499–505
pure virtual interface, 503
PYTHON primary, 158–159

Q

quadrigraphs, 143
quiet build systems, 54, 173

R

readme-alpha option, 148
recursive build systems, 38
recursive extension targets, 505–508
Red Hat Package Manager (RPM), 67,
    70, 101, 377, 428
redundancy, 47
reject parameter, 122
Remnant, Scott James, 180
remote build functionality, 28–29,
    111–112
repository revision numbers, 508–510
Ritchie, Dennis, 80
root permissions, 71
Rossum, Guido van, 293
RPM (Red Hat Package Manager), 67,
    70, 101, 377, 428
runtime directories, 70

S

salutations, 324
Savannah Gnulib git repository,
     354–355, 528–529
SCons, 14
SCRIPTS primary, 159
sed expressions, 90–91
semicolon (;) character, 40, 42
setlocale function, 296–297
shared libraries
     benefits of, 180–181
     installing, 187
     interfaces, 209
     pthread libraries, 126–131
     tables, 182
     use of, 181–187
shared object name (soname), 211
shared option, 193
shell commands, 42–43
shell condition, 381
shell variables, 114
side-by-side cache (SxS), 214–215
Simon, Peter, 528
Solaris systems, 122
Solaris Versioning, 210–212
source archives
     building, 4–7
     downloading, 2
     installing, 9–11
     testing, 7–9
     unpacking, 3–4
source distribution archives, 50–54
spec files, 70
src_b64_CPPFLAGS directive, 363
$(srcdir), 93
staged installations, 67
stamp targets, 403–404
standard C library, 296–297
standard targets and variables,
     64–66, 389
static option, 193
stdbool module, 358
strfmon function, 315–316
strftime function, 307–308
string module, 358
strip program, 458
Stroustrup, Bjarne, 293
strxfrm function, 311–314
subdir-objects option, 149
SUBDIRS variable, 224–225, 343, 505
substitution variables, 114
Subversion project, 509, 539

sudo, 11
suffix rules, 47–49
SVNREV variable, 509–510
SxS (side-by-side cache), 214–215

T

TAB characters, 39–40, 537–538
tag argument, 106
Tanner, Thomas, 180
tar utility, 3–4
tarballs, 50–54
target systems, 517
targets, 40, 44, 64–65, 536–537
TDs (transitive dependencies), 401–403
telephone numbers, 324
templates
     header files, 107–108
     Makefile.in template, 99
     .pc.in templates, 282–283
     portable object template (*.pot*) files,
          334–337
test groups, 248–249
TESTS variable, 236–237
TESTSOURCES variable, 243
testsuite program, 240–241
TESTSUITE variable, 243–247
TEXINFOS primary, 160, 537
text replacement, 523–528
thesaurus library, 184–185
time and date, 307–308
tool chain options, 452–453
transitive dependencies (TDs),
     401–403
translation units, 195
transparency, 14
Tromey, Tom, 145
troubleshooting, 449–450
two-pass systems, 531–533
type and structure definitions, 138–141

U

Ubuntu, 461
uintN_t definitions, 139–140
uninstallation, 60–62, 285–286
unit testing, 55–56, 162–164, 257–260
Unix compilers, 45
UNQUOTED versions of macros, 115
unzip utility, 3
--update option, 364
url argument, 101

Autotools, 2nd Edition is set in New Baskerville, Futura, Dogma, and TheSansMono Condensed. This book was printed and bound at Sheridan Books, Inc. in Chelsea, Michigan. The paper is 50# Finch Opaque, which is certified by the Forest Stewardship Council (FSC).

The book uses a layflat binding, in which the pages are bound together with a cold-set, flexible glue and the first and last pages of the resulting book block are attached to the cover. The cover is not actually glued to the book's spine, and when open, the book lies flat and the spine doesn't crack.

RESOURCES

Visit *https://www.nostarch.com/autotools2e/* for resources, errata, and more information.

More no-nonsense books from **NO STARCH PRESS**

LINUX BASICS FOR HACKERS

Getting Started with Networking, Scripting, and Security in Kali

by OCCUPYTHEWEB

DECEMBER 2018, 248 PP., $34.95

ISBN 978-1-59327-855-7

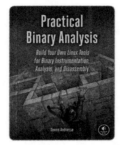

PRACTICAL BINARY ANALYSIS

Build Your Own Linux Tools for Binary Instrumentation, Analysis, and Disassembly

by DENNIS ANDRIESSE

DECEMBER 2018, 456 PP., $49.95

ISBN 978-1-59327-912-7

HOW LINUX WORKS, 2ND EDITION

What Every Superuser Should Know

by BRIAN WARD

NOVEMBER 2014, 392 PP., $39.95

ISBN 978-1-59327-567-9

AUTOMATE THE BORING STUFF WITH PYTHON, 2ND EDITION

Practical Programming for Total Beginners

by AL SWEIGART

FALL 2019, 504 PP., $39.95

ISBN 978-1-59327-992-9

THE LINUX COMMAND LINE, 2ND EDITION

A Complete Introduction

by WILLIAM SHOTTS

MARCH 2019, 504 PP., $39.95

ISBN 978-1-59327-952-3

YOUR LINUX TOOLBOX

by JULIA EVANS

AUGUST 2019, 7 ZINES, $29.95

ISBN 978-1-59327-977-6

PHONE:
1.800.420.7240 OR
1.415.863.9900

EMAIL:
SALES@NOSTARCH.COM

WEB:
WWW.NOSTARCH.COM